MW00667583

FORENSIC SCIENCE

Advanced Investigations

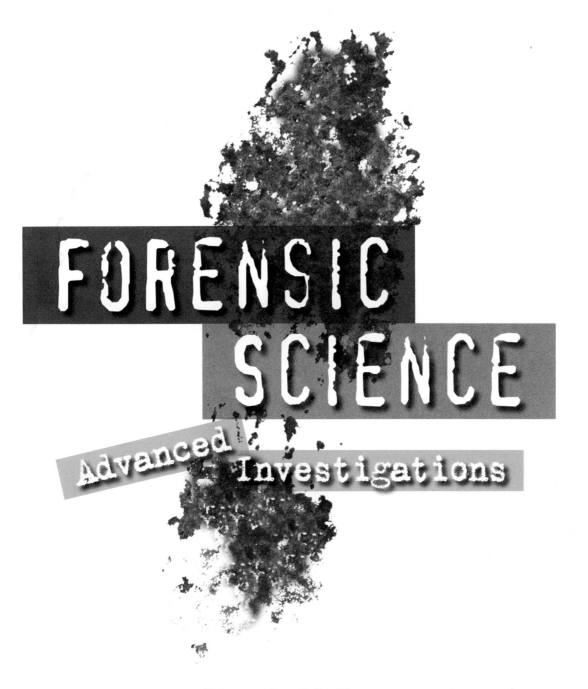

Rhonda M. Brown
Lake County Schools, East Ridge High School, Clermont, FL

Jackie S. Davenport
Lake County Schools, Tavares High School, Tavares, FL

SOUTH-WESTERN
CENGAGE Learning™

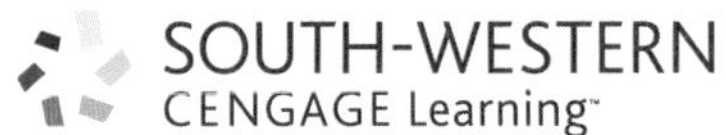

Australia • Brazil • Japan • Korea • Mexico • Singapore • Spain • United Kingdom • United States

Forensic Science: Advanced Investigations
First Edition

Rhonda M. Brown, Jackie S. Davenport

Vice President of Editorial, Business: Jack W. Calhoun

Vice President/Editor-in-Chief: Karen Schmohe

Senior Acquisitions Editor: Mike Schenk

Senior Developmental Editor: Dave Lafferty

Marketing Manager: Robin LeFevre

Senior Content Project Manager: Diane Bowdler

Digital Project Manager: Lysa Kosins

Senior Website Project Manager: Ed Stubenrauch

Senior Frontlist Buyer: Kevin Kluck

Copyeditor: Jeanne R. Busemeyer

Production Service: Integra Software Services Pvt. Ltd.

Senior Art Director: Tippy McIntosh

Cover and Internal Design: Joe Devine, Red Hangar Design, LLC

Cover Image: ©Fotosearch

Senior Rights Specialist, Photo: Deanna Ettinger

Photo Researcher: Bill Smith Studios

ISBN-13: 978-0-538-45089-8
ISBN-10: 0-538-45089-4

South-Western Cengage Learning
5191 Natorp Boulevard
Mason, OH 45040
USA

Cengage Learning products are represented in Canada by Nelson Education, Ltd.

For your course and learning solutions, visit www.cengage.com/school

Visit our company website at www.cengage.com

Dedication

To our families and friends who supported us throughout this incredible journey.

Printed in the United States of America
1 2 3 4 5 6 7 15 14 13 12 11

Preface

WELCOME TO *FORENSIC SCIENCE: ADVANCED INVESTIGATIONS*

Introducing a new textbook that provides more advanced science behind forensics as well as labs and activities appropriate for high school students! ***Forensic Science: Advanced Investigations*** is student *and* teacher friendly. ***Forensic Science: Advanced Investigations*** integrates science, mathematics, and writing skills by using real-life applications and case studies, providing complete flexibility for any science program. Teachers can conduct a full-year's study of forensics or select topics for a half-year course. As another option, teachers may use the textbook in any science class to motivate students to learn science concepts through forensic applications. ***Forensic Science: Advanced Investigations*** is the new standard in high school forensic science . . . Case Closed!

Forensic Science: Advanced Investigations is meant as a companion book to ***Forensic Science: Fundamentals and Investigations*** by Bertino (ISBN 978-0-538-44586-3). For some classroom situations, this text can be used independently.

GETTING STARTED

Forensic Science: Advanced Investigations explains the science used in forensic-science techniques. It provides a chapter-by-chapter description of specific types of evidence and the techniques to collect and analyze the evidence. As students progress through the course, they refine the techniques and apply them to other areas of study. The topics covered in the 15 chapters include interrogation and reporting, laboratory techniques, arson and fire investigations, explosions, body systems, physical trauma, autopsy and truthful reporting, physiology of alcohol and poisons, advanced studies of DNA, forensic odontology, entomology, crime and accident reconstruction, cyber crimes, and criminal profiling.

The textbook is designed to motivate students. Forensic topics are introduced in case studies taken from headlines and popular media. These features engage students as they describe the historical development of forensic-science techniques. Inexpensive, easy-to-use labs provide students with opportunities for successful laboratory experiences as well as an appreciation of the true nature of forensic science problem-solving techniques. Suggestions for research projects extend and enrich student learning and interest.

CHAPTER FEATURES

Each chapter of ***Forensic Science: Advanced Investigations*** begins with a true story, student objectives, key vocabulary, and a ***Key to Science Topics***. The ***Key to Science Topics*** identifies biology, earth science, chemistry, physics, psychology, or mathematics concepts integrated into chapter topics.

Special features include ***Did You Know?*** margin notes that provide additional facts and information, and ***Digging Deeper*** research activities that refer students to the free online InfoTrac® database from Gale Publishing called the **Forensic Science e-Collection database.**

At the end of each chapter, a ***Chapter Summary*** reviews the main points of the chapter. A series of short ***Case Studies*** offer high-interest topics for critical thinking, writing, and classroom discussion. ***Career Profiles*** describe occupations related to forensic science. A ***Chapter Review*** contains both objective and short-answer questions to assess student understanding. A chapter bibliography lists the research sources.

Each chapter has ***Activities*** that provide hands-on experiences with forensic-science techniques. Each activity has clear, step-by-step directions for students of all reading levels. For teachers, they offer easy, quick preparation and minimal materials expense. Each activity includes objectives, materials, safety precautions, procedures, and other learning tools.

New features include Crime Scene S.P.O.T. (Student Prepared Original Titles) and Preventing Adolescent Crime Together (P.A.C.T.™). ***S.P.O.T.*** offers short stories written by actual students followed by critical-thinking questions. These also provide interdisciplinary instructions integrating reading and writing throughout the text. ***P.A.C.T.*** provides service-learning opportunities through projects addressing issues such as anti-bullying and social responsibility.

Capstone Projects give students the opportunity to apply key topics learned throughout the course.

FOR THE TEACHER

The **Wraparound Teacher's Edition** (ISBN 978-0-538-45090-4) contains teaching strategies and tips to engage students. It provides clarification of science content and forensic-science procedures, ideas to help stimulate student interest, evaluation opportunities, additional questions, and suggestions for further exploration and research.

An **Instructor's Resource CD-ROM** is available to teachers who adopt a classroom set of ***Forensic Science: Advanced Investigations*** (ISBN 978-0-538-45091-1). The CD contains additional activities, Power Point© presentations, student activity forms, rubrics, content blueprints, and enrichment materials.

Teachers can purchase a flexible, easy-to-use **Exam***View*® **Electronic Test Bank and Test Generation Software CD-ROM** that contains objective questions that cover textbook content (ISBN 978-538-45092-8). The test bank includes questions for each chapter and a final exam. The **Exam***View* software enables teachers to modify questions from the test bank or add their own questions to create customized tests.

An electronic **eBook** is also sold separately (ISBN 978-1-111-74685-8 PAC and 978-0-8400-6441-7 IAC). The eBook is a digitized version of ***Forensic Science: Advanced Investigations.*** It offers the same rich photographs, illustrations and other graphics, and easy-to-read fonts as the printed text. Students may view the PDF files on their computers.

CourseMate is also available separately for this text. CourseMate brings course concepts to life with interactive study and exam-preparation tools that support the printed textbook. This includes the interactive eBook, and interactive teaching and learning tools including quizzes, flashcards, videos, and more. It also includes ***Engagement Tracker***, a tool that monitors student engagement in the course. See http://www.cengagesites.com/academic/?site=4625 for more information.

Available separately, the Forensic Science ***Virtual Labs*** give real-world lab experience within an online environment. Eighteen forensic-science lab activities based on two crime scenes are included. The labs can be used with any text on forensic science. See www.cengage.com/school for more information.

South-Western maintains a **website** to support this text. Both students and teachers using ***Forensic Science: Advanced Investigations*** may access the website at www.cengage.com/school/forensicscienceadv. The site provides teacher resources and information about related products. Student resources on the site include forms, additional projects, and links to related sites. In addition, a link is provided to the ***Gale Forensic Science e-Collection*** database which allows free online research in various journals and the Gale Virtual Reference Library.

ABOUT THE AUTHORS

Rhonda Brown is the department chair in the Science Department at East Ridge High School in Clermont, Florida. She teaches forensic science as well as biology and zoology. Brown developed the Florida state curriculum for Forensic Sciences I and co-authored the curriculum for Forensic Sciences II. She received her Master's degree in Education from Indiana Wesleyan and Bachelor of Science degree from the University of Southern Indiana. In 2004 she received National Board Certification, the East Ridge High School Teacher of the Year Award, and the 2005 Disney Teacherrific Award for Outstanding Educational Program. In 2010, she was named Central Florida STEM (Science, Technology, Engineering and Math) Teacher of the Year by the Air Force Association. Rhonda Brown has been a presenter at several local, state, and national education conferences including the National Science Teachers Association (NSTA) and Florida Association for Science Teachers (FAST).

Jackie Davenport teaches earth science, honors chemistry, and forensic science at Tavares High School in Tavares, Florida. She is the Small Learning Community Lead teacher for the Science, Technology, and Health Sciences community. Davenport is co-author of the Forensic Science II curriculum for the state of Florida. She earned a Master's degree in Educational Leadership from National Louis University and a Bachelor's in Science Education degree from East Carolina University. Her awards include Tavares High School Teacher of the Year, Lake County Teacher of the Year finalist, the 2005 Disney Teacherrific Award for Outstanding Educational Program, and Disney Teacherrific Special Judges Award in 2003. Jackie Davenport has served as a consultant for the National Education Program. She also has presented and facilitated at several state and national science teaching events, including the National Science Teachers Association (NSTA), Florida Association for Science Teachers, and Florida Association of Science Supervisors.

Brief Contents

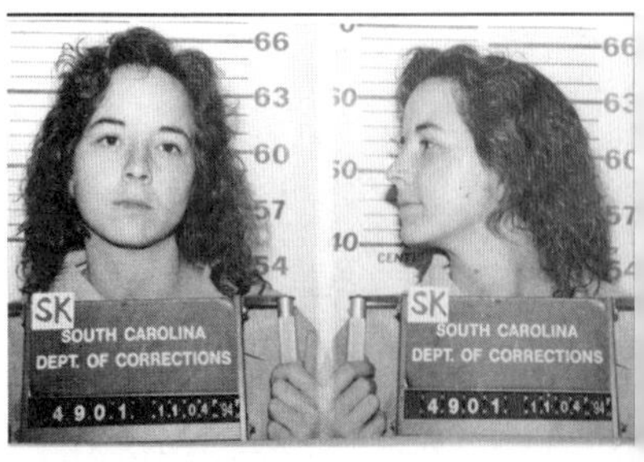

REUTERS/LOU DEMATTEIS/FILE LD/GN

AP PHOTO

Contents

Chapter 1
Overview of Forensics 2

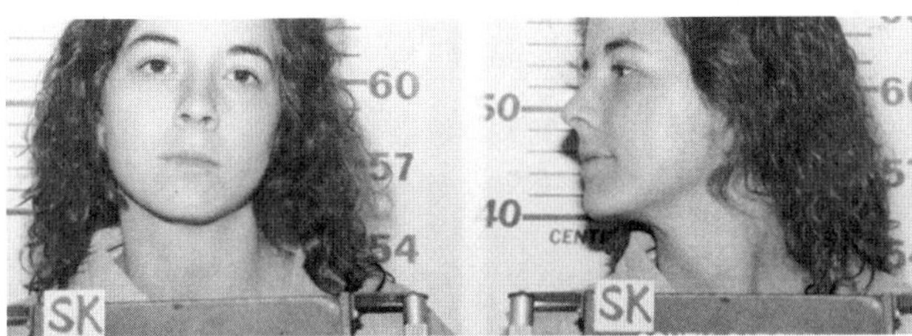

Chapter 2
Interrogation and Forensic Reporting 36

Chapter 3
Forensic Laboratory Techniques 70

Chapter 4
Arson and Fire Investigation 104

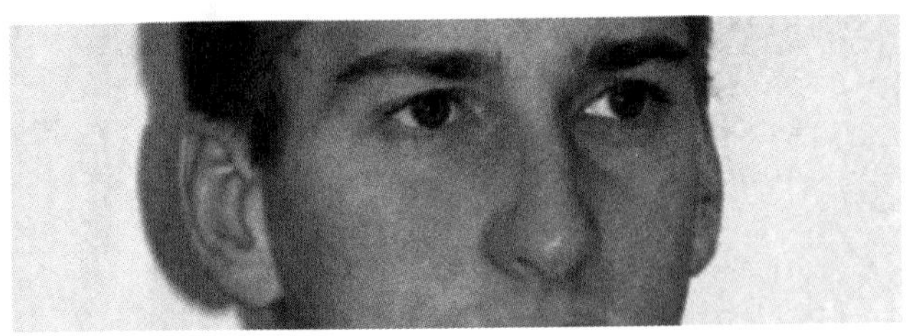

Chapter 5
Explosions 132

Chapter 6
Body Systems 162

Chapter 7
Physical Trauma 194

Chapter 8
Autopsy 218

Chapter 9
Physiology of Alcohol and Poisons 252

Chapter 10
Advanced Concepts in DNA 286

Chapter 11
Forensic Odontology 318

Chapter 12
Forensic Entomology 342

Chapter 13
Crime and Accident Reconstruction 368

Chapter 14
Cyber Crimes 398

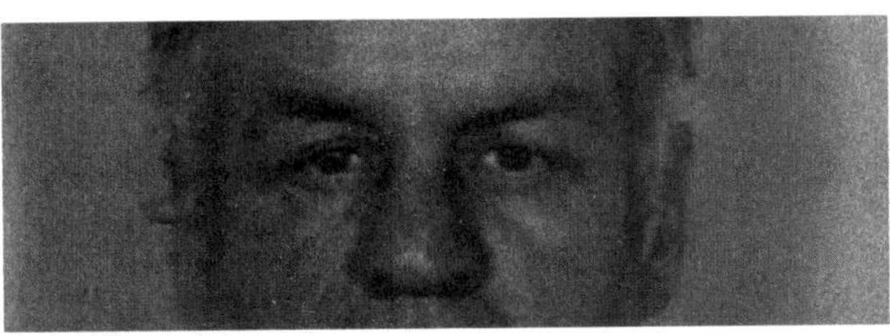

Chapter 15
Criminal Profiling 428

Capstone Projects

Reviewers

Dukagjin Binishi, Ph.D.
Harry S Truman High School
Bronx, NY

Joseph Bomba
Ellet High School
Akron, OH

Rebecca Collins
Granada High School
Livermore, CA

Karen Lynn Cruse
The Summit Country Day School
Cincinnati, OH

Sheryl Fischer
Simon Kenton High School
Independence, KY

David R. Foran, Ph.D.
Michigan State University
East Lansing, MI

Myra Frank
Marjory Stoneman Douglas High School
Parkland, FL

Daniel Hartung
Plant City High School
Plant City, FL

Robert D. Hall, Ph.D., J.D.
University of Missouri
Columbia, MO

Glennis Kaplansky
East Brunswick High School
East Brunswick, NJ

Lawrence Kobilinsky, Ph.D.
Chairman and Professor, Department of Sciences
John Jay College of Criminal Justice
New York, NY

Mari Knutson
Lynden High School
Lynden, WA

Barbara E. Llewellyn, Ph.D.
American Academy of Forensic Sciences
Elmhurst, IL

Madeline Loftin
Wingfield High School
Jackson, MS

Danielle Ristow
Rancho High School
Las Vegas, NV

Carol A. Robertson
Fulton High School
Fulton, MO

Mary Robinson
Rio Rancho High School
Rio Rancho, NM

Ruth Smith, Ph.D.
Michigan State University
East Lansing, MI

Donald C. Snyder, Jr.
South Philadelphia High School
Philadelphia, PA

Mike Tracy
Publishing Consultant
Austin, TX

Zach Tracy
Cleveland Heights High School
Cleveland Heights, OH

Sue Trammell, Ph.D.
John A. Logan College
Carterville, IL

Eva A. Vincze, Ph.D.
George Washington University
Department of Forensic Sciences
Washington, DC

Jeff Roth-Vinson
Cottage Grove High School
Cottage Grove, OR

Leesa Wingo
South Anchorage High School
Anchorage, AK

James Zlomke
Cheyenne East High School
Cheyenne, WY

STRAIGHT FROM THE SOURCE, NO ASSUMPTIONS

THE CONTENT YOU REQUESTED

EVIDENCE: Fingerprints of Expert Educators, Like You, Throughout.

This program's balance of relevant content and hands-on lab activities are a direct result of your input at every stage! No other forensic science program delivers precisely what you need for your students and your course.

- **Review board of more than 70 educators, focus groups, and ongoing educator feedback** guided each decision to ensure the program meets the educational needs of your students.
- **Student and teacher supplements support the workflow with time-saving tools.** Many resources represent "firsts" for the high school course, such as the innovative **Forensic Science e-Collection™ database** and the new **Virtual Labs.**
- **Forensic Science combines topics from math, chemistry, biology, physics, psychology, and earth science** into a single course with all materials clearly aligned with the **National Science Education Standards**. Distinctive icons identify topics in the chapter opener and throughout the text.

DR. CARL COPPOLINO

As science continues to advance, so do the tests
analyze and process evidence. In 1963, William F
his home in New Jersey. Doctors thought that Fa
failure. Two years later, the wife of Mrs. Farber's l
in her home in Florida. At first, the family ph
Coppolino had died of a heart attack. However
prosecutors that Dr. Carl Coppolino had poiso
Carmela Coppolino. Both bodies were exhumed
separate trials. The first trial, in New Jersey, was
the second trial, for the murder of Coppolino's
Coppolino was acquitted of the first murder. The
however, admitted evidence from the toxicologis

BIOLOGY · CHEMISTRY · EARTH SCIENCES · MATHEMATICS · PHYSICS · PSYCHOLOGY

FORENSIC SCIENCE
Fundamentals & Investigations
Anthony J. Bertino

EQUIP STUDENTS WITH REAL-WORLD FIELD EXPERIENCE

ENGAGE THEM WITH HANDS-ON ACTIVITIES

The variety of high-interest lab activities saves you time, while giving students the hands-on experience needed to integrate their knowledge of science and related subjects. These activities outline the objectives, materials, safety precautions, scenarios, background information, detailed procedures and data tables, follow-up questions, and research opportunities. Plus, the affordable materials allow labs to be completed with minimal investment.

Capstone Crime Scene projects give students the opportunity to apply key topics learned throughout the course.

ACTIVITY 12-2 *Ch. Obj: 12.2, 12.3, 12.6*
A WORLD OF INSECTS

Objectives:

By the end of this activity, you will be able to:
1. Research insects native to various parts of the world.
2. Assess the importance of insects to forensics.
3. Produce a multimedia project illustrating a research topic.

Materials:

Computers with Publisher®, PowerPoint®, and/or Flash® Movie

Procedure:

1. After choosing a geographical area, research insects of forensic importance in that part of the world. Information should include the anatomy, life cycle, habitat, where they are found, food source(s), position in the food chain (predators, prey, etc.), forensic importance, and any other "fun facts" you may find.
2. Choose the format to showcase your research:
   a. PowerPoint presentation
   b. Flash Movie
   c. Brochure

Figure 1. World map.

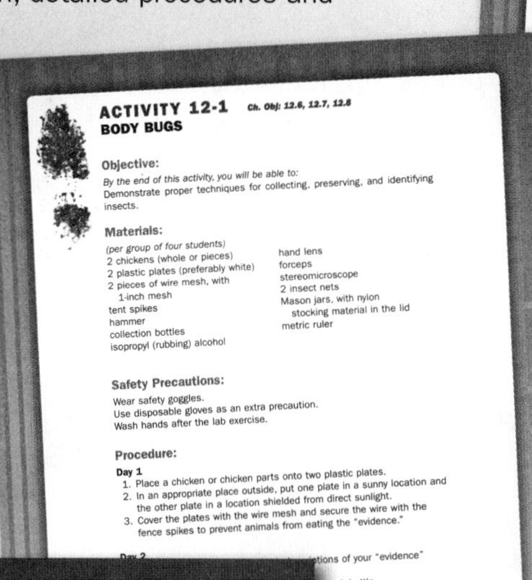

ACTIVITY 12-1 *Ch. Obj: 12.6, 12.7, 12.8*
BODY BUGS

Objective:

By the end of this activity, you will be able to:
Demonstrate proper techniques for collecting, preserving, and identifying insects.

Materials:

(per group of four students)
2 chickens (whole or pieces)
2 plastic plates (preferably white)
2 pieces of wire mesh, with 1-inch mesh
tent spikes
hammer
collection bottles
isopropyl (rubbing) alcohol
hand lens
forceps
stereomicroscope
2 insect nets
Mason jars, with nylon stocking material in the lid
metric ruler

Safety Precautions:

Wear safety goggles.
Use disposable gloves as an extra precaution.
Wash hands after the lab exercise.

Procedure:

Day 1
1. Place a chicken or chicken parts onto two plastic plates.
2. In an appropriate place outside, put one plate in a sunny location and the other plate in a location shielded from direct sunlight.
3. Cover the plates with the wire mesh and secure the wire with the fence spikes to prevent animals from eating the "evidence."

Day 2

Preventing Adolescent Crime Together™ (P.A.C.T)

Introduction:

Every year, children and adults are injured when using fireworks in celebrations. In this activity, you will develop a public awareness brochure to educate others about the dangers and safe handling procedures of explosives, particularly fireworks.

Procedure:

Part 1
1. Research the following information regarding fireworks safety:
   - Current statistics on fireworks injuries/deaths
   - Tips for preventing accidents
   - Laws and regulations regarding the use of fireworks in your area. Be sure to look for local as well as statewide regulations.
2. After completing the research, your group will design an informational brochure for other students. The brochure must be easy to read, informative, creative, and well organized.

Part 2
1. Find out who the legislators are for your area.
2. Using the information from your research, write a persuasive letter to a legislator to:
   a. Request more specific regulations for use of explosive fireworks.
   b. Request stricter or more lenient legislation regarding use of explosive fireworks.
   c. Suggest a safety campaign or contest for students across the country. The winning entry could be presented before the next national celebration involving fireworks.

VAKHRUSHEV PAVEL/GETTY IMAGES

Figure 5-22. *Fireworks.*

Explosions

P.A.C.T.™ Activities (Preventing Adolescent Crime Together) provide service learning opportunities through projects addressing issues such as anti-bullying and social responsibility. Available with *Forensic Science Advanced Investigations*.

"I found this program to be the most student friendly. The well-written laboratory activities capture the students' interest immediately and enhance the forensic science curriculum by encouraging the students to think critically."

Glennis Kaplansky, East Brunswick High School

RIVETING DESIGN & FEATURES OFFE

Tracing the Clues to Student Success!

From a dynamic design to captivating images, practical applications, intriguing case studies, and glimpses into actual crime and lab scenes, every part of this text appeals to students. The program presents information the way students learn best.

Chapter-opening scenarios highlight intriguing news shaping forensic science today. These exciting stories, taken directly from gripping headlines, draw students into the chapter information. Scenarios feature well-known cases, such as Chris Brown, Michael Jackson, Timothy McVeigh, and more.

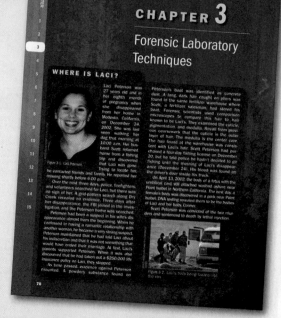

CHAPTER 1 2 3 4 5 6 7 8 9 10 11 12 13 14 15

CHAPTER 3

Forensic Laboratory Techniques

WHERE IS LACI?

Laci Peterson was 27 years old and in her eighth month of pregnancy when she disappeared from her home in Modesto, California, on December 24, 2002. She was last seen walking her dog that morning at 10:00 a.m. Her husband Scott returned home from a fishing trip and discovered that Laci was gone. Trying to locate her, he contacted friends and family. He reported her missing shortly before 6:00 p.m.

Figure 3-1. Laci Peterson.

Over the next three days, police, firefighters, and volunteers searched for Laci, but there was no sign of her. A grid-pattern search along Dry Creek revealed no evidence. Three days after her disappearance, the FBI joined in the investigation, and the Peterson home was searched.

Peterson had been a suspect in his wife's disappearance almost from the beginning. When he confessed to having a romantic relationship with another woman, he became a very strong suspect. Peterson maintained that he had told Laci about his indiscretion and that it was not something that would have ended their marriage. At first, Laci's parents supported Peterson. When it was also discovered that he had taken out a $250,000 life insurance policy on Laci, they stopped.

As time passed, evidence against Peterson mounted. A powdery substance found on Peterson's boat was identified as concrete dust. A long, dark hair caught on pliers was found in the same fertilizer warehouse where Scott, a fertilizer salesman, had stored his boat. Forensic scientists used comparison microscopes to compare this hair to hair known to be Laci's. They examined the cuticle, pigmentation, and medulla. Recall from previous coursework that the cuticle is the outer layer of hair. The medulla is the center core. The hair found at the warehouse was consistent with Laci's hair. Scott Peterson had purchased a four-day fishing license on December 20, but he told police he hadn't decided to go fishing until the morning of Laci's disappearance (December 24). His blood was found on the driver's door inside his truck.

On April 13, 2002, the body of a fetus with the umbilical cord still attached washed ashore near Point Isabel in Northern California. The next day, a female's body was discovered in a park near Point Isabel. DNA testing revealed them to be the bodies of Laci and her baby, Conner.

Scott Peterson was convicted of the two murders and sentenced to death by lethal injection.

Figure 3-2. Laci's body being loaded into the van.

70

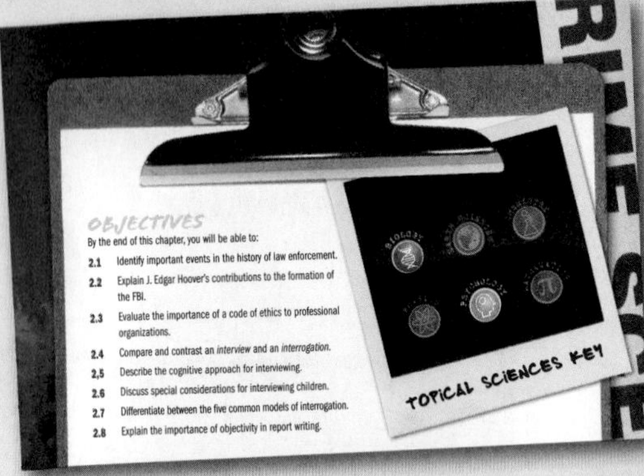
OBJECTIVES

By the end of this chapter, you will be able to:

2.1 Identify important events in the history of law enforcement.
2.2 Explain J. Edgar Hoover's contributions to the formation of the FBI.
2.3 Evaluate the importance of a code of ethics to professional organizations.
2.4 Compare and contrast an interview and an interrogation.
2.5 Describe the cognitive approach for interviewing.
2.6 Discuss special considerations for interviewing children.
2.7 Differentiate between the five common models of interrogation.
2.8 Explain the importance of objectivity in report writing.

Clear Learning Objectives guide how students' study, helping them focus on the key topics. Chapter content and assessment materials, tagged to specific learning objectives, provide a strong framework for mastering key concepts.

Scientific Terms and Vocabulary, highlighted in each chapter, introduce key terms, ensuring students are able to understand their meaning.

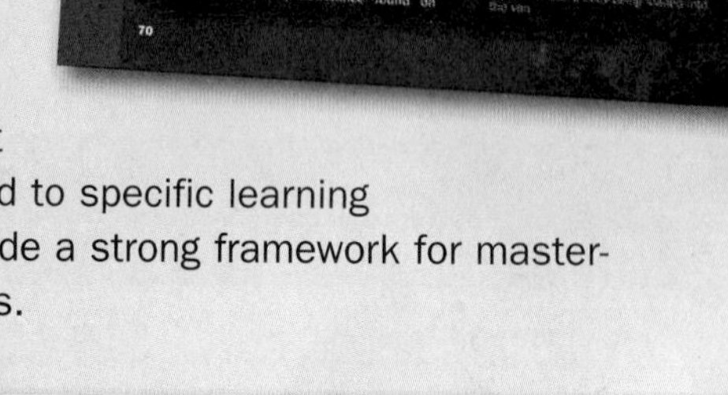
VOCABULARY

ethics - a set of rules that define appropriate behavior in a situation

interrogation - official questioning of a suspect or witness by law enforcement

interview - a question and answer session that does not accuse but is instead intended to gather information concerning a case and/or a suspect

interviewer - a trained individual who questions witnesses or suspects and is able to interpret cues in verbal and physical behavior

objectivity - judgment that is not influenced by personal feelings or bias, focused on fact

suspect - an individual under investigation for his or her alleged involvement in a crime

End of Chapter Review Questions, including multiple choice, short answer, and critical thinking, highlight cross-curricular connections and ensure students thoroughly comprehend principles before moving ahead.

CHAPTER 9 REVIEW

Matching

Match the following poisons with their method of entry into the body. Some answers may be used more than once and some review items may have more than one answer.

1. carbon monoxide *Obj. 9.3, 9.4* a. ingestion
2. ricin *Obj. 9.3, 9.4* b. inhalation
3. anthrax *Obj. 9.3, 9.4* c. injection
4. snake venom *Obj. 9.3, 9.4* d. absorption

Multiple Choice

NEW WAY TO EXPERIENCE SCIENCE

Dr. Park Dietz: Forensic Psychiatrist

Dr. Park Dietz (see Figure 2-17) knew at a very young age that he had a deep interest in criminology. However, he had family pressures leading him in the direction of medicine. In an effort to find a satisfying compromise, Dr. Dietz decided on forensic psychiatry. He graduated from Cornell University where he earned a double major in psychology and genetics. After completing his bachelor's degree, he continued his education at Johns Hopkins University. He earned M.D., Ph.D., and Master of Public Health degrees from Johns Hopkins University. During this time, he published several works and made a number of connections in the field. He continued his residency in the Clinical Scholars Program in Psychiatry at Johns Hopkins. He spent his third year of residency as the Chief Fellow of forensic psychiatry at the University of Pennsylvania.

AP PHOTO/D.J. PETERS, POOL

Figure 2-17. Dr. Park Dietz.

Dietz has testified or consulted in all 50 states. As a forensic psychiatrist, Dietz evaluates the defendant's state of mind. He is asked to determine whether the defendant is mentally competent to stand trial. He is also sometimes asked to evaluate the defendant's mental state when the crime took place. In other words, he is asked whether the defendant was legally insane at the time of the crime.

Dietz has worked on a number of high-profile ... D.C. Sniper, and Columbine cases. Dietz firmly stands on the premise of full disclosure and believes in being as honest as possible in preparing reports and presenting expert opinions at trial. "The responsibility of all forensic scientists is to be as objective as possible, to uncover the truth, and report it clearly and accurately," states Dietz. He notes the importance of clearly stating all sources of data and explaining the basis of all opinions.

When Dietz interviews a defendant soon after the crime, he videotapes the interview to preserve the evidence. If the crime has occurred months or even years prior to the interview, Dietz will read the case records before interviewing the defendant. During an interview session, Dietz states that to remain unbiased, he stays centered on the facts and focused on justice.

Today, Dietz owns a private practice and serves as an expert witness 6 to 12 times a year. On a typical day, he handles 10 to 12 cases, consults with attorneys to prepare them for trial, reviews records, visits crime scenes, prepares reports, conducts interviews, and prepares exhibits for trial. Dietz is modest when asked about his achievements—he feels that nothing he has achieved was achieved alone.

Careers in Forensic Science sections focus on the hottest careers related to forensic science today, detailing job requirements, necessary preparation, and challenges.

Case Studies bring closure to the chapter concepts using intriguing case facts drawn from actual forensic investigations. Questions prompt discussion and further students' critical thinking skills.

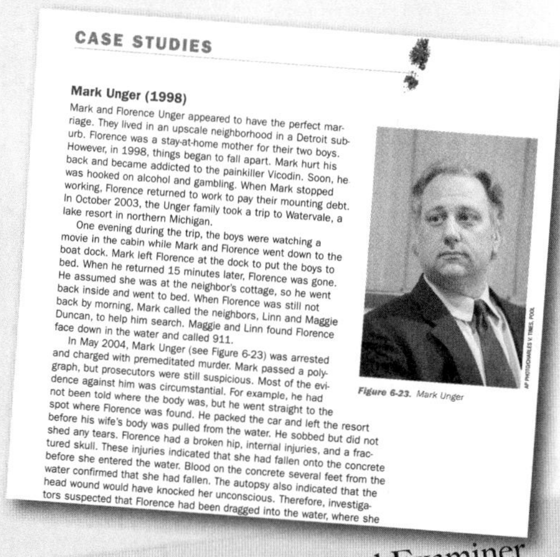

CASE STUDIES

Mark Unger (1998)

Mark and Florence Unger appeared to have the perfect marriage. They lived in an upscale neighborhood in a Detroit suburb. Florence was a stay-at-home mother for their two boys. However, in 1998, things began to fall apart. Mark hurt his back and became addicted to the painkiller Vicodin. Soon, he was hooked on alcohol and gambling. When Mark stopped working, Florence returned to work to pay their mounting debt. In October 2003, the Unger family took a trip to Watervale, a lake resort in northern Michigan.

One evening during the trip, the boys were watching a movie in the cabin while Mark and Florence went down to the boat dock. Mark left Florence at the dock to put the boys to bed. When he returned 15 minutes later, Florence was gone. He assumed she was at the neighbor's cottage, so he went back inside and went to bed. When Florence was still not back by morning, Mark called the neighbors, Linn and Maggie Duncan, to help him search. Maggie and Linn found Florence face down in the water and called 911.

In May 2004, Mark Unger (see Figure 6-23) was arrested and charged with premeditated murder. Mark passed a polygraph, but prosecutors were still suspicious. Most of the evidence against him was circumstantial. For example, he had not been told where the body was, but he went straight to the spot where Florence was found. He packed the car and left the resort before his wife's body was pulled from the water. He sobbed but did not shed any tears. Florence had a broken hip, internal injuries, and a fractured skull. These injuries indicated that she had fallen onto the concrete before she entered the water. Blood on the concrete several feet from the water confirmed that she had fallen. The autopsy also indicated that the head wound would have knocked her unconscious. Therefore, investigators suspected that Florence had been dragged into the water, where she

AP PHOTO/CHARLES V. TINES, POOL

Figure 6-23. Mark Unger

Crime Scene S.P.O.T. (Student Prepared Original Titles) offers short stories written by actual students, followed by critical thinking questions and a writing activity. S.P.O.T. provides interdisciplinary instruction integrating reading and writing throughout the text. Available with *Forensic Science Advanced Investigations*.

CRIME SCENE S.P.O.T. Student Prepared Original Titles

The Case of the Sleepy Driver
By: Kyle Banas, Kelsey Janos, and Melissa Pena
East Ridge High School
Clermont, Florida

The phone rang at 3:33 A.M., waking Gill suddenly. Gill was the crime-scene investigator in Blue Springs. "Hello?" said Gill sleepily, as he answered.

"Hey, Gill. It's Rick. We got one for you. We're at Joe's Bar."

Gill forced himself to his feet, threw on his uniform, grabbed his wallet and keys, and headed for the car. On the ride over to Joe's, he called Sarah, his partner. She answered on the third ring, and Gill told her what he knew. They agreed to meet at the bar as soon they could both get there.

Upon arrival at the scene, Gill spoke to paramedics while Sarah and Rick began their investigation. Rick ran the license plate and found that the car belonged to Jason Schwartz, the driver. They walked around to the front of the car. A man was pinned between the car and the outside wall of the bar.

"Looks like the driver fell asleep at the wheel. We don't have an ID on the second vic. He was apparently at the wrong place at the wrong time." Rick's assessment was as good as any, for now.

Gill slipped on his state-issued gloves in order to avoid contaminating evidence before he began his examination of the car. The first thing he noticed was the driver's clothing. Schwartz had either gotten dressed in a hurry or in the dark. His hair was disheveled. There was a cell phone lying on the ground next to the car. After Sarah finished the photographs, Gill picked up the phone and examined it. He hit the redial button. After a few rings, a woman answered.

"Logan! Are you on your way home yet?"

Gill answered, "This is CSI Gill Barkin. To whom am I speaking?"

The woman identified herself as Logan's girlfriend, Taylor. After a few questions, Gill was able to determine that the other victim (who was pinned to the wall) was most likely Logan Mefford. He also discovered that the two dead men worked together.

Witnesses at Joe's agreed that Jason and Logan had been having drinks with another man, their boss Ryan Little. Little also happened to be Taylor's brother. According to rumors, the boss was not happy that both of the men had been dating his baby sister, and he had been trying to convince them to leave her alone. The glasses the three had been drinking from were still on the table.

Continuing their investigation, the team pop... trunk and searche... of the interior of ... gym bags were c... evidence. One w... other was a blac... the initials TL o... time all of the ... been tagged a... the medical ex... up to transpor...

It took a c... to get the au... Schwartz ha... broken neck ... according to ... report, cryst... in the kidne... oxalate cry... consistent ... poisoning ... Gill had a...

MAURITIUS IMAGES GMBH/ALAMY

Figure 9-24. *Antifreeze is used in car radiators to prevent the water from freezing. When ingested by animals and humans, it is highly toxic.*

...e ME about this. According to the doctor, ...tifreeze is sweet and can easily be mixed ...to tea or some other sweetened beverage. ...e symptoms of antifreeze poisoning ...clude dizziness—a possible cause of the ...ash.

Sarah asked, "So you are saying he was ...oisoned?"

"Yes," replied Gill. "It looks as if our acci...nt has just become a homicide."

A search of Schwartz's apartment ...vealed evidence of a female visitor to the ...artment. A glass with lipstick on the rim ...as collected for possible DNA and finger...int analysis. Other than that, everything ...emed to be in order. Mefford's home ...vealed a little more. He was not as neat ...d tidy as Schwartz, and hidden under the ...chen sink was a nearly empty container ... antifreeze. The container was bagged and ...ken to be processed for trace evidence.

Weeks later, when all of the evidence had ...en processed, the team sat down for their ...al assessment of the case.

Gill addressed the group. "The anti...eze found in Mefford's home appears ... have been planted there. A hair stuck to the inside of the cap was analyzed fo... mitochondrial DNA. It showed a familial relationship to Taylor. The fingerprint on ... glass in Schwartz's apartment belonged ... Taylor. We believe Schwartz's murder wa... supposed to be pinned on Mefford, but, instead, Mefford ended up dead as well. We have an arrest warrant coming, and ... this afternoon, the real murderer will be behind bars."

Activity:

Answer the following questions based on information in the Crime Scene S.P.O.T.

1. Based on the information provided, whose name is on the arrest warrant? What evidence supports your theory?
2. If Schwartz died of a broken neck in the crash, why was his death ruled a homicide?
3. After reading the story, write a one- or two-paragraph conclus... to the story. Be sure to use correct terminology and logical, well-reasoned arguments.

Office of Chief Medical Examiner

Official Statement

Expert Testimony, taken from Mari Knutson Herbert, Lynden High School:

"The Forensic Science chapters are engaging and informative. Students are presented with an actual case that hooks them, clear explanations, thoughtful activities and assessment, all aligned to the national standards. Students will appreciate the pictures and examples of forensic techniques."

DOWN AND DIRTY

ENABLE STUDENTS TO EXPLORE THE TOOLS OF THE TRADE WITH DYNAMIC ONLINE RESOURCES

The **Forensic Science companion website** takes learning to a new level with a wealth of learning and teaching resources.
www.cengage.com/school/forensicscienceadv

- Interactive learning activities and labs
- Forensic Science e-Collection™ database
- Web links to other dynamic science sites
- Interactive flashcards and crossword puzzles to review key terms
- Lab forms to complete in-text lab activities
- Lesson Plans
- PowerPoint® presentation slides

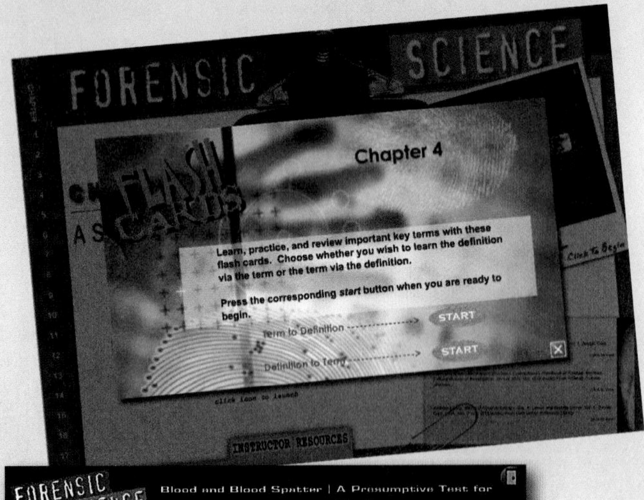

Digging Deeper with Forensic Science e-Collection guides students in exploring specific areas of interest related to forensic science for additional reinforcement.

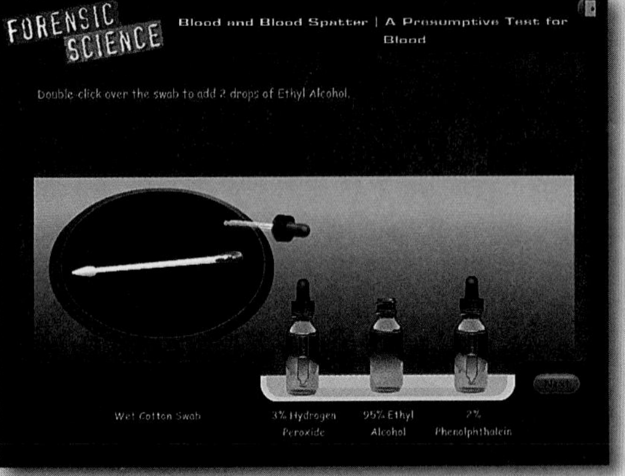

Digging Deeper
with Forensic Science e-Collection

Many people die each year because someone was dri
while intoxicated. Alcohol impairs reflexes, which makes dri
extremely dangerous. But how much is too much? For pe
under the age of 21, it is illegal to consume any alcohol. Sta
have laws that define legal blood alcohol levels for drivers over
Confirmatory tests for blood alcohol levels use gas chroma
raphy. New presumptive tests are very accurate and can
law-enforcement personnel the information they need m

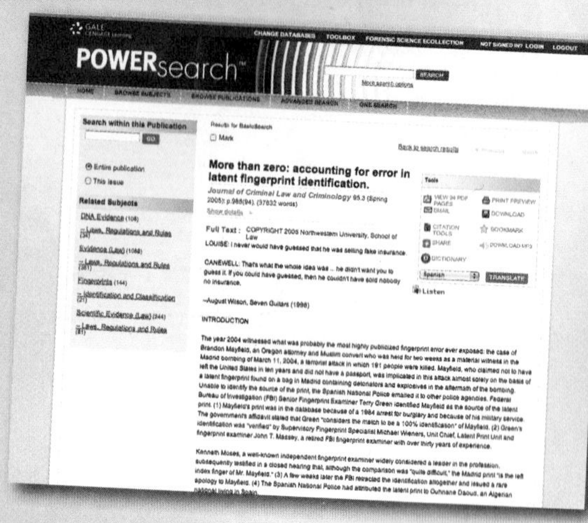

The **Gale Forensic Science e-Collection**™ allows you and your students to investigate the mysteries of forensic science in-depth with online access to hundreds of articles–from highly specialized academic journals to general science-focused magazines. Stay current with the latest scientific developments in this growing field. No other publisher offers such a complete, exclusive resource. Free student access on the textbook companion website.

COMPREHENSIVE INSTRUCTOR RESOURCES PROVIDE THE BACKUP YOU NEED

We've Got You Covered with the Finest Instructor Resources Around

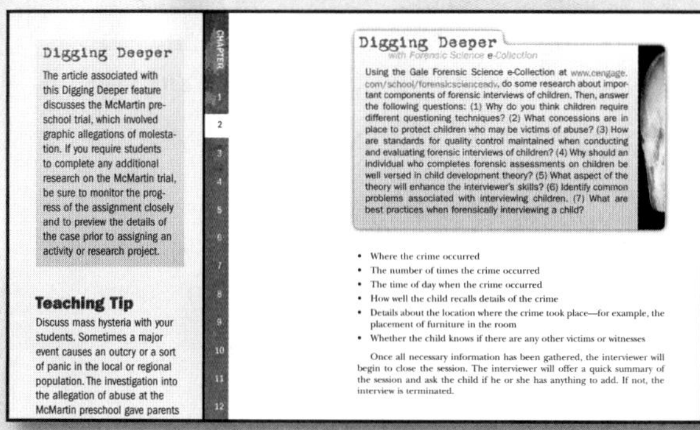

Digging Deeper

The article associated with this Digging Deeper feature discusses the McMartin preschool trial, which involved graphic allegations of molestation. If you require students to complete any additional research on the McMartin trial, be sure to monitor the progress of the assignment closely and to preview the details of the case prior to assigning an activity or research project.

Teaching Tip

Discuss mass hysteria with your students. Sometimes a major event causes an outcry or a sort of panic in the local or regional population. The investigation into the allegation of abuse at the McMartin preschool gave parents

CHAPTER 1 2 3 4 5 6 7 8 9 10 11 12

Digging Deeper with Forensic Science e-Collection

Using the Gale Forensic Science e-Collection at www.cengage.com/school/forensicscienceadv, do some research about important components of forensic interviews of children. Then, answer the following questions: (1) Why do you think children require different questioning techniques? (2) What concessions are in place to protect children who may be victims of abuse? (3) How are standards for quality control maintained when conducting and evaluating forensic interviews of children? (4) Why should an individual who completes forensic assessments on children be well versed in child development theory? (5) What aspect of the theory will enhance the interviewer's skills? (6) Identify common problems associated with interviewing children. (7) What are best practices when forensically interviewing a child?

- Where the crime occurred
- The number of times the crime occurred
- The time of day when the crime occurred
- How well the child recalls details of the crime
- Details about the location where the crime took place—for example, the placement of furniture in the room
- Whether the child knows if there are any other victims or witnesses

Once all necessary information has been gathered, the interviewer will begin to close the session. The interviewer will offer a quick summary of the session and ask the child if he or she has anything to add. If not, the interview is terminated.

Wraparound Teacher's Edition (WTE) provides reduced student pages with comprehensive related teaching tips and support that expand upon the topics presented on each page. Teaching notes and overviews include:

- The Big Ideas highlight the key science focus of the chapter content
- Teaching Resources
- Tips for Differentiated Learning
- Teaching Tips to engage, teach, explore, evaluate, and close the lesson
- Answers to end-of-chapter reviews and activities with additional details related to lab materials, safety precautions, procedures, data tables, and further student research
- Expanded cross-curricular information that connects biology, chemistry, earth science, physics, math, and English

Fundamentals & Investigations ISBN: 978-0-538-49403-8
Advanced Investigations ISBN: 978-0-538-45090-4

Instructor's Resource CD places key instructor resources at your fingertips, including Lab Activity pages in PDF format, Lab Activity forms, Additional Activities, Chapter Outlines, Tests in Microsoft Word, Instructor Notes, Lesson Plans, Objective sheets, PowerPoint® Presentation slides, and Rubrics.

Fundamentals & Investigations ISBN: 978-0-538-49743-5
Advanced Investigations ISBN: 978-0-538-45091-1

ExamView® Test Generator CD saves time in effectively assessing your students' understanding of chapter concepts by allowing you to easily create tests, worksheets, and study guides. Simply edit, add, delete, or rearrange questions.

Fundamentals & Investigations ISBN: 978-0-538-49744-2
Advanced Investigations ISBN: 978-0-538-45092-8

Instructor Companion Website provides online access to instructor resources and professional development webinars.
www.cengage.com/school/forensicscience
www.cengage.com/school/forensicscienceadv
Coming March 2011

Online Adobe eBook offers a rich, electronic textbook option that can be viewed on any device that uses Adobe Acrobat Reader®. The eBook looks exactly like the printed version, including content, photos, graphics, and rich fonts.

Fundamentals & Investigations (PAC) ISBN: 978-1-111-57570-0
Advanced Investigations (PAC) ISBN: 978-1-111-74685-8

CourseMate

Cengage Learning's **CourseMate** brings course concepts to life with interactive learning, study, and exam preparation tools that support the textbook. CourseMate includes an integrated eBook, interactive teaching and learning tools including quizzes, flashcards, and Engagement Tracker, a first-of-its-kind tool that monitors student engagement in the course.

Fundamentals & Investigations ISBN: 978-1-111-86202-2
Advanced Investigations ISBN: 978-1-111-86322-7

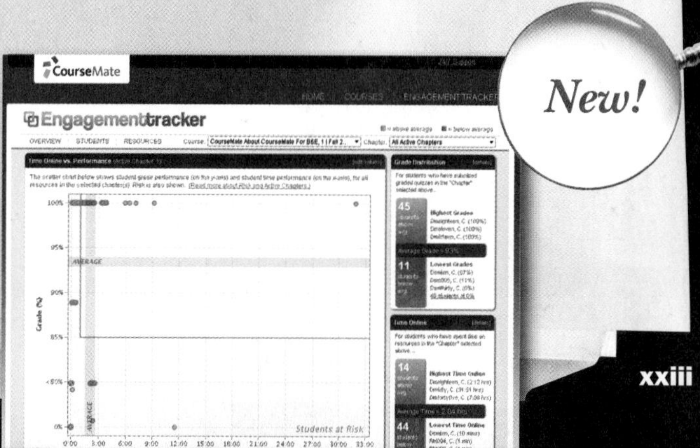

OBSERVE THE EVIDENCE

NEW! FORENSIC SCIENCE VIRTUAL LAB PROGRAM

Set the Scene for Online Learning Success!

Now you can give your students real-world lab experience within an online environment! The new virtual lab program includes 18 forensic science lab activities between two crime scenes.

Each lab activity includes:

- Background information
- 3-D crime scene
- Clear instructions
- Toolkit
- Post lab assessment
- Critical thinking questions
- Research activities

Instructors can choose whether the items below will count as practice or graded assignments (entered in instructor grade book):

- Notebook
- Worksheet
- Post Lab Assessment
- Additional Research
- Mini Lab Assessment
- Individual Labs
- Crime Scene

Information about each activity includes: the time spent on the task, the number of attempts, and the amount completed. Instructors can arrange the graded items so that several assessments feed a single grade or each item will have an individual grade.

Forensic Science Virtual Lab Crime Scene 1
ISBN-13: 978-1-111-48253-4 (Printed Access Code)

Forensic Science Virtual Lab Crime Scene 2
ISBN-13: 978-1-111-57717-9 (Printed Access Code)

Coroner's Office

Laboratory Data Sheet

The Basic Crime Lab includes 8 labs that let students experience:

☑ Blood Spatter Analysis
☑ Pollen/Spore Lab
☑ Hair Analysis
☑ Fiber Analysis
☑ Glass Analysis
☑ Document Analysis
☑ Fingerprint Analysis
☑ Ballistics Lab

The Advanced Crime Lab features 8 labs focused on enabling students to apply more sophisticated tools, such as:

☑ Toxicology
☑ Death /Autopsy
☑ Soil Examination
☑ Forensic Anthropology
☑ DNA Fingerprinting
☑ Tool Marks
☑ Casts and Impressions
☑ Fire and Explosives

Both Labs include activities and assessment for:

☑ Crime Scene Investigation/Evidence Collection
☑ Crime Scene Observation Skills

EADS STUDENTS TO THE RIGHT CONCLUSION

Blood Spatter Lab

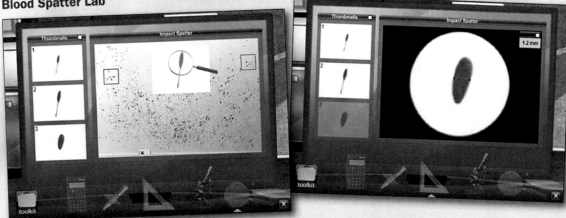

Exploring the Crime Scene Labs

The Crime Scene and Lab activities can be done together or independently.

1. Enter a 3–D Crime Scene with 360 degree view of the surroundings.
2. Accompanying text narrates the scene and prompts the user to take action.
3. Investigate and collect evidence using a virtual forensic toolkit.
4. Identify victim and determine suspects.
5. Take notes and save to a personal forensic file.
6. Enter Virtual Forensic Lab (Lab environment/ tools will change based on the type of lab being performed).
7. Step-by-step instructions guide students through the lab.
8. Worksheets filled out during lab can be saved to forensic file or submitted to instructor.
9. Complete Post Lab Assessment.

Lab Worksheet

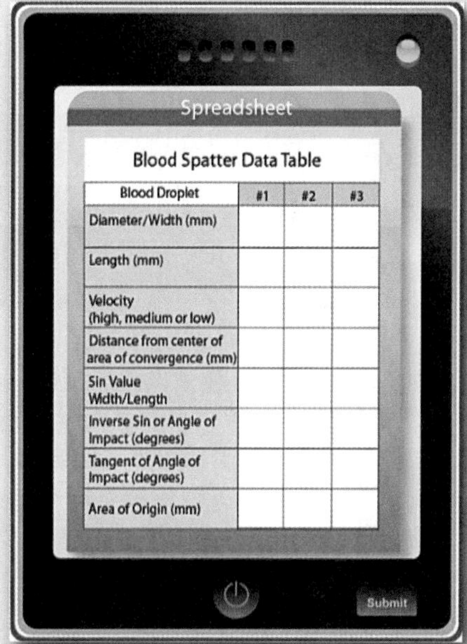

Spreadsheet

Blood Spatter Data Table

Blood Droplet	#1	#2	#3
Diameter/Width (mm)			
Length (mm)			
Velocity (high, medium or low)			
Distance from center of area of convergence (mm)			
Sin Value Width/Length			
Inverse Sin or Angle of Impact (degrees)			
Tangent of Angle of Impact (degrees)			
Area of Origin (mm)			

Submit

Post Lab Assessment

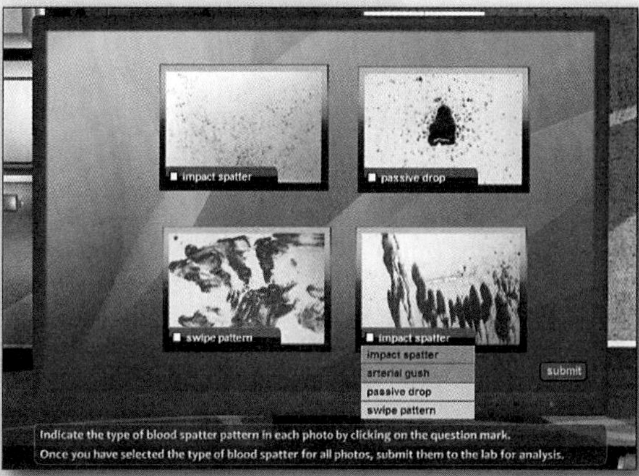

CHAPTER 1

Overview of Forensics

MURDERS AT THE MANSION

MELISSA TERAOKA

Figure 1-1. The Menendez Mansion.

On August 20, 1989, Lyle Menendez, 21, and his brother Erik Menendez, 18, went out for the evening. When they returned, they found that their parents, Jose and Kitty, had been shot and killed in the living room. At 11:47 p.m. Lyle called 911. The police arrived shortly afterward. There was no evidence of forced entry, and nothing had been stolen from the home—an indication Jose and Kitty likely knew their attackers. A witness told police she had seen two men enter the home at around 10:00 p.m. The brothers were questioned at the scene, but they were not considered suspects. No gunshot residue tests were administered.

On August 28, the brothers began cashing in on their $650,000 life insurance policy. Jose and Kitty had owed money on their mortgage and several other loans. After those had been paid, the brothers were left with a total inheritance of approximately $2 million. By the end of the year, they had spent more than $1 million. This behavior drew suspicion from police.

The brothers were arrested in March 1990. With little physical evidence, the investigators were hoping to find a link between the brothers and the guns used in the killings. Investigators searched the firearms records of a Big 5 store and uncovered the sale of two shotguns on August 18, 1989, to Donovan Goodreau of San Diego. Goodreau had an alibi for August 18 and August 20, and the signature for the firearms did not match his. A court order was issued for handwriting samples from Lyle and Erik. Erik refused to provide a handwriting sample.

On December 8, 1992, the Menendez brothers were indicted by the Los Angeles Grand Jury. The trial began on July 20, 1993. The defense admitted the brothers killed their parents; they argued that the brothers had been sexually, physically, and emotionally abused for years. Under California law, jurors had to believe the brothers feared for their lives in order to acquit them. The case resulted in a mistrial. A second trial in 1995 resulted in guilty verdicts for Lyle and Erik. They were convicted of first-degree murder and conspiracy to commit murder. On April 17, 1996, the brothers were sentenced to life in prison without the possibility of parole.

TED SOQUI/SYGMA/CORBIS

Figure 1-2. Lyle and Erik Menendez.

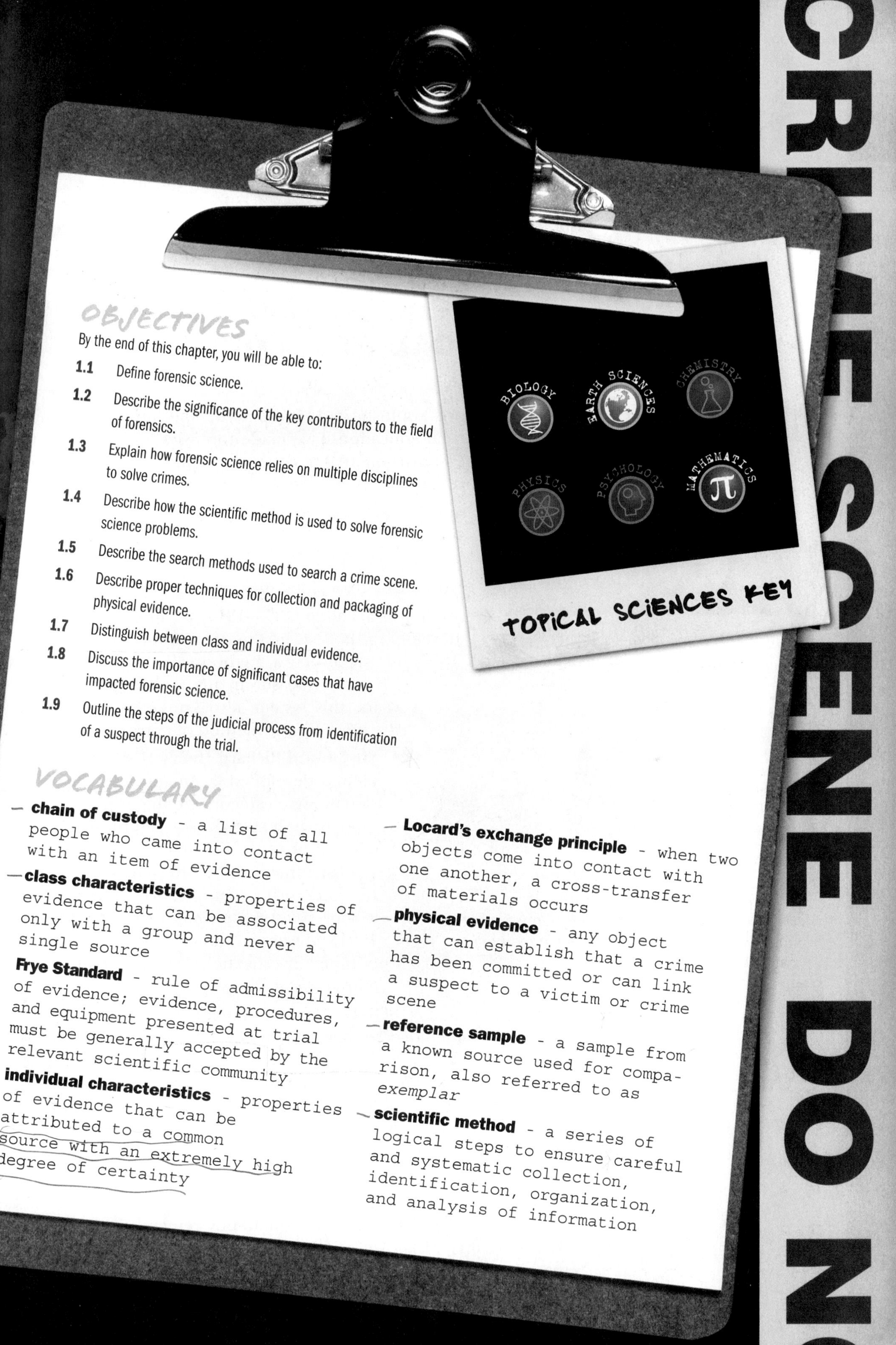

OBJECTIVES

By the end of this chapter, you will be able to:

1.1 Define forensic science.

1.2 Describe the significance of the key contributors to the field of forensics.

1.3 Explain how forensic science relies on multiple disciplines to solve crimes.

1.4 Describe how the scientific method is used to solve forensic science problems.

1.5 Describe the search methods used to search a crime scene.

1.6 Describe proper techniques for collection and packaging of physical evidence.

1.7 Distinguish between class and individual evidence.

1.8 Discuss the importance of significant cases that have impacted forensic science.

1.9 Outline the steps of the judicial process from identification of a suspect through the trial.

VOCABULARY

- **chain of custody** - a list of all people who came into contact with an item of evidence
- **class characteristics** - properties of evidence that can be associated only with a group and never a single source
- **Frye Standard** - rule of admissibility of evidence; evidence, procedures, and equipment presented at trial must be generally accepted by the relevant scientific community
- **individual characteristics** - properties of evidence that can be attributed to a common source with an extremely high degree of certainty
- **Locard's exchange principle** - when two objects come into contact with one another, a cross-transfer of materials occurs
- **physical evidence** - any object that can establish that a crime has been committed or can link a suspect to a victim or crime scene
- **reference sample** - a sample from a known source used for comparison, also referred to as *exemplar*
- **scientific method** - a series of logical steps to ensure careful and systematic collection, identification, organization, and analysis of information

Obj. 1.1 INTRODUCTION

Forensic science has a rich history dating back to humble and meager beginnings. *Forensic science* is the application of science to the law. The first recorded case ever solved using forensic evidence took place in China during the 13th century. Flies were attracted to traces of blood on a suspect's sickle. The suspect confessed. Over the past 150 years, extraordinary advancements in forensic science have led to a more sophisticated approach to solving crime.

Obj. 1.2 HISTORICAL DEVELOPMENT

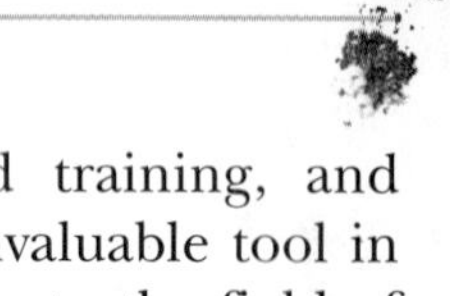

Figure 1-3. *Anthropometry relied on measurements of such features as the nose.*

More sophisticated equipment, specialized training, and formal education have made forensics an invaluable tool in the courtroom. A few of the key contributors to the field of forensics are listed below:

- **Mathieu Orfila (1787–1853)** Orfila is often referred to as the father of toxicology. In 1812, he published a dissertation on the detection of poisons and their effects on animals. This work made toxicology an accredited scientific standard.
- **Alphonse Bertillon (1853–1914)** Bertillon is recognized for developing the first method of criminal identification. In 1879, he established a method called *anthropometry* using 11 bodily measurements (see Figure 1-3). For 20 years, this system, also known as *Bertillonage,* was the most precise method of personal identification.
- **Sir Edward Richard Henry (1850–1931)** In 1896, Henry developed a fingerprint identification system that categorized fingerprints by whorl, loop, or arch pattern (see Figure 1-4). His system allowed prints to be filed, traced, and searched against other prints. After the development of his fingerprint identification system, he wrote a book, *Classification and Uses of Finger Prints.* In 1901, he went to work for Scotland Yard in the criminal investigations unit where he developed a fingerprint bureau.

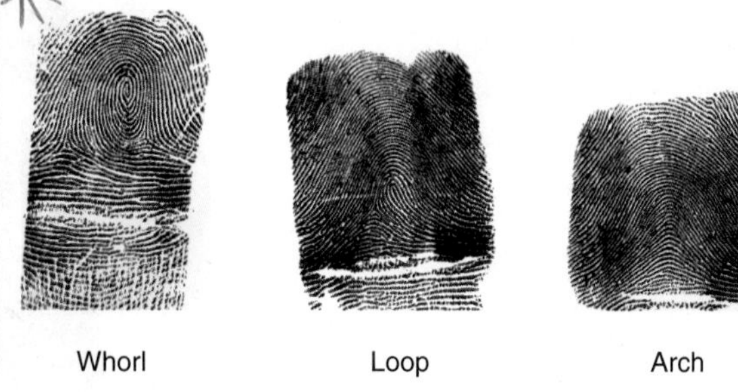

Figure 1-4. *Sir Edward Richard Henry found that fingerprints can be organized into three broad categories—whorl, loop, and arch.*

- **Karl Landsteiner (1868–1943)** Landsteiner's career spanned several decades and included studies in pathological anatomy, physiology, and immunology. In 1901, he discovered blood groups. His ABO system of blood typing is still used today. In forensics, blood typing can be used to narrow the list of possible suspects. Landsteiner received the Nobel Prize in Physiology or Medicine in 1930 for this discovery.
- **Edmond Locard (1877–1966)** Locard was the founder of the Institute of Criminalistics at the University of Lyon in Lyon, France. The Institute of Criminalistics is considered to be the flagship university in the field of forensic science. He is most prominently remembered for his cross-transfer principle known as **Locard's exchange principle**, which states that whenever two pieces of evidence come in contact with each other, there is always an exchange of materials.

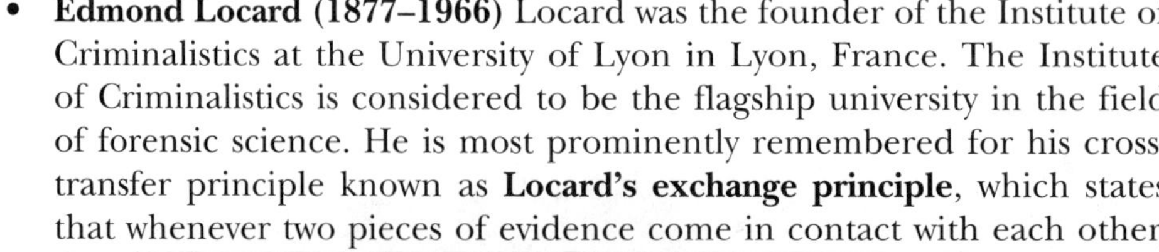

- **Albert Osborn (1858–1946)** In 1930, Osborn wrote a book called *Questioned Documents.* The published work is still in use today as a reference guide for crime-scene investigators. Osborn was widely regarded as an expert in document forgery. Initially, a questioned document was any handwritten or typed document or signature whose source was unknown or that needed to be authenticated for the purpose of a criminal investigation. Today, document examiners also study erasures, obliterations, physical characteristics of paper, watermarks, and impressions.

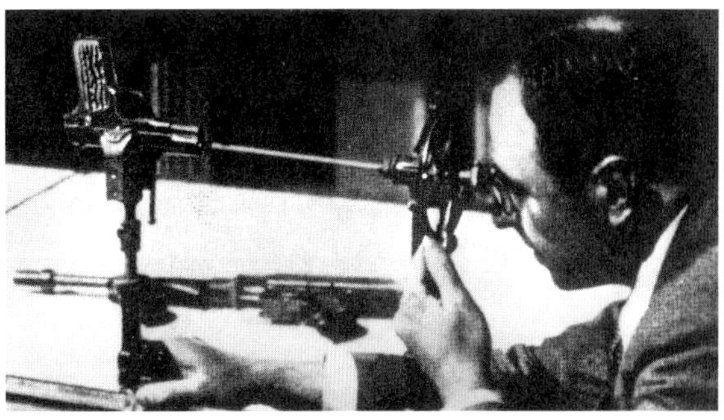

Figure 1-5. *Calvin Goddard with his comparison microscope.*

- **Calvin Goddard (1891–1955)** Goddard's interest in firearms led him to work for the Bureau of Forensic Ballistics in New York City in 1925. While working there, he invented the comparison microscope (see Figure 1-5). The comparison microscope has since become an essential tool necessary for firearms examination.
- **Rosalind Franklin (1920–1958), James Watson (1928–), and Francis Crick (1916–2004)** Franklin began studying DNA in the early 1950s. Her X-ray diffraction photograph of DNA provided significant clues about the molecule's structure. Her images of DNA suggested that the molecule looked similar to a twisted ladder. In 1953, Watson and Crick saw Franklin's photograph and were able to build a structural model of DNA. In 1962, Watson and Crick received the Nobel Prize for their discovery of the structure of DNA. This incredible contribution to the field of science made it possible for other scientists, such as Alec Jeffreys, to develop ways to use DNA to identify individuals.

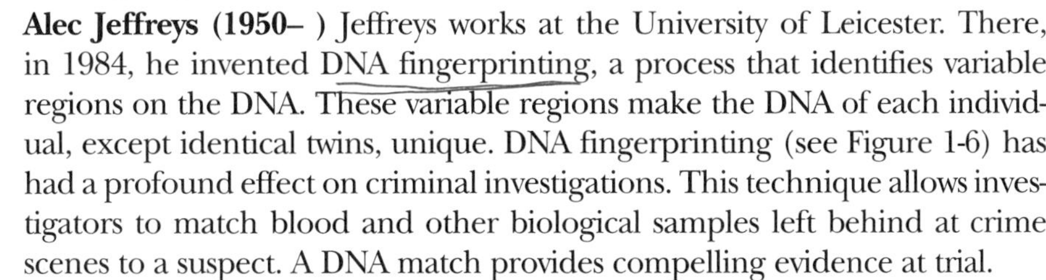

- **Alec Jeffreys (1950–)** Jeffreys works at the University of Leicester. There, in 1984, he invented DNA fingerprinting, a process that identifies variable regions on the DNA. These variable regions make the DNA of each individual, except identical twins, unique. DNA fingerprinting (see Figure 1-6) has had a profound effect on criminal investigations. This technique allows investigators to match blood and other biological samples left behind at crime scenes to a suspect. A DNA match provides compelling evidence at trial.

On February 14, 1929, seven gangsters were murdered by men dressed as Chicago police officers. Calvin Goddard was instrumental in proving that Al Capone's hit men were responsible for the St. Valentine's Day massacre.

Since the discovery of DNA fingerprinting, the advancements in DNA analysis have been extraordinary. Later in the course we will study, in depth, the current, state-of-the-art technology surrounding DNA and the implications it has had in the courtroom.

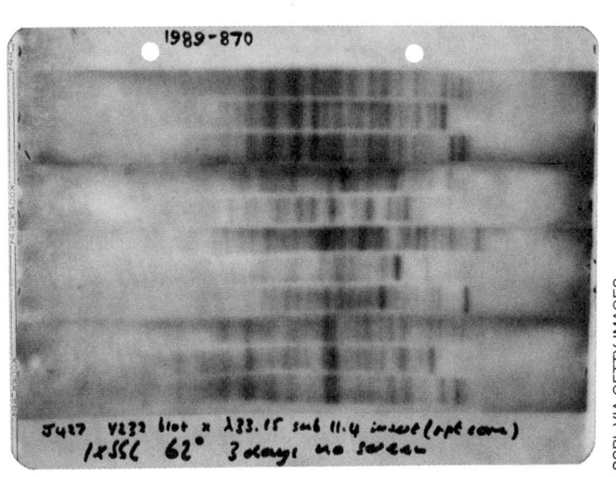

Figure 1-6. *Alec Jeffreys is shown here with a DNA fingerprint. At right is an autoradiograph of the first DNA fingerprint, which was prepared on September 19, 1984.*

Digging Deeper

with Forensic Science e-Collection

The Innocence Project is a non-profit legal organization dedicated to exonerating individuals who are wrongfully convicted and securing their release from prison. Although well intended, the Innocence Project has controversial and political issues. DNA testing or retesting of convicted criminals is allowed in some states but not in others. Using the Gale Forensic Science e-Collection at www.cengage.com/school/forensicscienceadv, research the related laws in several states. Then write a one-page position paper supporting one state's interpretation.

Obj. 1.3 A MULTIDISCIPLINARY APPROACH

Did You Know?

The first case ever solved using DNA fingerprinting was in 1988. Colin Pitchfork was convicted of murdering two young girls: one in 1983 and one in 1986 in Narborough, Leicestershire, England. Another suspect in this case was the first person ever exonerated using DNA analysis.

Forensic science is a multifaceted discipline, drawing on various subject areas such as history, math, technology, language, and reasoning skills (see Figure 1-7). Careers in forensics that require professionals to use multiple disciplines include:

- Forensic nurse
- Forensic chemist
- Forensic toxicologist
- Forensic meteorologist
- Forensic accountant

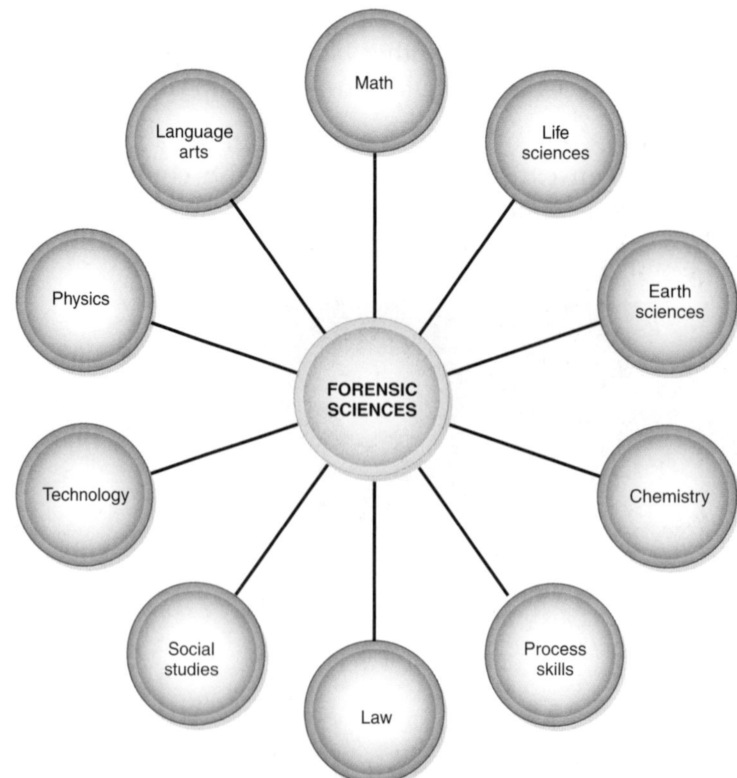

Figure 1-7. *Forensic science is a multidisciplinary subject.*

Each of these careers requires not only training and experience in the core field but also knowledge of the historical milestones of the field and the ability to use the most current technology and equipment to provide a descriptive and complete analysis of the evidence.

THE SCIENTIFIC METHOD

Obj. 1.4

Forensic science begins at the crime scene, the place where an incident took place. Investigators rely on the scientific method (see Figure 1-8). The **scientific method** is a series of logical steps used to solve a problem. Once the crime has been discovered and the police arrive at the scene, the *problem* has been established. Police detectives and crime-scene investigators use the scientific method to observe and collect the physical evidence found at the crime scene. The police and investigators carefully evaluate the scene and the surrounding areas. The police will then develop a hypothesis. The hypothesis is an attempt to answer the following questions: what happened, how did it happen, and when did it happen? The crime-scene investigator does not form a hypothesis. Instead, he or she sends the evidence to the crime lab for further analysis.

Once the physical evidence has been collected from the crime scene, it is important to establish a thorough and systematic procedure to process the evidence. Physical evidence will be sent to the lab for testing. Consider the following: Hair removed from a crime scene is processed. Tests reveal color, texture, and shape, and the toxicology report shows no use of

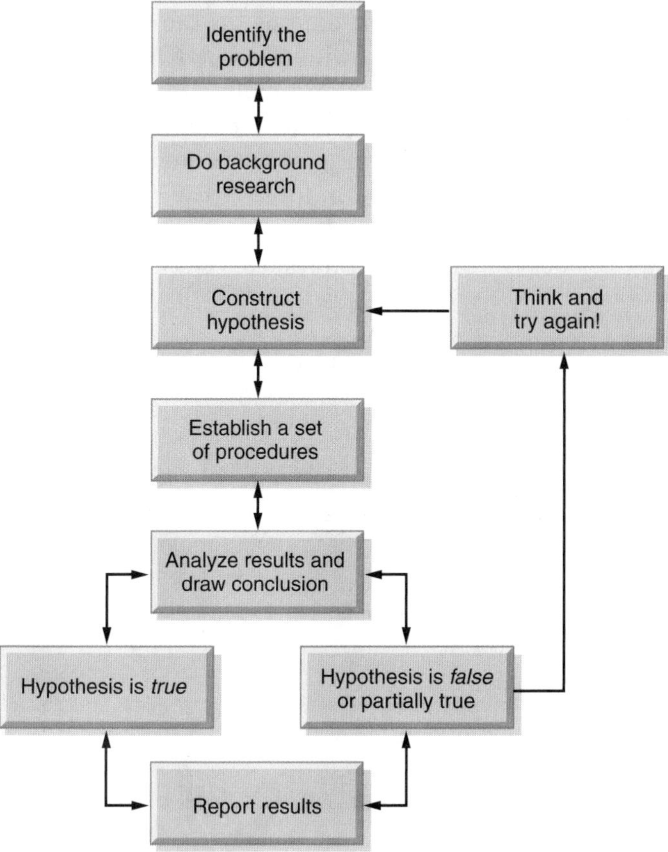

Figure 1-8. *The scientific method.*

drugs. No follicular tag is present. Therefore, there is no nuclear DNA. The tests are not likely to link the hair to one person. However, the tests can help the investigators reduce the number of potential suspects. After all of the physical evidence has been processed and evaluated, a conclusion is drawn. If a suspect can be linked to a crime, then the suspect is apprehended.

Digging Deeper
with Forensic Science e-Collection

There are actually two types of DNA—nuclear DNA and mitochondrial DNA. Nuclear DNA is found in all nucleated cells and is used in evaluating evidence such as blood and semen. Mitochondrial DNA is found in the mitochondria of eukaryotic cells and is passed down to offspring through maternal lineage. Using the Gale Forensic Science e-Collection at www.cengage.com/school/forensicscienceadv, conduct research and type a one-page summary answering the following questions: (1) What is mitochondrial DNA? (2) Compare and contrast mitochondrial DNA and nuclear DNA. (3) Complete an analysis of a case in which mitochondrial DNA was used to solve a crime.

Obj. 1.5, 1.6, 1.7

THE CSI TEAM AND CRIME LABS

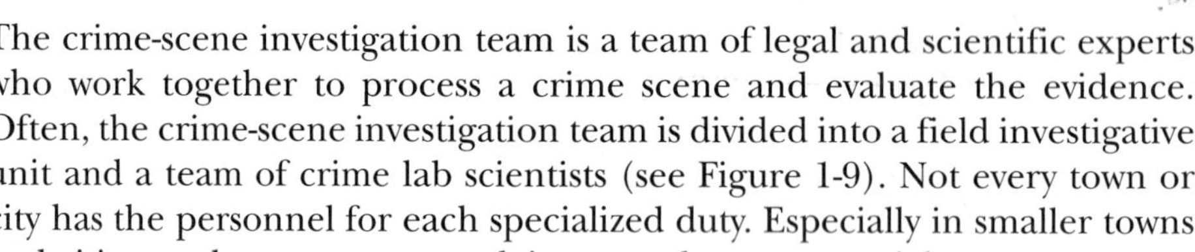

The crime-scene investigation team is a team of legal and scientific experts who work together to process a crime scene and evaluate the evidence. Often, the crime-scene investigation team is divided into a field investigative unit and a team of crime lab scientists (see Figure 1-9). Not every town or city has the personnel for each specialized duty. Especially in smaller towns and cities, each person may work in more than one specialty.

Police officers are usually the first to arrive at the scene. They secure the scene and provide first aid to anyone who needs it. Crime-scene investigators record details about the crime scene and collect physical evidence. Crime-scene investigators include photographers, sketch artists, and evidence collectors. Depending on the crime, medical examiners and district attorneys may also arrive at the scene. As a team, these experts are looking for clues that will link the evidence to a suspect or victim. In this way, they take a holistic approach. Collectively, they begin to piece together the whole picture.

Edger Allen Poe's *Murders in the Rue Morgue* was published in 1841 and is the first detective story on record.

Field Investigative Unit	Crime Lab Scientists
• Secure the crime scene	• Receive/sign for evidence
• Photograph the crime scene	• Review paperwork
• Search the crime scene	• Complete chemical and/or physical tests on evidence
• Properly collect and package evidence	• Complete an analysis of the findings
• Complete proper forms	• Provide expert testimony
• Evidence submission form	
• Chain of custody	
• Deliver or ship evidence to proper processing site	
• Provide expert testimony	

Figure 1-9. *Responsibilities of the crime-scene investigation team.*

Evidence collected at the scene is sent to the crime lab for analysis. Forensic scientists at the crime lab specialize in fields such as chemistry, toxicology, pathology, and firearms. These scientists remain completely neutral in their analysis. They do not form a hypothesis or draw conclusions about guilt or innocence. Information gathered from chemical and physical tests can be used to establish a timeline of events. It may also corroborate a witness's statement. Poor communication could lead to critical mistakes. Also, if mistakes are made, the entire process is in jeopardy. For example, if the evidence is not collected properly, investigators in the crime lab may not make the correct analysis. Proper training is essential for experts in the field as well as in the lab to reduce the risk of errors.

When necessary, experts in specialized fields are called in to assist with cases requiring very specific training. For example, forensic entomologists study insect evidence. Forensic odontologists gather evidence from dental records.

PROCESSING A CRIME SCENE

The investigators work together to collect all of the clues from the crime scene. The team must use a systematic and thorough approach. The first authorized personnel to arrive must secure the scene, usually with crime-scene tape, and move unauthorized people away from the area. If someone at the scene is in need of medical attention, the responding officer will help that person before doing anything else. Depending on the circumstances, people near and around the crime scene may be detained and questioned. Before any evidence is collected, the entire scene must be sketched and photographed (see Figure 1-10). Evidence must be left undisturbed in its original location. If circumstances warrant moving the evidence, to tend to an injured victim for example, it must be noted in the records. Measurements of the crime scene and locations of items of evidence must also be documented.

REUTERS/SAM MIRCOVICH

Figure 1-10. *Evidence being photographed.*

After the crime scene and the evidence have been photographed, the field investigator will begin to collect and package the **physical evidence**, material found at the scene that may be integral to solving a crime. The size and location of a crime scene will determine the search method used (see Figure 1-11). Examples of various search methods and the type of crime include:

- Zone—building or other structure—homicide, home invasion, robbery, sexual assault, etc.
- Spiral—large area, no barriers—open field—kidnapping, homicide
- Line search—large area looking for a large object in a single direction—site of a plane crash
- Grid—large area looking for a large object in two directions—arson investigation

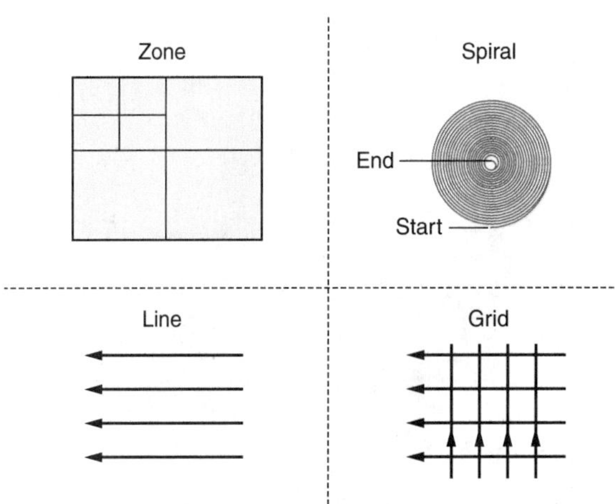

Figure 1-11. *Investigators choose the search method based on the size and location of the crime scene.*

EVIDENCE

The analysis of physical evidence and the presentation of findings in court are the most powerful resources available to a prosecutor. Often, the physical evidence presented at court can be the determining factor of guilt or innocence of the suspect. Two basic types of evidence exist—testimonial and physical. Testimonial evidence, gathered by law enforcement or crime-scene investigators, is the witness testimony used to build a timeline of events or to confirm a suspect's whereabouts.

Physical evidence may be any material collected or observed at a crime scene that could link potential suspects to a crime (see Figure 1-12). The evidence can range in size from a drop of blood to large objects, such as furniture or a door. Examples of physical evidence include a document, hair, fibers, fingerprints, soil, and blood. Each type of evidence must be collected using specific techniques and guidelines set up by the crime lab to preserve the integrity of the evidence.

The evidence must be collected and packaged properly so that a crime-scene technician can identify and compare physical evidence. Each piece of evidence is packaged differently based on the chemical and physical properties of the evidence. For example, bloody clothing is first air dried

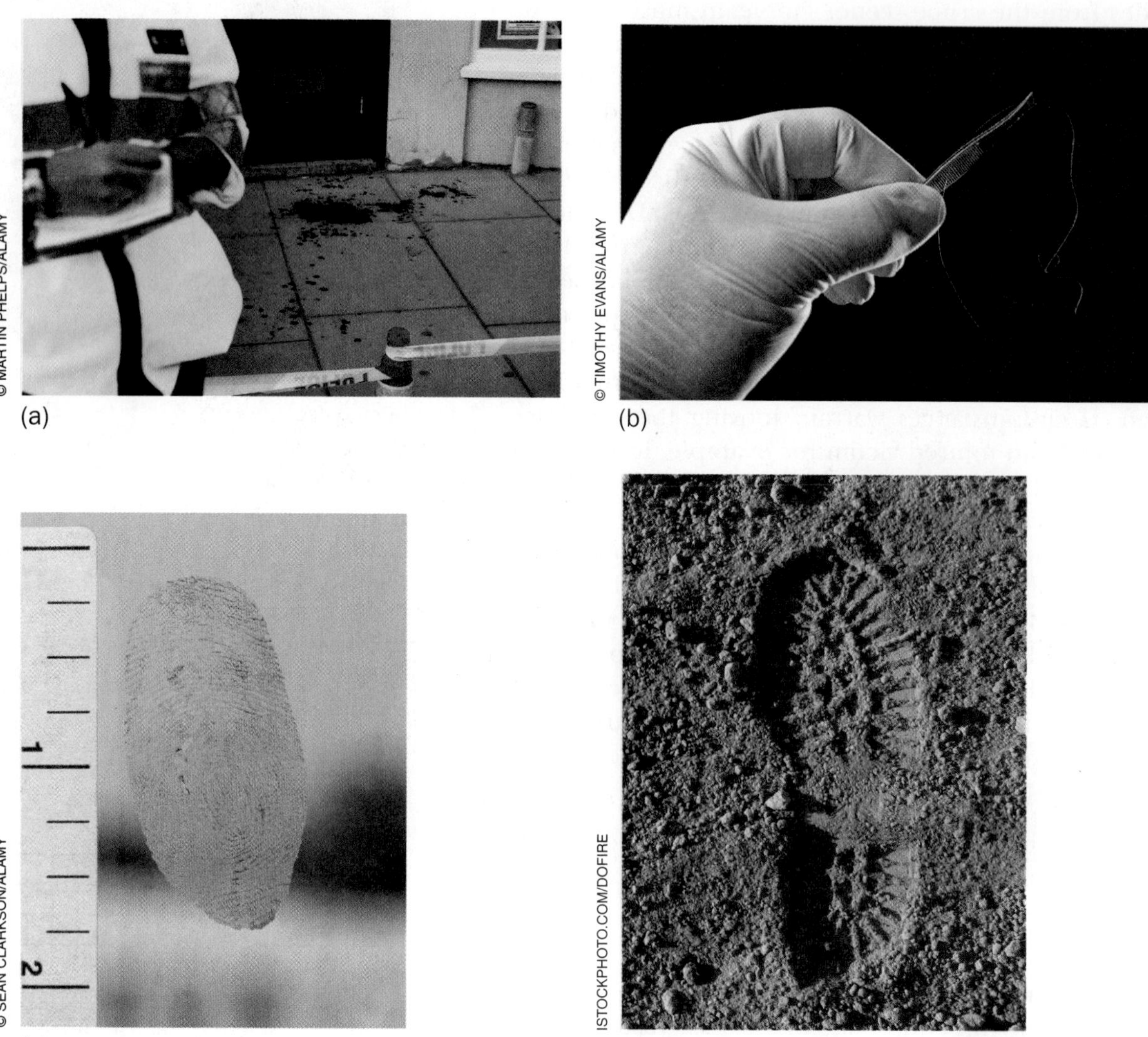

Figure 1-12. *Blood spatter (a), hair (b), fingerprints (c), and footprints (d) are all examples of physical evidence.*

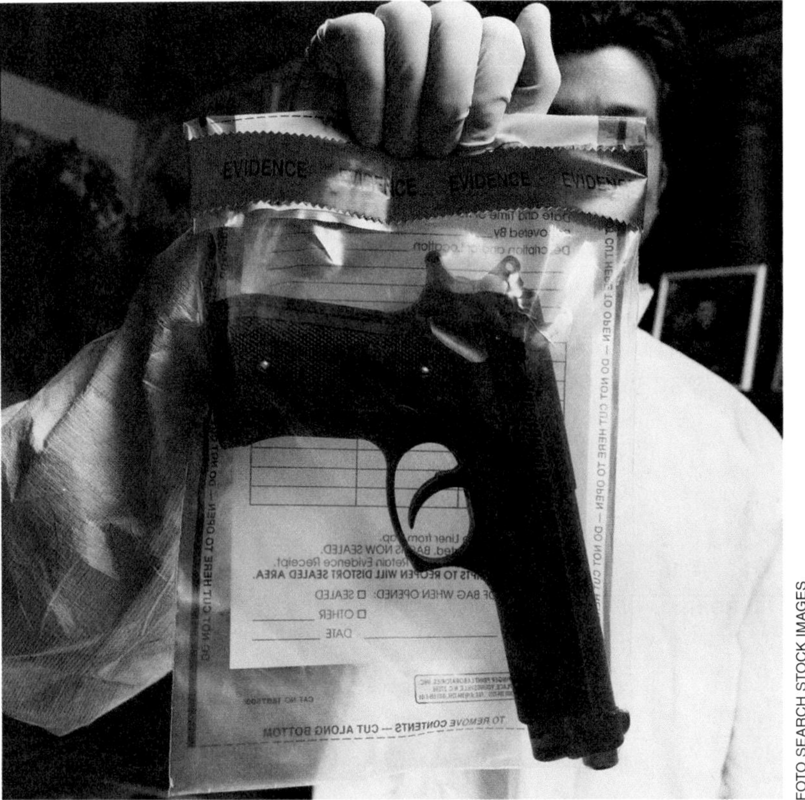

Figure 1-13. *A gun collected from a crime scene is packaged in a sealed plastic bag.*

and then packaged in a paper bag. A pool of blood will be swabbed. The swab is then air dried and placed in a plastic vile. In each case, the blood evidence must be dried prior to storage in order to prevent mold and bacterial growth. Each piece of evidence collected must be packaged separately to avoid cross-contamination (see Figure 1-13).

Once an item of evidence is collected, a **chain of custody** is established. The chain of custody provides documentation of every person who has come into contact with the evidence (see Figure 1-14 on page 12). The chain of custody acts as a "paper trail." The name of the person who collected and packaged the evidence is recorded. After that, the name of each person who has handled, analyzed, or transported the evidence is recorded. Every time the evidence is opened, a new seal is formed. The person who handled the evidence then signs over the new seal. The chain of custody demonstrates to the courts that the evidence that is being presented at trial is free of contamination, alteration, or substitution. The chain of custody preserves the integrity of the evidence and provides a mechanism of accountability.

Sometimes, very small amounts of certain physical evidence are found at the crime scene. This evidence is called *trace evidence* and can include gunshot residue, a tiny amount of blood, a single hair, or paint. According to Locard's exchange principle, contact between a victim and a suspect or between a suspect and a location results in a transfer of material. In an effort to prevent contaminating or destroying the evidence, field investigators do not remove trace evidence, if possible, from the object that bears it. Highly trained field investigators are able to collect and bag the most apparent evidence as well as any possible carriers of trace evidence. For example, a soda can may be collected and sent to the crime lab for testing because fingerprints were visible on the can (see Figure 1-15). Additional tests might reveal lip prints and saliva on the rim and drugs dissolved inside the can.

CHAPTER

1

2

3

4

5

6

7

8

9

10

11

12

13

14

15

West Virginia Department of Health and Human Resources
Bureau of Public Health, Office of Laboratory Services
Threat-Preparedness and Response Section

☐ 167th 11th Avenue
South Charleston, WV 25303
Phone: 304-558-3530 x 2301
Fax: 304-558-2006

☐ 4710 Chimney Drive, Suite G
Charleston, WV 25302
Phone: 304-558-0197
Fax: 304-558-4143

EVIDENCE/CHAIN OF CUSTODY DOCUMENT

1. SAMPLE:	2. DATE COLLECTED:	3. EOC NUMBER:
4. COLLECTED BY:	5. COUNTY:	6. OLS LAB NUMBER:

7. SAMPLE DESCRIPTION: (Number, Quantity, Type, Packaging, etc.)

☐ Clinical Sample
☐ Environmental Sample

8. INCIDENT LOCATION:

9. SUSPECTED BIOLOGICAL/CHEMICAL TERRORISM AGENT(S):

Bio-terrorism Agent(s)

☐ Anthrax, ☐ Plague, ☐ Tularemia, ☐ Brucella,

☐ Unknown, ☐ Other ____________________

Chemical Terrorism Agent(s)

☐ Nerve Agent, ☐ Blister Agent/Vesicant,

☐ Blood Agent, ☐ Choking Agent/Irritant Agent,

☐ Riot Control, ☐ Unknown, ☐ Other ____________

Rationale:

10. SAMPLE RECEIVED FROM:

Organization: ______________________ Date: __________ Time: ____________

Address: ______________________________________ Phone: ____________

Received from: ____ [Sign in Section 11]

Witnessed by: ______________________________________ Date: ____________

Received via: ☐ US Mail, ☐ Hand Delivered, ☐ Shipped via ____________________________

11. SAMPLE ACKNOWLEDGMENT:

Sample Received from: [signature]	Date/Time:	Sample Received by: [signature]	Sample Received by: [print name]	Remarks:

12. SAMPLE RELEASED TO:

Organization: ______________________ Date: __________ Time: ____________

Address: ______________________________________ Phone: ____________

Received by: ____ [Sign in Section 11]

Witnessed by: ______________________________________ Date: ____________

Transferred via: ☐ US Mail, ☐ Hand Delivered, ☐ Shipped via ____________, ☐ Sample Destroyed, Date: ________

13. SAMPLE STORAGE CONDITIONS:

January 22, 2004 Attach additional pages as required Page _____ of _____

WEST VIRGINIA DEPARTMENT OF HEALTH & HUMAN RESOURCES

Figure 1-14. *Sample chain of custody form.*

When investigators submit evidence to the lab for processing and analysis, they include an evidence submission form. The form lists each item of evidence, a brief description of each item, and each testing procedure requested. Items of evidence are also identified and compared in the lab. Identification determines the identity of a piece of evidence. Comparison determines whether the evidence can be linked to a victim, location, or suspect. Evidence may also be compared to known samples to determine the kind of weapon or to build a timeline of events. **Reference samples**, samples from a known source, are used as a basis of comparison. For example, a white powder is found at a crime scene. Upon processing, the white powder is identified as cocaine mixed with some inorganic substances. However, there is not yet enough information to link the substance to a suspect. If a suspect is apprehended and traces of white powder are present in his or her pocket, the powder can be compared to the cocaine mixture found at the crime scene. If the two samples are consistent, the suspect may be arrested. Essentially, identification answers the question "What is it?" and comparison answers the question "Where did it come from?"

Figure 1-15. *This soda may provide trace evidence.*

Analysis of crime-scene evidence will attempt to answer one of two questions:

1. Does the analysis provide enough information to reduce the number of potential suspects?
2. Does the analysis provide enough information to know, with certainty, who committed the crime in question?

If the evidence is determined to possess **class characteristics**, it may serve as a mechanism to reduce the number of suspects. Evidence with class characteristics cannot be directly connected to one person or source. Examples of class evidence include blood type, fibers, and paint. When evidence exhibits **individual characteristics**, it can be matched to a single source with a high degree of probability. Examples of individual evidence include anything that contains nuclear DNA, such as a hair follicle, blood cells, or semen. Tool marks and fingerprints are also classified as individual evidence because of the uniqueness of their patterns. The analysis of all of the evidence helps law enforcement personnel form, support, and even disprove hypotheses about what happened when and about who might have been involved.

All of the evidence can then be used to prove the investigators' conclusions in court. Figure 1-16 illustrates both individual and class characteristics.

LANDMARK CASES

Obj. 1.8

As forensic science has continued to evolve, how evidence is processed, what type of evidence is admissible in court, and the equipment used to analyze evidence have all been scrutinized. Several cases have led to modifications in how forensic experts process evidence and in the types of evidence presented at court.

FRYE V. UNITED STATES

In 1923, in the case of *Frye v. United States,* James Frye was convicted of second-degree murder. He had confessed to the murder, but later retracted his confession. Frye appealed his conviction on the grounds that the lie detector test he had taken proved he was telling the truth. He also had an

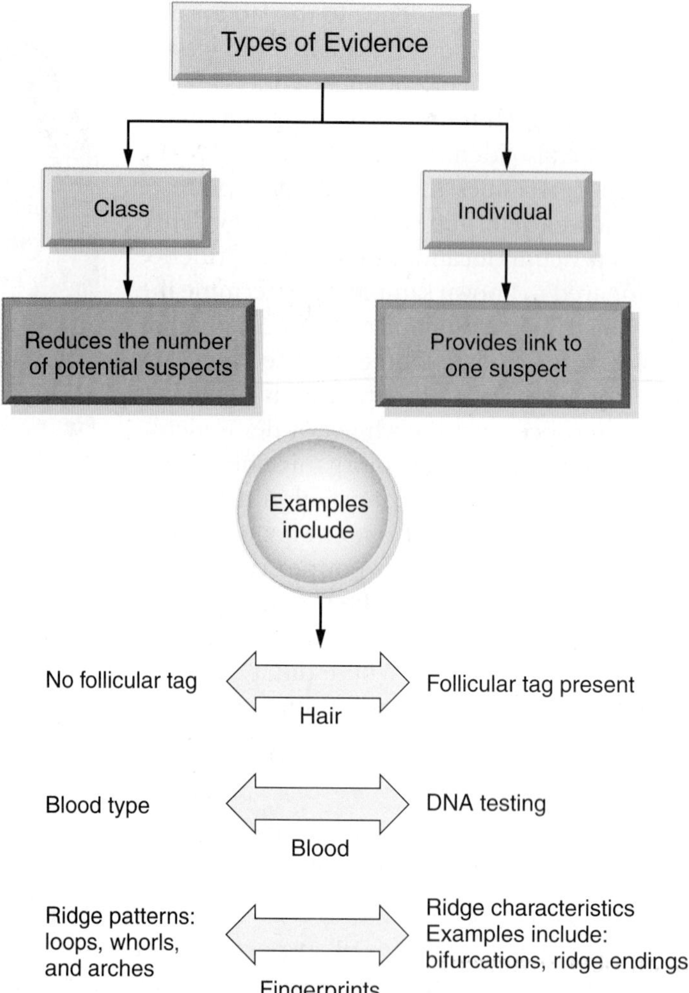

Figure 1-16. *Individual and class evidence.*

expert prepared to explain the scientific methods used to determine the outcome of his lie detector test. Nonetheless, the trial court refused to allow the expert's testimony. Frye's argument provided great detail of the science behind how the lie detector test functioned. The Supreme Court upheld the lower court's decision, stating one of the most famous opinions ever sent down by federal court. The opinion was as follows:

> "Just when a scientific principle or discovery crosses the line between the experimental and demonstrable stages is difficult to define. Somewhere in this twilight zone the evidential force of the principle must be recognized, and while courts will go a long way in admitting expert testimony deduced from a well-recognized scientific principle or discovery, the thing from which the deduction is made must be sufficiently established to have gained general acceptance in the particular field in which it belongs."

Essentially, *Frye v. United States* stated that lie detector tests lacked scientific backing and were not considered to be "generally accepted" by the relevant scientific community. This case led to what is now commonly called the **Frye Standard**. According to the Frye Standard, evidence, procedures, and equipment presented at trial must be generally accepted by the scientific community.

Digging Deeper
with Forensic Science e-Collection

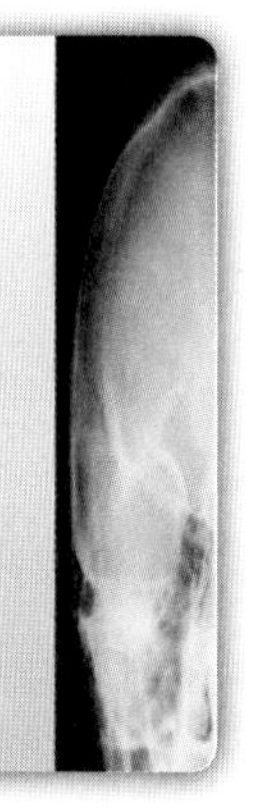

Historically, science and technology have always overlapped. Since the discovery of DNA fingerprinting, phenomenal technological advances have given investigators new tools necessary to identify and convict criminals. Using the Gale Forensic Science e-Collection at www.cengage.com/school/forensicscienceadv, research and write a one-page summary answering the following questions: (1) What are some branches of forensic science that have benefited from advancements in technology? (2) How has technology challenged the credibility of forensic science?

DAUBERT v. MERRELL DOW PHARMACEUTICALS

Daubert v. Merrell Dow Pharmaceuticals involved Jason Daubert and Erik Schuller, minor children who were born with birth defects. Their mothers had both been prescribed Bendectin, an anti-nausea medication, during their pregnancies. The women thought that the Bendectin had caused their sons' birth defects. The lawsuit against the manufacturers of Bendectin, Merrell Dow Pharmaceuticals, was filed in 1993 and named the children as the petitioners. The case was heard by the U.S. Supreme Court. The Court ruled that the trial judge had ultimate decision-making power regarding expert testimony at trial. The ruling suggests criteria for evaluating the science used to support evidence presented by an expert:

1. Has it been tested?
2. Has it been peer reviewed?
3. What is the rate of error?
4. Is it generally accepted?

Figure 1-17 compares the way the Frye and Daubert cases changed the way scientific evidence could be used in court.

Frye v. United States
1. Scientific evidence must meet "general acceptance" standards.
2. Any evidence, procedure, or equipment presented at trial must gain acceptance from the scientific community before becoming admissible in court.
1. Recognizes that science is constantly changing.
2. Decisions affected the use of expert testimony at trail.
Daubert v. Merrell Dow Pharmaceuticals
1. Trial judge has ultimate decison-making power regarding expert testimony used at trial.
2. Expert testimony is not automatically admissible. The judge must consider whether the underlying science is generally accepted and based on testable theory and whether the procedures have undergone peer review and have a reasonable error rate.

Figure 1-17. *The Frye and Daubert cases changed the way scientific evidence could be used in court.*

AP PHOTO/TOBY MASSEY

Figure 1-18. *Carl Coppolino is being led by a police officer after being convicted of second-degree murder.*

DR. CARL COPPOLINO

BIOLOGY

As science continues to advance, so do the tests and equipment used to analyze and process evidence. In 1963, William Farber was found dead in his home in New Jersey. Doctors thought that Farber had died from heart failure. Two years later, the wife of Mrs. Farber's lover was also found dead in her home in Florida. At first, the family physician thought Carmela Coppolino had died of a heart attack. However, Mrs. Farber convinced prosecutors that Dr. Carl Coppolino had poisoned William Farber and Carmela Coppolino. Both bodies were exhumed and Coppolino had two separate trials. The first trial, in New Jersey, was for the murder of Farber; the second trial, for the murder of Coppolino's wife, was held in Florida. Coppolino was acquitted of the first murder. The judge in the Florida case, however, admitted evidence from the toxicologist. The toxicologist used a newly developed test to screen for succinylcholine chloride, a paralytic. A toxicology analysis revealed elevated levels of succinic acid, the metabolite of succinylcholine chloride, in Carmela's brain. At trial, the defense argued that this new procedure had not yet been widely accepted by the scientific community. The toxicology screening completed was an experimental test with little scientific backing by the toxicology community. In the Coppolino case, the court ruled that the fact that a technique, test, or procedure is new does not necessarily mean its findings are inadmissible in court. The expert witness, however, is responsible for providing scientifically valid testimony to support the findings. As a result, Coppolino was convicted of the second-degree murder of his wife (see Figure 1-18).

Obj. 1.9

THE JUDICIAL PROCESS

The U.S. Constitution was signed in 1787. The Constitution set up the executive, legislative, and judicial branches of the U.S. government. In 1789, Congress added 10 amendments to the U.S. Constitution. These 10 amendments are called the Bill of Rights (see Figure 1-19). The sixth amendment ensures that a person will be tried by an impartial jury of his or her peers. The jury listens to arguments by both the defense and prosecution. The jury hears information about physical and circumstantial evidence and hears testimony from witnesses. The jury uses all of this information to make an informed decision about the guilt or innocence of the person on trial. The jury is instructed to assume that the defendant is innocent until proven guilty. In criminal cases, the prosecution must prove, beyond a reasonable doubt, that the defendant committed the crime for the jury to return a guilty verdict.

ARREST

Before a trial can begin, the suspect needs to be arrested and charged with the crime. The fourth amendment protects against unreasonable search and seizure. This means that law enforcement personnel must show probable cause before they can arrest or search a suspect or search the suspect's property. There are four broad categories of evidence gathering that may be used to show probable cause:

1. **Observation:** Law enforcement personnel gather information using their five senses. The information may provide a direct link to a crime, or law enforcement personnel may detect a familiar pattern.

First Amendment	Congress shall make no law respecting an establishment of religion, or prohibiting the free exercise thereof; or abridging the freedom of speech, or of the press; or the right of the people peaceably to assemble, and to petition the government for a redress of grievances.
Second Amendment	A well regulated militia, being necessary to the security of a free state, the right of the people to keep and bear arms, shall not be infringed.
Third Amendment	No soldier shall, in time of peace be quartered in any house, without the consent of the owner, nor in time of war, but in a manner to be prescribed by law.
Fourth Amendment	The right of the people to be secure in their persons, houses, papers, and effects, against unreasonable searches and seizures, shall not be violated, and no warrants shall issue, but upon probable cause, supported by oath or affirmation, and particularly describing the place to be searched, and the persons or things to be seized.
Fifth Amendment	No person shall be held to answer for a capital, or otherwise infamous crime, unless on a presentment or indictment of a grand jury, except in cases arising in the land or naval forces, or in the militia, when in actual service in time of war or public danger; nor shall any person be subject for the same offense to be twice put in jeopardy of life or limb; nor shall be compelled in any criminal case to be a witness against himself, nor be deprived of life, liberty, or property, without due process of law; nor shall private property be taken for public use, without just compensation.
Sixth Amendment	In all criminal prosecutions, the accused shall enjoy the right to a speedy and public trial, by an impartial jury of the state and district wherein the crime shall have been committed, which district shall have been previously ascertained by law, and to be informed of the nature and cause of the accusation; to be confronted with the witnesses against him; to have compulsory process for obtaining witnesses in his favor, and to have the assistance of counsel for his defense.
Seventh Amendment	In suits at common law, where the value in controversy shall exceed twenty dollars, the right of trial by jury shall be preserved, and no fact tried by a jury, shall be otherwise reexamined in any court of the United States, than according to the rules of the common law.
Eighth Amendment	Excessive bail shall not be required, nor excessive fines imposed, nor cruel and unusual punishments inflicted.
Ninth Amendment	The enumeration in the Constitution, of certain rights, shall not be construed to deny or disparage others retained by the people.
Tenth Amendment	The powers not delegated to the United States by the Constitution, nor prohibited by it to the states, are reserved to the states respectively, or to the people.

Figure 1-19. *The Bill of Rights.*

2. **Expertise:** Law enforcement personnel use specialized skills and training to gather evidence. For example, a police officer may be trained to identify movements, gestures, or graffiti.
3. **Information:** Law enforcement personnel may gather information from witnesses, informants, or victims.
4. **Circumstantial evidence:** Law enforcement personnel may gather evidence that implies, but does not prove, a crime has been committed or that the suspect was involved.

Law enforcement personnel must convince a judge that the crime took place and that there is probable cause to suggest that the suspect was involved. If these two criteria are met, the judge will sign a search or arrest warrant (see Figure 1-20). It is sometimes necessary for law enforcement personnel to perform a search or make an arrest without a warrant. For example, the suspect, victim, or bystanders may be in immediate danger or there may be imminent risk of evidence being destroyed. Officers may also perform a search if the suspect consents.

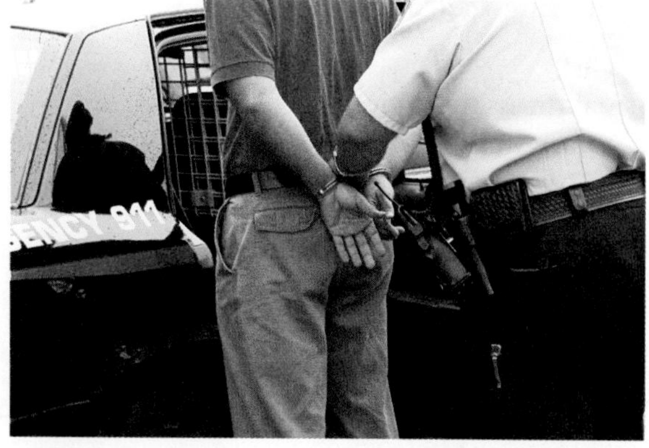

Figure 1-20. *Once an arrest warrant is served, the suspect is taken into custody.*

When a suspect is brought in for questioning, Miranda rights are read. This means that the officer must point out that anything the suspect says can be used in court and that the suspect has the right to remain silent. The suspect also has a right to an attorney. The suspect may hire an attorney or use a court-appointed attorney. After a suspect is arrested, an arraignment is held. At the arraignment, the suspect is formally charged and a plea of guilty or not guilty is entered. Depending on the type of crime, bail will be set. In capital murder cases, bail is usually not set and the suspect must remain in jail until the trial.

BEFORE THE TRIAL

After an arraignment, pretrial conferences take place between the defense and the prosecution. Both sides share information so that each side can prepare arguments. This is the discovery stage of the process. If the prosecution is going to offer a plea agreement, it will do so during discovery. A plea agreement often requires the defendant to plead guilty to a lesser charge in return for a lighter sentence.

At the preliminary hearing, the judge hears both sides and determines whether there is sufficient evidence to take the case to trial. If the judge determines enough evidence is present, then a court date is set for trial to begin. If the judge determines that there is not enough evidence, the case is dismissed. In several states and in federal cases, such as capital murder cases, a grand jury hears from the prosecution and determines whether there is enough evidence to justify a trial.

AT THE TRIAL

At the trial, the defense and prosecution present their cases to the judge and jury (see Figure 1-21). Each is given a chance to call witnesses and to cross-examine witnesses. During cross-examination, attorneys question, and try to discredit, the opposing side's witnesses. Jurors then leave the courtroom

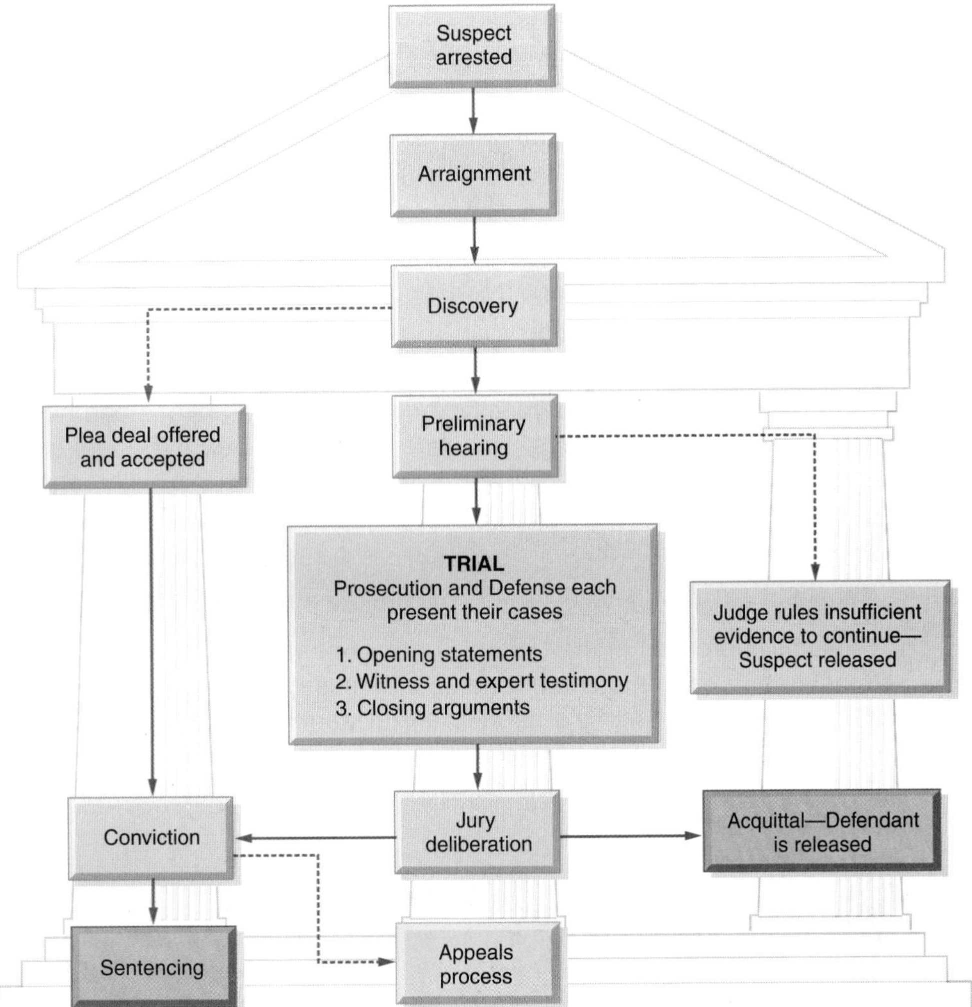

Figure 1-21. *An outline of court proceedings.*

to deliberate. In criminal cases, the verdict must be unanimous—everyone on the jury must agree with the verdict. If the verdict is "not guilty," the defendant is released. If the verdict is "guilty," the defendant must return to court on the date of sentencing. Upon sentencing, the defendant will be remanded to a state or federal correctional facility. If an appeal is filed, attorneys may not present new evidence in appellate court. The arguments must focus on legal and procedural errors made in the original court. Once a case has been appealed to the highest court in the state, it may move to the U.S. Supreme Court if the case involves interpretation of the U.S. Constitution. However, most are decided before they go that far.

CHAPTER SUMMARY

- Forensic science is the application of science to law. Forensic scientists rely on the scientific method for processing and evaluating evidence.
- Many disciplines—math, technology, history, language, and reasoning skills—are utilized in forensic science. Forensic scientists specialize in fields such as chemistry, toxicology, pathology, and firearms.

CHAPTER 1 2 3 4 5 6 7 8 9 10 11 12 13 14 15

- Processing a crime scene requires a thorough and systematic approach. The size and location of the crime scene determine the search method used.
- Physical evidence is any substance that can link a potential suspect to a victim or crime. Examples can include footprints, tire tracks, hair, fiber, fingerprints, or blood.
- When processing evidence, investigators use identification and comparison tests to attempt to make connections between victims, suspects, and crime scenes.
- Evidence that possesses class characteristics can be associated with a group rather than with a specific individual. This evidence can help narrow the field of suspects. Evidence that possesses individual characteristics can be associated with a single person with a high degree of certainty. This sort of evidence can be used to show that a specific individual was involved.
- A chain of custody is necessary to maintain the integrity of the evidence.
- The judicial process in the United States is based on the U.S. Constitution. Law enforcement personnel must prove probable cause before a judge will sign a search or arrest warrant. In court, a jury of the suspect's peers hears evidence from both sides. The jury is instructed to assume that the suspect is innocent until proven guilty.

CASE STUDIES

Ernesto Miranda (1963)

In 1963, Ernesto Miranda (see Figure 1-22) was arrested and charged with kidnapping and raping an 18-year-old woman. The Constitution gives citizens the right to remain silent and to have an attorney present when being questioned by police. However, when Miranda was arrested, there was no rule in place requiring police to remind people of these rights. Miranda was brought in for questioning and confessed to the crime. He was never reminded of his right to remain silent or to hire an attorney. At trial, Miranda's attorney tried to get Miranda's confession thrown out. The court denied the request, and Miranda was convicted. In 1966, the case was heard by the U.S. Supreme Court. The lower court's ruling was overturned. The Supreme Court ruled that citizens should be told their rights before they are questioned. Later, physical evidence was used to convict Miranda.

BETTMANN/CORBIS

Figure 1-22. *Ernesto Miranda.*

Think Critically

1. **What impact did the Miranda case have on the judicial process?**
2. **Outline the steps that took place between Miranda's confession and the ruling by the U.S. Supreme Court.**
3. **Which Constitutional amendment(s) were relevant to this case?**

The Patrick Gilligan Family (1980)

On January 14, 1980, Patrick Gilligan and his wife Teresa found an intruder in their Evansville, Indiana, home. The intruder, who was surprised by the Gilligans, tied Patrick, Teresa, and their two young children up. He then shot and killed Patrick, Teresa, Lisa (5 years old), and Gregory (4 years old). Before leaving the scene, the intruder stole guns, a police scanner, a CB radio, and some other property from the home.

Evidence led investigators to Donald Ray Wallace (see Figure 1-23). An eyewitness identified him. Upon searching his home, investigators found blue jeans that were covered with blood. Lab tests confirmed that the blood was that of the victims. Several of Wallace's acquaintances testified in court that Wallace had admitted to killing the Gilligans. The items he stole from the Gilligan's home were ultimately recovered from his home or traced back to him.

Wallace was declared incompetent to stand trial. He spent nearly two years in a mental institution. Then, in 1982, a judge ruled that Wallace was mentally competent to stand trial. On October 10, 1982, he was sentenced to death by lethal injection. In the weeks prior to his execution, he admitted to killing the family and to faking his mental illness in order to avoid prosecution.

Figure 1-23. *Donald Ray Wallace.*

Think Critically

1. **What significance did the bloodstained jeans have on this case?**
2. **Do bloodstains have individual or class characteristics? Explain.**
3. **What steps must the investigators have taken prior to searching Wallace's home? What evidence do you think they had before the search?**

Bibliography

Books and Journals

Lerner, K. Lee, and Brenda Wilmoth Lerner, eds. (2006). *The World of Forensic Science*. Farmington Hills, MI: Gale Cengage Learning.

Newton, Michael. (2008). *The Encyclopedia of Crime Scene Investigation*. New York: Info Base Publishing, pp. 125–126.

Web Sites

http://aafs.org
www.cengage.com/school/forensicscienceadv
www.clarkprosecutor.org
www.crime-scene-investigator.net/dutydescription.html
www.enotes.com/forensic-science/henry-edward-richard
www.enotes.com/forensic-science/osborn-albert-sherman
www.essortment.com/all/dnascientists_rchz.htm
www.in.gov/judiciary/opinions/previous/wpd/01130501.ad.doc
www.innocenceproject.org
www.invent.org/hall_of_fame/235.html
www.law.cornell.edu/constitution/constitution.billofrights.html
www.law.cornell.edu/supct/html/92-102.ZO.html
www.mobar.org/journal/1997/novdec/bebout.htm
http://nobelprize.org/nobel_prizes/medicine/laureates/1930/landsteiner-bio.html
www.usconstitution.net/miranda.html

CAREERS IN FORENSICS

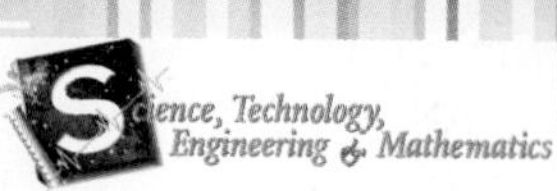

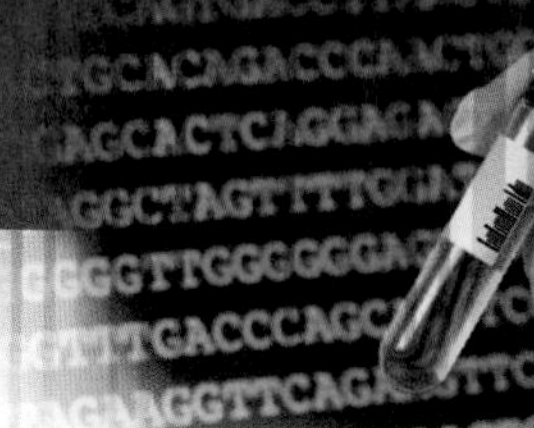

Jason Byrd, Ph.D.: Forensic Entomologist

Jason Byrd (see Figure 1-24) has been interested in forensic science and insects since the sixth grade. However, he didn't think of combining these two interests until he was in college. Dr. Byrd earned Bachelor of Science and Master of Science degrees in entomology with a graduate minor in criminology and law. He also earned a Ph.D. in entomology with an emphasis on forensic botany. Dr. Byrd is a board-certified forensic entomologist; an administrative officer with the Disaster Mortuary Operational Response Team (DMORT), a branch of the Department of Health and Human Services (DHHS); and the logistics chief for the Florida Emergency Mortuary Response System (FEMORS). He worked on his first case in 1993.

AP PHOTO/POOL, CHAD PILSTER

Figure 1-24. *Dr. Jason Byrd.*

Every day is different for Dr. Byrd. He teaches forensic science classes in the Hume Honors College at the University of Florida. He develops and conducts workshops for law enforcement officials, death investigators, attorneys, and medical examiners. He often serves as a consultant in criminal and civil legal cases. As if all of that didn't keep him busy enough, he has published several scientific articles and two books on forensic entomology. Forensic entomology is the study of insects in legal situations, such as criminal cases. The insects found on and around a body provide clues to investigators about the time of death. This information is often vital in solving crimes.

Dr. Byrd enjoys the educational aspect of his job the most. "I like being able to pass knowledge to my students," he says. His recommendation for someone considering a career in a field of forensic science is to "take all the chemistry classes you can!"

It is very important for scientists to remain unbiased. A bias is a preference or preconceived idea that is based on personal preference rather than on scientific analysis. Scientists take steps to avoid allowing their biases to affect their analysis of data. For example, forensic scientists performing lab analysis are not shown unnecessary personal information about the victim or suspect. When working a case, Dr. Byrd says it is easy for him to remain unbiased. "You simply focus on the science, and report what the science tells you as an expert. Ethics is of critical importance. Without high personal ethical standards, you are compromised as an expert witness. Your ethics provide the validity to your statements as much as the science behind your statements."

Learn More About It

To learn more about forensic science careers, go to www.cengage.com/school/forensicscienceadv.

CHAPTER 1 REVIEW

Matching

Match the scientist on the right with the appropriate contribution to forensic science.

1. established anthropometry as a system of personal identification *Obj. 1.2*
2. discovered blood groups *Obj. 1.2*
3. invented the comparison microscope *Obj. 1.2*
4. expert in document analysis *Obj. 1.2*
5. discovered structure of DNA *Obj. 1.2*

a. Karl Landsteiner
b. Watson and Crick
c. Albert Osborn
d. Alphonse Bertillon
e. Calvin Goddard

Multiple Choice

6. ________ is the application of science to law. *Obj. 1.1*
   a. Anatomy
   b. Physiology
   c. Forensic science
   d. Physical science

7. A series of logical steps to solve a problem is called the ________. *Obj. 1.4*
   a. scientific method
   b. hypothesis
   c. theory
   d. experiment

8. The process of DNA fingerprinting was discovered by ________. *Obj. 1.2*
   a. James Watson
   b. Francis Crick
   c. Edmond Locard
   d. Alec Jeffreys

9. ________ developed a fingerprint identification system. *Obj. 1.2*
   a. Calvin Goddard
   b. Sir Edward Richard Henry
   c. Albert Osborn
   d. Watson and Crick

10. Before any evidence is collected, the entire scene must be ________. *Obj. 1.6*
    a. processed
    b. organized
    c. cleaned up
    d. photographed

11. Gunshot residue, hair, and paint are all examples of ________. *Obj. 1.6*
    a. biological evidence
    b. trace evidence
    c. chemical evidence
    d. circumstantial evidence

12. A ________ protects the integrity of and provides a paper trail for collected evidence. *Obj. 1.6*
    a. chain of events
    b. chain of evidence
    c. chain of custody
    d. all of the above

Short Answer

13. Identify the following as having either class or individual characteristics: *Obj. 1.7*
    a. pry marks on a window pane
    individual
    b. fibers from a football jersey left at a crime scene
    class
    c. type A blood found on a knife
    class
    d. semen found on a bedsheet at the scene of a sexual assault
    individual
    e. striations on a spent bullet
    class

14. List and explain three additional disciplines useful in forensic science. *Obj. 1.3*

15. Compare and contrast the various methods of searching a crime scene. Give an example of a search area requiring each search method. *Obj. 1.5*

zone- homicide; building
spiral- open field, no barriers, + kidnapping
line- large area for large object in 1 direction; plane crash
grid- large area for large obj. in 2 directions; arson

16. Explain how Locard's exchange principle is useful in discovering trace evidence. *Obj. 1.6*

Trace evidence is small amounts of evidence found @ a crime scene that could be linked back to the suspect.

17. In your own words, describe what happens from the time a suspect is brought in for questioning to the trial. *Obj. 1.9*

18. Describe how the scientific method is used to solve forensic science questions. *Obj. 1.4*

Think Critically **19. Evaluate the three landmark cases discussed in the chapter and explain the forensic significance of each.** ***Obj. 1.8***

20. Make a concept map that illustrates the various directions the judicial process can take. ***Obj. 1.9***

CRIME SCENE S.P.O.T. Student-Prepared Original Titles

Jack and Jill: Forensics Style

By: **George Foley**

East Ridge High School
Clermont, Florida

When Jack fell down, it wasn't by accident. Jill was a nefarious sort of girl, but she did not exhibit these characteristics openly. Always jealous of Jack for being the center of attention, she envied her brother. One day, it seemed apparent that their ill mother would leave Jill out of her will. This made Jill furious, and for months she plotted to restore her name in the will. One afternoon, their ailing mother asked the children to fetch a pail of water from the well atop the hill.

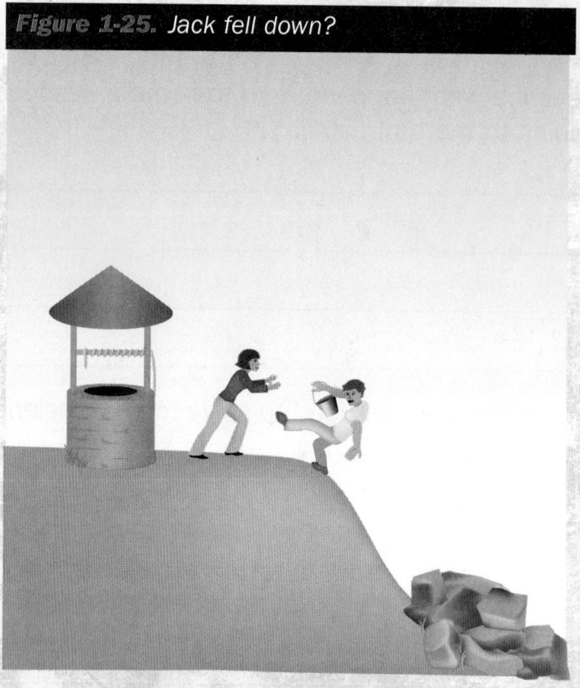

Figure 1-25. Jack fell down?

Jill didn't think her plan all the way through, but she decided to act. As soon as Jack finished filling the bucket of water, Jill distracted him by pointing to something out over the hill. With Jack distracted, Jill gave him a firm push that knocked Jack off balance. He tumbled down the hill (see Figure 1-25).

Jack fell down the hill head over feet until he reached a ledge where he hit his head on a rock. Finally coming to rest at the bottom of the hill, Jack's lifeless body lay across the green bloodstained grass.

Satisfied, Jill ran home. She avoided her mother and went straight to her room to clear her head and began to write about the deed in her diary. Meanwhile, her mother was becoming very thirsty and, in her ailing state, died.

Several days later, neighbors became worried that they had not seen Jack or his mother and called the police. The police found the mother deceased in her bed and no one else home. The medical examiner arrived and determined the cause of death of the woman to have been the disease that ravaged her body, but the mechanism of death was dehydration. The manner of death was ruled accidental. Since the death was unattended, local law required a law enforcement investigation. After the body was removed, crime-scene investigators were asked to look at the scene. A note on the nightstand in the woman's bedroom read: "If I am asleep, Jack, please pour me a glass of water from the pail that you brought back from the well."

Because the woman was dehydrated, and there was no evidence of water anywhere in the house, investigators decided to find out where Jack was. Legal records confirmed the family consisted of two children, Jack and Jill, and their mother. Investigators tried to locate Jack but to no avail. They then decided to search around the well mentioned in the note.

Upon arrival, the crime-scene technicians noticed a pail on its side close to the top of the hill. They were able to lift prints off the pail and take a red thread as trace evidence.

On one side of the hill, there was crushed grass that showed where someone or something had rolled down the hill. Halfway down the hill, a boulder was found spattered with a red substance. The CSI took pictures and swabbed the substance to test for blood. The results of the test were positive. The team moved further down the hill and found a decomposing body.

The medical examiner processed the body. An autopsy was performed. It was determined to be the body of a young boy. Comparisons of dental records to the teeth of the deceased proved that it was Jack. The medical examiner determined his cause of death to be blunt-force trauma to the head (possibly from the rock with blood found at the scene). The mechanism of death was a fractured skull. The manner of death was ruled accidental—temporarily.

The investigation then concentrated on Jill, who, after the death of her mother, could inherit all the possessions because of the death of her brother. This made Jill a prime suspect in both mysterious deaths. Of course, Jill provided an alibi to where she was at the time of both deaths. She claimed to be staying at her friend's house.

Meanwhile, back at the crime lab, examination of the red thread collected at the scene revealed it was from a specific type of red sweater because of its special weave and material. Having no other suspects in the deaths of Jack and his mother, and having probable cause, investigators searched Jill's residence for this specific garment. Indeed, the investigators found a red sweater that matched the one described in the lab report. To make matters worse for Jill, the sweater was slightly torn. Crime lab technicians were able to confirm that the thread found at the scene of Jack's death was consistent with the sweater found at Jill's residence. The police arrested Jill on two counts of first-degree murder after investigators found her diary with entries about her foul deeds and her obsession to inherit her mother's fortune. No amount of apology from Jill could sway the jury in her favor.

Activity:

Answer the following questions based on information in the Crime Scene S.P.O.T.

1. Identify several key pieces of evidence within the story. How were they useful to investigators?
2. What evidence in the story is class evidence and what evidence is individual? Explain.
3. How can Locard's exchange principle help investigators in this case?
4. WRITING Choose a fairy tale or nursery rhyme as the basis for your own forensic story. Be sure to include details about evidence, evidence collection, and the judicial process. Your story should be 800 words or less.

Preventing Adolescent Crime Together™ (P.A.C.T)

Introduction:

In this activity, you will be given an opportunity to research crime trends in your community and discuss their impact and consequences.

Materials:

Newspaper articles
Map of your area

Procedure:

1. Over the course of two weeks, collect at least 10 newspaper articles pertaining to bullying and/or teen crime.
2. Summarize each article. The summaries need to include the date and a description of the crime, the age(s) and gender of the suspect(s), the number of victims, and the area where the crime occurred. Keep the summaries and articles in a portfolio. Be sure to include sources.
3. Work in a group of two or three students to share and compare articles. Combine the data and record in a data table similar to the one shown in Figure 1-26. If there are duplicates, record each article only once.
4. Plot the crimes on a map. It may be helpful to number each crime in your data table to help you label the map.

Figure 1-26. *Data Table 1.*

Date	Crime	Gender of Suspect(s)	Age of Suspect(s)	Number of Victims	Location

Questions:

1. What was the most commonly committed crime? The least?
2. Were the majority of crimes committed by males or females?
3. Is there a difference in the types of crime committed by males and females? If so, describe the differences.
4. What information can you gather from the map? Did the crimes tend to take place in the same area?
5. What do you think may account for these trends in crime?
6. What would you do if you lived in a high-crime area (assume moving out is not an option)?

ACTIVITY 1-1 *Ch. Obj: 1.3, 1.4*
PICK A PENNY

Objectives:

By the end of this activity, you will be able to:

1. Describe the importance of observation skills in identification.
2. Improve your attention to detail.
3. Explain the importance of ethics in law enforcement.

Materials:

hand lens or stereomicroscope
penny

Procedure:

1. Using a hand lens or stereomicroscope, carefully examine the penny you are given.
2. Draw and label two circles for your drawings. Label one circle "Heads" and the other "Tails."
3. Sketch both sides of the penny.
4. Document as many characteristics as possible on the data sheet.
5. When you have completed your drawings, return the penny to your instructor.
6. After all of the pennies have been collected, your instructor will allow you to examine all of the pennies.
7. Based on your sketches, try to identify your penny.

Questions:

1. Were you able to identify your penny? If so, how are you sure?
2. What system or procedure did you use to narrow down the possibilities? Be very specific.
3. Did the accuracy and detail of your sketch help you in finding your penny? Explain.
4. Make a list of the class characteristics of a penny.
5. Make a list of the individual characteristics of your penny.
6. How does this activity relate to forensic science and crime-scene investigation?

ACTIVITY 1-2 *Ch. Obj: 1.3, 1.4*
PUZZLE ACTIVITY

Objectives:

By the end of this activity, you will be able to:

1. Work as a team.
2. Rely on collaboration between teams.

Materials:

(per group of 3 or 4)
job description cards
one set of puzzle pieces

Procedure:

1. Each person in the group should select one job description card.
2. Empty contents of bag onto the table.
3. As a team, put the entire puzzle together. Each member of the team should perform according to the job description card.

Questions:

1. Were you able to piece together the entire puzzle? Why or why not?
2. Could you tell what the picture on the puzzle was before putting it together? Explain.
3. What might have made the task easier?
4. How does the activity relate to forensic science and a crime scene?

ACTIVITY 1-3 *Ch. Obj: 1.3, 1.4, 1.6, 1.7*
SOCK LAB

Objectives:

By the end of this activity, you will be able to:

1. Collect and identify trace evidence.
2. Illustrate proper documentation of data.
3. Identify individual and class characteristics of the socks.

Materials:

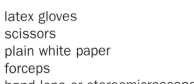

(per group of two students)
pair of clean, white socks
brown paper bag
transparent tape
permanent marker
latex gloves
scissors
plain white paper
forceps
hand lens or stereomicroscope

Procedure:

Day 1:

1. At home, wear a pair of clean, white socks around the house for a couple of hours. Choose a time when you are likely to be moving around inside the house quite a bit rather than when you plan to be sitting still for a long period of time. Take care not to walk in anything wet and do not walk outside.
2. Carefully take off the socks and place them in the brown paper bag (the evidence bag). Be very careful not to lose any of the trace evidence.
3. Seal the bag with tape. Write your name, the date, the "item of evidence," and where it was "collected" on the outside of the bag across the seal.
4. Give the "evidence bag" to your teacher.

Day 2:

1. Put on a pair of latex gloves. Use scissors to carefully open the evidence bag without breaking the seal.
2. Remove the socks and lay them out over a piece of paper.
3. Before removing anything from the socks, observe and record any visible evidence (see Data Table 1).
4. Examine the socks for any obvious wear patterns. Note the size of the socks.
5. Using forceps and/or transparent tape, remove hairs, fibers, and particulate matter from the socks. Record your findings.
6. Using a hand lens or stereomicroscope, look at the trace evidence found on the socks. Draw what you see.
7. Repackage the evidence and make the necessary notations on the bag.

Data Table 1. Trace evidence observations.

Item	Description/Drawing

Questions:

1. Why are you required to wear gloves to handle the evidence?
2. Why was it necessary to lay the socks on paper before processing them for evidence?
3. Do the socks you examined belong to a male or a female? How do you know?
4. What can you assume about the person to whom the socks belong?
5. What is the purpose of the written notations on the bag?

USING PRIOR KNOWLEDGE

PROJECT 1 *Ch. Obj: 1.3, 1.4*
HOT CASE GONE COLD

Objectives:

By the end of this activity, you will be able to:

1. Research an unsolved missing person or murder victim.
2. Predict the outcome of the case.
3. Recognize potential forensic resources that may not have been utilized in the case.
4. Design a presentation of the material.

Materials:

Materials may vary depending on student choices. They may include the following:

computers with Internet access and word processing
and presentation software
computer paper
poster board
markers
old magazines
scissors
glue

Background:

What happened to Jon Benet Ramsey, Natalie Holloway, Jennifer Kesse, Trenton Duckett, and Lisa Stebic? Due to the high-profile nature of these cases, enormous amounts of manpower and taxpayer dollars have been dedicated to solving these cases.

Procedure:

Your group has been given the opportunity to review, consult, and collaborate on a designated cold case. As a group, select a case or review the case assigned by your teacher. Research as much as you can about the case and then develop a proposal as a group. Your proposal must provide detailed and in-depth background information on the case, including dates and times of events, key players, and the current status of the case. Your proposal should also answer the questions listed on the next page.

- What forensic issues surround this case? What is considered the crime scene? When was it processed? What was collected? What were the results?
- What forensic resources have not been utilized? Justify your answers.
- What additional questions do you think key players in this case need to be asked?
- What is your analysis of the manner in which this case has been handled by all parties, including law enforcement, media, the attorneys, and the families of the victims?

Note that these questions provide a good starting point for your research. Your research may provide information beyond the answers to these questions. This information should also be included in your proposal.

Finally, predict the outcome of this case. Based on the evidence and the information released to the public, and all the research you have done, what do you think happened to the victim in your case? Be sure to include supporting statements.

Medium:

As a group, you should decide the best way to present your proposal. You can do a computer presentation, posters, a documentary, or anything you think presents the material best. The presentation should be 10–20 minutes long. Be sure to include a source page. It is very important to document sources clearly and completely. Use only reliable sources that meet the strict standards of scientific investigation.

Each student in the group should also complete a one-page reflection paper. In this paper, answer the following questions: Did you like the project? Why or why not? What would you do differently? What did you learn?

Grading:

Your grade will be determined as follows:

How well you utilized forensic terminology and cited key issues relating to law and law enforcement (20%)
How well you worked as a group (10%)
Your content (30%):
Content must be accurate.
Do you have a strong background on this case?
How well did you organize the timeline of events and key players?
How clearly and logically was your analysis presented?
Were your predictions backed up by evidentiary value and research?
What questions have not been asked? Experts, forensic resources, how the case has been handled, etc.?
Creativity (20%)
Source page (10%)
Reflection paper (10%)

PROJECT 2 *Ch. Obj: 1.3, 1.4*

CREATE YOUR OWN CRIME SCENE: KEEP EVERYONE GUESSING!

Objectives:

By the end of this activity, you will be able to:

1. Develop a crime scene by using your prior knowledge.
2. Draw a crime-scene sketch.
3. Build a scale model of a crime scene.
4. Apply forensic terminology correctly to a scenario.
5. Evaluate the probative value of evidence in a scenario.

Materials:

Materials will vary depending on student choices.

Procedure:

Create a crime-scene scenario. Be sure to include the following:

- Background information—who, what, when, where?
  - Include a little information about your victim and any possible suspects.
  - Use correct forensic terminology.
- A small-scale model of your crime scene—The crime can take place anywhere your imagination carries you.
- Actual samples of evidence—Remember, you should include known and unknown samples.
- A crime-scene sketch
- A solution to your crime-scene scenario—The solution should make sense and be aligned well with all of the information and evidence you supplied. Any surprises should be clearly explained and justified with the evidence and background information provided.

Complete a one-page reflection paper. In this paper, answer the following questions: Did you like the project? Why or why not? What would you do differently? What did you learn? Did this project help to refresh your memory of information learned in the previous course?

Grading:

Your grade will be determined as follows:

How well you utilized forensic terminology and concepts and how detailed you were in your background information (20%)

How well you worked as a group (10%)

Your crime scene (30%):

- Was your small-scale crime scene to scale and aligned with your background information as well as actual evidence samples included?
- Was your crime-scene sketch neat, clean, and courtroom ready? Did it include all the key components of a complete crime-scene sketch?

Creativity (20%)

Solution page (10%)

Reflection (10%)

CHAPTER 2

Interrogation and Forensic Reporting

THE LAST DRIVE

At around 9:00 P.M. on October 25, 1994, Susan Smith stood screaming on Shirley McCloud's porch: "A black man has got my kids and my car! Please help me!" The McClouds called 911. When police arrived, Sheriff Wells took Susan's statement.

According to Susan, she had driven to a local park but stayed in her car. When she stopped at a red light on Monarch, she had been carjacked. Investigators were suspicious of the story. One troubling detail was that the light at the Monarch intersection would not have been red. It was set to remain green unless there was a car on the cross street. Susan had said that there were no other cars at the intersection. On October 27, Susan was asked to go over the details of her story three separate times. Wells asked her to describe that day from the time she woke up to the carjacking. Her story changed in several ways over the course of the day. For example, she first told investigators that she had gone shopping. She later admitted that she had merely been driving around for hours. Additionally, investigators noticed that she looked and sounded as if she was crying, but she never shed any tears.

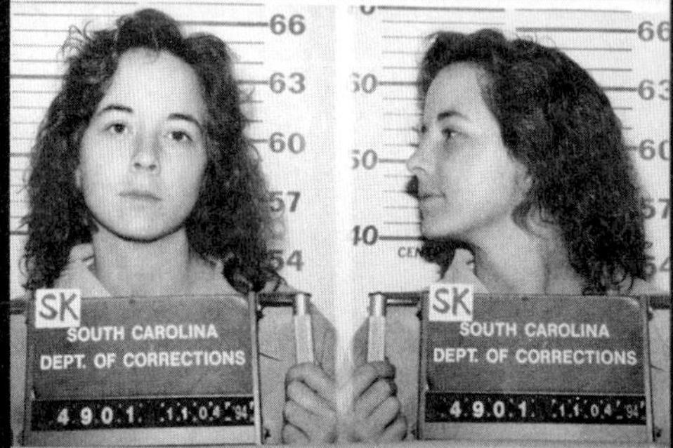

Figure 2-1. Susan Smith will be eligible for parole in 2025, after serving 30 years.

Investigators asked the FBI Behavioral Sciences Unit for a profile for a homicidal mother. Susan Smith shared many of the characteristics described in the profile. As the investigation continued, the sheriff's department and the FBI developed questions and strategies to try to get Susan to confess. For example, Sheriff Wells told Susan that he knew she was not carjacked at the red light at the Monarch intersection. He led her to believe that undercover officers were working a drug case at a nearby intersection and would have witnessed the crime. He threatened to tell the community that she had lied. Susan asked him to pray with her. At the close of the prayers, Wells added, "Lord, we know that all things will be revealed to us in time." He then looked at Susan and said, "Susan, it is time."

Susan immediately dropped her head and cried. She then confessed to letting her car roll into the lake with 3-year-old Michael and 14-month-old Alex strapped in the backseat. Investigators recovered the car, with the boys' bodies still inside. Autopsies confirmed that the boys were alive when their mother sent her car into the lake.

Smith was arrested and charged with two counts of murder. She was convicted and sentenced to life in prison.

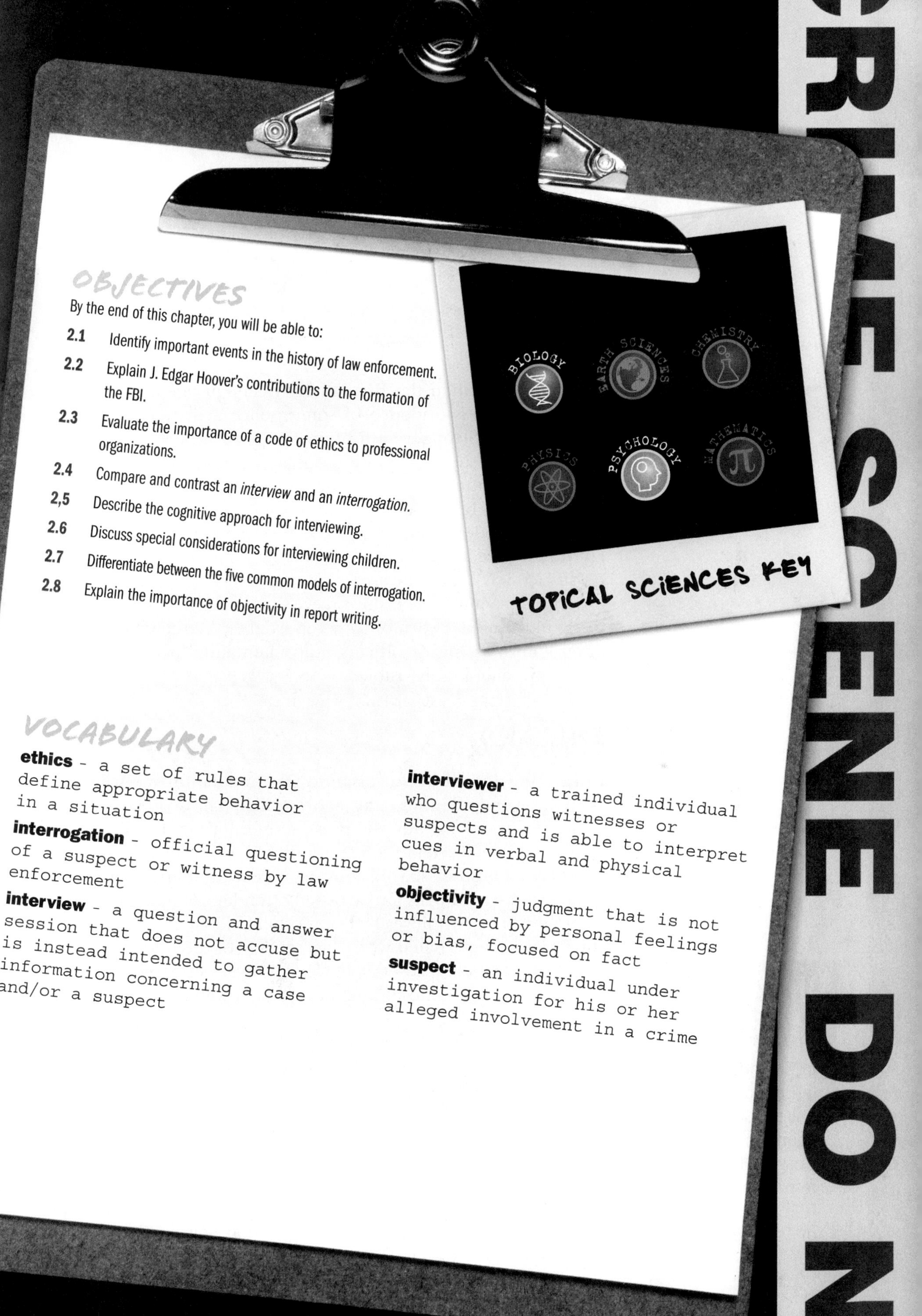

OBJECTIVES

By the end of this chapter, you will be able to:

2.1 Identify important events in the history of law enforcement.

2.2 Explain J. Edgar Hoover's contributions to the formation of the FBI.

2.3 Evaluate the importance of a code of ethics to professional organizations.

2.4 Compare and contrast an *interview* and an *interrogation.*

2,5 Describe the cognitive approach for interviewing.

2.6 Discuss special considerations for interviewing children.

2.7 Differentiate between the five common models of interrogation.

2.8 Explain the importance of objectivity in report writing.

VOCABULARY

ethics - a set of rules that define appropriate behavior in a situation

interrogation - official questioning of a suspect or witness by law enforcement

interview - a question and answer session that does not accuse but is instead intended to gather information concerning a case and/or a suspect

interviewer - a trained individual who questions witnesses or suspects and is able to interpret cues in verbal and physical behavior

objectivity - judgment that is not influenced by personal feelings or bias, focused on fact

suspect - an individual under investigation for his or her alleged involvement in a crime

INTRODUCTION

Have you ever noticed police officers working at a public event? Do you know that they are there to maintain order and promote public safety? Can you think of other places where police work hard to protect your safety? The earliest police forces were actually formed as peacekeepers. It wasn't until the 18th century that law enforcement began to evolve into modern crime prevention and investigation units.

Obj. 2.1, 2.2

HISTORICAL DEVELOPMENT

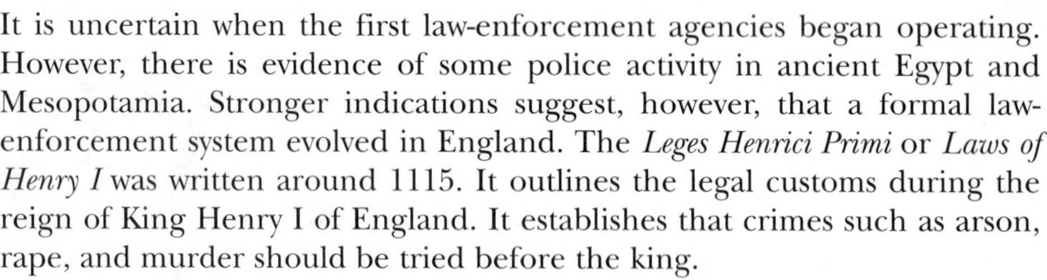

It is uncertain when the first law-enforcement agencies began operating. However, there is evidence of some police activity in ancient Egypt and Mesopotamia. Stronger indications suggest, however, that a formal law-enforcement system evolved in England. The *Leges Henrici Primi* or *Laws of Henry I* was written around 1115. It outlines the legal customs during the reign of King Henry I of England. It establishes that crimes such as arson, rape, and murder should be tried before the king.

At that time in England, the chief law-enforcement officer in the shire, or district, was called the *reeve.* The reeve was responsible for carrying out orders from a judge. The shire reeve was primarily responsible for keeping peace in the area. Sworn officers reported to the reeve. These officers were often known as "peace officers" and maintained peace in the community simply by being highly visible.

ENGLAND

During the Industrial Revolution, the population in British cities grew. With larger populations came more crime. The scope of responsibility of officers began to include the prevention of crime. In 1748, London's first magistrate, Henry Fielding, was appointed to prosecute criminals. Fielding and his half-brother started the Bow Street Runners, a band of eight constables who served writs and arrested offenders. Fielding also implemented programs to catch thieves by using other thieves. Those who provided the "tips" were compensated with a percentage of the fines collected when the other thieves were prosecuted. The Bow Street Runners quickly gained a reputation for honesty and began to travel the country in search of criminals.

In 1822, Sir Robert Peel was appointed Home Secretary. His primary purpose was to make policing more effective. He is considered the father of modern policing—in part, because he called on experts in the fields he was investigating. In 1829, Peel introduced the Metropolitan Police Act into Parliament. Parliament passed the act, establishing the Metropolitan Police of London. The Metropolitan Police replaced a less centralized system of constables and night watchmen. The London force became the first modern police force and has served as the model for urban police forces around the world. In honor of Sir Robert "Bobby" Peel, all London police officers are often called bobbies (see Figure 2-2).

In England, the land is divided into shires, similar to counties in the United States. The term *sheriff* comes from the contraction of "shire reeve."

THE UNITED STATES

In 1783, British forces were withdrawing from occupied lands in the newly established United States of America. On November 25, a watch system was put in place in New York City. In this way, New York became the first city in the United States to create a system of policing and enforcing laws. By 1790, New York had a fully operational daytime paid police force. In 1845, the police force was organized into a single organization called the *New York Police Department.* Many other large cities, such as Boston, Chicago, and Philadelphia, established police forces during this same time period.

© JON ARNOLD IMAGES LTD/ALAMY

Figure 2-2. The "bobby."

U.S. Department of Justice

In 1789, Congress established the position of attorney general. For almost 100 years, the attorney general functioned without a staff. In 1870, Congress established the U.S. Department of Justice, which was headed by the attorney general. The Department of Justice was to standardize legal opinions and manage legal proceedings. At the time, the Department of Justice did not include an investigative department. The agency hired people from outside the department and used informants to complete investigations. Sometimes, the Department of Justice "borrowed" personnel from other federal agencies to complete investigations. When Theodore Roosevelt became president, he and his attorney general, Charles Joseph Bonaparte, requested that the Congress set up a "force of permanent police" within the Department of Justice. Congress refused his request. In July 1908, Roosevelt allowed his attorney general to organize a group of 34 Special Agents within the Department of Justice. In 1909, this group was named the *Bureau of Investigation.*

Hoover and the Bureau of Investigation

In 1923, President Calvin Coolidge appointed Harlan Stone as the attorney general. Stone's primary goal was to get rid of corruption in the Bureau of Investigation. He hired J. Edgar Hoover (see Figure 2-3) to be the director. Prior to Hoover's appointment, appointments to the Bureau were based on political connections rather than on experience and training. Hoover set new standards for employment and promotion in the Bureau. All new hires would be required to have a college degree in law or in accounting. Additionally, new hires would be subject to an extensive background check. Employment within the department was no longer dependent on political connections. Promotion had been based on seniority. Hoover demanded that promotion be based on merit.

Hoover made significant changes in law enforcement processes as well. For example, prior to 1924, there was no central location for storing fingerprints. The Department of Justice maintained a collection of fingerprint cards at Leavenworth Federal Penitentiary in Kansas. The International Association of Chiefs of Police kept another collection. Hoover thought that it would be more efficient and logical to store all the prints in one centralized location, at the Bureau. In 1924, 800,000 prints were relocated to the Bureau of Investigation's Identification Division.

LIBRARY OF CONGRESS

Figure 2-3. J. Edgar Hoover, the first FBI director.

Figure 2-4. *J. Edgar Hoover FBI Building in Washington, DC.*

Hoover also thought forensic science would be a powerful tool for solving crimes. He established a crime lab at the Bureau in 1932 to help federal investigators process evidence. State and local police departments needing assistance also sent evidence to the crime lab. This common professional courtesy continues today. In 1935, the Bureau of Investigation changed its name to the *Federal Bureau of Investigation (FBI)* (see Figure 2-4).

The FBI

Today, various other federal investigative agencies, including the Drug Enforcement Administration and Immigration and Naturalization Service, exist within the Justice Department. The Treasury Department has jurisdiction over the Bureau of Alcohol, Tobacco, Firearms, and Explosives (ATF); the Internal Revenue Service (IRS); and the Secret Service. The Department of Defense manages internal investigations of the military. It is important to note that federal agencies are governed by federal statutes. These statutes are rules that determine which agency has jurisdiction over a specific federal law. For example, if a child is abducted and taken into another state, the case is under the jurisdiction of the FBI because the case is no longer under the jurisdiction of only one state. However, if a child is abducted by someone who keeps the child within the state, the investigation belongs to the local city or county policing agency.

Digging Deeper

with Forensic Science e-Collection

All stakeholders in the judicial process are expected to maintain the highest integrity and ethics. Criminal investigators, trial lawyers, deputy sheriffs, judges, psychiatrists, and everyone else involved in the process are held to a standard of professional ethics. Some trial lawyers have a unique ability to focus on the facts of the case while still connecting with the jury. Using the Gale Forensic Science e-Collection at www.cengage.com/school/forensicscienceadv, read about how well some trial lawyers connect with juries. Then complete the following activities: (1) Identify ways in which trial lawyers connect with the jury and respect the courtroom. (2) Explain why it is ethically significant for a trial lawyer to refrain from providing an opinion in the courtroom. (3) Do you think that the "likability" of an attorney can carry as much weight with a jury as the lawyer's presentation of the facts can? Explain.

ETHICS

Obj. 2.3

Doctors, educators, police officers (see Figure 2-5), and attorneys all follow specific codes of ethics. **Ethics** is a set of guidelines—written and unwritten—that explain appropriate conduct for a particular situation or profession. A written code of ethics describes a process for enforcing appropriate conduct by a profession's members. A written code usually also establishes a system for the investigation and resolution of alleged unethical conduct.

All scientists, including forensic scientists, must follow the scientific method to ensure the validity of their testing techniques and procedures. Scientists have an ethical responsibility to report their data and analysis accurately. Their scientific procedures must also be verified through peer review.

Figure 2-5. *When cadets become police officers, they agree to follow a code of ethics.*

The American Academy of Forensic Sciences (AAFS) (see Figure 2-6) has set forth ethics guidelines. These guidelines outline proper behavior for a forensic scientist and include responsible conduct, honesty, and integrity in all aspects of the profession. The credibility of forensic science and its related professions depends on the confidence the public has in the reliability and accuracy of the work done by individuals in those professions. A code of ethics outlines the responsibilities of the forensic scientist. These responsibilities include close attention to detail when collecting evidence, accurate communication of findings, and objectivity when conducting interviews or interrogations.

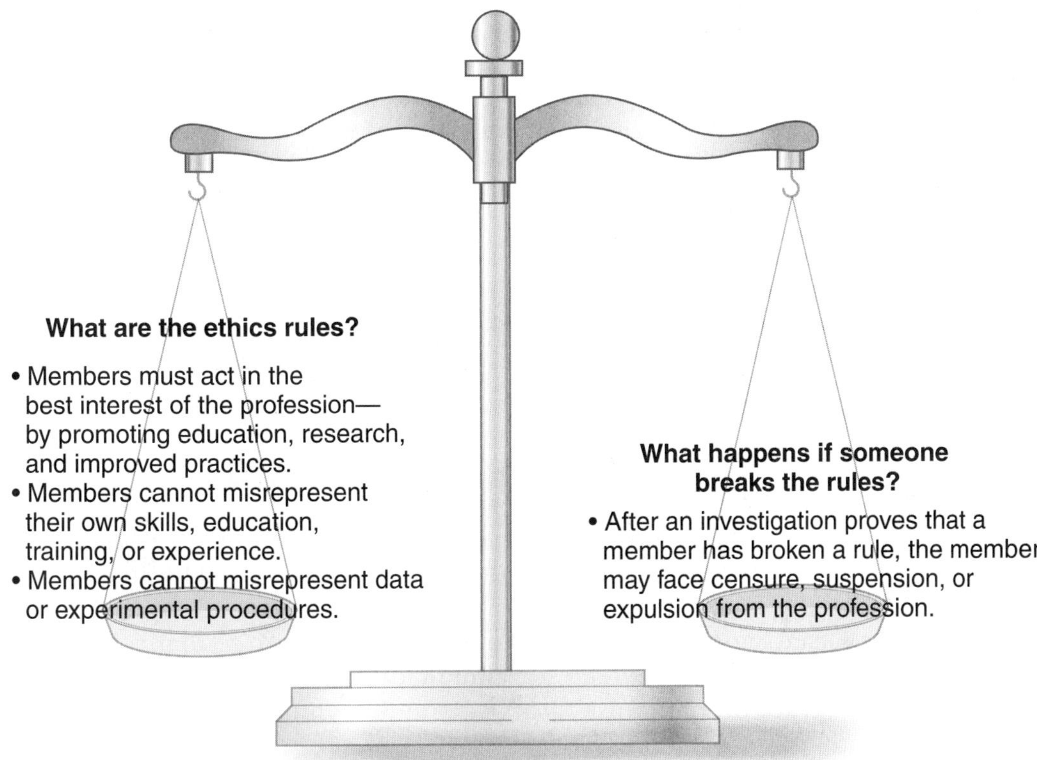

What are the ethics rules?
• Members must act in the best interest of the profession—by promoting education, research, and improved practices.
• Members cannot misrepresent their own skills, education, training, or experience.
• Members cannot misrepresent data or experimental procedures.
What happens if someone breaks the rules?
• After an investigation proves that a member has broken a rule, the member may face censure, suspension, or expulsion from the profession.

Figure 2-6. *The code of ethics outlined by the AAFS includes the rules and the consequences of breaking the rules.*

Obj. 2.3 BIAS

Throughout the criminal investigation, law-enforcement personnel and forensic scientists must do what they can to avoid bias. This means that they try to keep any preconceived ideas about a case, suspect, victim, or group from influencing their approach to the case. To avoid bias, cases are given unique identification numbers so that forensic scientists performing the lab analysis are not shown unnecessary personal information. Unnecessary details, especially those details personal in nature, may influence the work and analysis negatively.

Obj. 2.4, 2.5, 2.6 INTERVIEWING TECHNIQUES

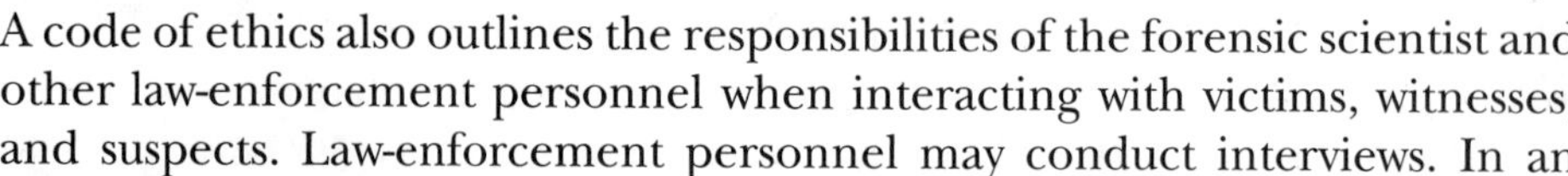

A code of ethics also outlines the responsibilities of the forensic scientist and other law-enforcement personnel when interacting with victims, witnesses, and suspects. Law-enforcement personnel may conduct interviews. In an **interview**, the dialogue is intended to gather information about a case and/or a suspect. A **suspect** is an individual under investigation for alleged involvement in a crime. The investigator completing the interview session is called the **interviewer**. As the interviewer establishes rapport with the victim, suspect, or witness, he or she will ask general questions. Because the interviewer is not expecting the subject to make incriminating statements and is not trying to elicit a confession, there is no need to issue a Miranda warning. The questions are very broad and open-ended to allow a free flow of information. As the victim, suspect, or witness becomes more comfortable, the interviewer will begin to ask more specific questions to gather more detail. For example, an open-ended question such as "What were you doing on Tuesday?" will lead to very general answers. The more specific and closed-ended question, "What were you doing at 3 o'clock on Tuesday afternoon?" will yield a very definite answer. Miranda rights must be presented before a suspect is interviewed (see Figure 2-7).

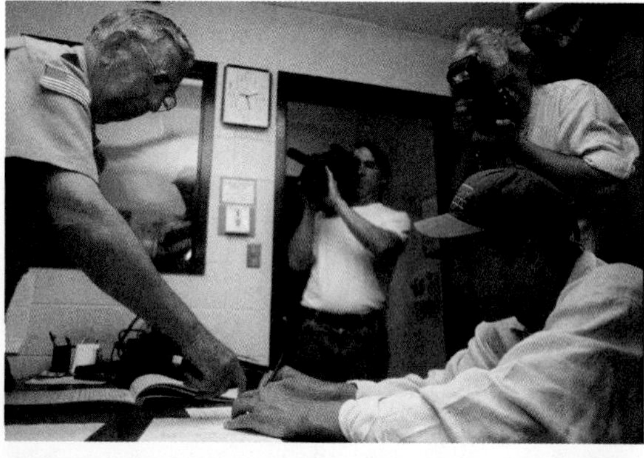

AP PHOTO/LEXINGTON HERALD-LEADER, MICHELLE PATTERSON

MIRANDA WARNING

1. You have the right to remain silent.
2. Anything you say can and will be used against you in a court of law.
3. You have the right to talk to a lawyer and have him present with you while you are being questioned.
4. If you cannot afford to hire a lawyer, one will be appointed to represent you before any questioning if you wish.
5. You can decide at any time to exercise these rights and not answer any questions or make any statements.

WAIVER

Do you understand each of these rights I have explained to you?
Having these rights in mind, do you wish to talk to us now?

Figure 2-7. *By signing a Miranda card, such as the one shown above, actor Woody Harrelson affirms that he understands his rights prior to an interrogation.*

THE COGNITIVE APPROACH

Interviewers use several different techniques to gather the information they need. The most common interview technique is called the *cognitive approach* (see Figure 2-8). This technique is designed to enhance the person's recollection of the details of the incident. It encourages the subject to reconstruct the circumstances surrounding the crime or incident in different ways. The interviewer may then probe previous topics or choose to end the interview and resume questioning at a later date.

The same suspect, victim, or witness may be interviewed several times. The interviewer takes careful notes each time. Often, the subject provides a

Goals of the Cognitive Approach	Examples of Interview Questions
Recall emotions or retrieve memory of the surroundings.	Describe the room in which the assault took place.
Report even the smallest details.	Can you tell us what was on the wall?
Recall the events in a different order.	What did you do when you entered the room? What did you do when you found the body? What were you doing before you entered the room?
Alter the suspect's or witness's perspective.	Close your eyes and describe the scene as if you are watching it on television.

Figure 2-8. *The cognitive approach.*

lot of information in the first interview. As time passes, the person may have time to "get the story straight." In subsequent interviews, the interviewer is trained to discover any deceptions and inconsistencies. Inconsistencies in the story can lead to a shift from an interview to an interrogation. Legally, an **interrogation** is a questioning session in which the subject is likely to make incriminating statements or to confess. An interrogation often takes on an accusatory tone. Before an interrogation, the suspect must be read the Miranda rights.

SPECIAL CONSIDERATIONS WHEN INTERVIEWING CHILDREN

In every interview and interrogation, the primary objective of the interviewer is to determine the truth. Sometimes, law-enforcement personnel need to interview children. Children may be the victims of crime or they may have witnessed a crime. In either case, they can provide important information for investigators. Interviewers take special considerations when interviewing a child. The interviewer needs to help the child feel as safe and comfortable as possible so that the child is able to provide the best information. The techniques vary depending on the age of the child and on the nature of the crime. When interviewing a young child, the interviewer follows certain established and accepted guidelines. For example, all of the interviewer's questions are prepared in advance. This allows the interviewer to focus on the child rather than on developing questions during the interview. The interview always occurs in a private location, and the child is offered the opportunity to bring a toy or blanket into the room for added comfort and security. Additionally, a parent or advocate for the child is usually nearby. The interviewer will eventually compile notes of the interview in report form to be used in court. The interview itself is likely to be videotaped. Again, this allows the interviewer to focus directly on the child. Perhaps more importantly, the tape can be used to show the defense that the child was not coerced in any way.

When the interview begins, the interviewer makes introductions and explains to the child what is happening. The interviewer is likely to begin by asking who the child's legal guardian is and where that person was when the crime took place. Also, in cases of child abuse, it is important for the interviewer to learn what the child calls major body parts. By the end of the interview, the interviewer should be able to determine:

CHAPTER

1

2

3

4

5

6

7

8

9

10

11

12

13

14

15

Digging Deeper

with Forensic Science e-Collection

Using the Gale Forensic Science e-Collection at www.cengage.com/school/forensicscienceadv, do some research about important components of forensic interviews of children. Then, answer the following questions: (1) Why do you think children require different questioning techniques? (2) What concessions are in place to protect children who may be victims of abuse? (3) How are standards for quality control maintained when conducting and evaluating forensic interviews of children? (4) Why should an individual who completes forensic assessments on children be well versed in child development theory? (5) What aspect of the theory will enhance the interviewer's skills? (6) Identify common problems associated with interviewing children. (7) What are best practices when forensically interviewing a child?

- Where the crime occurred
- The number of times the crime occurred
- The time of day when the crime occurred
- How well the child recalls details of the crime
- Details about the location where the crime took place—for example, the placement of furniture in the room
- Whether the child knows if there are any other victims or witnesses

Once all necessary information has been gathered, the interviewer will begin to close the session. The interviewer will offer a quick summary of the session and ask the child if he or she has anything to add. If not, the interview is terminated.

AP PHOTO/AKRON BEACON JOURNAL, ROBIN TINAY SALLIE

Figure 2-9. *Playing videotaped testimony in court.*

CHILDREN IN THE COURTROOM

Child witnesses present several unusual circumstances in court. Research shows that children are naturally very truthful in reporting crimes. However, the trial judge must determine whether the child understands the responsibilities of a witness before the child's testimony can be admissible. The judge must also weigh the emotional effect that acting as a witness may have on the child against the importance of the information provided by the child witness. Additionally, a person being accused of a crime usually has the right to face the person making the accusation. However, several factors, considerations, and limitations go into deciding whether a child witness will have to face the alleged perpetrator. Special technology, such as closed circuit television (see Figure 2-9), has been used. Children have been important witnesses in many cases. Often the child's testimony is invaluable. At other times, the information gathered by investigators has provided enough evidence to allow the child to avoid the courtroom.

PROPER PROCEDURES FOR INTERROGATION

When a crime has occurred, investigators must secure the crime scene, compile evidence, and interview witnesses in an effort to narrow the search for potential suspects. Potential suspects may be interviewed. Once law-enforcement personnel have gathered enough information, they will interrogate a suspect (see Figure 2-10). A detective usually conducts the interrogation. The primary purpose of the interrogation

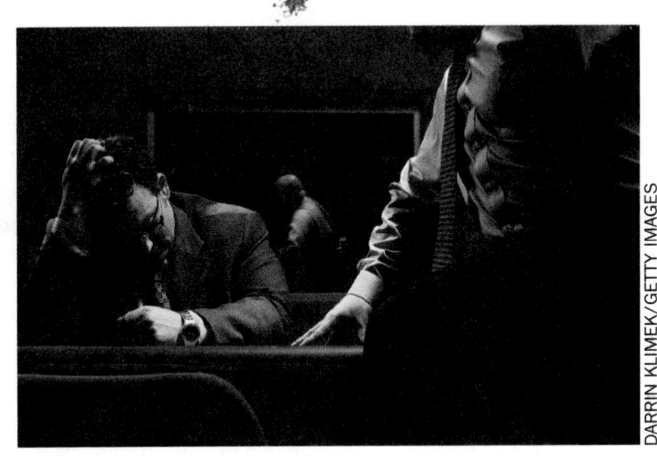
DARRIN KLIMEK/GETTY IMAGES

Figure 2-10. *Suspect being interrogated.*

Digging Deeper
with Forensic Science e-Collection

Khalid Sheik Mohammed was allegedly the mastermind of the 2001 attacks on the World Trade Center and the Pentagon. The Bush administration was questioned for allowing torturous methods of interrogation, including 183 allegations of simulated drowning. Using the Gale Forensic Science e-Collection at www.cengage.com/school/forensicscienceadv, research the topic of terrorist interrogation. Then answer the following questions: (1) Do you think the interrogation methods were appropriate? (2) Do you think investigators would have secured a confession utilizing traditional interrogation techniques? (3) What may be the long-term political repercussions of these actions? Write a position paper weighing the importance of securing a confession against the implications of denying constitutional rights to certain criminals.

is to gather evidence from an uncooperative source. Detectives hope to determine innocence or guilt of an individual. If the detective determines that the suspect is guilty, the primary objective is then to obtain a legally admissible confession.

PREPARING FOR AN INTERROGATION

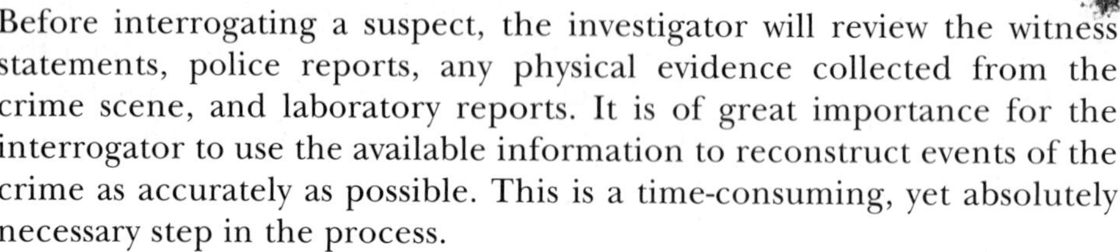

Before interrogating a suspect, the investigator will review the witness statements, police reports, any physical evidence collected from the crime scene, and laboratory reports. It is of great importance for the interrogator to use the available information to reconstruct events of the crime as accurately as possible. This is a time-consuming, yet absolutely necessary step in the process.

It is impossible to determine exactly how long an interrogation will take. An interrogation can last several hours. It will end immediately if the suspect requests an attorney or invokes the right to remain silent. However, a large percentage of suspects waive their right to silence or to an attorney. This is almost always a bad decision on the part of the suspect.

There are guidelines to ensure that investigators do not use unreasonable interrogation techniques to wear down the suspect. For example, interrogators cannot deny use of the restroom or access to medication. If and when a suspect confesses to a crime, the investigator must be sure not to show any emotion, hostility, or anger. The investigator must remain calm. Often, the investigator will videotape the confession and ask the suspect to write out and sign the confession.

A polygraph measures physiological responses during an interview or interrogation. A trained examiner uses sensors to measure respiration rate, heart rate, blood pressure, and perspiration.

THE PHYSICAL SETTING OF THE INTERROGATION

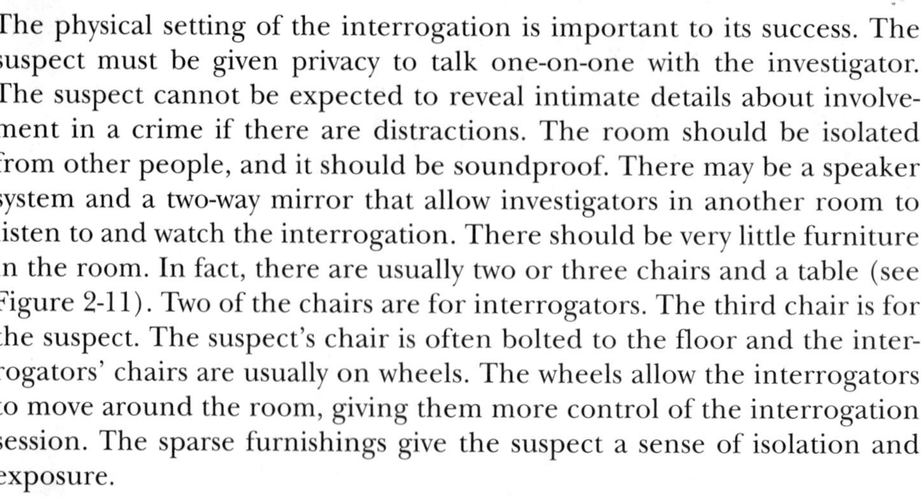

The physical setting of the interrogation is important to its success. The suspect must be given privacy to talk one-on-one with the investigator. The suspect cannot be expected to reveal intimate details about involvement in a crime if there are distractions. The room should be isolated from other people, and it should be soundproof. There may be a speaker system and a two-way mirror that allow investigators in another room to listen to and watch the interrogation. There should be very little furniture in the room. In fact, there are usually two or three chairs and a table (see Figure 2-11). Two of the chairs are for interrogators. The third chair is for the suspect. The suspect's chair is often bolted to the floor and the interrogators' chairs are usually on wheels. The wheels allow the interrogators to move around the room, giving them more control of the interrogation session. The sparse furnishings give the suspect a sense of isolation and exposure.

© A. T. WILLETT/ALAMY

Figure 2-11. *An interrogation room is usually sparsely furnished.*

ORGANIZING THE INTERROGATION SESSION

When an interrogator walks into an interrogation room, the session will occur in very specific steps. First, the interrogator will establish control of the session by directing the suspect to be seated. The interrogator will sit down and "roll" into the suspect's space. At this point, the interrogator will explain the suspect's Fifth and Sixth Amendment rights. Recall from Chapter 1 that the Fifth Amendment guarantees a person's right to remain silent and to the due process of law. The Sixth Amendment guarantees a person's right to a trial by an impartial jury of his or her peers.

Once the interrogation begins, the interrogator will show evidence to support how the suspect is directly involved in the crime in question. The interrogator's body language is very deliberate during an interrogation. When the interrogator wants to show interest in the suspect's story, he or she may lean forward and offer direct eye contact. On the other hand, if the interrogator thinks the suspect is lying, the interrogator may tell the suspect directly that the story is not believable.

Regardless of whether the interrogator believes the suspect, he or she must allow the suspect to finish answering each question. If the interrogator approaches a lie very early in the session, it may impede further dialogue between the suspect and the interrogator. Also, if the suspect is lying, the statements can often be disproved by witness statements or physical evidence. These inconsistencies can be used against a suspect at trial.

In 2006, John Mark Karr confessed to the murder of JonBenet Ramsey. Janine Driver, a body language expert, cites several indications that Karr was actually lying. He would shrug his shoulders (indicates uncertainty) while answering with a definitive statement. He also mixed up tenses of the verbs in his sentences, indicating he was making up a story.

INTERROGATION MODELS

Obj. 2.7

There are many different models of interrogation. The investigator will decide which model (or models) to use based on factors including the suspect's age, education and professional experience, and prior

Interrogation Model	Questioning Strategies	Why It Works
Suspect Decision-Making Model	Interrogator offers the options available and the consequences of each.	Suspect weighs pros and cons of confessing; considers all options and consequences; decision to confess lies in perceived consequences
Cognitive-Behavioral Model	Interrogator continues questioning over a period of time until the suspect is less resistant.	Factors such as exhaustion, overwhelming guilt, and isolation may cause someone to confess; relies on fear of isolation or being arrested, or on the suspect's assumption that guilt can be proved
Psychoanalytical Model	Interrogator tells suspect that confessing will lift the burden of guilt and that the suspect will feel better.	Individual feels the need to confess to punish himself/herself; model is unlikely to be used successfully on suspects who are career criminals, as they are unlikely to feel guilt for their actions
Emotional Model	Interrogator allows suspect to shift the blame without escaping legal responsibility.	Suspect simply does not tell the truth to avoid the consequences, such as loss of freedom or social status
Interaction Process Model	Investigator determines the best method to approach the interrogation based on the suspect's background and history and on the facts of the case.	Allows the interrogator to evaluate the suspect and then proceed using one or more of the above models to obtain a confession

Source: Zulawaski, David E., and Douglas E. Wicklander. (2002). *Practical Aspects of Interview and Interrogation*. Boca Raton, FL: CRC Press.

Figure 2-12. *Interrogation models.*

experience with the criminal justice system. In each case, the primary objectives of any interrogation are to determine the truth and to obtain a legally admissible confession. Five common models are outlined in Figure 2-12.

SIGNS OF DECEPTION

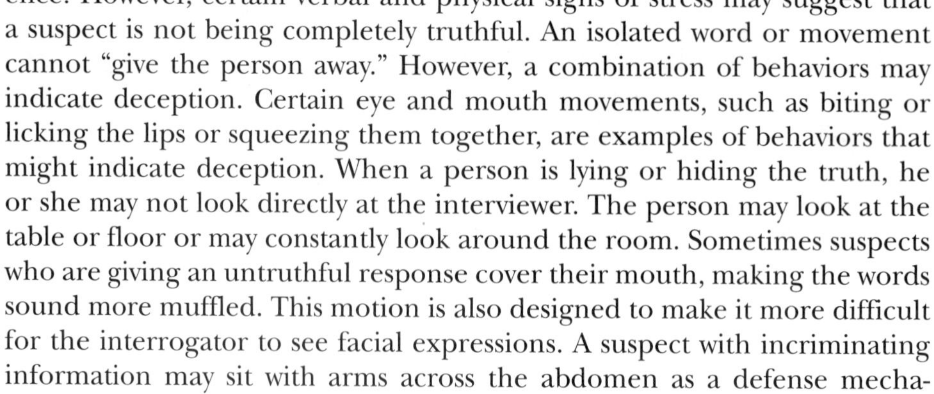

Even if someone has nothing to hide, an interrogation is a stressful experience. However, certain verbal and physical signs of stress may suggest that a suspect is not being completely truthful. An isolated word or movement cannot "give the person away." However, a combination of behaviors may indicate deception. Certain eye and mouth movements, such as biting or licking the lips or squeezing them together, are examples of behaviors that might indicate deception. When a person is lying or hiding the truth, he or she may not look directly at the interviewer. The person may look at the table or floor or may constantly look around the room. Sometimes suspects who are giving an untruthful response cover their mouth, making the words sound more muffled. This motion is also designed to make it more difficult for the interrogator to see facial expressions. A suspect with incriminating information may sit with arms across the abdomen as a defense mechanism. When an investigator asks an emotionally relevant question, suspects may shift their posture abruptly. The interviewer may be able to use these cues within the interrogation to generate dialogue that leads to a confession. In other cases, these cues may be used in later interrogations. Review Figure 2-13 to compare different body language.

Figure 2-13. *Notice the different body language. In the image on the left, the woman is open and engaged. On the right, her arms are crossed and she is looking away. This may indicate deception.*

REPORTING INFORMATION

Obj. 2.8

As discussed earlier, forensic scientists must follow a set of ethics guidelines. One significant piece of these guidelines is accurate and truthful reporting. When reporting details of an interview or interrogation, the investigator must maintain objectivity. **Objectivity** is simply reporting the facts of the case, without presenting the opinion of the investigator. In the report, details include, but are not limited to, the name and address of the person interviewed and whether the person is regarded as a witness, a suspect, or a victim. Thorough straightforward reporting demonstrates professionalism and integrity for the field.

Reports are filed throughout the investigation process, not just after an interview or interrogation. All reports are stored so that investigators and attorneys can review all of the information related to the case (see Figure 2-14). The types of reports vary depending on the purpose of the report and the person generating the report. For example, a field investigator fills out chain of custody reports as well as evidence submission forms. A forensic scientist in the lab reports the findings of the observations and testing procedures performed on the evidence of a case. In each case, the reports and forms must be thorough, straightforward, and accurate. Objectivity and ethics also are key elements to all investigative and forensic procedures—from evidence collection and analysis to courtroom testimony. You will learn more about these procedures throughout this course.

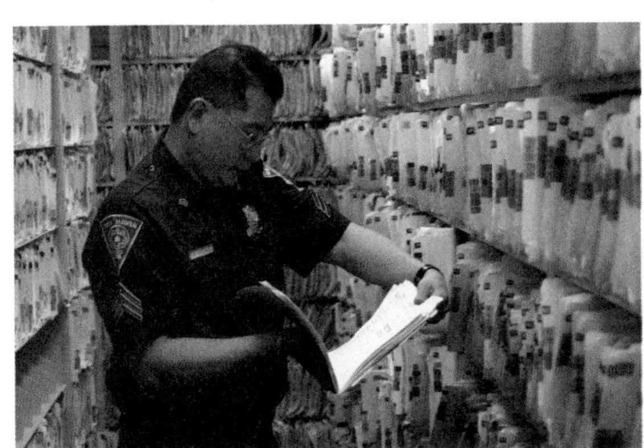

Figure 2-14. *Records room.*

CHAPTER SUMMARY

- As cities grew during the Industrial Revolution, the focus of policing shifted from peacekeeping to crime prevention.
- The U.S. Department of Justice, established by Congress and headed by the attorney general, initially had no investigative department and relied on outside organizations to perform investigations.
- In 1924, J. Edgar Hoover was named director of the Bureau of Investigation, which was later renamed the Federal Bureau of Investigation, or the FBI. Hoover set standards for employment and promotion, consolidated the fingerprint databases, and created a crime lab to help federal, state, and local agencies process and analyze evidence.
- A code of ethics is a set of guidelines for appropriate behavior as well as a system to enforce professional conduct.
- During an interview, an investigator asks questions designed to gather information. During an interrogation, the investigator expects to gather incriminating statements or a confession.
- The most common interview technique is the cognitive approach, which enhances recollection of information.
- Special considerations are always given when interviewing children, but the ultimate goal is to obtain the truth.
- The five common models of interrogation are suspect decision-making, cognitive-behavioral, psychoanalytical, emotional, and interaction process.
- Accurate, truthful, and objective reporting is essential to maintaining professionalism in forensic science fields. Objectivity is a focus on facts without influence from opinion or bias.

Ruby Ridge (August 1992)

Randy Weaver was indicted on federal firearms charges stemming from the sale of sawed-off shotguns to an undercover agent for the Bureau of Alcohol, Tobacco, Firearms, and Explosives. Weaver and his wife were involved with federal agents in an altercation that led to Randy's arrest. After posting bail, Randy was released. However, he did not appear in court on his assigned date; and U.S. Marshals were issued an arrest warrant for him. In August, U.S. Marshals began a surveillance of the Weaver's property. On August 21, 1992, there was a standoff between the Marshals and the Weavers. During the standoff, U.S. Marshal Bill Degan and Weaver's 14-year-old son were killed. Weaver and a family friend, Kevin Harris, were both injured. The next day, an FBI sniper killed Weaver's wife. Negotiators were called in and, after three days, Randy Weaver surrendered to agents. Agents confiscated the guns and ammunition found on the property (see Figure 2-15).

Randy Weaver and Kevin Harris were arrested and charged with murder, aiding and abetting murder, conspiracy, and assault. During the trial, witnesses for the prosecution contradicted one another, and government agents admitted to evidence tampering and providing "reenacted" crime-scene photos. Additionally, Marshals had not followed typical procedure from the beginning. The agents knew that Weaver had a lot of weapons on his property. The Marshals told the judge that they had never considered knocking on the door and arresting Weaver. Instead, they had begun an armed surveillance of his property. The jury acquitted Weaver and Harris of the murder and conspiracy charges. Weaver was found guilty of lesser charges.

An investigation into the government's involvement in the Weaver case resulted in letters of reprimand and 10- to 15-day suspensions for many of the agents involved. Prosecutors were ordered to apologize in open court and to pay for part of the defense attorney's fees.

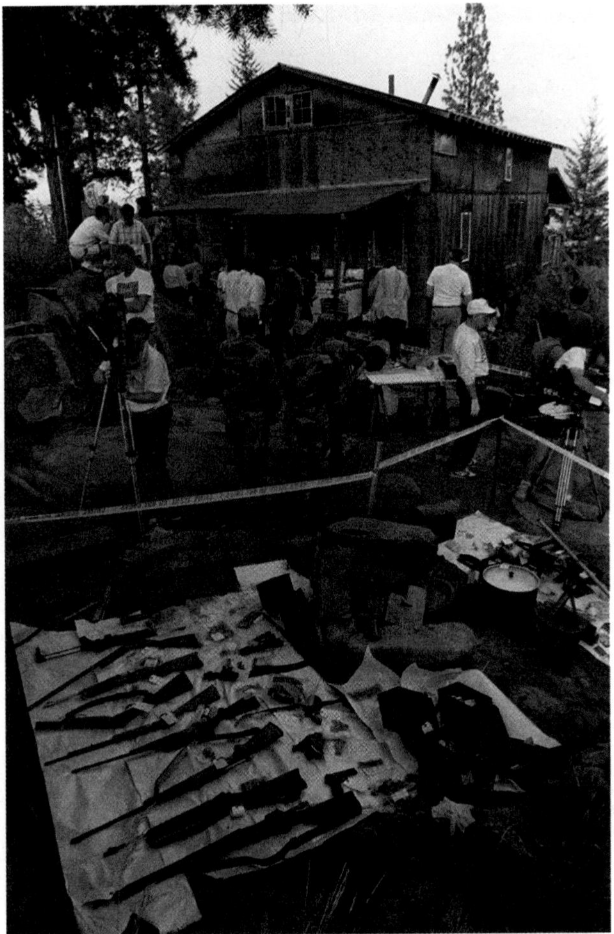

AP PHOTO/GARY STEWART

Figure 2-15. *Weapons and ammunition confiscated from the Weaver home after the Ruby Ridge standoff.*

Think Critically **Using the Gale Forensic Science e-Collection at www.cengage.com/school/forensicscienceadv, research the Ruby Ridge case. Describe the ethics violations in one or two paragraphs.**

AP PHOTO/MARCIO JOSE SANCHEZ

Figure 2-16. *John Stoll spent 20 years in jail after a Child Protective Services employee and a sheriff's investigator allegedly pressured children into lying about the events.*

John Stoll (1984)

In 1984, John Stoll was accused of abusing six boys in his Bakersfield, California, home. He was convicted of 17 counts of child abuse and received a 40-year sentence. When questioned by investigators for the Innocence Project nearly 20 years after Stoll's convictions, the alleged victims (who are now adults) admitted that they had lied. Four of the six testified that investigators had manipulated and nagged them until they had made up the stories. A fifth has no memory of the events at all. Only one, Stoll's son, maintains that Stoll abused him.

A review by the attorney general cited flawed interrogation techniques and poorly trained personnel. Medical experts had never examined the children, and no physical evidence had been found to support the allegations. The interviews had been long, and law-enforcement personnel had asked leading and suggestive questions. At least one child had been interviewed alone, and another had been promised ice cream if he told the investigators what they wanted to hear. Stoll was exonerated and released from prison in 2004. He is shown in Figure 2-16 about one year after his release from prison.

Think Critically

1. **What are some specific ethical responsibilities when interrogating children?**
2. **Why is it important to make special concessions when interviewing children?**
3. **How does the John Stoll case illustrate the importance of careful interviewing of children?**

Bibliography

Books and Journals

Barnett, Peter D. (2001). *Ethics in Forensic Science: Professional Standards for the Practice of Criminalistics.* Boca Raton, FL: CRC Press.

Boetig, Brian Parsi, and Arnold R. Belmer. (October 2008). "Understanding Interrogation." *FBI Law Enforcement Bulletin*.

Burnstein, Harvey. (1999). *Criminal Investigation*. Upper Saddle River, NJ: Prentice Hall.

Inbau, Fred E., John E. Reid, Joseph P. Buckley, and Brian C. Jayne, eds. (2004). *Criminal Interrogation and Confessions.* An Asphen Publication.

Zulawaski, David E., and Douglas E. Wicklander. (2002). *Practical Aspects of Interview and Interrogation*. Boca Raton, FL: CRC Press.

Web Sites

www.aafs.org

www.brookings.edu/papers/2009/0510_interrogation_law_wittes.aspx?more=rc

www.cengage.com/school/forensicscienceadv

www.fbi.gov

www.innocenceproject.org/Content/314.php

www.nycpolicemuseum.org

http://nypd.police-memorial.com/george.htm

CAREERS IN FORENSICS

Science, Technology, Engineering & Mathematics

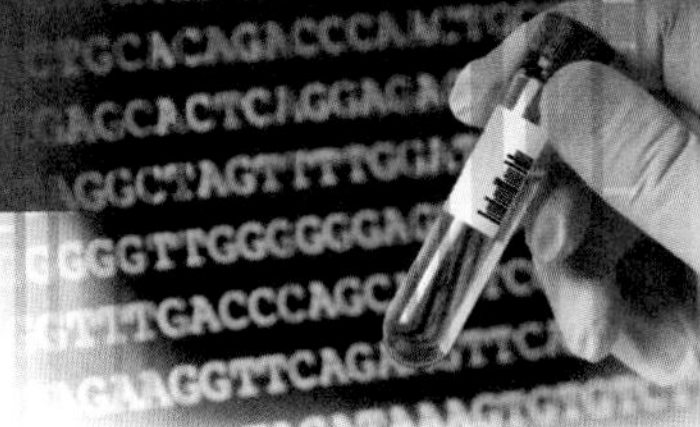

Dr. Park Dietz: Forensic Psychiatrist

AP PHOTO/D.J. PETERS, POOL

Figure 2-17. *Dr. Park Dietz.*

Dr. Park Dietz (see Figure 2-17) knew at a very young age that he had a deep interest in criminology. However, he had family pressures leading him in the direction of medicine. In an effort to find a satisfying compromise, Dr. Dietz decided on forensic psychiatry. He graduated from Cornell University where he earned a double major in psychology and genetics. After completing his bachelor's degree, he continued his education at Johns Hopkins University. He earned M.D., Ph.D., and Master of Public Health degrees from Johns Hopkins University. During this time, he published several works and made a number of connections in the field. He continued his residency in the Clinical Scholars Program in Psychiatry at Johns Hopkins. He spent his third year of residency as the Chief Fellow of forensic psychiatry at the University of Pennsylvania.

Dietz has testified or consulted in all 50 states. As a forensic psychiatrist, Dietz evaluates the defendant's state of mind. He is asked to determine whether the defendant is mentally competent to stand trial. He is also sometimes asked to evaluate the defendant's mental state when the crime took place. In other words, he is asked whether the defendant was legally insane at the time of the crime.

Dietz has worked on a number of high-profile cases, including the John Hinkley, Jeffrey Dahmer, Susan Smith, Menendez Brothers, Unibomber, D.C. Sniper, and Columbine cases. Dietz firmly stands on the premise of full disclosure and believes in being as honest as possible in preparing reports and presenting expert opinions at trial. "The responsibility of all forensic scientists is to be as objective as possible, to uncover the truth, and report it clearly and accurately," states Dietz. He notes the importance of clearly stating all sources of data and explaining the basis of all opinions.

When Dietz interviews a defendant soon after the crime, he videotapes the interview to preserve the evidence. If the crime has occurred months or even years prior to the interview, Dietz will read the case records before interviewing the defendant. During an interview session, Dietz states that to remain unbiased, he stays centered on the facts and focused on justice.

Today, Dietz owns a private practice and serves as an expert witness 6 to 12 times a year. On a typical day, he handles 10 to 12 cases, consults with attorneys to prepare them for trial, reviews records, visits crime scenes, prepares reports, conducts interviews, and prepares exhibits for trial. Dietz is modest when asked about his achievements—he feels that nothing he has achieved was achieved alone. "By standing on the shoulders of our predecessors we reach a bit higher."

Learn More About It

To learn more about forensic psychiatry, go to www.cengage.com/school/forensicscienceadv.

CHAPTER 2 REVIEW

Multiple Choice

1. Strong evidence suggests the first formal system of law enforcement began in _____. *Obj. 2.1*
   a. the United States
   b. Mesopotamia
   c. England
   d. Egypt
2. _____ is considered to be the father of modern policing. *Obj. 2.1*
   a. Robert Peele
   b. Henry Fielding
   c. J. Edgar Hoover
   d. Calvin Coolidge
3. The first city in the United States to create a system of policing and enforcing laws was _____. *Obj. 2.1*
   a. Chicago
   b. Philadelphia
   c. New York City
   d. Boston
4. A code of ethics is important because _____. *Obj. 2.3*
   a. it provides guidelines for appropriate behavior
   b. it gives credibility to the profession
   c. it outlines rights and responsibilities
   d. all of the above
5. Which of the following is *not* a special consideration when interviewing a child? *Obj. 2.6*
   a. videotaping the child's testimony to show there was no coercion
   b. allowing a special toy or blanket to be held by the child as comfort
   c. allowing the child to face the defendant in court
   d. conducting the interview in a private location
6. The purpose of an interrogation is to _____. *Obj. 2.4*
   a. gather information about a suspect or crime
   b. develop a timeline of events
   c. elicit an incriminating statement or a confession
   d. create a detailed crime-scene sketch
7. The model of interrogation focusing on exhaustion, guilt, and isolation to secure a confession is the _____. *Obj. 2.7*
   a. suspect decision-making model
   b. emotional model
   c. psychoanalytical model
   d. cognitive-behavioral model
8. Which of the following is *not* a sign of deception? *Obj. 2.4*
   a. maintaining eye contact
   b. biting the lip
   c. covering the mouth
   d. crossing the arms over the abdomen
9. Reporting only the facts of a case illustrates _____. *Obj. 2.8*
   a. bias
   b. objectivity
   c. partiality
   d. subjectivity

True/False

Determine whether each of the following statements is true. If the statement is false, explain why the statement is false, and correct the statement.

10. The Internal Revenue Service is under the jurisdiction of the Treasury Department. *Obj. 2.1*

____

____

____

11. In the cognitive approach, the suspect weighs the pros and cons of a confession. *Obj. 2.5*

____

____

____

12. An investigator will know precisely how long an interrogation will last when questioning a suspect. *Obj. 2.4*

____

____

____

Short Answer

13. Discuss the role of industrialization in the need for broadening the scope of law enforcement. *Obj. 2.1*

____

____

____

14. What were some of the initial problems with the Department of Justice that ultimately led to the establishment of the Bureau of Investigations? *Obj. 2.2*

____

____

____

15. Explain J. Edgar Hoover's role in overhauling the Bureau of Investigation. *Obj. 2.2*

16. When would an investigator be more likely to use an interview approach when questioning a suspect? When would an interrogation method be used? *Obj. 2.4*

17. Briefly describe the psychoanalytical model of interrogation. When might an investigator choose to use this model and what kinds of questions might the investigator ask? *Obj. 2.7*

18. Explain how objectivity is related to ethics in forensic science. *Obj. 2.8*

Think Critically

19. Justify the importance of objectivity when processing a crime scene, evaluating evidence, and interrogating suspects. *Obj. 2.8*

20. Evaluate the effect bias can have on a report. *Obj. 2.8*

CRIME SCENE S.P.O.T. Student-Prepared Original Titles

Sibling Rivalry

By: **Casey Langfeld**

Tavares High School
Tavares, Florida

The nightmare began when I woke up in an interrogation room. My vision was blurred and I felt as though I might pass out. I was handcuffed to the chair. A well-dressed man entered the room; he was carrying a cup of coffee in one hand and photographs in the other. Watching me closely, he placed four pictures in front of me.

I was confused. My boyfriend, my sister, and both of my parents—all shot in the head. Detective Ryan started to ask me questions about the previous night, but I couldn't remember. I just sat there staring at the pictures through my tears and praying it was just a bad dream (see Figure 2-18).

Figure 2-18. *The interrogation.*

Figure 2-19. *The blood-stained clothes.*

"Why did you kill them?" Ryan asked. I felt my face grow red with anger, but said nothing. He asked me the question again.

"I could never kill anyone, especially the people I loved the most." I didn't kill them . . . at least I hoped I hadn't. "The last thing I remember is having lunch with my sister yesterday afternoon," I told him. "We were planning a girls' night out, so I went home to take a nap."

"Anything else?" he asked. I just stared at him. Ryan stood up and yelled, "I know you killed them! Your fingerprints were all over the murder weapon! So play stupid or tell us what happened. Either way you are going to jail."

For the first time, I noticed my blood-stained clothes. I must have killed them—my own family—but I couldn't remember any of it. I began crying again, and this time when he asked why I killed them, I could only reply, "I don't know." With that "confession," my attorney advised me to plead guilty to avoid a lengthy trial and a possible death sentence.

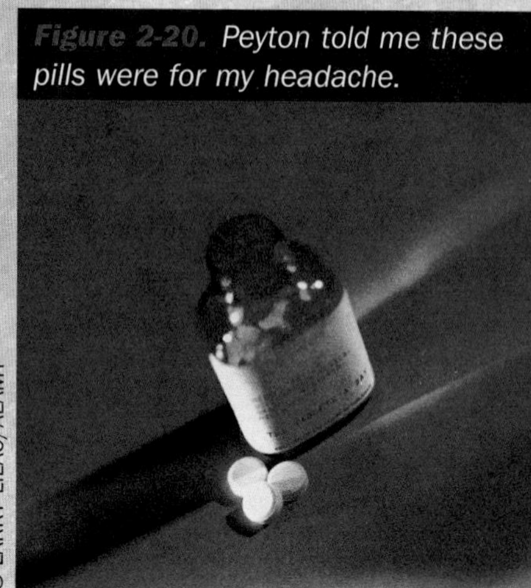

Figure 2-20. Peyton told me these pills were for my headache.

After six years, I began to remember bits and pieces of the nightmare. At lunch, my sister had given me something for my headache because I wanted to cut the afternoon short. I realize now that my sister must have drugged me. When I told Detective Ryan my theory, he laughed. "Why would your sister drug you, then murder her family and kill herself? It doesn't make sense." I couldn't believe what I was hearing. How could they not know? "Not my younger sister. I had lunch with Peyton, my older sister." Peyton wasn't actually my sister, but my parents had raised her from the age of eight.

When Ryan brought Peyton in for questioning, she just laughed. She admitted to drugging me, killing everyone I loved, and then framing me for the crime. Why? Pure and simple—jealousy.

Activity:

Answer the following questions based on information in the Crime Scene S.P.O.T.

1. In the story, what question or questions alerted you to the fact that the narrator was being interrogated rather than interviewed?
2. Describe any ethical issues involved in the interrogation techniques described in the story.
3. Which interrogation model did Detective Ryan use?
4. WRITING Write your own story in which someone is falsely accused of a crime. Be sure that any interviews or interrogations within your story follow at least one of the models described in the chapter. Your story should be 800 words or less.

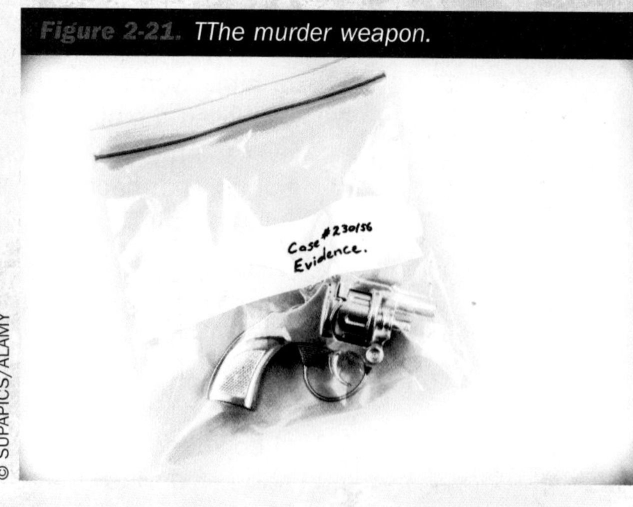

Figure 2-21. TThe murder weapon.

Preventing Adolescent Crime Together™ (P.A.C.T)

Introduction:

In this activity, you will research bullying. Focus on why kids bully, the emotional and psychological effects of bullying, and healthy conflict resolution.

Procedure:

1. In groups of three or four, use the Internet or other media resources to research bullying.
2. Develop two information packets similar to the one shown in Figure 2-22. The first packet should be directed at students several years younger than you. Your teacher will specify the age group. The second packet should be directed to their parents.
3. The student publication should provide information about healthy conflict resolution. It should also answer the following questions:
   - What is bullying?
   - Why do kids bully?
   - What are some signs that someone might be the victim of bullying?
   - What are the psychological and social effects of bullying and being bullied?
   - Where can students go for help if they are being bullied?
   - How can your target audience discourage bullying behavior?

Bullying hurts...

Preventing
Adolescent
Crime
Together

COURTESY OF RHONDA BROWN

Figure 2-22. *Sample P.A.C.T. brochure cover.*

4. The parent publication should give parents information about teaching healthy conflict resolution skills. It should also answer the following questions:
   - What is bullying?
   - How can a parent recognize the signs that their child is being bullied or is being a bully?
   - How can parents address bullying issues with their child?
   - What support groups are available to help parents and families deal with bullying?

ACTIVITY 2-1 *Ch. Obj: 2.4, 2.8*
DRIVE-BY SHOOTING

Objectives:

By the end of this activity, you will be able to:

1. Complete different types of reports properly.
2. Reconstruct a chain of events from witness and police statements.

Materials:

(per pair of students)
blank report forms
drawing paper or poster board
metric rulers

Procedure:

Working with a partner:

1. Read through the following scenario.
2. From the information in the scenario, fill out the reports and draw a sketch of the crime scene. Be sure to meet the following guidelines:
   - Include the positions of each of the suspects inside the car.
   - Place and label all of the evidence in the proper location at the scene.
   - Write a complete incident report detailing the chain of events, including information about who fired the shots. Be sure to explain the paint on the bat, the blood, and any other relevant evidence.

Scenario:

At 9:47 P.M., the 911 operator answered a call. The caller reported hearing multiple gunshots in the area of Main Street and Oak Boulevard. Police were immediately dispatched to the area.

Officers Barber and Coley arrived on the scene at 9:59 P.M. After carefully assessing the scene to make sure there were no injured victims, Coley began interviewing potential witnesses. As she did so, Barber placed a call to crime-scene investigators.

At 10:15 P.M., investigators began a thorough search of the area. Investigators flagged, photographed, and collected evidence, including

- Blood spatter on the side of the corner building facing Main Street; gravitational drops leading across the street, up a set of stairs, and into an apartment
- Bullet casings from multiple weapons—.45 and .357 calibers
- Tire tracks in mud on the side of the road and similar tracks and skid marks in the middle of Oak Street
- A piece of red cloth, consistent with a t-shirt, on the street near the skid marks
- Some dark blue paint on a baseball bat on the ground near the blood

Barber radioed for backup officers to help search the apartment, and paramedics were called to the scene.

At 10:45 P.M., additional officers and paramedics arrived. Officers Moses and Moore followed the path of blood drops to Apartment C and knocked on the door. An elderly woman answered the door and invited the officers in. She allowed them to search the rooms. In the bathroom, the officers found an unconscious white male on the floor; he had a blood-soaked towel wrapped around his leg. Paramedics transported him to the nearest hospital. The injured male was identified by his grandmother as Neil Marshall.

After interviewing witnesses, police were able to piece together the following chain of events:

- At 9:30 P.M., a group of young men in a late-model black car pulled up and began talking to two males—one black, one white.
- The discussion got louder as it progressed, and one of the young men standing next to the car hit the driver's door with a bat.
- The other young man outside the car pulled out a knife and stabbed at the passenger in the backseat on the driver's side, cutting the passenger and ripping his shirt.
- The passengers in the car began shouting, pulled out guns, and began shooting as the car sped away, tires squealing.
- Sara Fuller was watching television in an apartment across the street and heard the shots fired. She immediately reported the incident.

Several hours later, a late-model dark blue sedan was found abandoned several blocks from the shooting. Officers impounded the car. Scientists from the forensic laboratories found traces of gunpowder on the back door behind the driver and on the upholstery in the middle of the backseat, a small smear of blood—type O—on the same back door, and a dent to the driver's door. Multiple fingerprints were found, but none matched any of the database records. The car had been reported stolen three days earlier.

When the injured male regained consciousness, he identified the young men in the car as Bart Lambert, Calvin Davis, Joseph Rivera, and Garrison Hastings. The young men were brought in for questioning. Officers read the young men their Miranda rights before questioning them.

- B. Lambert, a white male, born 12/13/80: tested positive for gunshot residue, had no visible wounds, had a Colt .45 in his possession
- C. Davis, a white male, born 6/01/81: negative for gunshot residue, no visible wounds, but confessed to being present at the scene, had a knife, a 9 mm Browning and a .22 Berretta in his possession
- J. Rivera, a Hispanic male, born 4/27/82: positive for gunshot residue, a superficial cut to the left shoulder, wearing a red t-shirt with a piece of the sleeve missing, had a knife in his possession
- G. Hastings, a black male, born 7/04/89: negative for gunshot residue, no visible wounds

During the interview, it was determined the four were indeed involved in the incident. Each was then interrogated. The front-seat passenger, fearing jail, confessed and made a complete statement naming the driver and both shooters.

Questions:

1. Why was the original questioning considered an interview?
2. When did the interviews become interrogations? How is an interview different from an interrogation?
3. Who was driving the car? How do you know?
4. Who were the shooters? Where were they seated in the car?
5. Can it be determined who shot Marshall? Why or why not?
6. Can it be determined who wounded Rivera? Why or why not?
7. If you were part of the investigation team, what questions would you still need to ask?

Extension:

Build a three-dimensional scale model of the crime scene. You may use clay, wood, computerized technology, or any other medium approved by your teacher.

ACTIVITY 2-2 *Ch. Obj: 2.3, 2.8*
ETHICAL ERRORS

Objectives:

By the end of this activity, you will be able to:

1. Evaluate ethical behavior of law-enforcement personnel in various scenarios.
2. Explain the importance of proper record keeping and report writing.

Procedure:

1. Read your group's assigned case.
2. Determine the ethical error and answer the questions for each situation.
3. As a group, prepare a presentation regarding the details of the case. Be sure to include the answers to the questions in your presentation.

CASE 1

Officer Peacock observed a car pass his car on the highway. The driver was a black male smoking a cigar. The officer thought it was a "blunt" (cigar stuffed with marijuana) and signaled for the vehicle to pull over. The driver took his time, moving across several lanes of traffic before stopping. Officer Peacock pulled his car next to the driver's side, pointed his handgun at the driver, and began yelling instructions. Officer Peacock yelled vulgar profanities and expletives at the driver. The officer grabbed the cell phone from the driver's hand and threw it into the car with adequate force to chip the windshield. The officer asked if the driver had flicked his "blunt" at him. The driver told him that it was merely a cigar. Officer Peacock terminated the stop without documenting the traffic stop in any fashion. The department has a strict policy of detailed documentation. The officer also did not write the required use-of-force report.

Questions:

1. Did the officer have probable cause for the "stop"?
2. Explain the mistakes the officer made in this case.
3. Identify reasons you think the department requires detailed documentation of all traffic stops.
4. What do you think is the importance of a "use-of-force" report? Why do you think Officer Peacock did not file one?

CASE 2

Deputy Torres reported that her department-issued handgun, stun gun, and three magazines of ammunition had been stolen from her home. She wrote a memo detailing the theft in which she falsely claimed that the equipment had been secured in the closet of a locked bedroom and that she was the only one with access to the bedroom. During a criminal investigation of the theft, it was determined that neither the closet door nor the bedroom door had a lock on it.

Questions:

1. Why is it important for law-enforcement officers to report missing equipment as soon as they recognize an item is missing?
2. What motive(s) do you think Deputy Torres would have to create such an elaborate plan falsely detailing the theft?
3. What was the likely outcome of the criminal investigation? What did the investigation clearly indicate?

CASE 3

Officer Saunders was part of a team investigating a large drug operation. After shutting down the operation and arresting the drug dealers, he failed to impound various items of found and confiscated property, including two cell phones and a GPS. He took them home and kept them. He then sold one of the phones online for $50.

Questions:

1. What are the proper procedures for recording confiscated property?
2. What was the likely way in which Officer Saunders got caught?
3. Why must all property be recorded and retained by law enforcement?

CASE 4

Officer Evans lied to his supervisor when he said he had finished a traffic accident report that he had not yet finished. He then finished the report and backdated it to confirm his earlier statements. A department review of reports and procedures revealed that he had also neglected to complete reports on several other cases. A few months later, Officer Evans's girlfriend made a complaint about threats she had received from Evans by phone and text message. Among the messages were the following statements: "Woman, I can take care of you and I won't even have to lift a finger." "I know where you live, and you might have forgotten, but I am a police officer, don't worry, I've got you, I'll make you pay."

Questions:

1. What are the ethical issues with backdating a report?
2. What is the connection between the first infraction and the second one a few months later?
3. Do you think the girlfriend's complaint is valid? Was she threatened?
4. Do you think Officer Evans abused his power in law enforcement? Explain.

CASE 5

Deputy Palanski received information that a possible suspect in numerous commercial burglaries might be at a particular residence. Palanski arrived at the residence along with a detective and another deputy. They were met in the driveway by the homeowner. Deputy Palanski explained his reasons for being there and asked if the suspect he was searching for was in the home. The homeowner said "no," and refused to give consent to search the home. The homeowner was placed in handcuffs and detained in the backseat of a marked patrol car. Deputy Palanski then went to the front door and knocked several times. He opened the door and announced who he was, and then entered the home without consent and without a search warrant. Palanski acknowledges that he should not have entered the home under these circumstances, but at the time he believed he was justified in making entry.

Questions:

1. What amendment right was violated in this case? Explain.
2. What plausible justification would have allowed Deputy Palanski to search the home without a warrant?
3. How could the Deputy Palanski have handled this case differently?
4. Did Deputy Palanski exhibit professional and ethical misconduct in this case? Explain.

CASE 6

Officer Hayes chronically failed to complete reports and properly impound evidence. He lied to numerous citizens, to his supervisors, and eventually to the professional standards bureau to cover up his failures. He informed investigators that he had no property relating to the 150 or so cases with unfinished reports. He later returned with three boxes and several trash bags containing various paperwork, evidence, and property.

Questions:

1. What would happen if law-enforcement officers commonly failed to complete reports and impound evidence?
2. Based on the information provided in this case, is there any way to "undo" the mess created by this officer?
3. Identify three characteristics that are important for law-enforcement officers to have in order to generate proper reports and documentation.
4. What could be some possible negative outcomes of Officer Hayes' behavior?
5. Is this an ethical concern, a "getting organized" concern, or both? Explain your answer.

ACTIVITY 2-3 *Ch. Obj: 2.8*
THE MISSING MASCOT

Objectives:

By the end of this activity, you will be able to:

1. Use deductive reasoning to solve a case.
2. Explain the importance of organization and proper documentation.

Procedure:

1. Read the background information.
2. Record the data from the background information in Data Table 1.
3. Read the witness notes gathered by police during their investigation.
4. Record important information in Data Table 2.

Background Information:

Upon arriving at the school on Friday at 6:15 A.M., Principal Schwengel noticed that the locks had been cut off and the gate was standing open. He immediately alerted the campus resource officers, who discovered that the school mascot, a plaster statue of a knight on a black horse, was missing from the courtyard. The officers called in additional police officers to help with their investigation.

It was the morning of the big football game between the Harrison High School Knights and their long-time, cross-town rival, the Central High Bulldogs. The Harrison mascot was to be transported to the football field later that afternoon. A pep rally the previous evening had ended at around 9.00 P.M. and the principal had locked the gate at precisely 9:45, just after the custodians had finished clearing the litter. It had rained heavily from about 10:00 until 11:30.

As investigators scanned the crime scene and interviewed witnesses, they made the following notes:

- A paper cup and sandwich wrapper from the nearby fast-food restaurant—the Burger Barn—littered the courtyard.
- Graffiti in the courtyard said "GO BULLDOGS!!" and "Slay the KNIGHTS!"
- An empty can of red spray paint and two partial cans of gray spray paint were found behind the bushes near the parking lot.
- One of the cans had a smudge of a slippery black substance on the side.
- A recently smoked cigarette butt was found near the water fountain.
- Prints from three different pairs of athletic shoes were found in the soil between the parking lot and the fence, going in both directions. The prints were made by a man's size 12 shoe, a man's size 10 shoe, and a woman's size 6 shoe. Another set of tracks, from a man's boot, size 9 1/2, was found in the courtyard.
- A man's 10-speed bicycle was parked under a tree a few feet outside the gate. It had a flat tire.
- Car tracks were found on the shoulder of the road near the school's entrance.

Investigators thought students from Central High School stole the mascot to taunt the Knights. After questioning students and neighbors near the school property, investigators discovered there had been a disturbance of some sort around 12:30 A.M.

After interviewing several people who attended the pep rally, investigators made the following additional notes:

- Jeff works part-time at the Burger Barn. He worked until 12:45 A.M. last night because it was his night to close the restaurant. His best friend and lab partner is Kevin.
- Wayne is a mechanic who recently began working the early morning weekend shift doing oil changes for a local cab company.
- Mikayla and Ethan went to the movies after the pep rally. They still had the ticket stubs in their pockets.
- Addison is a flyer for the cheerleading squad. She is so light weight that the other girls are able to toss her higher than any of the other flyers.
- Ethan and Kevin are football players at Central. Kevin is also the starting center for the basketball team.
- Matthew and Kyle are football players for Harrison. They live in the same neighborhood, down the street from school.
- Addison is Matthew's sister. Mikayla is Kyle's sister. Addison and Mikayla are cheerleaders for their school.
- Wayne lives alone in a modest house across the street from the football field.
- Addison is a very petite girl who wears size 4 blue jeans.
- Mikayla is a tumbler—her specialty is a back-handspring series. She takes her sport seriously, so she swore off sodas years ago.
- Wayne is 5 feet, 5 inches tall.
- In his spare time, Kyle rides his bike to keep in shape.
- Ethan was wearing his red and gray football jersey when he was interviewed.
- Wayne hates football because the games are noisy.
- Addison and Matthew came back to the school after the pep rally to pick up Kyle because his bike got a flat tire. They all walked over to the fence because they thought they heard noises from the courtyard. They didn't see anything, so they left.

Data Table 1. Background information.

Description of crime scene	
Physical evidence	
Circumstantial evidence	
Any other evidence that may help investigators solve the crime	

Data Table 2. Witness notes.

	Size of the Person	Occupation or Hobbies	School Affiliation	Inferences That Can Be Made
Jeff				
Kevin				
Matthew				
Kyle				
Mikayla				
Ethan				
Addison				
Wayne				

Questions:

1. Why did investigators initially believe the Bulldogs committed the crime?
2. Who took the mascot?
3. What information helped you identify the perpetrator?
4. Cite key pieces of evidence that caused you to believe the other seven suspects were innocent.
5. How did organizational and documentation skills help in solving the case?
6. Write a two-paragraph scenario explaining what you think happened.

Extension:

Use the methods outlined in this activity to develop an outline of events in a current, local criminal case. Include all possible suspects, motives, and witnesses. Then, based on the facts you accumulate, complete a scenario of what you think occurred in the case.

ACTIVITY 2-4 *Ch. Obj: 2.4, 2.5, 2.7*
INTERROGATION STATION: A ROLE-PLAYING ACTIVITY

Objectives:

By the end of this activity, you will be able to:

1. Demonstrate proper questioning and note-taking procedures.
2. Utilize interviewing and interrogation skills.

Cast of Characters:

victim
suspect
eyewitnesses
interviewer
interrogator

Procedure:

1. In a small group, create a fictitious criminal scenario in which each person in the group will portray one of the cast of characters.
2. Make sure the behavior of each character is portrayed realistically. If necessary, research details about the role.
3. Practice the skit several times.
4. Perform the skit in front of the class.

The requirements for the role-play activity include the following:

- Provide clear details regarding the crime that occurred, location, time, etc.
- Use proper interviewing and interrogation techniques. (For example, you should begin with open-ended questions.)
- Characters must act and behave as they would if the crime had truly occurred, within reason.
- Scenario may be creative and have a twist at the end that the audience is not expecting, but it should be logical and realistic.

CHAPTER 1 2 3 4 5 6 7 8 9 10 11 12 13 14 15

CHAPTER 3

Forensic Laboratory Techniques

WHERE IS LACI?

Figure 3-1. Laci Peterson.

Laci Peterson was 27 years old and in her eighth month of pregnancy when she disappeared from her home in Modesto, California, on December 24, 2002. She was last seen walking her dog that morning at 10:00 a.m. Her husband Scott returned home from a fishing trip and discovered that Laci was gone. Trying to locate her, he contacted friends and family. He reported her missing shortly before 6:00 p.m.

Over the next three days, police, firefighters, and volunteers searched for Laci, but there was no sign of her. A grid-pattern search along Dry Creek revealed no evidence. Three days after her disappearance, the FBI joined in the investigation, and the Peterson home was searched.

Peterson had been a suspect in his wife's disappearance almost from the beginning. When he confessed to having a romantic relationship with another woman, he became a very strong suspect. Peterson maintained that he had told Laci about his indiscretion and that it was not something that would have ended their marriage. At first, Laci's parents supported Peterson. When it was also discovered that he had taken out a $250,000 life insurance policy on Laci, they stopped.

As time passed, evidence against Peterson mounted. A powdery substance found on Peterson's boat was identified as concrete dust. A long, dark hair caught on pliers was found in the same fertilizer warehouse where Scott, a fertilizer salesman, had stored his boat. Forensic scientists used comparison microscopes to compare this hair to hair known to be Laci's. They examined the cuticle, pigmentation, and medulla. Recall from previous coursework that the cuticle is the outer layer of hair. The medulla is the center core. The hair found at the warehouse was consistent with Laci's hair. Scott Peterson had purchased a four-day fishing license on December 20, but he told police he hadn't decided to go fishing until the morning of Laci's disappearance (December 24). His blood was found on the driver's door inside his truck.

On April 13, 2002, the body of a fetus with the umbilical cord still attached washed ashore near Point Isabel in Northern California. The next day, a female's body was discovered in a park near Point Isabel. DNA testing revealed them to be the bodies of Laci and her baby, Conner.

Scott Peterson was convicted of the two murders and sentenced to death by lethal injection.

Figure 3-2. Laci's body being loaded into the van.

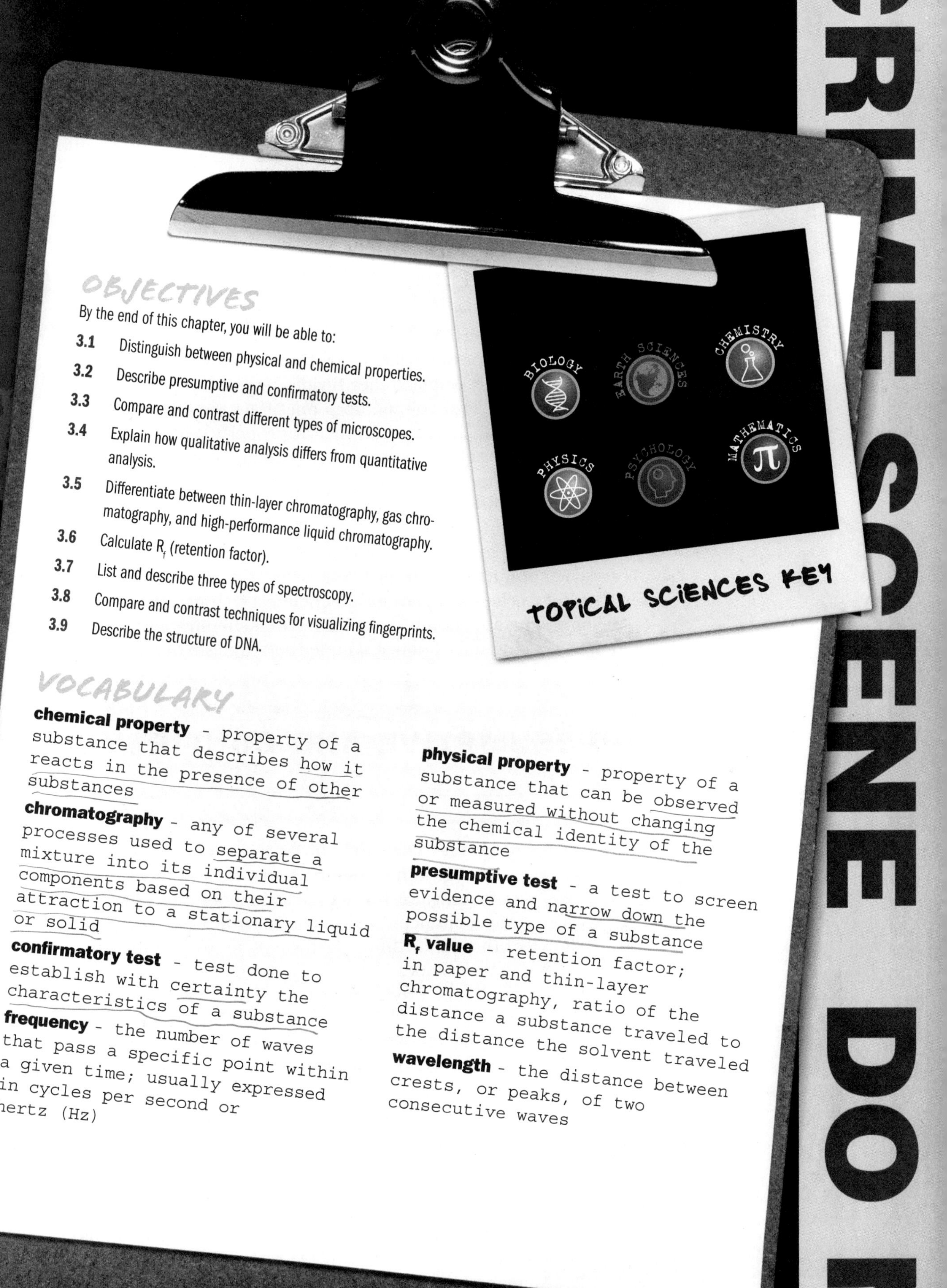

OBJECTIVES

By the end of this chapter, you will be able to:

3.1 Distinguish between physical and chemical properties.
3.2 Describe presumptive and confirmatory tests.
3.3 Compare and contrast different types of microscopes.
3.4 Explain how qualitative analysis differs from quantitative analysis.
3.5 Differentiate between thin-layer chromatography, gas chromatography, and high-performance liquid chromatography.
3.6 Calculate R_f (retention factor).
3.7 List and describe three types of spectroscopy.
3.8 Compare and contrast techniques for visualizing fingerprints.
3.9 Describe the structure of DNA.

VOCABULARY

chemical property - property of a substance that describes how it reacts in the presence of other substances

chromatography - any of several processes used to separate a mixture into its individual components based on their attraction to a stationary liquid or solid

confirmatory test - test done to establish with certainty the characteristics of a substance

frequency - the number of waves that pass a specific point within a given time; usually expressed in cycles per second or hertz (Hz)

physical property - property of a substance that can be observed or measured without changing the chemical identity of the substance

presumptive test - a test to screen evidence and narrow down the possible type of a substance

R_f value - retention factor; in paper and thin-layer chromatography, ratio of the distance a substance traveled to the distance the solvent traveled

wavelength - the distance between crests, or peaks, of two consecutive waves

Obj. 3.1 INTRODUCTION

Hairs and fibers found at a crime scene can give investigators a great deal of information. Microscopy and other analytical techniques can be used to determine whether the hairs came from a human or another animal. These techniques also help investigators determine whether the fibers came from carpet, clothing, or something else.

In the crime lab, forensic scientists compare samples of evidence from the crime scene to known samples. The evidence samples are called *questioned samples.* The known samples are called *controls.* Ultimately, investigators hope to identify the evidence samples through specific physical and chemical properties.

Physical properties are properties that can be measured without changing the identity of the evidence. For example, when forensic scientists calculate the density of glass, they divide the mass of the glass by its volume. Measuring mass and volume does not affect the chemical makeup of the glass. Therefore, density is a physical property. Other physical properties include color, melting point, boiling point, odor, and viscosity. Changes to substances that do not alter the chemical makeup of the substance—cutting, shredding, melting, or freezing—are physical changes.

Chemical properties determine how a substance behaves in the presence of other substances. For example, iron will react with oxygen in the presence of water to produce rust, or iron oxide. Changes to a substance that alter its chemical identity are chemical changes. Rusting, burning, and decomposing are chemical changes. When chemical testing is done on evidence, the original evidence sample is often destroyed.

Obj. 3.2 PRESUMPTIVE AND CONFIRMATORY TESTS

© PABLO PAUL/ALAMY

Figure 3-3. Presumptive test for blood.

At a crime scene, field investigators must make immediate decisions regarding potential items of evidence. For example, if an investigator finds a red stain at a homicide scene, he or she must conduct initial tests to narrow down the possible identity of the stain. Although it is easy to assume that the red stain is human blood, it could also be paint, ketchup, or blood from an animal. **Presumptive tests** allow a field investigator to screen evidence to reduce the number of possibilities and to get a preliminary identification. If presumptive tests at the scene show that the red stain is blood, investigators will collect additional samples. The tests do not, however, tell crime-scene investigators whether the blood is from a human or another animal. These samples are sent to the lab for **confirmatory tests.** Confirmatory tests are used to make a more specific identification. A confirmatory test would determine whether the blood belonged to a human or some other animal (see Figure 3-3).

Presumptive tests exist for saliva, semen, blood, urine, and vaginal secretions. There are also presumptive tests for many kinds of drugs. Presumptive tests screen for chemicals in each fluid. At the crime scene, investigators might use a

UV (ultraviolet) light, or another alternative light source, to determine whether a stain is a body fluid. If so, a sample will be collected and sent to the lab. At the lab, forensic scientists may perform presumptive tests to determine what kind of body fluid. For example, semen contains the enzyme acid phosphatase. Although this enzyme is found in other fluids, including vaginal secretions, it is found in much higher concentrations in semen. The presumptive test for semen is actually screening for acid phosphatase. To definitively identify the evidence as semen, confirmatory testing would be completed at the crime lab. For example, the fluid may be viewed under a microscope. If sperm is present, the fluid is semen. In this case, microscopy is a confirmatory test. To identify the person who produced the semen, the sample must undergo more specialized confirmatory tests, such as DNA profiling. Presumptive testing reduces costs and aids field investigators in collecting evidence essential to the case. Confirmatory tests are necessary to identify, with certainty, a piece of evidence. Each test plays an integral role in forensic investigations.

MICROSCOPY

Obj. 3.3

Prior to the mid-1600s, microscopes could magnify a specimen only about six to ten times its original size. In 1665, Robert Hooke published a book called *Micrographia*. In it, he described cork cells. Hooke was the first to observe cells; he used the term *cells* because the dead cork cells looked like small rooms. In 1676, Antoine van Leeuwenhoek was the first to observe single-celled microscopic organisms. Today, there are several different kinds of microscopes. Some can magnify an object hundreds of thousands of times.

Some of the earliest microscopes were simply magnifying glasses. Looking at small insects like gnats and fleas was very common. For that reason, microscopes were sometimes referred to as *flea glasses*.

COMPOUND LIGHT MICROSCOPE

The compound light microscope is probably the most widely used microscope today. This microscope has a light source and multiple lenses to obtain high magnification. The compound microscope usually has a magnification between 40× (40 times) and 1,000× (1,000 times). Compound microscopes are powerful enough to view hair, fibers, and cells. Figure 3-4 shows a cross-section of an artery through a compound light microscope at different magnifications.

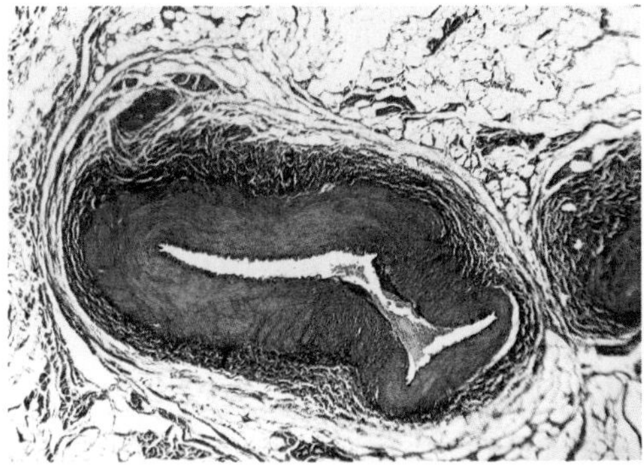

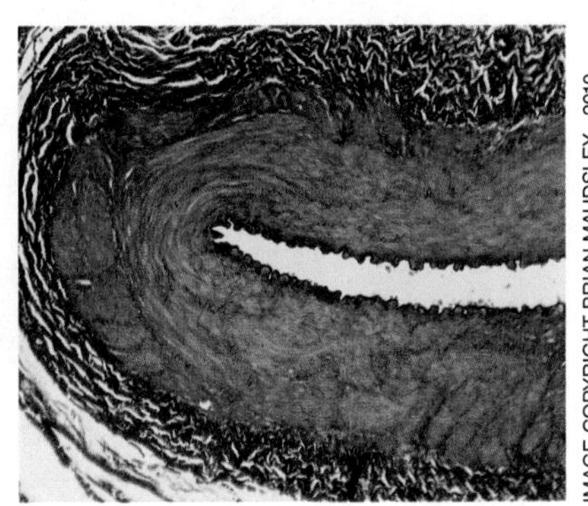

Figure 3-4. *The image on the left has been magnified 40×. The image on the right has been magnified 100×.*

CHAPTER 1 2 3 4 5 6 7 8 9 10 11 12 13 14 15

OMAR TORRES/AFP/GETTY IMAGES

Figure 3-5. Comparison microscope.

STEREOMICROSCOPE

A compound microscope works by sending light through the specimen. Sometimes, a specimen is too thick or opaque to be seen through a compound microscope. The light of a stereomicroscope, or dissecting microscope, is reflected from the surface of the specimen. Because the light is reflected, the stereomicroscope produces a three-dimensional image useful for dissecting. Surface details are also more visible with the stereomicroscope. Forensic investigators use a stereomicroscope to examine insect larvae, paint chips, and other small items of evidence.

COMPARISON MICROSCOPE

The comparison microscope is another useful tool (see Figure 3-5). The comparison microscope is actually two microscopes connected to one eyepiece. When the investigator looks through the eyepiece, he or she sees a circular, split-view window. The image on the right is of the specimen under the microscope on the right and can be compared side-by-side to the image on the left. The comparison microscope is particularly useful when comparing bullet striations, fibers, and hair samples. Investigators are able to make comparisons while viewing two samples at the same time (see Figure 3-6). Usually, the investigators will compare a known sample to a questioned sample.

PHILIPPE PSAILA

Figure 3-6. Images of two bullet casings as seen through a comparison microscope. One casing was found at the crime scene. The other was taken from the suspect's gun.

ELECTRON MICROSCOPES

Compound microscopes, stereomicroscopes, and comparison microscopes all use light. Electron microscopes, on the other hand, use beams of electrons to form images. These microscopes can magnify materials up to 500,000× with good resolution, but the image is in black and white. The *transmission electron microscope (TEM)* passes a beam of electrons through a thin slice of a specimen. This produces images of internal structures. The *scanning electron microscope (SEM)* passes a beam of electrons over the surface of a sample to produce a three-dimensional image. This image provides details about the surface of the sample. Forensic investigators use electron microscopes to analyze small specimens and to view tiny surface or internal details. SEM is also an important step in the determination of the identity of trace materials, such as gunshot residues.

CHROMATOGRAPHY

Obj. 3.4, 3.5, 3.6

Most analytical techniques can be classified as either quantitative or qualitative. A quantitative analysis will result in a measurable amount—a quantity. Qualitative analysis, on the other hand, will result in a description or identification of the components of a mixture. In chemistry, a *mixture* is the combination of two or more substances. The substances in a mixture do not react chemically. Therefore, they can be separated based on their physical properties.

Qualitative tests are based on the physical and chemical properties of the sample. **Chromatography** separates substances within a mixture based on their physical properties. Different substances will adhere, or stick, to solid surfaces or dissolve in a solvent differently. In paper chromatography, a small amount of a liquid mixture is placed near the bottom of a piece of paper. Usually, a scientist will place a drop of an unknown mixture and a drop of a known mixture several millimeters apart. The known mixture acts as the control. The bottom of the paper is placed into a liquid solvent. The solvent must be lower on the paper than the samples. The solvent moves up the paper and is called the *mobile phase.* The paper itself is called the *stationary phase* because it does not move. You have probably seen a liquid move along a piece of paper. If you spill a little water on a paper towel, for example, the water will spread out. In paper chromatography, the liquid solvent spreads across the paper in much the same way. As the liquid solvent moves up the paper, different components of the mixture will adhere to the paper at different places. These components leave marks on the paper.

The result of any chromatography is a called a *chromatogram* (see Figure 3-7). A chromatogram shows substances that were dissolved in the original mixture. The chromatogram also shows how far the solvent traveled. Investigators can identity components of the original

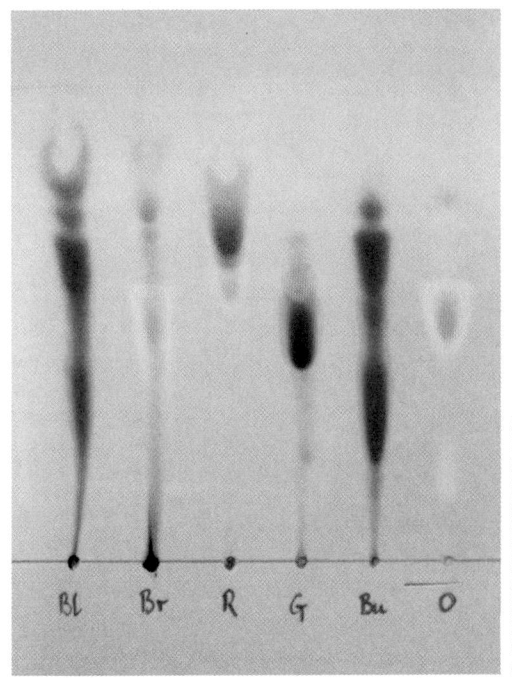

Figure 3-7. *Chromatogram from paper chromatography.*

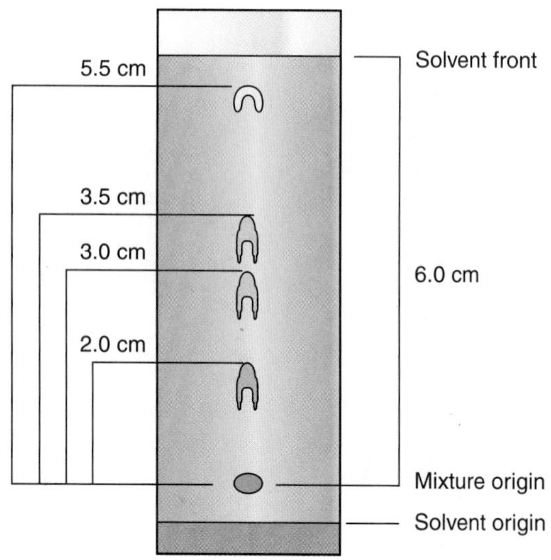

Figure 3-8. Chromatogram used for calculating R_f.

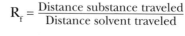

mixture by calculating the retention factor, or the R_f value. The R_f value is a qualitative comparison between the length of time the substance is in the mobile phase and in the stationary phase. In paper chromatography, the **R_f value** is the ratio of the distance the substance traveled to the distance the solvent traveled. Chromatography can be done with different solvents to establish the identity of an unknown substance. The R_f value for each substance will depend on the solvent being used. First, the investigator measures how far the solvent traveled. The line that shows where the solvent stopped moving is called the *solvent front.* The investigator then measures how far each dissolved substance traveled. For example, if the substance traveled 3.0 cm and the solvent traveled 6.0 cm, as in Figure 3-8, the R_f is calculated as follows:

MATHEMATICS

$$R_f = \frac{\text{Distance substance traveled}}{\text{Distance solvent traveled}}$$

$$R_f = \frac{3.0\text{ cm}}{6.0\text{ cm}}$$

$$R_f = 0.5$$

Chromatography is used in forensic science to analyze dyes in fibers, test for explosives or accelerants, and to check body fluids for the presence of drugs. More sophisticated forms of chromatography have replaced paper chromatography in forensic laboratories in recent years. Most other chromatographic techniques pass liquid or gas through a column or tube packed with a porous solid material. Thin-layer chromatography (TLC), gas chromatography (GC), and high-performance liquid chromatography (HPLC) are commonly used in forensic laboratories. TLC is similar to paper chromatography, but the stationary phase is a thin layer of gel-like material on a glass or plastic plate. TLC is faster and produces clearer separation than paper chromatography. Both paper and thin-layer chromatography are useful for separating dyes and inks. GC is performed at high temperatures and is useful for separating mixtures that contain large molecules, such as the proteins found in blood. HPLC uses high pressure to force mixtures through a column of liquid. Unlike GC, HPLC can take place at room temperature. Therefore, HPLC can be used to test for the presence of flammable materials, such as explosives or accelerants, which may be found during an arson investigation. Figure 3-9 shows a sample of a gas chromatogram.

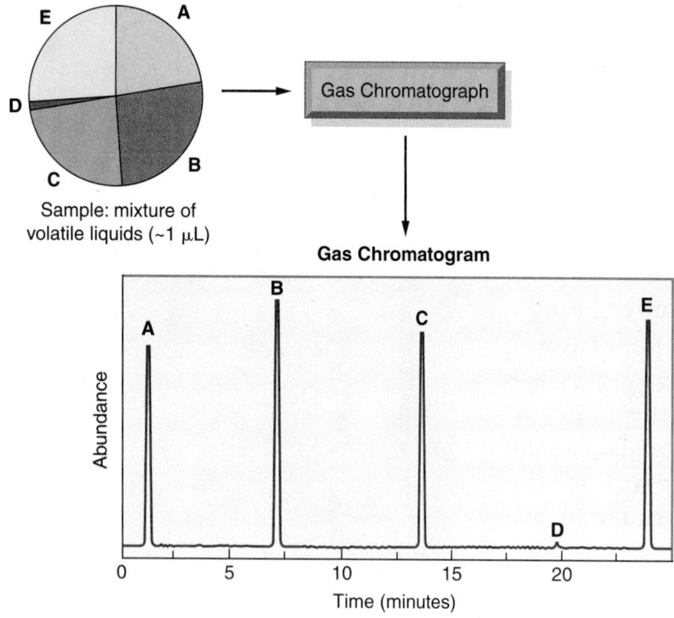

Figure 3-9. Gas chromatography helps forensic scientists determine the components of a mixture. Each spike represents a different component of the mixture. The relative heights tell the scientist the relative abundance of each component.

Digging Deeper

with Forensic Science e-Collection

Many people die each year because someone was driving while intoxicated. Alcohol impairs reflexes, which makes driving extremely dangerous. But how much is too much? For people under the age of 21, it is illegal to consume any alcohol. States have laws that define legal blood alcohol levels for drivers over 21. Confirmatory tests for blood alcohol levels use gas chromatography. New presumptive tests are very accurate and can give law-enforcement personnel the information they need much more quickly. Using the Gale Forensic Science eCollection at www.cengage.com/school/forensicscienceadv, research current techniques used to get quicker results for law enforcement.

ELECTROMAGNETIC RADIATION

Light travels in electromagnetic waves. The highest point in the wave is called the *crest*. The distance between two consecutive crests is the **wavelength. Frequency** refers to how many waves pass a specific point within a given time. Therefore, a wave with a high frequency will have a short wavelength (see Figure 3-10). Visible light, X-rays, radio waves, and microwaves are all electromagnetic waves. Electromagnetic waves can be organized into the electromagnetic spectrum, based on their wavelengths and frequencies. Forensic scientists use the electromagnetic spectrum (see Figure 3-11) to search for latent fingerprints, examine articles of clothing for trace evidence, or determine the structure of a molecule.

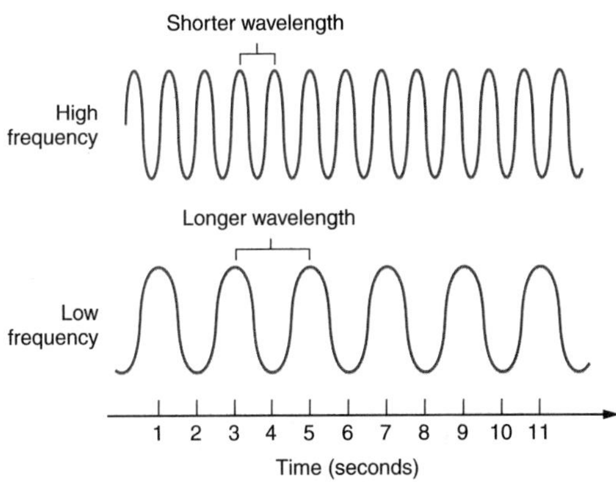

Figure 3-10. *Higher frequency waves have shorter wavelengths.*

high freq. = short wl
low freq = long wl.

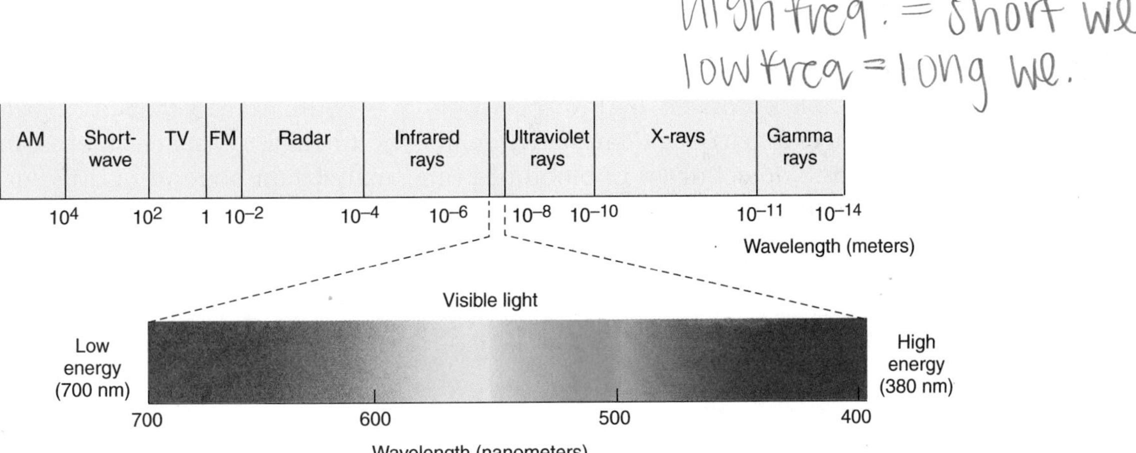

Figure 3-11. *The electromagnetic spectrum is divided into wavelength ranges, or bands. Visible light has a wavelength between 380 and 760 nm (nm = 10^{-9} m).*

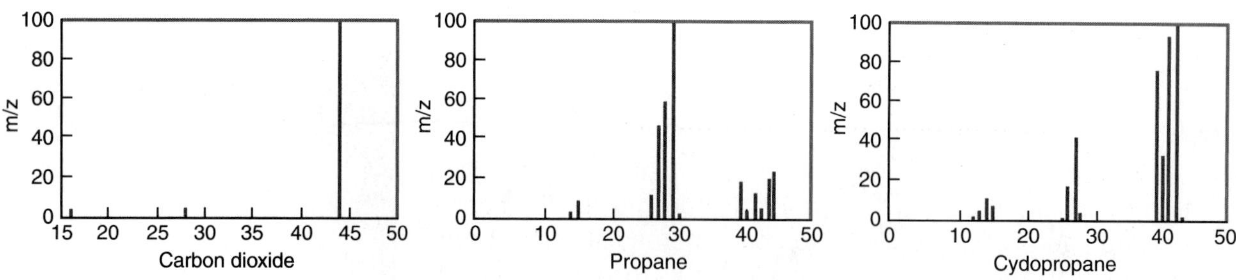

Figure 3-12. *Spectrographs of carbon dioxide, propane, and cyclopropane.*

Obj. 3.7 SPECTROSCOPY

When a substance is heated, it emits light at a specific wavelength. Electromagnetic spectroscopy uses this chemical property to determine what elements are present in a sample. Spectroscopy can be used to identify fibers and other trace evidence. It can also be used to detect contaminants in various materials. Spectroscopy can detect accelerant and explosive residue. An electromagnetic spectrograph measures the wavelengths of light emitted and captures a spectral image on photographic film. The spectral image is a series of lines (see Figure 3-12). Each element produces a unique line in the spectral image. So, the pattern of lines tells the scientists which elements are found within the sample. Spectroscopy also measures the amount of light absorbed, which can be used to determine the concentrations. There are several forms of spectroscopy. Mass spectroscopy is often combined with gas chromatography to identify atoms and molecules by their masses. A sample is loaded into the mass spectrometer and vaporized and ionized, forming charged particles called *ions*. The ions are then sent through a magnetic or electric field. The path of the ion depends on the ratio of its mass to its charge. The results are recorded on a photographic plate. Every chemical has a unique mass spectrum, making mass spectroscopy useful as a confirmatory test.

In June 2008, a partial fingerprint on a painting called "Young Girl in Profile in Renaissance Dress" confirmed that the painting was a da Vinci original. The fingerprint was consistent with two other fingerprints found on da Vinci works. In 1995, the painting sold for $19,000. The painting is now worth $160 million.

Atomic absorption spectroscopy (AAS) measures the amount of light of a specific wavelength absorbed by atoms of a particular substance. This technique is especially useful in determining heavy-metal contaminants in air, water, and soil samples. It is also useful when analyzing paint chips. This technique can help forensic scientists determine whether soil or paint at the crime scene can be linked to another location. A link may help connect a suspect or victim to the crime.

Ultraviolet (UV) spectroscopy measures wavelengths of light and can be used to determine the concentration of different elements in a solution. The graph produced by UV spectroscopy is compared to that of known substances as part of a quantitative analysis of the data. UV spectroscopy can be used to detect drugs in blood or urine, analyze components of dyes and food additives, and monitor air and water quality.

Obj. 3.8 FINGERPRINT-DEVELOPING TECHNIQUES

CHEMISTRY

Often, crime-scene investigators find objects and surfaces at a crime scene that were touched by the victim or perpetrator of the crime. The person may have left fingerprints behind. These fingerprints can be a very important piece of evidence. Most of the time, the fingerprints are latent, or invisible. Certain chemical and physical properties of fingerprints make it possible for a latent fingerprint examiner to use a variety of techniques to lift and visualize these prints. For example, most body fluids can be seen with a high-intensity UV light. This light helps investigators detect the presence of fingerprints. Another well-known technique for visualizing fingerprints is called *dusting for fingerprints* (see Figure 3-13). Investigators apply finely ground powders to the surface with a soft brush. The powders stick to fingerprint residues, making the print visible. The print can then be lifted with adhesive tape and placed on a labeled card. The powders come in a variety of colors. The color of the object or surface being analyzed for prints determines the color of the powder used.

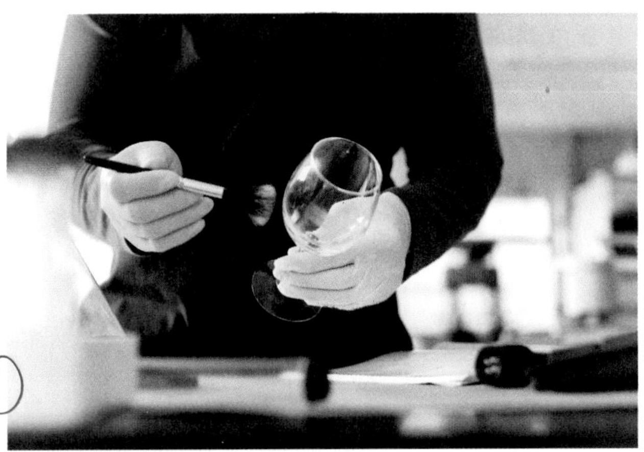

SEAN JUSTICE/CORBIS

Figure 3-13. *Dusting for fingerprints.*

CYANOACRYLATE FUMING

Dusting works best on fresh fingerprints. Sometimes, however, latent fingerprints are not found right away. To view prints found on nonporous substances, such as glass and many plastics, crime-lab technicians may need to use cyanoacrylate, or Super Glue®. When heated, the cyanoacrylate reacts with traces of proteins and fatty acids in the fingerprint. The technique is called *cyanoacrylate fuming* or *Super Glue fuming*. The item being analyzed is placed or suspended inside an enclosed fuming chamber (an aquarium with a lid will work) along with warmed water. The glue is placed in a small container inside the chamber and heated. The fumes react with the fingerprints, and the fingerprints become whitish in color. Although the technique is not difficult, it is potentially hazardous. The fumes are irritating to mucous membranes and should never be inhaled.

Super Glue fuming was first used in 1978 by the Criminal Identification Division of the Japanese National Police Agency.

SILVER NITRATE

Crime-lab technicians may use silver nitrate ($AgNO_3$) solution instead of cyanoacrylate to detect fingerprints left on nonporous surfaces. The solution reacts with salts in perspiration in the fingerprint to form silver chloride ($AgCl_2$). Under an ultraviolet light source, the silver chloride will appear black.

NINHYDRIN

If the latent prints are found on porous surfaces, such as wood, fabrics, and concrete, technicians may use ninhydrin. Ninhydrin reacts with the amino acids in the fingerprints to produce a purple fingerprint. Ninhydrin may be sprayed on the item or the item may be dipped into the ninhydrin. It takes up to 48 hours to develop fingerprints in this manner. However, heating the surface will increase the rate of the reaction.

Digging Deeper

with Forensic Science e-Collection

DFO is a relatively new method used to enhance fingerprints. Using the Gale Forensic Science eCollection at www.cengage.com/school/forensicscienceadv, research DFO and investigate a case in which DFO was instrumental. Discuss the key elements of the case. Additionally, describe how this case relates to the Frye standard introduced in Chapter 1.

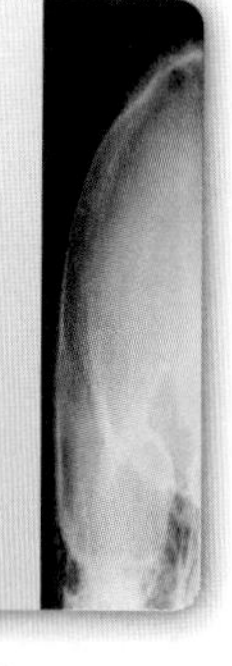

DIAZOFLUOREN

CHEMISTRY

Ninhydrin is often used to develop prints found on paper. However, diazofluoren (DFO) is even more effective than ninhydrin. The paper is dipped into the DFO for 10 seconds and then allowed to dry. DFO reacts with amino acids in the fingerprints. For the best results, the process may need to be repeated. The developed fingerprints may not be visible to the naked eye. However, they will be visible when viewed under a blue-green light.

VACUUM METAL DEPOSITION

Another technique uses a vacuum metal deposition chamber to evaporate gold and zinc. The gold is attracted to the thin layer of tissue left by the latent fingerprint. The zinc condenses on the gold coating and between the fingerprint ridges. This technique works best on nonmetal surfaces. It has also been somewhat successful in developing fingerprints on finely woven fabrics. Figure 3-14 summarizes some of the fingerprint-developing techniques described here.

Many people who have fingerprint cards on file have never been accused or convicted of a crime. For example, the fingerprints of teachers, military personnel, civil servants, and adoptive parents are included in the database.

FINGERPRINT DATABASES

Once latent fingerprints have been developed, they must be photographed. The photograph is scanned and digitized. Forensic scientists can then compare the print with other prints stored in a database. The database provides a set of potential matches. The scientist compares the potential matches to make the final identification. The FBI maintains the Integrated Automated Fingerprint Identification System (IAFIS). Some state and local agencies maintain databases as well.

Process	Surfaces	Adheres to
Dusting	Nonporous	Residues
Cyanoacrylate fuming	Nonporous	Proteins, fatty acids
Ninhydrin	Porous	Amino acids
Diazofluoren	Porous	Amino acids
Silver nitrate	Nonporous	Salts
Vacuum metal deposition	Nonmetal	Gold is attracted to the tissue; zinc adheres to the gold

Figure 3-14. *Summary of some fingerprint-developing techniques.*

DNA ANALYSIS

Obj. 3.9

Deoxyribonucleic acid (DNA) is found in the cells of all organisms. It can be detected in blood, saliva, semen, tissues, hair, and bones. With the exception of identical twins, each person's DNA is unique. Therefore, DNA samples provide individual evidence to tie a suspect to the crime or to identify a victim. DNA analysis can also be used to determine paternity or to detect genetic disease.

DNA is a very long double-stranded molecule made up of units called *nucleotides*. Each nucleotide is composed of a sugar, a phosphate, and a nitrogenous base. The sugar is called *deoxyribose*. There are four different bases—adenine, guanine, thymine, and cytosine. The structure of DNA is sometimes referred to as a twisted ladder (see Figure 3-15). The sugar and phosphate form the sides of the ladder. The bases make up the rungs of the ladder. Adenine always pairs with thymine; cytosine always pairs with guanine. The uniqueness of each person's DNA comes from the sequence of the base pairings along the ladder.

A *DNA fingerprint* is a DNA pattern that distinguishes one individual from another. A DNA fingerprint can be used to determine whether two samples are from the same person, related people, or unrelated people. Gel electrophoresis is one technique involved in the production of a DNA fingerprint. The process is a form of chromatography. Segments of DNA are separated by size through a gel (the stationary phase) by a mild electric current (the moving phase). Chapter 10 discusses in detail the various techniques for extracting and analyzing DNA.

3DRENDERINGS/SHUTTERSTOCK

Figure 3-15. *Three-dimensional structure of DNA.*

CHAPTER SUMMARY

- Physical properties are characteristics that can be observed and measured without changing the chemical identity of the substance.
- Chemical properties are characteristics that determine the way a substance interacts with other substances.
- Presumptive tests screen evidence for possible identification. Confirmatory tests determine the identity of evidence.
- The comparison microscope is one of the most useful tools in forensic investigations because it allows for the side-by-side comparison of samples.
- Most analytical techniques can be classified as either quantitative or qualitative. Quantitative analysis will always produce a number or a proportion. Qualitative analysis will always provide a description or statement.
- Chromatography is used to separate a mixture into its individual components.
- Wavelength is the distance between peaks of consecutive waves.

- Frequency is the number of waves that pass a specific point in a given amount of time.
- Electromagnetic spectroscopy measures the wavelength of light emitted when a substance is heated. Mass spectroscopy separates atoms and molecules according to mass.
- Fingerprint-developing techniques include dusting, cyanoacrylate fuming, silver nitrate, and vacuum deposition chambers.
- Repeating patterns of DNA base pairs are used to produce a DNA fingerprint. A DNA fingerprint can be used to determine whether two DNA samples are from the same person or related people.

CASE STUDIES

Gerald Fit Mason

On July 21, 1957, four teenagers were parked near oil fields in Hawthorne, California. A man walked up to the car and pointed a gun at the driver. He robbed the teens and raped one of the girls. He then stole the car. About 30 minutes later, he was pulled over for running a red light. As the two officers were walking away, he shot them. Both officers were killed. Several hours later, an abandoned 1949 Ford was found. The crime-scene unit took photos of the vehicle and dusted for fingerprints. Two partial latent prints of a thumb were found on the steering wheel, but investigators were unable to find a match. Unfortunately, the case went cold.

Forty-six years later, the LA County Criminalistics Unit reexamined the two partial prints. Investigators used cutting-edge digital fingerprinting technology to make digital copies of the prints. They stitched together the prints collected from the steering wheel and ran the print through the Integrated Automated Fingerprinting Identification (IAFIS) database. Investigators were able to determine that the print was consistent with the fingerprints of Gerald Fit Mason (see Figure 3-16).

On January 24, 2003, an official complaint was filed. Later, detectives arrested Mason for rape, murder, and robbery. On March 21, 2003, Mason confessed. Forty-six years after he committed these crimes, Mason was sent to jail. He is serving two life sentences.

AP PHOTO/ANNE CUSACK, POOL

Figure 3-16. *Gerald Fit Mason*

Think Critically

1. **Discuss some reasons why you think this case may have gone cold.**
2. **Why do you think a criminalistics department would reopen a case after nearly 50 years?**
3. **Had this crime taken place today, what additional evidence may have been collected and how would it have been processed?**

Roger Reece Kibbe: The I-5 Strangler

Over the course of a decade, several women were murdered along Interstate 5 near Sacramento, California. Each of the victims was strangled and bound in a similar manner. The killer had cut the victims' hair and cut their clothes in unusual patterns. These similarities suggested that the same person had murdered all the victims.

Eventually, Roger Reece Kibbe (see Figure 3-17) was charged with killing Darcie Frackenpohl, a 17-year-old runaway. Extensive microscopic and fiber analysis was presented at trial. Investigators had used a comparison microscope to determine that a rope found in Kibbe's car was the same kind of rope used to strangle Darcie. Electron microscopy showed that the ropes had 10 elements in common. The electron microscope analysis also showed that paint had been sprayed near each rope. Both had traces of the same air contaminants.

On May 10, 1991, Kibbe was sentenced to 25 years to life in prison. He was not eligible for parole for a minimum of 16 years and 8 months.

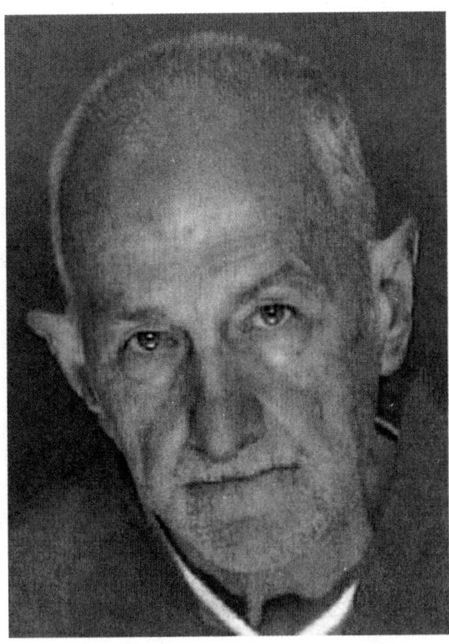

Figure 3-17. *Roger Reece Kibbe*

Think Critically **1. Was the analysis described in this case study quantitative or qualitative? Explain.**

2. Hair that was consistent with Kibbe's two cats was found on the clothing of one of the victims. Was this class or individual evidence? What is the relevance of the fact that the hair was from two different cats?

CHAPTER 1 2 3 4 5 6 7 8 9 10 11 12 13 14 15

Bibliography

Books and Journals

Lerner, K. Lee, and Brenda Wilmoth Lerner, eds. (2006). *The World of Forensic Science*. New York: Gale Group.

Newton, Michael. (2008). *The Encyclopedia of Crime Scene Investigation*. New York: Info Base Publishing, pp. 125–126.

Siegel, Jay. (2006). *Forensic Science: The Basics*. Boca Raton, FL: CRC Press, pp. 317–318.

Websites

www.biologymad.com/cells/microscopy.htm
www.cengage.com/school/forensicscienceadv
www.fbi.gov/hq/cjisd/takingfps.html
www.forensic-medicine.info/fingerprints.html
http://inventors.about.com/od/mstartinventions/a/microscope.htm
www.microscope-microscope.org/basic/microscope-history.htm
www.nfstc.org/pdi/Subject02/pdi_s02_m02_01_a.htm
www.redwop.com/download/dfo.pdf
www.rpi.edu/dept/chem-eng/Biotech-Environ/CHROMO/chromtypes.html
www.trutv.com

CAREERS IN FORENSICS

Science, Technology, Engineering & Mathematics

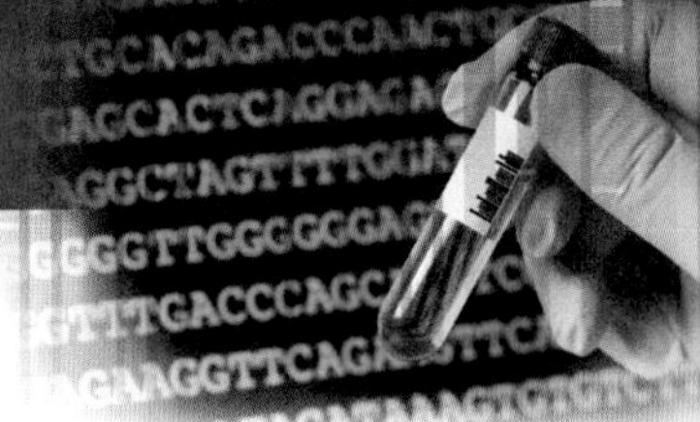

Gene Cushing: Latent Fingerprint Expert

Gene Cushing (see Figure 3-18) has been a crime-scene investigator for the Lake County, Florida, Sheriff's Office (LCSO) since November 1996. He is a court-certified expert in processing scenes and items of evidence for latent fingerprints. He is also the bloodstains expert for the LCSO and is often called in to document and interpret bloodstains at crime scenes. He has taken extensive training in basic crime-scene procedures, forensic science, tire and shoe documentation and recovery, evidentiary photography, medicolegal investigation of deaths, and blood pattern analysis.

PHOTO TAKEN BY L. LINDA DRESCHER, LCSO

Figure 3-18. *Gene Cushing.*

Currently, the LCSO requires only that applicants for crime-scene investigator positions have a high-school diploma. Cushing points out, however, that most recently hired investigators have four-year college degrees in Forensic Science or in Criminal Justice and Forensic Justice.

Cushing says that most people think that what they see on popular television crime dramas is a portrayal of what he does every day. For example, jurors often expect crime and evidence to be handled precisely and expertly in an hour. Additionally, these televisions shows misrepresent automated fingerprint identification systems. These systems do not give a person's name, address, or place of employment, and they do not determine matches. Actual confirmation is achieved manually.

Gene's typical day involves processing crime scenes, following up on pending cases, and processing evidence for latent prints. He usually receives subpoenas for latent print evidence about twice a month, but he rarely appears in court. Because fingerprints are individual evidence, defendants often settle the case without going to trial.

Gene's favorite part of his job is that LCSO has had the good fortune to receive some of the best equipment available. The sophisticated equipment lessens the likelihood of closing cases without prosecution. To him, the most challenging part of the job is having cases with more questions than answers. One such case would be Gene's investigation into the disappearance of still-missing two-year-old Trenton Duckett from his Lake County home in 2007.

Learn More About It

To learn more about forensic science careers, go to www.cengage.com/school/forensicscienceadv.

CHAPTER 3 REVIEW

Matching

1. the distance between peaks of consecutive waves *Obj. 3.7*
2. property described by behavior in the presence of another substance *Obj. 3.1*
3. the number of waves that pass a specific point within a given amount of time *Obj. 3.7*
4. property of substance that can be observed or measured without changing the composition of the substance *Obj. 3.1*

a. frequency
b. wavelength
c. physical property
d. chemical property

Multiple Choice

5. Which of the following is a physical property? *Obj. 3.1*
   a. density
   b. mass
   c. melting point
   d. all of the above

6. A fiber found at a crime scene is burned during a burn test. The investigator observes several things, including the rate of the burn, the color of the flame and ashes, and the smell released, to determine the identity of the fiber. What kind of change is taking place during a burn test? *Obj. 3.1*
   a. physical
   b. chemical
   c. complete
   d. none of the above

7. Which of the following microscopes shines light through the specimen, allowing observation of cells? *Obj. 3.3*
   a. comparison microscope
   b. dissecting microscope
   c. compound light microscope
   d. electron microscope

Short Answer

8. Distinguish between physical properties and chemical properties. Give examples of each. *Obj. 3.1*

physical = property than can be observed w/o change; mass + color
chemical = property described by behavior
w/ another substance; rusting + burning

9. Why is it important to reserve some of the evidence sample when completing chemical tests? *Obj. 3.1*

Because the original evidence sample
is often destroyed.

10. Differentiate between presumptive and confirmatory tests. *Obj. 3.2*

Presumptive = reduces the possibilites to get preliminary ideti..

Confimatory = make more specific identification;
ex = blood from human or animal

11. Why is a comparison microscope a useful tool in forensic investigations? *Obj. 3.3*

You can compare to samples at the
same time; ex: comparing blood
samples or bullet striations

12. Compare and contrast the scanning electron microscope and the transmission electron microscope. *Obj. 3.3*

13. What types of materials can be analyzed using chromatography? Be specific about which type of chromatography is best for each substance. *Obj. 3.5*

dyes in fibers = TLC (thin layer: paper)
explosives = HPLC (high performance: room temp.)
body fluids for drugs = GC (gas: good for separating
proteins found in blood)

14. List and describe two forms of spectroscopy. *Obj. 3.7*

1. Mass=Combined w/ GC to identify atoms by their masses; good for chemicals
2. Atomic Absorbtion= measures the amount of light absorbed; paint & soil

15. Distinguish between the six kinds of fingerprint-developing techniques described in this chapter. *Obj. 3.8*

1. Cyanoacrylate=

16. In a paper chromatography experiment, the solvent traveled 5.0 cm. Substance A traveled 3.0 cm. What is the R_f value of substance A? *Obj. 3.6*

sub/solv

3.0/5.00

Rf value=0.6

17. Describe the structure of DNA. *Obj. 3.9*

like a twisted ladder

18. Explain the difference between quantitative and qualitative analysis. Give an example of a technique that aids in each. *Obj. 3.4*

Quantative=#

Qualitative= description

Think Critically **19. A woman was found dead in her office one Saturday morning. She had been shot in the back of the head. It was clear that the perpetrator had opened the metal drawers of the woman's desk and searched through several files. Which fingerprint-development techniques would the latent fingerprint expert be most likely to use to visualize any fingerprints found at the scene?** ***Obj. 3.8***

20. In the scene described in question 19, the investigators found a bullet lodged into the window frame above the woman's desk. What kinds of equipment are forensic scientists most likely to use when examining the bullet? ***Obj. 3.3***

A Tale of Two Brothers

By: **Meagan Gallant and Corey Stillwaggon**

Tavares High School
Tavares, Florida

John Calabash, a tall, handsome 21-year-old, was eating lunch with his beautiful girlfriend, Casey Perez, a 20-year-old social butterfly with blue eyes and dark curly hair. They began to talk about their relationship and about his business. Casey wanted to be more involved with the business, but John, knowing what was at stake, told her "no." John and his older brother, Carl, were always competing for the top spot in the family business.

John and Casey had a serious relationship, but marriage was never discussed between them. John, however, had been talking to Carl a lot and was going to ask Casey to marry him soon. Carl was envious of his brother's relationship with Casey. Above everything, though, he wanted to be in charge of the family business. Their father had recently died and, if anything ever happened to John, and John and Casey were married, the business would go to Casey, not Carl. The family business was the largest known drug business throughout the United States.

Later that night, John and Carl got together at John's house to talk about a big business deal coming up in a couple of weeks. Casey was reading in the other room, but she was also listening to the deal. At 9:42 p.m., as Carl was getting ready to leave, he heard a muffled scream coming from the direction of the kitchen. He didn't think much of it and left the house. Later, Carl returned to John's house, planning to pretend that he had forgotten his wallet. His intention, though, was to put a plan into action to discredit Casey. However, when he arrived at 10:33 p.m., he found his brother dead in the kitchen; John had been stabbed twice in the chest. Carl went to look for Casey and found her dozing in a recliner with the book still open on her lap. After being told that her boyfriend was dead, she and Carl called 911.

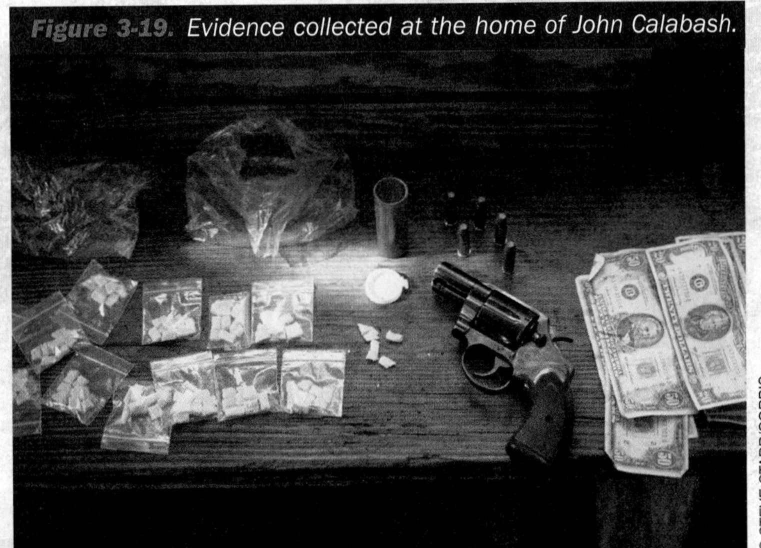

Figure 3-19. *Evidence collected at the home of John Calabash.*

When the police arrived, the crime scene was secured, and close-up and long-shot photos of the body and evidence were taken. Police began to collect the evidence, including a large butcher knife found in the kitchen. They also found evidence of a drug-related business, including weapons, large amounts of cash, and drug paraphernalia (see Figure 3-19). They found hair resembling Casey's and blood. All of the evidence was sent to the lab for further analysis. They dusted for fingerprints and found some footprints that were cast and photographed. There was no sign of a break-in—no sign of tool marks or broken glass at entry points. As the police began to investigate an open window, they also began questioning Casey and Carl.

Casey stated that she had been talking with John when Carl came over to visit, but that she had gone into the study to leave them alone to discuss business. She had been reading a book, but eventually fell asleep in the chair. When she awoke, Carl told her that he had forgotten something and had returned to find John, stabbed to death, lying in the kitchen.

Carl claimed that he left the house at about 9:45 p.m., and returned at about 10:30 p.m., to pick up his wallet. When asked how he got into the house, he admitted he had a key. He said he walked into the kitchen, where he believed he had left his wallet, and that was when he saw John lying on the floor. He instantly ran to find Casey.

Activity:

Answer the following questions based on information in the Crime Scene S.P. O.T..

1. Identify the evidence in the story and classify it as class or individual.
2. What motives did Carl and Casey have for killing John?
3. Hair that resembled Casey's was found at the crime scene. Is that important to the case? Why or why not?

WRITING

4. Write your own ending to this mystery. Be sure to include details about the analysis that would be done in the lab. You can be creative, but be sure that your ending is logical and that you demonstrate an understanding of various analysis techniques. Your ending should be 500 words or less.

Preventing Adolescent Crime Together™ (P.A.C.T)

Introduction:

Each person—even an identical twin—has unique fingerprints. However, the pattern of fingerprints is a class characteristic because there are only loops, whorls, and arches. The minute detail of the ridges within the patterns gives fingerprints their individual characteristics. When comparing fingerprints, investigators look at the number and location of ridge endings, bifurcations, and several other specific details. Some law-enforcement departments use fingerprint cards and child identification kits to record individuals' fingerprints.

Materials:

wet wipes
ink (strips or pads)
fingerprint 10-card

Procedure:

1. Wash hands and use a wet wipe to make sure all residues and dirt are removed from hands and fingers.
2. Place the ink pad or strip on the table. Stand at a comfortable distance from the table.
3. Grasp the hand of the person being fingerprinted carefully but firmly. Be sure not to touch your partner's fingertips.
4. Place the index finger on the ink pad or strip. Roll the bulb of the finger from one edge of the fingernail to the other edge. Make sure the ink is on the finger from the tip to just below the first crease (see Figure 3-20). **Note:** Using too much ink will make the ridge characteristics indistinguishable. Using too little ink will cause the print to be too light to see detail.
5. Roll the index finger from nail to nail on the designated area of the labeled card. Lift the finger after rolling to avoid smearing the print detail.
6. Repeat steps 2–5 until you have printed each finger on both hands.

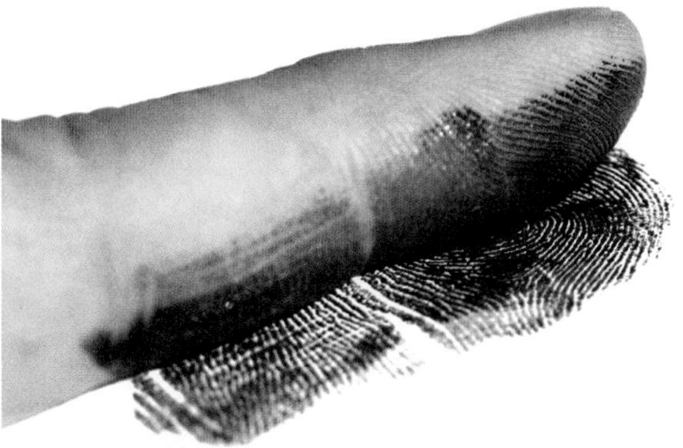
ISTOCKPHOTO

Figure 3-20. *Proper inking for fingerpriniting.*

7. Wash hands thoroughly to remove all traces of the ink.
8. Trade roles with your partner, and repeat steps 1–7.
9. Complete the card with as much detail as possible. If necessary, use the notations described in Figure 3-21 on your card.

Abnormality	Acceptable Notations
One or more fingers, thumbs, or hands are missing.	AMP (for an amputation) MAB (missing at birth) XX
An individual has a bandage or cast on a finger, thumb, or hand.	Unable to Print UP
Permanent tissue damage exists to a finger, thumb, or hand.	Take the prints as they exist; you may include the notation "scarred," but it is not necessary.
Injury, birth defect, or disease has caused abnormal structure of a finger, thumb, or hand.	Special equipment may be needed, but every attempt should be made to utilize techniques previously mentioned.
Individual has more than five fingers on one or both hands.	Print the thumb and the next four fingers.
Two or more fingers are grown together.	Roll as completely as possible and make a note on the fingerprint card.

Figure 3-21. *The FBI-recognized notations above should be made to describe abnormalities of the fingers, thumbs, or hands. All notations should be made in the corresponding block of the fingerprint card.*

ACTIVITY 3-1 *Ch. Obj: 3.1, 3.4*
SEPARATING MIXTURES

Objectives:

By the end of this activity, you will be able to:

1. Determine physical properties of various materials.
2. Design an experiment to separate a mixture.
3. Evaluate the success of the experiment.

Materials:

(per group of three or four students)

bag of sand (A)
bag of poppy seeds (B)
bag of salt (C)
bag of iron filings (D)
bag of mixture (E)
aluminum foil
filter paper
forceps
funnels
hot plate
magnets
paper plates
paper towels
rubber band
sifter
spoons
stirring rod
tissue paper
variety of glassware
water

Safety Precautions:

Wear safety goggles when handling glassware.
Wear apron.
Be careful when handling hot plate.
Do not handle heated glassware with bare hands.

Procedure:

1. Examine each of the samples (A–E). Note that sample E is a mixture of the other four.
2. Using the equipment and materials available to you, determine any properties that distinguish one sample from the others. Record your observations in your data table.

Data Table 1. Properties of samples.

Sand	Iron Filings	Salt	Poppy Seeds

3. Based on your observations, design an experimental procedure for separating sample E into its individual components. Be sure to include:
   a. The hypothesis and materials you used
   b. The steps of your procedure
   c. Qualitative and quantitative data
   d. An analysis of the data
   e. A conclusion

Questions:

1. Were you successful in separating the components of the mixture? Explain.
2. If you had done the steps of your procedure in a different order, would the outcome have been the same? Why or why not?
3. Are there supplies that were not provided but that might have made the task easier? Explain.
4. What would you do differently if you were given an opportunity to do this lab again?

Bonus:

1. The mixture in Sample E included iron filings. How would your experimental procedure have changed if the iron filings had been replaced by aluminum filings?
2. The mixture in Sample E included salt. How would your experimental procedure and results have changed if the salt had been replaced by sugar?

ACTIVITY 3-2 *Ch. Obj: 3.3*
MICROSCOPY

Objectives:

By the end of this activity, you will be able to:
1. Operate a compound microscope and a stereomicroscope.
2. Sketch samples in the field of view.

Materials:

(per group of two or three students)
prepared slides
compound microscope
stereomicroscope
protective gloves (optional)

Safety Precautions:

Always carry the microscope with both hands.
If you are allergic to latex, alert your teacher so that you may use alternative gloves.

Procedure:

1. Review the parts of the microscope with your teacher.
2. Place a prepared slide on the stage of the compound microscope.
3. Beginning with the lowest magnification, focus on the sample.
4. Once the field of view is in focus, rotate the nosepiece to a higher magnification.
5. Continue until you get the best image.
6. Draw what you see. Pay attention to detail.
7. View the slide under the stereomicroscope. Draw what you see, paying attention to detail.
8. Repeat with each of the additional slides provided.

Questions:

1. To get the best image, was the magnification the same for every slide? Why or why not?
2. Were some slides better viewed under one microscope than under the other? Explain.
3. If you had several samples of the same thing but from different sources (example: cat hair, dog hair, and human hair or dyed fibers and bloody fibers), how were you able to distinguish between the samples?
4. Based on the samples provided by your teacher, from what types of crime scenes could you expect to find those items of evidence?
5. Determine whether each item is class or individual evidence. Explain.

ACTIVITY 3-3 *Ch. Obj: 3.5*
PAPER CHROMATOGRAPHY

Objectives:

By the end of this activity, you will be able to:

1. Perform paper chromatography.
2. Determine whether the document had been altered.

Materials:

(per group of two students)
4 prepared chromatography strips
scissors
50 mL beakers (2)
10 mL distilled water
10 mL isopropanol
paper towels
stapler or tape

Safety Precautions:

Wear safety goggles, an apron, and gloves.
Consult MSDS sheets for specific instructions for handling and disposal of alcohol.
If you are allergic to latex, alert your teacher so that you may use alternative gloves.

Background:

Bill Evans owns a construction company. He has become very successful in the last several years. To save money, Bill and his wife have been keeping their own books rather than hiring an accountant. The business-related receipts and invoices are useful when they fill out their tax forms each year.

Last week, Bill was told that his business would be audited. The IRS wants to see business records for the last four years. The purpose of the audit is to make sure that Bill and his wife have been paying the taxes they owe. The auditor has asked Bill to bring all of his business-related receipts and invoices to the auditor's office.

Bill and his wife panicked and began collecting every piece of paper they could find. They found some of the receipts from the purchase of computer equipment for the new business, most of the invoices, the payroll records, and the majority of the receipts for materials purchased. However, they could not find all of the receipts. Bill was not sure he had all the documents he needed to support the numbers written in the business records books. He considered changing the amounts on several of the receipts. He thought, "I'll use this black pen and change some of those 3s to 8s. No one will ever know the difference."

Procedure:

1. Your teacher will give you four pieces of chromatography paper. Each of the papers will have an ink sample cut from different areas of Bill's receipts.
2. Trim the bottom corners of each strip to form a V. Be sure you do not cut the ink sample.
3. Add about 0.5 cm of water to one beaker and 0.5 cm of isopropanol to the other. Label each beaker.
4. Place paper strips 1 and 2 into the beaker labeled "water." Be sure the ink dots stay above the water, as shown in the following figure.

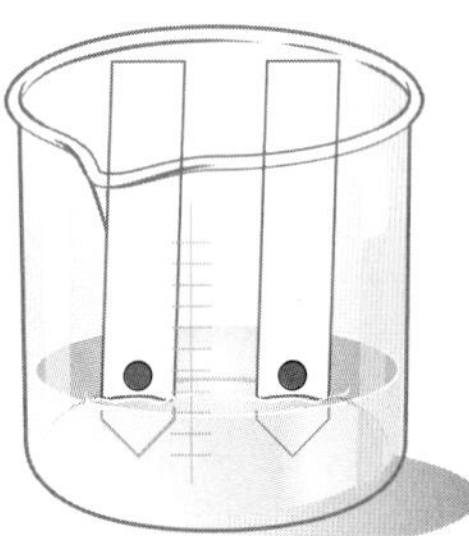

Figure 1. *Experiment set-up.*

5. Place paper strips 3 and 4 into the beaker labeled "isopropanol." Be sure the ink dots stay above the isopropanol.
6. Leave the beakers and strips undisturbed until the solvent (water or isopropanol) has dampened most of the length of the strip.
7. Carefully remove the strips from the beakers. Place them on a paper towel to dry. Do not allow the strips to touch each other.
8. Staple or tape the dry strips to your post-laboratory answers.

Questions:

1. Based on your chromatography results, do you think that Bill changed his receipts? Support your answer.
2. Is water always a good solvent for separating inks? Explain.
3. How could a teacher use chromatography to determine whether a student's answers have been changed after a test has been graded and returned?

ACTIVITY 3-4 *Ch. Obj: 3.4, 3.5, 3.6*
SMOOCH!

Objectives:

By the end of this activity, you will be able to:

1. Use chromatography to distinguish between the four similar shades of lipstick.
2. Calculate R_f values.
3. Differentiate between the lip patterns.

Materials:

(per group of two or three students)
chromatography paper
pencil
metric ruler
scissors
cotton swabs (5)
lipstick samples, labeled 1–5 and E
solvent
beaker
aluminum foil or plastic wrap
evidence envelope
suspect envelope
hand lens

Safety Precautions:

Wear safety goggles, an apron, and gloves.
Consult MSDS sheet for the proper handling and disposal of the solvent.
If you are allergic to latex, alert your teacher so that you may use alternative gloves.

Background:

The president of your school's student government needs your help. He has received several notes from a secret admirer and would like to know more about her. All of the notes are in envelopes with a lip print—a smooch—on the outside. The student president was able to get lipstick and lip-print samples from each of the girls he suspects. Your task is to determine which of the girls wrote the notes.

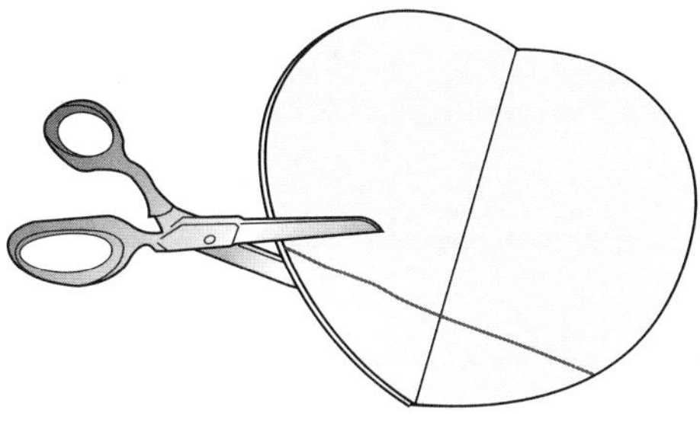

Figure 1. Cut the chromatography paper along the pencil line.

Procedure:

1. Gently fold the chromatography paper in half vertically. Open the paper, and use a pencil to draw a dot in the fold about 2.5 cm from the edge of the paper.
2. Using a pencil, draw a line through the dot perpendicular to the fold. Cut the paper along this line, as shown in Figure 1.
3. With your pencil, mark two points that are 1 cm from the cut edge of your paper. Draw a line through these points, all of the way across the page.
4. Label the paper across the top (in pencil) with your group number.
5. Write the number or letter for each lipstick sample across the top of your chromatography paper, as shown in Figure 2.
6. Use a cotton swab to place a small dot of lipstick sample 1 on the line below the number 1.
7. Repeat step 6 for each of the samples provided. Use a clean cotton swab for each sample.

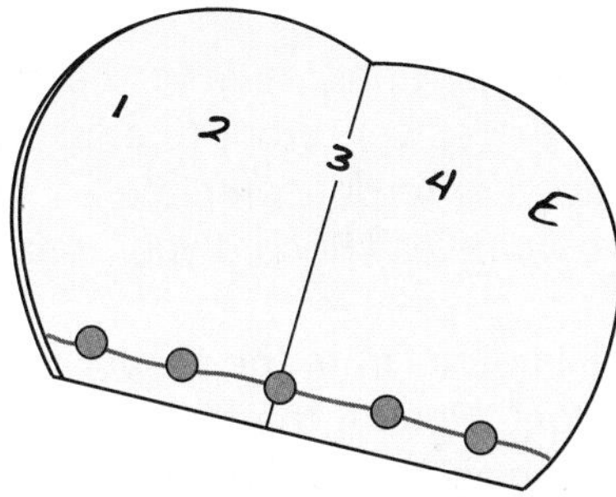

Figure 2. Place a small dot of each lipstick sample along the pencil line near the bottom of your chromatography paper.

8. Pour a small amount of solvent into the bottom of the beaker. The solvent should just cover the bottom of the beaker. Measure and record the depth of the solvent.

Data Table 1. Lip-print observations.

Sample	Lip-Print Sketches	Observations
Evidence		
1		
2		
3		
4		

9. Carefully place the prepared chromatography paper into the beaker with the flat side resting on the bottom. Make sure the solvent does not touch the lipstick dots. It may help to fold the paper slightly along the crease so it will stand alone in the beaker.
10. Cover the beaker with aluminum foil or plastic wrap. Leave the beaker undisturbed for approximately 30 minutes or until the solvent reaches the top of the chromatography paper.
11. While you are waiting, obtain the suspect and evidence lip prints from your teacher. Record the numbers and corresponding samples in Data Table 1.
12. Using a hand lens, compare the suspect prints with the one on the evidence envelope. Divide each lip print into quadrants and sketch the pattern one quadrant at a time. You may also take notes on your sketches.
13. When the chromatograms are finished separating, remove the paper from the beaker and mark with pencil the solvent front (the farthest distance traveled). Allow the chromatogram to dry completely.
14. Calculate the R_f value for each of the samples. Record your data in Data Table 2.
    a. Measure the distance in millimeters from the line of origin (pencil line) to the sample front. Record the distance.
    b. Measure the distance of the solvent front from the original depth of the solvent. Record.
    c. Calculate the R_f value using the following equation:

$$R_f = \frac{\text{Distance traveled by the sample (in cm)}}{\text{Distance traveled by the solvent (in cm)}}$$

    d. Repeat for all of the samples.

Data Table 2. Chromatography data.

Sample	Distance Solvent Traveled	Distance Sample Traveled	R_f Calculations
Evidence			
1			
2			
3			
4			

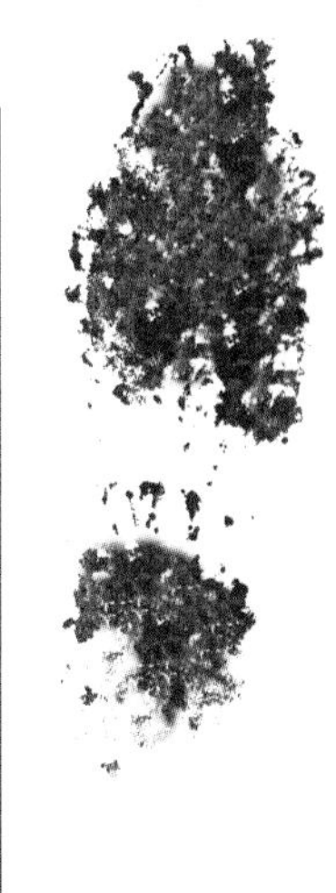

Questions:

1. Which suspect wrote the notes? How do you know?
2. If you were called in as an expert witness, what types of qualitative and quantitative data could you offer to identify the secret admirer?
3. Do you think a lip print could be used to convict or exonerate a suspect? Explain.
4. If the chromatography paper were doubled in size and the solvent were allowed to travel twice as long, would the R_f values change? Why or why not?

ACTIVITY 3-5 *Ch. Obj: 3.8*
FINGERPRINTING

Objective:

By the end of this activity, you will be able to:
Develop fingerprints using silver nitrate.

Materials:

(per group of three or four students)
3 pieces of paper
newspaper
small spray bottle
silver nitrate solution
watch or clock
UV light (optional)
stereoscope or hand lens

Safety Precautions:

Wear goggles, gloves, and an apron when working with silver nitrate.
If you are allergic to latex, alert your teacher so that you may use alternative gloves.
Silver nitrate will stain clothing. It will also leave skin discolored for several days.

Procedures:

1. Before putting on your gloves, pass three pieces of paper around your group. Be sure that at least one person places his or her fingers firmly in the center of the paper. These papers will now be your samples.
2. Cover your work space with newspaper and put the three samples on top of the newspaper. Put on your gloves.
3. Dampen all three samples with the silver nitrate solution. The samples should be damp but not completely wet.
4. Allow the samples to dry for 20 minutes.
5. Place the samples in direct sunlight or under an ultraviolet light. Check the development of your prints every few minutes until they turn dark gray or black. Record how long it took for the prints to develop.
6. Observe the prints under the stereoscope or magnifying lens. Note the differences among the different prints. Draw two different prints from each sample.

Questions:

1. On average, how long did it take for the fingerprints to develop?
2. What characteristics of the prints were visible? What characteristics might have helped you determine which prints belonged to each person in your group?

Extension:

Design an experimental procedure for identifying an unknown fingerprint. Be sure to include a list of the materials, the steps of your procedure, and the qualitative and quantitative data you would collect. Your experimental procedure should include a questioned sample and three or four known samples.

WRITING

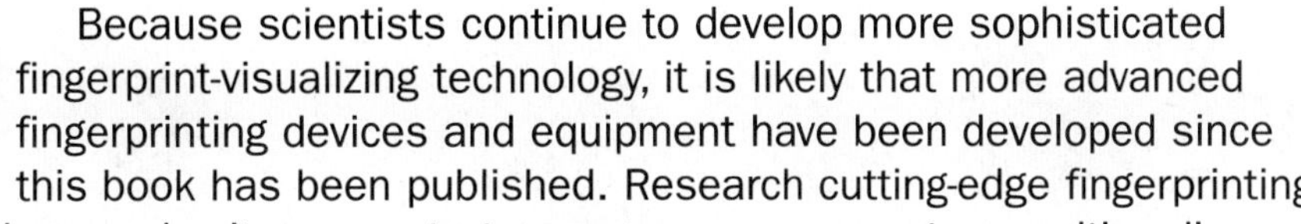

Because scientists continue to develop more sophisticated fingerprint-visualizing technology, it is likely that more advanced fingerprinting devices and equipment have been developed since this book has been published. Research cutting-edge fingerprinting technology, and write a one- to two-page paper or create a multimedia presentation describing the technology. Explain any controversy currently surrounding the technology. If the technology has been used to solve a case, describe the case and discuss how the technology helped investigators solve the case.

CHAPTER 4

Arson and Fire Investigation

HAPPY LAND FIRE

AP PHOTO

Figure 4-1. Happy Land social club after the fire.

Julio Gonzalez and Lydia Feliciano had dated off and on for six years. On March 25, 1990, Feliciano wanted to end the relationship. They argued at her workplace, a club in the Bronx called Happy Land. When Gonzales grabbed Feliciano's arm, the club's bouncers asked Gonzalez to leave. As he was being removed from the club, he shouted, "I'll be back, and I'll shut this place down!"

As Gonzalez walked away, he found an empty gasoline can. He filled it at a nearby gas station. The entrance to the club was a hallway that led to a staircase. The bar, the DJ, and most of the customers were at the top of those stairs. Gonzalez poured the gasoline along the hallway and on the steps. He used two matches to light the gasoline as he walked away. The gas ignited immediately, but the fire was initially contained to the hallway. Feliciano and a club customer noticed almost immediately and escaped. Most of the people in the club were upstairs listening to music. The DJ noticed the fire and announced it to the crowd. He then dived down the stairs and got out of the building.

When the DJ opened the door to get out, the rush of air provided more oxygen for the fire and pushed the fire up the stairs. A fire caused by gasoline on a wooden staircase produces thick black smoke. There were no windows or other ventilation upstairs. One of the gases in the smoke was carbon monoxide. Carbon monoxide is easily absorbed by blood cells. Victims die quickly. The DJ was the last person to escape. In less than 10 minutes, 87 people were dead.

Later that morning, Feliciano went to the police station to tell her story. She told them about her argument with Gonzalez and about his threat. The police were suspicious and went to talk to Gonzalez. The home smelled of gasoline. Gonzalez put on his gasoline-soaked shoes before leaving with the officers. Almost as soon as Gonzalez was read his Miranda rights, he confessed. He stated he started the fire as revenge against his ex-girlfriend who would not take him back. Later that day he was arraigned on 87 counts of murder. On August 19, 1991, he was convicted of arson and one of the largest mass murders in United States history.

AP PHOTO

Figure 4-2. The arrest of Julio Gonzalez.

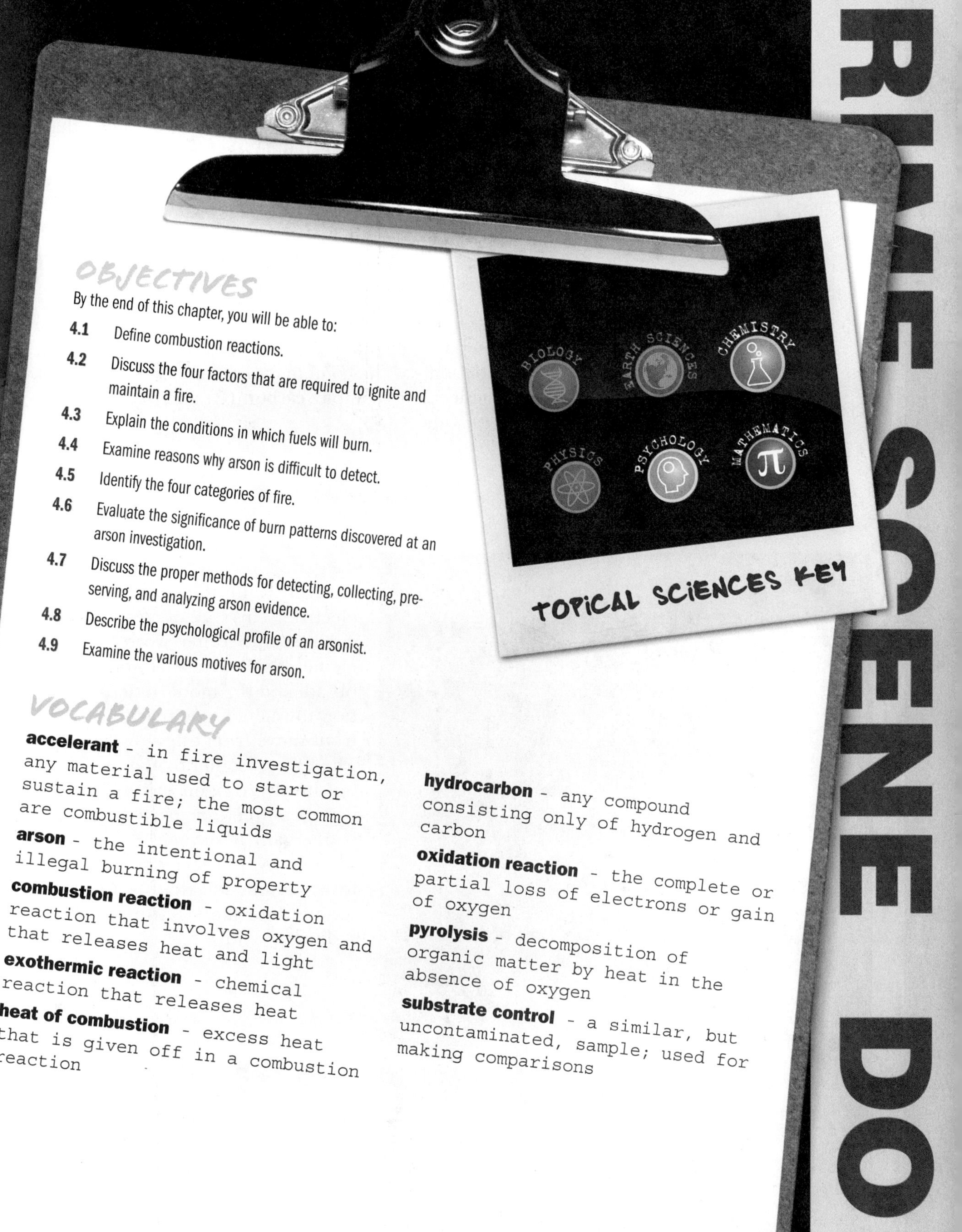

OBJECTIVES

By the end of this chapter, you will be able to:

4.1 Define combustion reactions.

4.2 Discuss the four factors that are required to ignite and maintain a fire.

4.3 Explain the conditions in which fuels will burn.

4.4 Examine reasons why arson is difficult to detect.

4.5 Identify the four categories of fire.

4.6 Evaluate the significance of burn patterns discovered at an arson investigation.

4.7 Discuss the proper methods for detecting, collecting, preserving, and analyzing arson evidence.

4.8 Describe the psychological profile of an arsonist.

4.9 Examine the various motives for arson.

VOCABULARY

accelerant - in fire investigation, any material used to start or sustain a fire; the most common are combustible liquids

arson - the intentional and illegal burning of property

combustion reaction - oxidation reaction that involves oxygen and that releases heat and light

exothermic reaction - chemical reaction that releases heat

heat of combustion - excess heat that is given off in a combustion reaction

hydrocarbon - any compound consisting only of hydrogen and carbon

oxidation reaction - the complete or partial loss of electrons or gain of oxygen

pyrolysis - decomposition of organic matter by heat in the absence of oxygen

substrate control - a similar, but uncontaminated, sample; used for making comparisons

INTRODUCTION

Fire has always fascinated people. Greek philosophers believed fire was one of the four basic elements—earth, wind, water, and fire. They thought that everything was made of these four elements. Now we know that fire is a chemical reaction that produces heat, light, and gas.

Obj. 4.1, 4.2, 4.3

CHEMISTRY OF FIRE

Fire is a rapid oxidation reaction that involves a combustible material. Combustible material is anything that will burn. For a long time, chemists thought that oxidation was the combination of any element with oxygen to form a new substance. For example, carbon (C) and oxygen gas (O_2) react to form carbon dioxide (CO_2).

$$C + O_2 \longrightarrow CO_2$$

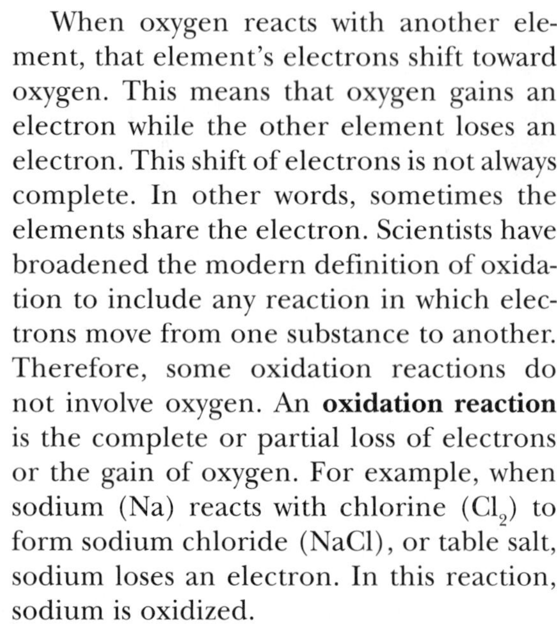

When oxygen reacts with another element, that element's electrons shift toward oxygen. This means that oxygen gains an electron while the other element loses an electron. This shift of electrons is not always complete. In other words, sometimes the elements share the electron. Scientists have broadened the modern definition of oxidation to include any reaction in which electrons move from one substance to another. Therefore, some oxidation reactions do not involve oxygen. An **oxidation reaction** is the complete or partial loss of electrons or the gain of oxygen. For example, when sodium (Na) reacts with chlorine (Cl_2) to form sodium chloride (NaCl), or table salt, sodium loses an electron. In this reaction, sodium is oxidized.

$$2Na + Cl_2 \longrightarrow 2NaCl$$

Figure 4-3. *Flames are the light released during combustion.*

COMBUSTION

Some oxidation reactions are called combustion reactions. **Combustion reactions** are oxidation reactions that involve oxygen and that produce flames (see Figure 4-3). Combustion reactions release energy in the form of heat and light. The excess heat energy is referred to as the **heat of combustion**. Chemical reactions that release heat are called **exothermic reactions**.

THE FIRE TETRAHEDRON

Four ingredients (see Figure 4-4) are needed to start a fire and to keep it burning:

1. *Oxygen.* Because oxygen is a major gas in the air, it is usually available. Oxygen is an important ingredient in any combustion reaction.
2. *Fuel.* The fuel is the material that is burning. If present, the accelerant burns first. In a fire investigation, an **accelerant** is any material used to start or maintain a fire. Soon, other nearby materials, such as wood or paper, ignite and become fuel for the continuation of the fire.
3. *Heat.* When a heat source is present, the temperature of a substance rises. Different fuels react with oxygen at different temperatures. The temperature at which a fuel will react with oxygen, or burn, is called the *ignition temperature.* At or above the ignition temperature, the fuel will continue to burn even after the heat source is removed.
4. *Chain reaction.* The fire itself usually releases enough heat to keep the fire burning. The fire will continue to burn until all of the oxygen or the fuel is used or removed.

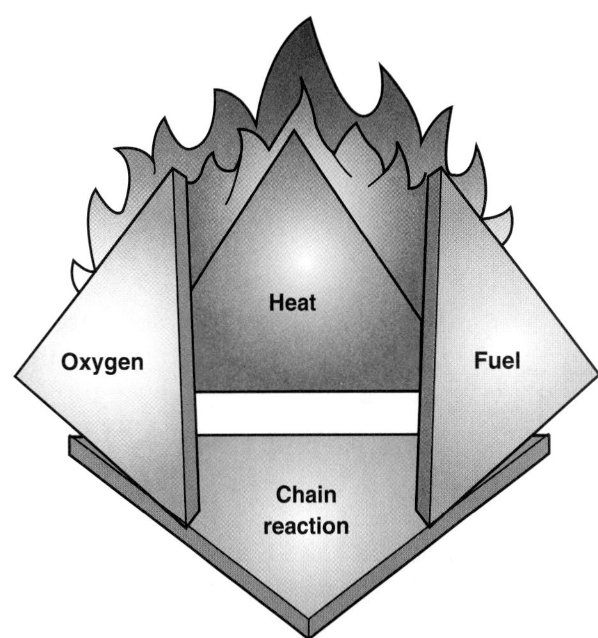

Figure 4-4. *The fire tetrahedron is a model used to describe how a fire starts and keeps burning.*

FLASH POINT AND IGNITION TEMPERATURE

Most accelerants are hydrocarbons. **Hydrocarbons** are compounds that are made of only hydrogen and carbon atoms. Gasoline, kerosene, and lighter fluid are hydrocarbons that can be used as accelerants. To react with oxygen, most accelerants must be in the gas state.

In a gas, the molecules are much further apart than in a liquid. Additionally, molecular bonding is weaker in the gas state. For example, liquid gasoline will not burn, but gasoline vapors will. When a liquid changes to the gas state, the process is called *vaporization.* For a liquid to burn, there must first be enough heat to vaporize the liquid. The lowest temperature at which this happens is called the *flash point.* At the flash point, the accelerant will continue to burn only as long as there is a heat source. The flash point of gasoline is −46°C. The flash point of gasoline, therefore, is 46°C *less than* the freezing point of water. Flash points and ignition temperatures of some commonly used accelerants are shown in Figure 4-5.

Accelerant	Flash Point (°C)	Ignition Temperature °C
Acetone	−20	465
Gasoline	−46	257
Kerosene	52–96	257
Mineral spirits	40–43	245
Turpentine	32–46	253

SOURCE: IAAI. (1999). *A POCKET GUIDE TO ACCELERANT EVIDENCE COLLECTION* (2ND ED.). MASSACHUSETTS CHAPTER.

Figure 4-5. *The flash points and ignition temperatures of several commonly used accelerants.*

At any temperature above the flash point, some of the liquid is becoming a vapor. These vapors can ignite. Gasoline's ignition temperature is 257°C. Once it reaches that temperature, gasoline will burn and continue burning.

PYROLYSIS

Figure 4-6. *Wood fire.*

Solid fuels, such as wood, are not flammable. However, vapors given off from the resins in the wood are flammable and will burn. As the wood itself decomposes in the fire, additional flammable vapors are released (see Figure 4-6). The process of decomposition caused by heat in the absence of oxygen is called **pyrolysis**. Charcoal, another common solid fuel, will not burn unless a liquid fuel is applied. The liquid fuel, often called *charcoal starter* or *lighter fluid*, is poured onto the charcoal. As the liquid vaporizes, it becomes flammable. Lighting it with a match causes a flame, and the heat from the flame decomposes the charcoal (see Figure 4-7). As the charcoal undergoes pyrolysis, more flammable vapors are released to sustain the fire.

Digging Deeper
with Forensic Science e-Collection

Certain behaviors and circumstances are considered "red flags" for arson investigators. For example, if the well-insured property of a person in debt burns, investigators might become suspicious. Sometimes, these red flags help investigators solve a crime. Sometimes, however, these red flags have innocent explanations. Cases involving fire are not always what they seem. Using the Gale Forensic Science eCollection at www.cengage.com/school/forensicscienceadv, read about how some investigators use these red flags. Then answer the following questions: (1) What red flags were described in this case? (2) How are burn patterns important in determining the cause of a fire? (3) Where was the fire's point of origin, and what did that indicate? (4) How are burn patterns and red flags different?

DIFFICULTIES IN ARSON DETECTION

Obj. 4.4

According to the National Fire Protection Association, fires caused more than $15 billion worth of property damage and killed more than 3,000 people in the United States in 2008. Accidents, such as faulty wiring or carelessness, cause many of the fires. For example, if someone falls asleep with a lighted cigarette or throws a lighted cigarette on dry grass, a fire can start. Unattended candles can also cause fires. Still other fires start when a fuel is stored in a warm place where there is little airflow. For example, as organic material decomposes, it releases heat. Decomposition takes place at the bottom of haystacks and in compost piles. If the temperature around the organic material gets high enough, the material might burn. Fires have also started when vapors from oily rags stored in a hot garage or storage building have been ignited by very high temperatures.

Figure 4-7. *Charcoal fire.*

Unfortunately, some fires are started on purpose. It is a crime to intentionally start a fire that damages property. This crime is called **arson.** The person who commits this crime is called an *arsonist.*

Arson is a very difficult crime to prove. The crime is usually carefully planned. The arsonist has usually left the scene well before anyone else notices the fire. The fire often destroys evidence from the crime, and what evidence is left is often destroyed as firefighters extinguish the fire. Therefore, investigators are left with little physical evidence to analyze. Combustible liquids are often used as accelerants in an arson fire. Any remaining traces of chemical accelerants usually evaporate very quickly. Detection of accelerants will be discussed later in this chapter.

Did You Know?

Arsonists often escape punishment. Only 16 percent of arson offenses lead to arrest, and only 2 percent of those arrested are convicted.

FUNCTION OF A FIRE INVESTIGATOR

Obj. 4.5

Firefighters are the first people to respond to a fire. Their priority has to be putting out the fire and saving any victims. As soon as a fire is out, the investigation begins quickly. Otherwise, evidence will be lost. At first, the main focus of the fire investigation is finding the fire's point of origin. In other words, investigators search for the point where the fire started. Once investigators find the point of origin, they examine the area for possible causes of the fire. Investigators look for accidental causes as well as evidence of arson. Investigators use four broad categories to classify fires (see Figure 4-8).

Did You Know?

In 2006, 10 firefighters died and 7,200 others were injured while responding to intentional fires.

Category	Description
Natural	A fire caused by acts of nature: a lightning strike or intense sunlight
Accidental	A fire that was unintentional and explainable; causes may include faulty wiring, malfunctioning appliances, or human carelessness
Undetermined	The cause of the fire is unknown and cannot be identified
Deliberate	A fire that was intentionally set (Not all deliberate fires are arson. For example, a campfire might spread out of control.)

Figure 4-8. *Fire categories.*

If investigators determine that the fire was not natural, they will try to find the person who started the fire. Investigators might prosecute someone who starts a fire, even if that person started the fire by accident.

DETERMINING CAUSE

Obj. 4.6

The point of origin can provide clues about the cause of a fire. The point of origin will be marked by a burn pattern. There often is a V-shaped burn near the point of origin. This shape is caused as the fire travels up from the point of origin. The V-shaped burn pattern is usually present in cases of natural or accidental fires. However, in cases of arson, other

Digging Deeper
with Forensic Science e-Collection

At the scene of a fire, investigators work hard to gather facts that will help them determine the cause of the fire. They might need to examine and analyze debris found at the scene. Additionally, investigators interview any witnesses to the fire. Eventually, fire investigators develop a report of their findings. Go to the Gale Forensic Science eCollection at www.cengage.com/school/forensicscienceadv to learn more about fire investigation. Use information from the article to write a brief essay describing fire investigation. Include answers to the following questions in your essay: (1) How are burn patterns important in determining the cause of a fire? (2) Explain the four causes of fire. (3) How can a fire be deliberate but not arson? (4) What guiding questions help investigators determine the origin of a fire?

Figure 4-9. Alligatoring.

Figure 4-10. Concrete spalling.

burn patterns may be found. Instead of a V-shaped burn, the point of origin might be surrounded by a burn pattern resembling the scales of an alligator. This burn pattern is called *alligatoring* (see Figure 4-9). The point of origin might also be surrounded by melted materials or concrete spalling. *Concrete spalling* is the breaking away of layers of concrete (see Figure 4-10). Both alligatoring and spalling are caused by intense heat. Because accelerants often cause fires that burn at higher temperatures than natural and accidental fires, these burn patterns are sometimes an indication of arson. Figure 4-11 on the next page lists several burn patterns and describes the clues they give investigators.

The burn pattern found near the point of origin might suggest that a fire was set on purpose. For example, some fires have more than one point of origin. It is possible for an accidental fire to have more than one point of origin, but it is not likely. Also, the way a fire appears to have spread can also suggest that the fire was set on purpose. More burning on the floor than on the ceiling and more burning on the underside of furniture indicate that an accelerant might have been used. Burn patterns that look like puddles are an indication that an accelerant was poured. These are called *pools* or *plants. Streamers,* which often connect pools, are trails of accelerants used to spread the fire. Streamers and other unnatural burn patterns also indicate the use of accelerants.

Obj. 4.7 COLLECTING THE EVIDENCE

Because accelerants often evaporate quickly, investigators must collect evidence from a fire as soon as possible. Investigators do not need to wait for a search warrant. The United States Supreme Court has upheld that a search warrant is unnecessary when evidence is likely to be lost. However, investigators need to get a search warrant for any later searches.

During the initial search, investigators often collect 3–4 L of ash and debris from the point of origin (see Figure 4-12 on the next page). They

Pattern	Description	Indication
Classic V	Burn pattern narrower at bottom and spreads outward as it rises	Ordinary burn pattern, no accelerant used
Inverted cone	Burn pattern is wider along the floor and narrower as it burns upward	May be caused by an accident or by accelerant poured along the floor
Alligatoring	Burn pattern resembling the scales of an alligator	Possible use of an accelerant, but not absolute
Spalling	The breaking away of layers of concrete due to exposure to high temperatures	Possible use of an accelerant, but not absolute
Streamers	Burn pattern that shows a trail from one area to another	Accelerant used to spread the fire from one area to another
Arc damage	Spark caused by a release of electricity	Electrical fire; may also be a result of a fire that has burned through wire insulation
Pool or plant	A burn pattern in a puddle configuration	Poured accelerants accumulated in a pool at the lowest point

Figure 4-11. *Burn patterns.*

also collect anything that could have traces of accelerant. Accelerants tend to flow down. This means that there might be traces of unburned accelerant in cracks in the floor or in upholstery. Investigators use portable vapor detectors, or sniffers, to search for debris that might hold traces of an accelerant. The sniffers detect molecules and particles in the air. Data from the sniffer can help investigators identify the area where accelerants may have been poured. Some investigators may even use arson dogs (see Figure 4-13 on page 112). Dogs have a very good sense of smell and can cover a wide area quickly. The dogs are trained to alert their handler when certain scents are present. These abilities make dogs very useful in detecting quantities of accelerant too small for the mechanical sniffers.

Did You Know?

The bedroom is the most common point of origin for intentionally set fires in the home. In public places, such as schools or offices, bathrooms are the most common point of origin.

COLLECTING THE CONTROL

Investigators often take several evidence samples from the scene. Each sample is packaged in its own container. Investigators also take a debris sample that has not been contaminated by the accelerant. This uncontaminated sample is called a **substrate control.** The substrate control must be packaged separately. The substrate control is compared to the other samples. These comparisons might help investigators prove that an accelerant was used. The burned substrate control gives information about products formed during the burning process. These products will also be

Figure 4-12. *Collecting fire debris.*

CHAPTER 1 2 3 4 5 6 7 8 9 10 11 12 13 14 15

found in the contaminated sample. However, any additional chemicals found on the contaminated sample can help investigators determine what kind of accelerant was used to start the fire. For example, some carpeting might be covered in flammable residues from carpet-cleaning products. Linoleum flooring breaks down into materials that might be mistaken for accelerants. If the suspect hydrocarbon is found in the fire debris but not in the substrate control sample, it is more likely to be the trace of an accelerant. If, on the other hand, the suspect hydrocarbon is found in both samples, the hydrocarbon was probably not used as an accelerant.

BILLY HUSTACE

Figure 4-13. *An arson dog helps an investigator search for traces of accelerant.*

PACKAGING THE DEBRIS

Fire debris must be packaged in an airtight container. Otherwise, vapor from accelerants might be lost. Investigators often use new, clean paint cans with airtight lids (see Figure 4-14) for the collection. Investigators do not fill the cans completely. The sample might release vapors. Therefore, investigators leave a few inches of space, called *headspace*, below the lid.

© INGRAM PUBLISHING/ALAMY

Figure 4-14. *Clean, unused paint cans are often used for collecting fire debris.*

FINDING THE IGNITER

Once fire investigators determine that an accelerant was used to start a fire, they need to find out what was used to light the accelerant. Matches are very common, but they usually are destroyed in the fire. Cigarette lighters are usually carried away from the scene. Sometimes, the arsonist uses things that leave evidence behind. For example, a Molotov cocktail will leave glass fragments behind. A *Molotov cocktail* is a homemade firebomb made with a glass bottle, fuel, and a wick. An accidental fire might be caused when electrical wiring sparks—causing an arc. This arc will cause a predictable pattern (see Figure 4-15). Electrical sparking devices, devices used to produce a spark, will also leave evidence behind. Faulty wiring might have produced the spark in an accidental fire. An arsonist might use a spark plug, fuse, or other device to make a spark to ignite the fire. Some part of this device is likely to be found during an investigation. Discovering the igniter, coupled with the accelerant used to start the fire, gives investigators information about the arsonist. For example, the arsonist's method of operation, or the methods the arsonist used to start the fire or to leave the scene, can sometimes be used to make a criminal profile. The *criminal profile* is a list of likely characteristics, such as age, gender, or motives. The profile helps the investigators narrow the list of suspects. By narrowing the list of suspects, investigators might be able to learn the identity of the arsonist. If an arsonist is responsible for more than one fire, the method of operation might help investigators connect the arsonist to all of the fires.

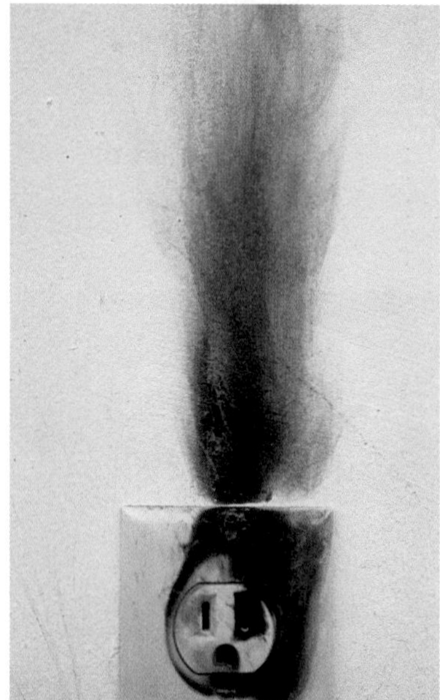
© FINE ART/ALAMY

Figure 4-15. *The charred electrical outlet and the surrounding burn pattern indicate an electrical fire.*

LAB ANALYSIS

Obj. 4.7

CHEMISTRY

Fire investigators will send all of the evidence to the lab for analysis. At the lab, forensic scientists will try to determine exactly what accelerant was used to start the fire. One method used to analyze accelerant residue from the collected debris is called the *direct headspace extraction procedure.* Remember that headspace is the space between the collected fire debris and the top of the container. When the container is heated, the vapors in the debris will collect in the headspace. The vapors are removed directly from the headspace with a syringe. The collected vapor is then analyzed using gas chromatography (GC). GC separates the accelerant vapors into individual components. Each component is recorded as a separate peak on a chromatogram (see Figure 4-16). Remember that in GC, a gaseous mixture is passed through the chromatograph. When a liquid sample is injected into the GC, the liquids vaporize and are carried through the system as a gas. Different components of the original mixture will take different amounts of time to reach the recording device, or the detector. First, forensic scientists compare the chromatograms of the fire debris and the substrate control. This comparison shows whether an additional chemical is found in the fire debris. This additional chemical would produce additional peaks on the chromatogram. Then, forensic scientists compare the additional peaks on the chromatogram from the fire debris to chromatograms of known hydrocarbons, such as gasoline, kerosene, or lighter fluid. This comparison allows the scientists to identify the accelerant. For a more detailed review of chromatography, turn to Chapter 3.

In the direct headspace extraction procedure, the amount of vapor collected is limited by the size of the syringe. Therefore, low concentrations of accelerant may not be detected. For this reason, many crime labs now use a technique called *passive headspace extraction procedure* to obtain traces of residue from the debris for analysis.

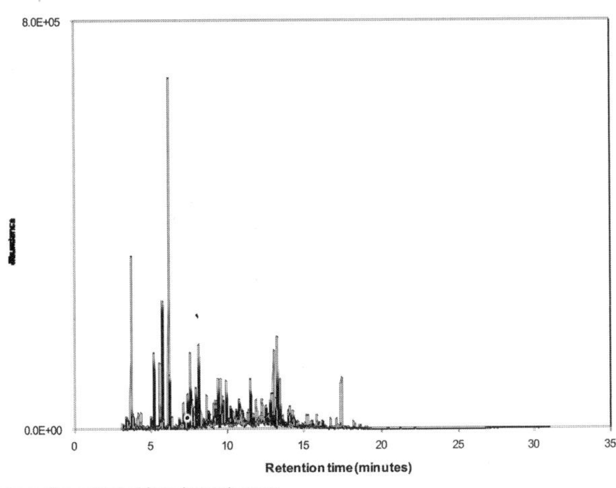

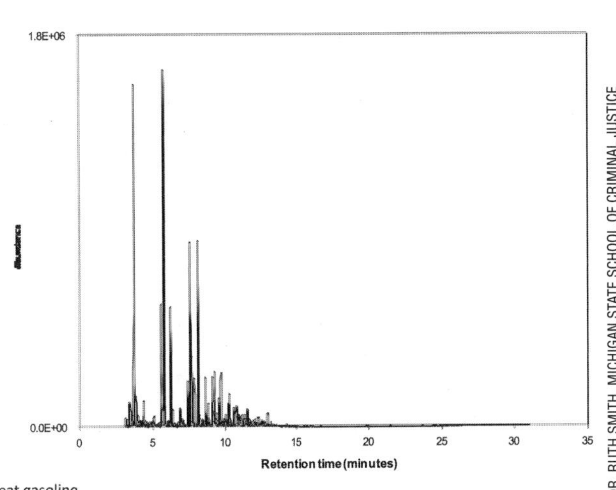

DR. RUTH SMITH, MICHIGAN STATE SCHOOL OF CRIMINAL JUSTICE

Figure 4-16. *The chromatogram on the left is the chromatogram from residues from a burned carpet. The chromatogram on the right is the chromatogram of gasoline. Notice that the two chromatograms share some peaks, indicating that there was gasoline on the carpet.*

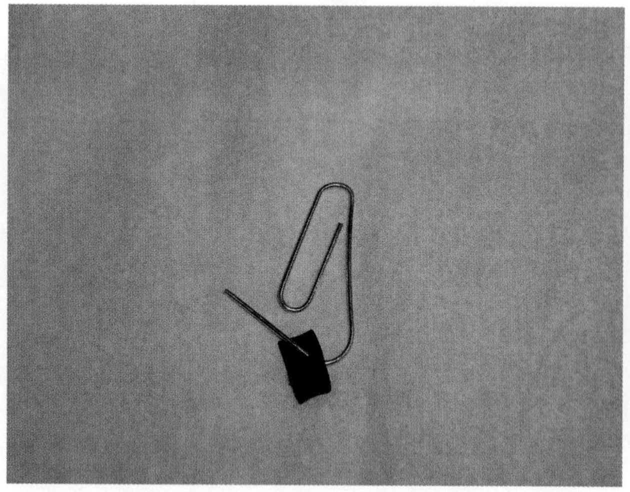

DR. RUTH SMITH, MICHIGAN STATE SCHOOL OF CRIMINAL JUSTICE

Figure 4-17. *Passive headspace extraction procedure.*

In the passive headspace extraction procedure (see Figure 4-17), a charcoal-coated strip is suspended in the headspace above the collected debris. The entire container, with the lid in place, is heated for four to 16 hours at 50–80°C. Any accelerant in the debris vaporizes and the charcoal absorbs the vapor. The charcoal strip is then removed from the container and washed with a small amount of organic solvent such as carbon disulfide. The solvent dissolves the accelerant but not the charcoal, thus removing the accelerant from the charcoal. The recovered accelerant is then analyzed using a gas chromatograph in the same manner as the direct headspace extraction procedure. The passive headspace extraction procedure is about 100 times more sensitive than the direct headspace extraction procedure. Lower concentrations of accelerant are more easily detected.

Obj. 4.8

PSYCHOLOGY OF AN ARSONIST

Psychologists think that arsonists often start fires for the sense of power they feel from watching their "work." The power gives the arsonist a physical and emotional "high" that is similar to a person's reaction to illegal drugs. Although there is not a typical arsonist, profilers have determined a set of characteristics often exhibited by arson suspects. This list of characteristics includes the following:

- Less than 25 years old
- Father not in the home
- Domineering mother
- Academically challenged
- Emotionally and/or psychologically disabled
- Unmarried, possibly still living at home with parents
- Feelings of inadequacy and insecurity
- Fascination with fire
- Alcoholism
- Parental neglect or abuse

These are common characteristics that have been linked to arsonists after psychological analysis of the suspect. Arsonists typically exhibit several, but not all, of the characteristics. Additionally, some arsonists do not display any of these characteristics; and not all people with these characteristics are arsonists. This list of characteristics is just one tool investigators can use to solve arson cases.

Research shows that adult serial arsonists often displayed certain behaviors in childhood. Behavior such as aggression, cruelty to humans and animals, destruction of property, deceit, and theft can indicate a conduct disorder. If a child has a conduct disorder, he or she often violates the rights of other people. If a child who has a conduct disorder starts a fire, it is usually with the clear intent to harm others or to damage property. Typically, children with this conduct disorder are male and very impulsive. Parents should seek professional help if a child exhibits several of the following signs:

- At the age of three, the child begins playing with matches.
- The child behaves as a "daredevil," especially when it comes to fire.
- The child makes his or her own mixtures to cause fires and explosions.
- The child is excited by fire.

Did You Know?

There appears to be a growing link between arson and illegal drug activity. Preliminary results of a new study by the National Fire Protection Association suggest that between 20 and 25 percent of reported arson cases in major American cities are drug related.

MOTIVES FOR ARSON

Obj. 4.9

People start fires for various reasons. Establishing motive is essential to the criminal investigation. The motives for arson tend to fall into six broad categories. Arsonists might start fires for financial gain, revenge, excitement, vanity, to conceal a crime, or as an act of vandalism.

FINANCIAL GAIN

Some people start fires for financial gain, usually through insurance fraud. Most homeowners have insurance for their house and its contents. If a house or business is damaged in an accidental fire, the insurance company gives the owner money to replace the building and its contents. Sometimes, people deliberately burn the house to collect the insurance money (see Figure 4-18). Other people burn vehicles for similar reasons. Arsonists might also burn a competitor's business. In this case, the arsonist might hope to get money by taking the competitor's customers. For this reason, insurance companies instigate many arson investigations. Most insurance companies employee professional arson investigators who are trained to determine whether a fire was caused by arson.

DALE A STORK/SHUTTERSTOCK

Figure 4-18. *Sometimes an arsonist hopes to receive insurance money after a fire.*

Digging Deeper
with Forensic Science e-Collection

Fires are very dangerous. They kill thousands of people each year. A person who starts a fire that kills someone may be convicted of murder. In some states, the arsonist faces the death penalty. Go to the Gale Forensic Science eCollection at www.cengage.com/school/forensicscienceadv to read about controversy surrounding some arson convictions. Review and discuss the article with a partner. Together, write a position paper or present a debate for your class. Be sure to address the significant ethical issues in the cases. Discuss the role of the Innocence Project and whether the group should be allowed to review all arson cases. Finally, explain your opinions about the outcome in the Willingham case.

REVENGE

Revenge arson, also referred to as *spite arson,* is a fire started to destroy an organization or a person for the sake of a cause. For example, some animal rights groups have burned animal research facilities to protest the use of animals for research purposes. Fires started as a result of jealousy or prejudice also fall into this category. The Happy Land Fire discussed at the beginning of this chapter fell into this category.

Figure 4-19. *Some arsonists are excited by fire itself.*

EXCITEMENT

Some people think that setting fires is exciting. People who set these fires might be bored or seeking attention. These arsonists are particularly dangerous—they generally have no specific target and the fires are random. They will burn anything at any time, just for the excitement (see Figure 4-19).

VANITY

Sometimes, the person who sets the fire is able to make himself or herself look like a hero. For example, the arsonist may pose as a passerby and report the fire. She or he might then get attention or even a reward for protecting the property or saving lives. People who create situations that make them appear to be brave suffer from hero syndrome. Basically, these arsonists are setting the fires out of vanity. They crave the positive attention they expect when they "save the day." Excitement and vanity are often overlapping motives. In other words, the person may find the fire and the false bravery to be exciting.

CRIME CONCEALMENT

Sometimes people start a fire to destroy evidence of another crime. This motive is called *crime concealment.* For example, a bank robber might steal a car to leave the bank. The robber might then burn the car to destroy evidence that might connect the robber to the car. A body burned inside a car or house to cover up a murder is also a form of crime concealment.

VANDALISM

The last category of arson is *vandalism.* School fires and fires in trash cans or dumpsters (see Figure 4-20), wooded areas, and abandoned buildings are often acts of vandalism. People under the age of 18 commit most of these crimes. Vandalism fires may also be categorized as revenge or excitement arsons.

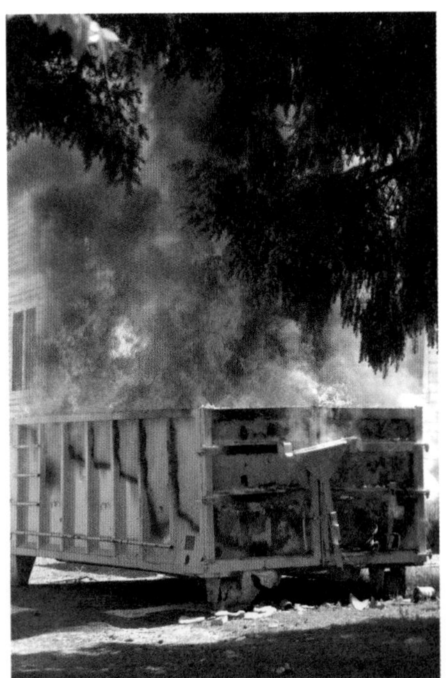

TFOXFOTO/SHUTTERSTOCK

Figure 4-20. *A dumpster fire may be an example of vandalism.*

CHAPTER SUMMARY

- Fire is an oxidation reaction that involves a combustible material. Not all oxidation reactions produce fire.
- The fire tetrahedron represents the four requirements for sustained fire—oxygen, fuel, heat, and a chain reaction.
- Most fuels are hydrocarbons. They typically have very high ignition temperatures. The ignition temperature is the temperature at which a fuel will light and continue to burn even if the heat source is removed.
- The initial focus of a search at the scene of a fire is to find the fire's point of origin.
- The burn pattern helps investigators determine whether the fire was natural, accidental, or deliberate.
- Three to four liters of ash and debris are generally collected from the point of origin and any area suspected of having traces of accelerant. Each sample of debris is placed in a separate airtight container.
- For comparison purposes, investigators also collect a substrate control.
- The direct headspace extraction procedure and vapor concentration are two methods of collecting accelerant vapors for analysis.
- Gas chromatography is used to analyze the accelerant residue. The resulting chromatogram is compared to chromatograms of known hydrocarbons in order to determine what type of accelerant, if any, is present.
- Psychologists have determined a set of characteristics that arsonists tend to exhibit. This set of characteristics helps investigators focus their search.
- Arsonists set fires for many reasons. Most of the motives fit into six broad categories—financial gain, revenge, excitement, vanity, crime concealment, and vandalism.

CASE STUDIES

Berns Triplets (1988)

At 6:00 p.m., on September 21, 1988, the Berns triplets were asleep in their crib. They were 17 months old. Their father, Scott Berns, was asleep on the couch. He woke up when he noticed smoke filling the room. He immediately ran outside of the home and shattered the triplets' window. The three girls were pulled from the burning house. Their mother, Patti Berns, died from smoke inhalation. The girls had third-degree burns, but they survived. Police suspected arson from the beginning. Suspicious burn patterns suggested that the fire had started near the couch where Scott Berns had been sleeping. Figure 4-21 shows the home after the fire was extinguished.

After almost two years, Berns was indicted. Berns abused prescription and illegal drugs. Witnesses said that he had shot at his wife's car earlier that day as she was leaving the home. However, it is very difficult to determine exactly what happened in cases of arson. The jury decided that there was reasonable doubt in the case against Scott Berns. Three strangers had been seen near the home the day of the fire. It was possible that the strangers had started the fire. Scott Berns was acquitted.

Think Critically

1. **What does it mean that the "jury decided that there was reasonable doubt"?**
2. **Why is it often very difficult to prove arson cases?**

COURTESY OF THE FORT WORTH STAR – TELEGRAM COLLECTION, SPECIAL COLLECTIONS, THE UNIVERSITY OF TEXAS AT ARLINGTON LIBRARY

Figure 4-21. *The Berns' home after the fire.*

John Orr (1984)

Figure 4-22. *John Orr.*

For several years, California was plagued by a series of arson fires. One of the fires resulted in the loss of 65 homes. A fire at a hardware store in Southern California killed four people. Investigators combed the areas and searched for clues; however, no clear pattern was found. Arson and fire investigations expert Marvin Casey noticed that the first fires were in Los Angeles. Each fire after that was further north. The pattern was moving toward Fresno, California. Casey also noted the odd coincidence that Fresno was the host city of an arson investigators convention. With this tenuous lead, Casey came up with 55 names of possible suspects. After several more years and another convention in Los Angeles, the suspect pool was finally narrowed down to just one individual—John Orr (see Figure 4-22). Orr was a respected arson investigator. People who knew him and worked with him were stunned by the accusation. Orr suffered from hero syndrome. He may also have been excited by the power he thought the fires gave him. Eventually, he was caught and convicted of four murders. He is serving a life sentence without the chance of parole. Arson investigators and criminal profilers think that Orr was one of the worst serial arsonists of the 20th century.

Think Critically

1. **Describe *hero syndrome* and explain how it relates to this case.**
2. **How did the arson investigator's attention to detail bring closure to this case?**
3. **Do you think the fact that Orr was an expert arson investigator contributed to the sentencing recommendation? Explain.**

Bibliography

Books and Journals

Conklin, Barbara, Robert Gardner, and Dennis Shortelle. (2002). *The Encyclopedia of Forensic Science*. Westport, CT: Onyx Press, pp. 5-6.

Gaensslen, R. E., Howard A. Harris, and Henry Lee. (2008). *Introduction to Forensic Science & Criminalistics*. New York: McGraw-Hill.

IAAI. (1999). *A Pocket Guide to Accelerant Evidence Collection* (2nd ed.). Massachussetts Chapter.

Lerner, K. Lee, and Brenda Wilmoth Lerner, eds. (2006). *The World of Forensic Science*. New York: Gale Group.

Redsicker, David R., and John J. O'Connor. (1997). *Practical Fire and Arson Investigation* (2nd ed.). Boca Raton, FL: CRC Press.

Websites

www.cbsnews.com/stories/2002/03/13/60II/main503634.shtml
www.cengage.com/school/forensicscienceadv
www.msnbc.msn.com/id/21475495
www.msnbc.msn.com/id/30965661
www.nfpa.org
www.njiaai.org/arson_facts.htm
www.officer.com/web/online/Investigation/Inside-an-Arsonists-Mind/18$38945

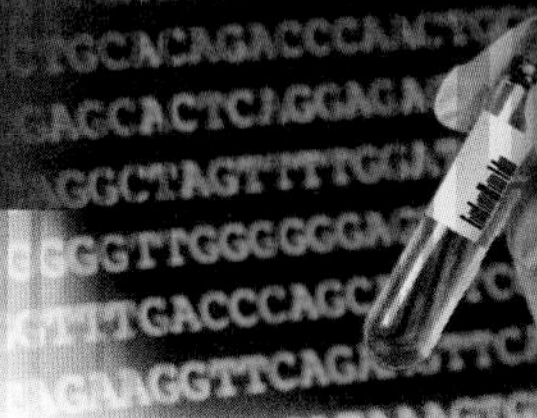

CAREERS IN FORENSICS

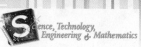

Science, Technology, Engineering & Mathematics

Jack Shannon: Arson Dog Handler/Trainer

COURTESY OF JACK SHANNON

Figure 4-23. *Jack Shannon, pictured here with one of his arson detection dogs.*

Jack Shannon (see Figure 4-23) works for the Vigo County Sheriff's Department in Terre Haute, Indiana. He is a K-9 handler and owns a dog-training school. Shannon has always loved dogs and began training dogs when he was 10 years old. He has been training dogs for more than twenty years! As an adult, he has graduated from several dog-training schools, including police dog-training schools. Shannon says that being able to read a dog's body language is essential for K-9 officers. "You have to give the dog time to work and not interfere with the dog."

Not all dogs make good arson detection dogs. An arson detection dog needs to have a strong *prey drive*, a desire to chase and catch. A dog who enjoys a game of fetch could be a candidate for detection training. An arson detection dog should also have a strong *hunt drive*. A dog who hunts enthusiastically for a toy hidden in weeds might be a good candidate. Shannon adds, "Throwing the toy into the weeds forces the dog to use his nose and not his eyes." A good arson detection dog will also be very possessive of the toy and reluctant to give it up.

To train the dog, the toy is scented with the target odor—an accelerant, for example. The trainer then hides the toy. The dog is taught to hunt for that odor by hunting for the scented toy. Once the dog finds the odor, the dog sits and waits for the handler. Many arson detection dogs are also trained to find explosives. Therefore, the dogs are trained not to scratch at the dirt or the debris. Scratching could detonate an explosive. Shannon adds, "A detection dog is really only searching for his toy. When he finds the target odor, he knows his toy will appear and playtime begins."

Learn More About It

To learn more about arson dog training, go to www.cengage.com/school/forensicscienceadv.

CHAPTER 4 REVIEW

Matching

1. heat produced when a substance is burned in oxygen *Obj. 4.1*
2. any compound consisting only of carbon and hydrogen *Obj. 4.3*
3. minimum temperature at which an accelerant will burn and continue to burn *Obj. 4.2*
4. any material used to start or maintain a fire *Obj. 4.7*
5. chemical reaction that involves a loss of electrons or a gain of oxygen *Obj. 4.1*
6. the intentional and illegal burning of property *Obj. 4.4*
7. decomposition of organic matter by heat *Obj. 4.3*
8. combination of oxygen with another substance, producing heat and light in the form of flames *Obj. 4.1*
9. minimum temperature at which liquid fuel will produce vapor *Obj. 4.2*
10. chemical reaction that releases heat *Obj. 4.1*

a. pyrolysis
b. combustion reaction
c. hydrocarbon
d. heat of combustion
e. flash point
f. accelerant
g. arson
h. oxidation reaction
i. exothermic reaction
j. ignition temperature

Multiple Choice

11. Which of the following is the most sensitive and reliable procedure for extracting combustible residues? *Obj. 4.7*
    a. high-performance liquid chromatography
    b. direct headspace extraction procedure
    c. gas chromatography
    d. passive headspace extraction procedure
12. The identity of an accelerant is determined by the pattern of its _________. *Obj. 4.7*
    a. burn
    b. debris
    c. atomic spectrogram
    d. gas chromatogram
13. _________ are trails of accelerant use to carry a fire from one location to another. *Obj. 4.6*
    a. Streamers
    b. Vapors
    c. Molotov cocktails
    d. Burn patterns

14. Most fuels cannot burn in the liquid state. What physical change must take place before a liquid fuel can burn? *Obj. 4.3*

    a. pyrolysis   c. vaporization
    d. solidification   b. decomposition

15. After the fire has been extinguished, the initial focus of the investigation at a fire scene is ________. *Obj. 4.6*

    a. determining whether the fire is arson
    b. identifying the accelerant
    c. finding the igniter
    d. locating the point of origin

16. Which of the following burn patterns most strongly suggests that a fire was either accidental or natural? *Obj. 4.5, 4.6*

    a. a "V" burn pattern   c. alligatoring of wood
    b. concrete spalling   d. streamers

17. Which of the following is a true statement about arson investigation? *Obj. 4.3, 4.5, 4.7*

    a. Fire investigators need to obtain a search warrant before collecting fire debris.
    b. All deliberate fires are considered arson.
    c. Wood does not burn in a fire.
    d. In the direct headspace extraction procedure, a charcoal strip is inserted into the evidence container with the fire debris.

Short Answer

18. What are some indications of accelerant use in a fire? *Obj. 4.5*

    ____
    ____
    ____

19. Why must all materials suspected of containing hydrocarbon residues be packaged in airtight containers? *Obj. 4.7*

    ____
    ____
    ____

20. Explain the terms *oxidation reaction* and *combustion reaction.* How are these terms related? *Obj. 4.1*

    ____
    ____
    ____

21. Why is it necessary to leave headspace in a container of fire debris? *Obj. 4.6*

____

____

____

22. Describe the four categories of fires. *Obj. 4.5*

____

____

____

23. Why are arson crimes difficult to detect and prove? *Obj. 4.4*

____

____

____

24. The motives for arson fires usually fall into six broad categories. Describe each of the six categories. *Obj. 4.9*

____

____

____

25. Describe the psychological profile of an arsonist and how investigators use this information. *Obj. 4.8*

____

____

____

Think Critically

26. Does the absence of chemical residues rule out the possibility of arson? Why or why not? *Obj. 4.5*

27. Online or at the library, find information about three different arson cases. Find information about one case that resulted in a conviction, one that resulted in an acquittal, and one that involved a conviction that was later overturned. Write an essay that discusses the significant differences in the three cases. Explain the value of arson evidence in court. *Obj. 4.7*

28. Research the differences in training for an arson investigator and a fire investigator. Examine the importance of arson evidence from each perspective. *Obj. 4.7*

CRIME SCENE S.P.O.T. Student-Prepared Original Titles

The Wynndom Warehouse Fire

By: **Yessenia Ramos**

East Ridge High School
Clermont, Florida

Amery Duke noticed the fire as she was driving past and hysterically called 911. When firefighters arrived, her red dress was blackened from soot and was torn in several places from trying to get into the building. After firefighters extinguished the fire, investigators searched the area to find the point of origin. A charred body was found on the floor of the burned out warehouse office (see Figure 4-24).

Within minutes, the fire was determined to be arson. There were streamers from the entrance of the building to the office. Just inside the entrance, investigators found wood exhibiting the alligator pattern of intense heat—the point of origin of the fire. The next step included talking to possible witnesses.

Dental records were used to identify the body as Sandra Anderson. From the carbon monoxide in the lungs, the medical examiner determined that the cause of death was smoke inhalation, not the blow to the head that apparently rendered her unconscious. Upon further examination, a small red button was found clutched in her right hand.

Figure 4-24. *Warehouse fire.*

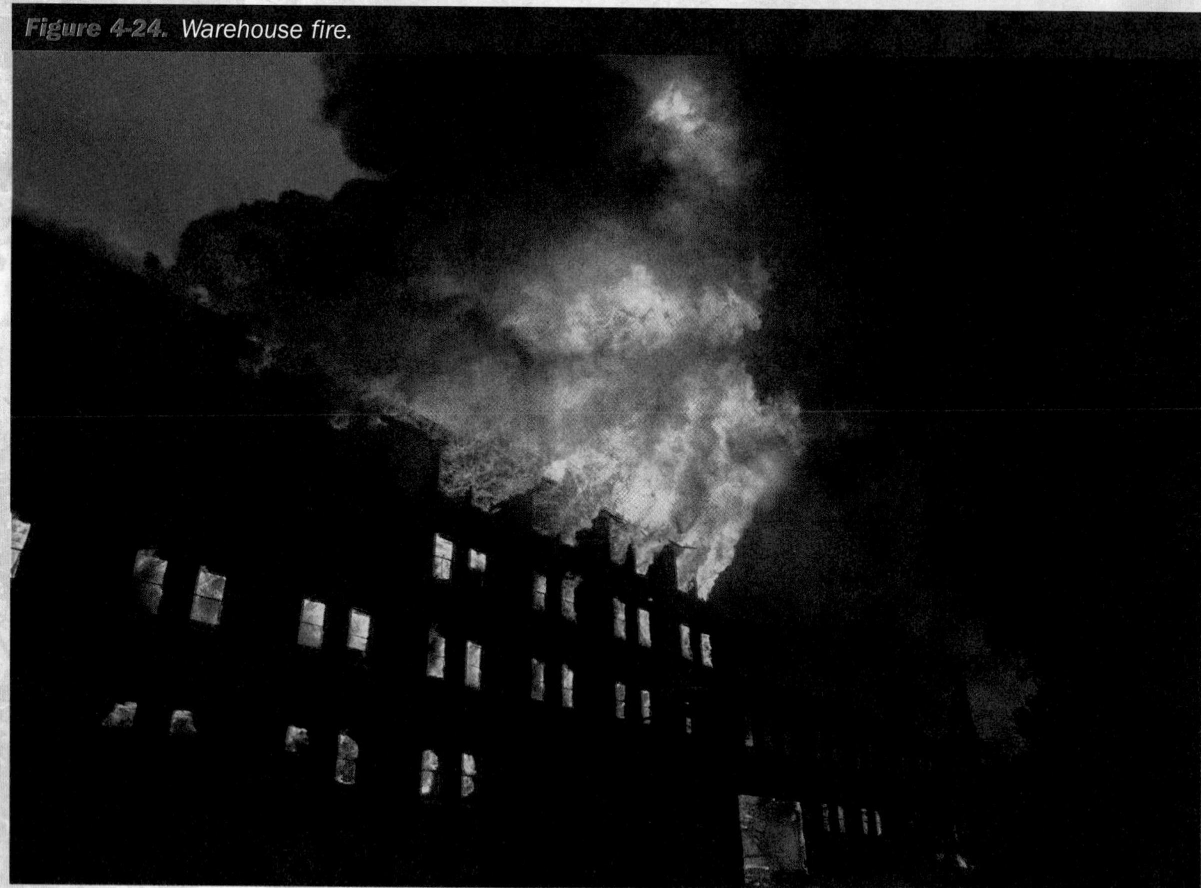

The following statements were obtained from witnesses:

- Sandra had been upset earlier in the day, but tried to cover her emotions. Several people overheard Sandra and Bryan Donahue arguing.
- Janine Adams, Sandra's neighbor, told investigators that she thought Paul Digiovanni, Sandra's boyfriend, was also dating someone else who worked at the warehouse. However, Paul had an alibi—he was making a delivery 120 km away when the fire started.
- Bryan Donahue, Amery's ex-boyfriend, had talked to Sandra earlier in the day to ask whether she had any idea who Amery was seeing. Sandra didn't know, but she said she was beginning to suspect someone. She wouldn't tell him who, so they had indeed argued about it.
- According to Amery, Sandra's coworker and friend, Sandra had called her to come to the warehouse. It had been Amery's day off. When she arrived, she saw the flames and called for help. Worried about her friend, she tried to enter the building to get to her friend, but she couldn't get in.

After investigators finished questioning the witnesses, police were called to arrest the murderer.

Activity:

Answer the following questions based on information in the Crime Scene S.P.O.T..

1. What evidence in the story indicates arson?
2. How would investigators process the evidence in this case?
3. Make a two-column chart. In the first column, list all of the possible suspects in the case. In the second column, describe motives they might have for setting the fire and killing Sandra Anderson.

WRITING

4. Using the clues in the story, write a conclusion to the story. In your conclusion, clearly describe the motive and explain how the evidence led to the suspect's arrest. You can be creative, but be sure that your conclusion is logical and that you demonstrate an understanding of arson investigation. Your conclusion should be 500 words or less.

Preventing Adolescent Crime Together™ (P.A.C.T)

Introduction:

Many arsonists are under the age of 18. In this activity, you will develop an arson awareness campaign for your school.

Procedure:

1. Organize the class into groups of three or four students.
2. At the library or online, research the answers to the following questions about juvenile arson:
    - What are the most current statistics about arson? How many arson fires happen each year? How many are committed by juveniles?
    - What are some common motives in juvenile arson cases?
    - What characteristics are included in the behavioral profile of an adolescent arsonist?
    - What tips do experts offer for preventing adolescent arson?
    - Where can students and parents go to for help? Are there websites or telephone hotlines?
3. Using your group's research, make two posters and a pamphlet describing the dangers of fire and promoting fire prevention.
    - Your group's posters should include a creative slogan that will capture the attention of your audience.
    - Your group's informational pamphlet must be easy to read, informative, creative, and well organized. The pamphlet must be typed. Include charts and illustrations. Your teacher may distribute your pamphlet to students throughout the school.

Figure 4-25. *Sample fire-prevention poster.*

ACTIVITY 4-1 *Ch. Obj: 4.7*
SUPER SNIFFERS

Objectives:

By the end of this activity, you will be able to:

1. Recognize the value of using dogs to detect accelerants.
2. Use your sense of smell to make careful observations.

Materials:

(per group of four students)
6 film canisters, labeled A, B, C, D, E, and F

Safety Precautions:

No safety precautions are needed for this lab. However, proper procedures for detecting odors should be observed. Do not hold nose directly over canister. Use hand to waft the odor toward the nose.

Background:

An arson dog was brought in to investigate a fire in a local warehouse. Fire investigators and law-enforcement personnel think the business owner intentionally started the fire for the insurance money. The dog immediately began moving about the fire scene. Within minutes, the dog alerted his handler. When arson dogs detect the presence of an accelerant, they will either sit or bark at the origin of the odor. This behavior is called an *alert*. The alert tells investigators where to collect samples to analyze for the accelerant.

Procedure:

Day 1:

1. With one hand, hold canister A slightly below your nose, about 7 cm in front of your face. With your other hand, waft the air toward your nose. Canister A represents the accelerant odor.
2. Pay attention to any familiar odors or sensations.

Day 2:

1. Take turns wafting the scents from each of the canisters labeled B, C, D, E, and F.
2. Without discussing it with your partners, decide which of the canisters contains the same odor as canister A.

Questions:

1. After your teacher identifies which canister contains the same substance as canister A, use the following formula to calculate the percentage of students who correctly identified the sample.

   $$\frac{\text{Number of correct answers}}{\text{Number of students in class}} \times 100 = \_\_\_\_\_ \%\text{ with correct answer}$$

2. How can you explain why not all students were able to correctly identify the scents? What factors made the process more difficult?
3. Why are dogs used to help investigators determine whether arson was a factor in a fire?

ACTIVITY 4-2 *Ch. Obj: 4.1*
EXOTHERMIC REACTIONS

Objectives:

By the end of this activity, you will be able to:

1. Determine the best proportion of chemicals to make an effective hot pack.
2. Explain exothermic reactions.
3. Properly collect, record, and interpret data.

Materials:

(per group of two students)
4 quart-size resealable freezer bags
permanent marker
triple beam balance
50 g $CaCl_2$
50 g $NaHCO_3$ (baking soda)
0.5 L distilled water
thermometer
watch or clock with a second hand

Safety Precautions:

Wear safety goggles, gloves, and an apron when working with chemicals.

If you are allergic to latex, alert your teacher so that you may use alternative gloves.

Make sure long hair is pulled back and dangling jewelry is removed.

Background:

Baking soda ($NaHCO_3$) and calcium chloride ($CaCl_2$) combine to produce calcium carbonate ($CaCO_3$), carbon dioxide (CO_2), table salt (NaCl), and water (H_2O). The balanced chemical reaction is shown here:

$$2NaHCO_3 + CaCl_2 \longrightarrow CaCO_3 + CO_2 + 2NaCl + H_2O$$

The reaction is exothermic. This means that, as the reaction takes place, heat is released.

Procedure:

1. Using the permanent marker, label the bags 1, 2, 3, and 4.
2. In bag 1, put 10 g of $CaCl_2$ and 10 g of $NaHCO_3$.
3. Place the thermometer in the bag and read the initial temperature of the mixture. Record the initial temperature in your data table.
4. Add 100 mL of distilled water to the bag. Seal the bag around the thermometer. Record the time it takes for the hot pack to reach the maximum temperature. Record the maximum temperature.
5. Record how long the mixture remains at the maximum temperature.
6. Predict what would happen if the ratios of the compounds were changed. Design the remaining hot packs with three different ratios of $CaCl_2$ and $NaHCO_3$. Be sure that your mixture includes a total of 20 g of $CaCl_2$ and $NaHCO_3$. Record the contents of each bag in Data Table 1.

Data Table 1. Observations.

	Amount of $CaCl_2$	Amount of NaH CO_3	Initial Temperature	Maximum Temperature	Time to Reach Max	Time Max Maintained
Bag 1						
Bag 2						
Bag 3						
Bag 4						

Questions:

1. How did you predict the ratios would affect the results of your experiment? Was your prediction correct? Explain.
2. The recommended therapeutic temperature for a hot pack is not more than 43°C. Based on your experimental results, what is the best ratio of $CaCl_2$ to $NaHCO_3$ for use in a homemade hot pack?
3. What was the purpose of determining the initial temperature?
4. Which of the hot packs maintained the maximum temperature for the longest time?
5. Explain why the temperatures changed when the water was added.

ACTIVITY 4-3 *Ch. Obj: 4.5, 4.6, 4.7, 4.9*
INVESTIGATIONS

Objectives:

By the end of this activity, you will be able to:

1. Identify the correct procedure for searching a fire scene.
2. Evaluate investigative procedures.

Procedure:

Read the details of each case and answer the questions that follow.

CASE 1

Firefighters arrived at the scene of a fire at 10:00 p.m., on a Saturday. After the fire was out, investigators began to search the building. Because it was so dark, they decided to return in the morning to complete the investigation. The next day, fire investigators were able to determine that the fire had started in a storage room at the back of the store. The owner told investigators that he stored extra shipping boxes and crates as well as solvents used for his business in the room. He said he was certain that the solvents (mostly mineral spirits) had vaporized. He assumed that heat from the overhead fluorescent bulbs must have ignited the fumes, setting the cardboard and wood on fire. Fortunately, he had insured his business.

Questions:

1. If you were investigating the fire, what questions would you ask the owner?
2. Use information from the chapter to explain whether the owner's explanation is reasonable. Could heat from the fluorescent lights have ignited the fumes from the solvents?
3. Do you think the fire was accidental or intentional? Explain.
4. If you think that the fire was intentionally set, what do you think the motive was?

CASE 2

On the morning after the fire, Jason Raines entered the building cautiously. This was his first investigation, and he was eager to get started. He systematically searched from room to room, taking notes, photographing the burn patterns, and collecting samples of debris. In the area where he found the most damage, he collected about 3 L of ash and debris. He placed the sample in an evidence envelope and properly documented the form. The fire had apparently started near the toaster in the kitchen. The plastic plug was melted, and there was severe damage around the outlet in a "V" pattern. The flames had apparently ignited the nearby curtains, and the fire had spread through the kitchen and into the dining room before firefighters arrived and extinguished the blaze.

Questions:

1. Should the fire be classified as deliberate, accidental, natural, or undetermined? Why?
2. Did Jason Raines follow proper procedure for searching the scene? Explain.
3. What mistakes did Raines make? What should he have done instead?

Extension:

WRITING

Write your own fire investigation scenario. Add three or four questions about evidence collection and analysis, motive, or fire categories. Trade your scenario and questions with a partner.

CHAPTER 5

Explosions

TIMOTHY MCVEIGH

AP PHOTO/HANDOUT FROM ATTORNEY STEPHEN JONES

Figure 5-1. Timothy McVeigh

At 9:00 A.M. on April 19, 1995, Timothy McVeigh loaded a rented truck with ammonium-nitrate fertilizer and several other chemicals. When mixed, these chemicals become explosives. He added two fuses and drove the truck to the front of the Alfred P. Murrah building in Oklahoma City. He lit the fuses and walked away. At 9:02 A.M., the 180 kg (4,000 pounds) of explosives detonated. The explosion demolished one-third of the seven-story building, and 168 people were killed. Nineteen of the victims were young children. The blast damaged buildings as far as 16 blocks away, and could be felt more than 80 km away.

Timothy McVeigh climbed into his getaway car and drove out of Oklahoma City. A state trooper stopped him about 120 km away because his car didn't have a license plate. The trooper did not yet know that McVeigh was involved in the explosion. However, McVeigh was taken into custody because he was carrying a gun that was not registered properly.

In the meantime, criminal profilers at the FBI were hard at work. The bombing in Oklahoma City had taken place exactly two years after a standoff with Branch Davidians in Waco, Texas. More than 70 people had died in Waco in the standoff with FBI agents. FBI profilers suspected that the person who built the bomb in Oklahoma City was trying to get revenge for that standoff. They thought he was a white man in his 20s and that he had a military background. These assumptions would be proven to be correct.

Meanwhile, investigators discovered that the truck carrying the bomb had an identification number on it. Investigators used the number to find the rental company that had rented the truck to McVeigh. McVeigh had used an alias—a false name—to rent the truck, but the employees were able to provide enough details for a police sketch. This sketch led investigators to McVeigh, who was still being held on the weapons charge. Eventually, McVeigh was indicted for the Oklahoma City tragedy. Evidence against him included phone cards that were used to trace his locations, fingerprints on the receipts for the chemicals, and explosives residue on his clothing and earplugs. McVeigh's trial started on April 24, 1997. He was convicted and sentenced to death. McVeigh was executed on June 12, 2001.

AP PHOTO/DAVID GLASS

Figure 5-2. *Alfred P. Murrah Federal Building.*

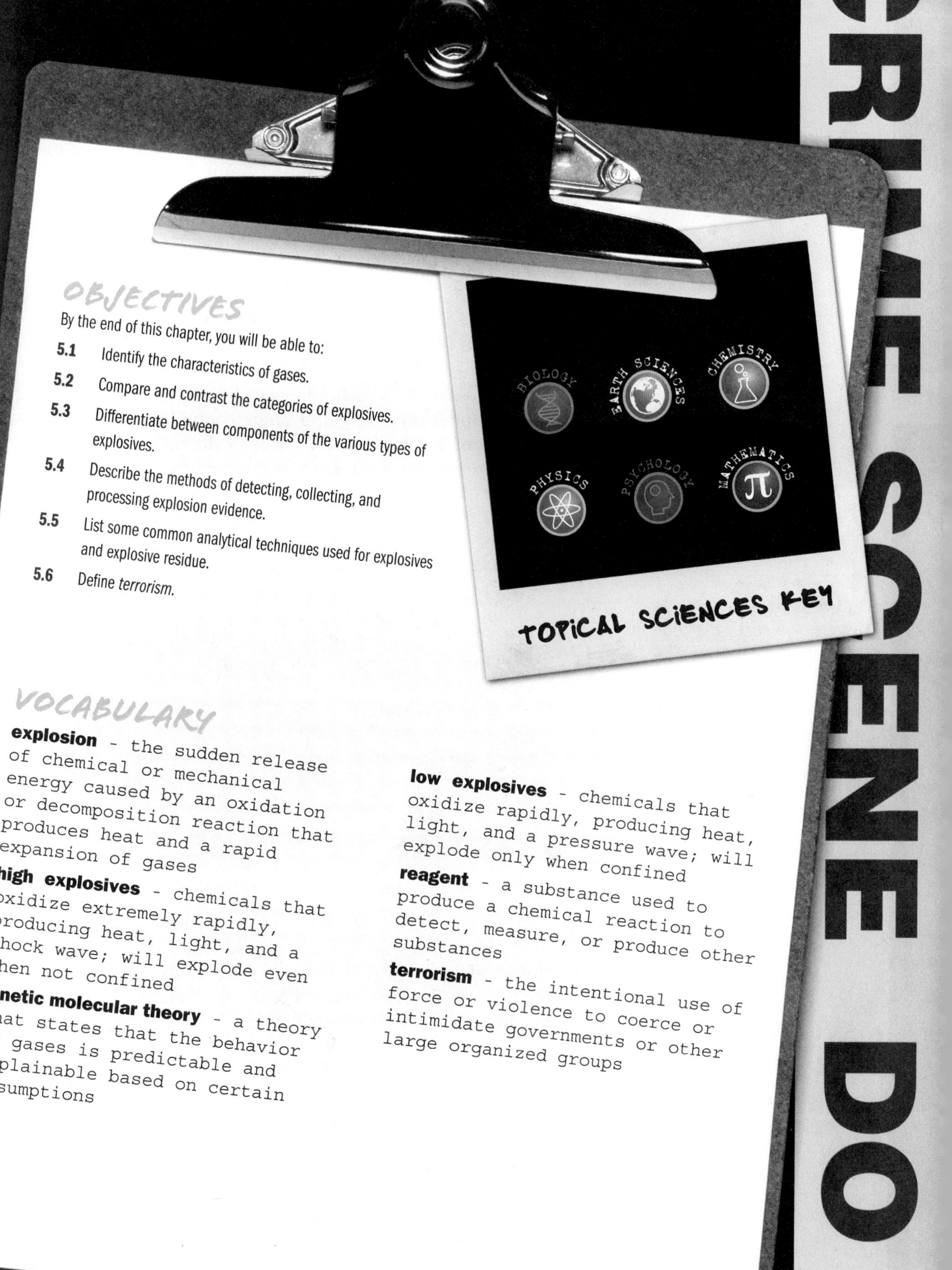

OBJECTIVES

By the end of this chapter, you will be able to:

5.1 Identify the characteristics of gases.

5.2 Compare and contrast the categories of explosives.

5.3 Differentiate between components of the various types of explosives.

5.4 Describe the methods of detecting, collecting, and processing explosion evidence.

5.5 List some common analytical techniques used for explosives and explosive residue.

5.6 Define *terrorism*.

VOCABULARY

explosion - the sudden release of chemical or mechanical energy caused by an oxidation or decomposition reaction that produces heat and a rapid expansion of gases

high explosives - chemicals that oxidize extremely rapidly, producing heat, light, and a shock wave; will explode even when not confined

kinetic molecular theory - a theory that states that the behavior of gases is predictable and explainable based on certain assumptions

low explosives - chemicals that oxidize rapidly, producing heat, light, and a pressure wave; will explode only when confined

reagent - a substance used to produce a chemical reaction to detect, measure, or produce other substances

terrorism - the intentional use of force or violence to coerce or intimidate governments or other large organized groups

Figure 5-3. *A chemical reaction is set off when a firecracker is lighted. When the fire reaches the materials inside the firework, the reaction releases nitrogen and carbon dioxide gases. The gases expand rapidly and explode the paper wrapping.*

INTRODUCTION

Explosions have been used commercially and in military settings for hundreds of years; they are useful in mining and demolition. Fireworks depend on controlled explosions (see Figure 5-3). Unfortunately, criminals and terrorists also use explosions to gain access, destroy property, and scare or hurt other people.

Like a fire, an explosion is caused by a chemical reaction. In an **explosion**, the chemical reaction releases a large amount of gas and a large amount of energy very quickly. The explosion sends a pressure wave through the surrounding materials. The chemical reaction is usually either an oxidation or decomposition reaction. Remember from Chapter 4 that oxidation reactions involve the combination of two substances to produce a new substance. In a decomposition reaction, a single compound is broken down into two or more simpler products. For example, carbonic acid (H_2CO_3) will decompose over time to produce carbon dioxide (CO_2) and water.

$$H_2CO_3 \longrightarrow H_2O + CO_2$$

Carbonic acid is added to sodas and other carbonated drinks. As the acid decomposes, it releases bubbles of carbon dioxide gas. This decomposition reaction provides the drink's "fizz."

Certain decomposition reactions release enough gas and energy that pressure and temperature can increase very quickly, causing an explosion.

Obj. 5.1

PROPERTIES OF GASES

Like all matter, gases are made up of tiny particles that are always moving. In general, gas particles move more quickly than particles in liquids and solids. Motion gives the particles energy. This energy is called *kinetic energy*. According to the **kinetic molecular theory**, the behavior of gases is predictable. The theory makes certain assumptions about the movement of gas particles:

- Gases are made up of many particles moving in rapid, random motion.
- The particles in gases are smaller than the distance between them. Therefore, most of the volume of a gas is empty space.
- When a gas particle collides with the container or with other particles, there is no net loss of energy.
- The average kinetic energy is directly proportional to the temperature of the gas. In other words, as the temperature increases, so does the average kinetic energy of its particles.
- There is no force of attraction between gas particles or between the particles and the walls of the container.

Because gas molecules are so far apart and move so much, a gas has no definite shape or volume. The gas will expand to fill the space. This property explains how smells diffuse through a room or down a hallway. A gas can also be contained and compressed. If the volume of a container decreases, the volume of the gas decreases. Although the volume of the gas decreases, the *amount* of gas—the number of particles—does not change. The gas particles get closer together, and the pressure inside the container increases. Figure 5-4 summarizes the relationship between temperature, pressure, and volume in gases.

If	And	Then
Temperature increases	Volume is constant	Pressure increases
Temperature increases	Pressure is constant	Volume increases
Volume increases	Temperature is constant	Pressure decreases
Volume increases	Pressure is constant	Temperature increases
Pressure increases	Temperature is constant	Volume decreases
Pressure increases	Volume is constant	Temperature increases

Figure 5-4. *Relationship between temperature, volume, and pressure in gases (assuming the amount of gas is constant).*

A single mathematical equation can be used to describe the relationship among temperature, volume, and pressure in confined gases. The relationship is called the *combined gas law*, and the equation is shown here:

$$\frac{P_1 \times V_1}{T_1} = \frac{P_2 \times V_2}{T_2}$$

In this equation, P is the pressure, measured in kilopascals (kPa); V is volume, measured in liters (L); and T is temperature, measured in Kelvins (K). You can use the combined gas law equation to predict how temperature, pressure, or volume changes when conditions change. For example, imagine that you have a helium-filled balloon on a warm day. The temperature is 300 K. The volume of the balloon is 25 L. The pressure is 101.3 kPa. Now, imagine that you carry the balloon into your air-conditioned home. The temperature changes to 293 K and the pressure remains constant. What is the new volume of the balloon?

$P_1 = 101.3$ kPa　　$P_2 = 101.3$ kPa

$V_1 = 25$ L　　$V_2 = ?$

$T_1 = 300$ K　　$T_2 = 293$ K

$$\frac{P_1 \times V_1}{T_1} = \frac{P_2 \times V_2}{T_2}$$

$$\frac{101.3 \text{ kPa} \times 25 \text{ L}}{300 \text{ K}} = \frac{101.3 \text{ kPa} \times V_2}{293 \text{ K}}$$

Rearrange the equation to isolate V_2, the new volume.

$$V_2 = \frac{101.3 \text{ kPa} \times 25 \text{ L} \times 293 \text{ K}}{300 \text{ K} \times 101.3 \text{ kPa}}$$

$$V_2 = 24.42 \text{ L}$$

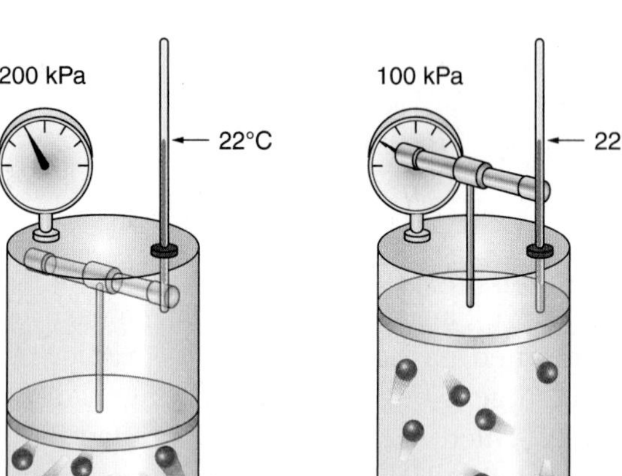

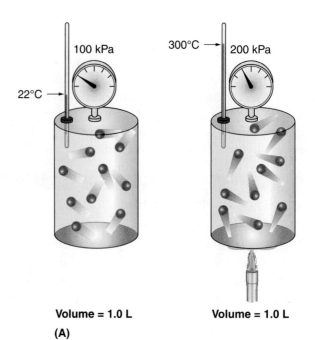

Figure 5-5. *(A) If the temperature of the enclosed gas increases and volume is held constant, the pressure increases. (B) If the volume of the enclosed gas increases and temperature remains constant, the pressure decreases.*

Based on the combined gas law, the volume of the balloon in the cooler room is 24.42 L.

Figure 5-5 illustrates the relationships between temperature, pressure, and volume in enclosed gases. Many explosive devices depend on the relationships between temperature, volume, and pressure in confined gases. Some devices heat the gases and hold the volume constant. Others increase the amount of gas in a container while keeping the volume constant. In both cases, the pressure increases. Eventually, the container will not be able to withstand the pressure and the container explodes. Sometimes, the chemical reaction releases such a large amount of heat and gas that the reaction causes an explosion even when the reaction doesn't take place within a container.

CHARACTERISTICS OF AN EXPLOSION

Obj. 5.2

When a fire burns, the reaction continues until one of the necessary components—fuel or oxygen—is used up. If oxygen can be removed from the fire, the fire will stop. For example, if a candle burns in a glass holder and someone covers the candleholder, the fire will stop as soon as the oxygen inside the holder is used up. An explosion does not require oxygen. Instead, energy and gases are released as the products of a chemical reaction. The expanding gases produce a pressure wave.

A *wave* is a disturbance in matter that carries energy. A pressure wave is a kind of wave called a *longitudinal wave*. You can produce a longitudinal wave with a spring toy (see Figure 5-6). If you hold the spring toy parallel to the ground and push or pull on one end of the toy, a wave forms. The coils get close together in some areas, called *compressions*. The coils are further apart in other areas, called *rarefactions*. The wave transfers energy from one end of the toy to the other.

The energy released in an explosion forces particles in the air to move back and forth. In some explosions, the gases are held in a container. In these explosions, the walls of the container will stretch until they burst and fragment. The debris, flying outward in all directions, is often the most dangerous result of the explosion. Flying debris is also referred to as *shrapnel* (see Figure 5-7).

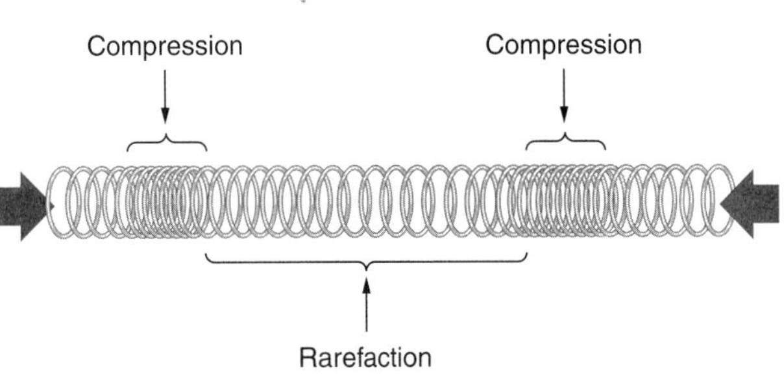

Figure 5-6. *Pressure waves move through the air in much the same way that this longitudinal wave moves through the spring toy.*

© PHOTOS 12/ALAMY

Figure 5-7. *Debris from an explosion is called* shrapnel.

TYPES OF EXPLOSIVES

Obj. 5.2, 5.3

Explosives are chemicals that react under certain conditions to cause explosions. Most explosives are solids. Explosives are generally placed into one of two categories based on their rates of reaction: low explosives and high explosives. Low explosives tend to react more slowly. When unconfined, low explosives burn rather than explode. Under the right conditions, low explosives produce less dramatic explosions than high explosives do.

The term *shrapnel* also means a bullet filled with lead pellets and a timing fuse. The fuse causes the bullet to explode in flight, spraying the pellets in all directions. Henry Shrapnel, an English artillery officer, invented this bullet in 1784.

LOW EXPLOSIVES

Materials that burn rapidly but explode only when confined in a container are called **low explosives**. These explosives produce a combustion reaction called *deflagration*. Deflagration is rapid, intense burning. Deflagration produces a pressure wave that travels at a speed of less than 340 meters per second (m/s),

Did You Know?

Natural gas is colorless and odorless. In 1937, approximately 300 people died in an accidental explosion when natural gas seeped into the New London School in New London, Texas. The gas ignited when it came into contact with an electrical source. Since that time, a small amount of t-butyl mercaptan, a chemical that smells like rotting cabbage, has been added to natural gas so that leaks can be detected.

the speed of sound. Low explosives are often used as *propellants.* The burning or small explosion of a low explosive produces enough gas and energy to push a bullet or other object away from the original explosion. A small amount of energy in the form of a spark or a burning fuse is required to ignite a low explosive. Black powder and smokeless gunpowder are some common low explosives.

One common black powder is a mixture of potassium nitrate, sulfur, and charcoal. This mixture will burn to produce potassium, sulfide, nitrogen gas, and carbon dioxide. The following equation illustrates the reaction of black powder combustion:

$$2KNO_3 + S + 3C \longrightarrow K_2S + N + CO_2$$

Because black powder will burn rather than explode in open air, it is used to make safety fuses. The powder is wrapped in a layer of fabric or plastic to form a fuse. The fuse carries the spark to another explosive, allowing an individual to remain at a safe distance from the explosion.

Smokeless gunpowder is the safest and most powerful low explosive. It is made of *nitrocellulose,* cotton treated with nitric acid.

Fireworks are made mostly of small amounts of black powder or smokeless gunpowder and a fuse. The firework is wrapped tightly in many layers of heavy paper. The reaction produces gases that put pressure on the paper. This pressure causes the explosion. Elements added to the firework emit different colors when they are heated. These elements are responsible for the various displays. Figure 5-8 shows a typical firework design.

Another low explosive forms when natural gas (usually from a furnace) escapes into an enclosed area and mixes with oxygen in the air. If the mixture is ignited, it will explode. If the explosion occurs inside a building, pressure waves will force walls and objects inside the building outward.

HIGH EXPLOSIVES

Materials that detonate are called *high explosives. Detonation* is an explosion that results in a violent disruption to the surrounding area. Unlike low explosives, high explosives do not need to be confined in order to detonate. Many high explosives, including TNT ($C_7H_5N_3O_6$), decompose upon detonation. The following equation shows the decomposition reaction that takes place as TNT explodes:

$$2C_7H_5N_3O_6 \longrightarrow 3N_2 + 5H_2O + 7CO + 7C$$

The carbon produced by the decomposition of TNT leaves black soot that indicates that TNT may have been involved in the explosion.

High explosives detonate easily and produce a pressure wave that moves as fast as 8,500 m/s. A pressure wave that moves this quickly is called a *shock wave.* When high explosives explode, the energy released causes the gas particles to move more quickly than the speed

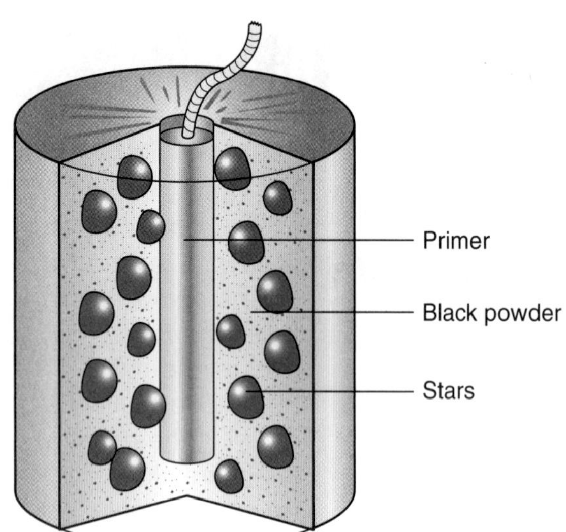

Figure 5-8. *The area labeled Stars refers to the elements that produce the colors and patterns when the firework explodes.*

of sound (340 m/s). The shock wave can cause widespread damage to the surrounding area. High explosives are used to blast or shatter a target. For example, workers in the mining industry use high explosives to blast through rock. High explosives can be placed into two categories: primary high explosives and secondary high explosives.

Primary High Explosives

Primary high explosives are extremely sensitive to heat, pressure, and movement. They are often used to detonate other explosives or as primers in shotgun shells or bullets. Lead styphnate and mercury fulminate are very sensitive to shock. This sensitivity makes them very useful in firearms cartridges. Nitroglycerin is probably the most well-known primary high explosive. Nitroglycerin was used for blasting into mountains, but its unpredictable nature makes it extremely dangerous to handle.

Secondary High Explosives

In small amounts, secondary high explosives can be handled relatively safely. They are not as sensitive to heat or shock as primary high explosives. However, secondary high explosives can cause very violent explosions. Most of the explosives used for military (see Figure 5-9) or commercial purposes are in this group. TNT (trinitrotoluene) and dynamite are well-known secondary high explosives. Often, primary high explosives are used to detonate secondary high explosives.

LUIS BENAVIDES

Figure 5-9. *Mortar grenades seized at a military base in Columbia in 2009.*

At one time, nitroglycerin was an important primary high explosive in the mining industry. However, it was very volatile and dangerous to handle. In 1866, Alfred Nobel discovered that a mixture of silica and nitroglycerin was much less volatile than nitroglycerin alone. The mixture was called *dynamite* and is still explosive under the right conditions. The discovery revolutionized the mining and construction industries.

The Nobel Company is the largest manufacturer of explosives in the world. Its founder, Alfred Nobel, used some of his company's enormous wealth to fund the Nobel prizes indefinitely.

CLASSIFICATION BY USE

Explosives are not only classified by how quickly they react. They can also be classified by how they are used. Commercial explosives are used in mining and road construction. Additionally, commercial explosives are used in demolition (see Figure 5-10). Explosives are placed near the support beams of a building. When these explosives are detonated, the building will collapse. Ammonium nitrate fuel oil (ANFO), black powder, and dynamite are typically used in controlled, commercial explosions.

Figure 5-10. *Explosives are used in a controlled demolition of a building.*

Military explosives include RDX (cyclotrimethylenetrinitramine), PETN (pentaerythritol tetranitrate), and TNT. RDX is very often formed into a pliable plastic known as C-4. It is also a component of military dynamite. PETN is mixed with TNT and used in grenades and small-caliber projectiles. The third type of explosive is the improvised explosive. Most improvised explosives are low explosives placed in a confining container, such as a pipe. Improvised explosives are illegal and often used by terrorists and in guerrilla warfare.

Obj. 5.4 COLLECTION OF EVIDENCE

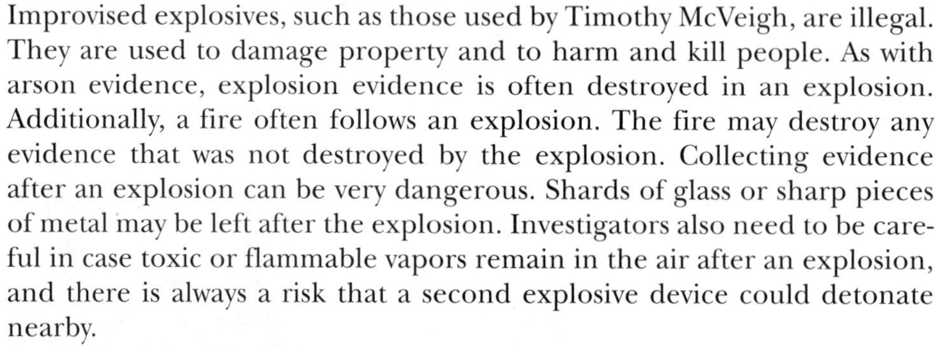

Improvised explosives, such as those used by Timothy McVeigh, are illegal. They are used to damage property and to harm and kill people. As with arson evidence, explosion evidence is often destroyed in an explosion. Additionally, a fire often follows an explosion. The fire may destroy any evidence that was not destroyed by the explosion. Collecting evidence after an explosion can be very dangerous. Shards of glass or sharp pieces of metal may be left after the explosion. Investigators also need to be careful in case toxic or flammable vapors remain in the air after an explosion, and there is always a risk that a second explosive device could detonate nearby.

Almost immediately after an explosion, investigators work to determine the cause. A large crater often marks the point of origin. Investigators collect soil samples and debris from inside the crater. They test objects near the crater for traces of explosive residue. The entire blast area must be carefully and systematically searched for residues, detonating devices, or pieces of the explosive device. Investigators may sift through the debris with wire mesh to find small pieces of physical evidence. Figure 5-11 shows examples of the sifting tools investigators use.

Figure 5-11. *Sieves with holes of different sizes are used to find small pieces of evidence.*

As they do with all physical evidence, investigators must maintain a chain of custody for each piece of evidence from the scene of an explosion. All evidence materials must be packaged separately to avoid cross-contamination. Soil and loose debris are packaged in metal containers, such as clean paint cans, or in glass jars. The cans or jars are labeled to record the location where the debris was found. Evidence from the scene of an explosion is not usually packaged in plastic bags. Gas residues from the explosion may diffuse through the plastic, and sharp-edged debris could puncture the bag.

Obj. 5.5 ANALYSIS OF EVIDENCE

Field investigators collect as much evidence as possible. This evidence is sent to the lab for further analysis. Forensic scientists in the laboratory try to identify the types of explosives and the mechanism used for detonation.

MICROSCOPIC EXAMINATION

A forensic scientist's first test is often low-power stereomicroscopy. The scientist examines the debris for unexploded material, such as dynamite or black powder. The shape and color of some low explosives, such as smokeless powders, make these explosives relatively easy to identify (see Figure 5-12). Smokeless powders are made of different combinations of nitrocellulose, nitroglycerin, and a third chemical called *nitroguanidine.* Pieces of the explosive device, such as part of a wrapper from dynamite, metal from a pipe, or pieces of wire, may also be visible in the debris. These items often provide important information in the investigation and are used in criminal prosecution.

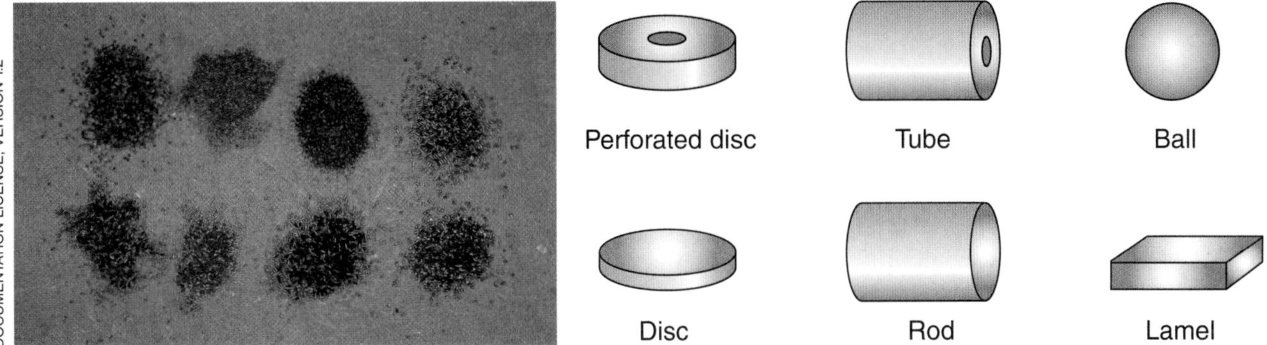

Figure 5-12. *Smokeless powder comes in very distinct colors (left) and shapes (right). The shape affects burn rate and pressure within the casing. These variables can be manipulated to produce the desired effect.*

Digging Deeper
with Forensic Science e-Collection

Before you get on an airplane or ship, your luggage and your person are scanned for explosives and explosive devices. Bombers have gotten more "creative" in their methods. Therefore, the federal government has hired some of the most highly trained and skilled explosion investigators to find the most sophisticated equipment and technology to detect explosive material and devices. Using the Gale Forensic Science e-Collection at www.cengage.com/school/forensicscienceadv, read about various explosive-detecting and analysis tools. Then write several paragraphs to address the following topics: (1) Describe a sniffer device and explain its various detection modes. (2) Discuss the two principle types of explosives. (3) Distinguish between the various types of analytical detection methods using explosion detectors. (4) Explain the value of walk-through scanning devices. (5) Why are dogs still considered some of the best detectors of explosives?

COLOR SPOT TESTS

After forensic scientists observe the samples under the microscope, they perform chemical analysis of the explosive residue. Scientists wash the debris with acetone. Acetone dissolves most explosives very easily. The acetone and explosive solution is concentrated and analyzed using a variety of techniques.

Forensic scientists use a presumptive color spot test to screen for the presence of various explosives. The scientist mixes a small amount of the concentrated acetone and explosive solution with a specific reagent and observes any reaction. A **reagent** is a substance used to produce a chemical reaction to detect, measure, or produce other substances. Based on the results of several of the spot tests, the scientist makes a preliminary identification of the explosive.

Scientists use the Modified Griess Test to check for the presence of nitrite compounds. Nitrite compounds contain the ion NO_2^-. Nitrite compounds are left behind when certain explosives, such as smokeless powders, burn. In the Modified Griess Test, the residues are rinsed in a solution of acetic acid and heated. The forensic scientist then adds the Griess reagent. If the original residue contained a nitrite, the solution turns bright orange.

Alcoholic potassium hydroxide (KOH) can also be used as a reagent in a presumptive test. Alcoholic KOH turns blue only in the presence of TNT or the residues left after an explosion of TNT. If TNT was used in an explosion, alcoholic KOH will turn blue or violet. Diphenylamine, another reagent, reacts in the presence of most explosive residues. However, it does not react with TNT or RDX. Figure 5-13 summarizes these three presumptive tests.

The Modified Griess Test and the diphenylamine test confirm that explosives are present. They do not show which explosive is present, but they narrow down the possibilities. Neither the Modified Griess Test nor the diphenylamine test detects TNT. Forensic scientists use the potassium hydroxide (KOH) test to detect TNT.

Digging Deeper

with Forensic Science e-Collection

Atlanta, Georgia hosted the Summer Olympic Games in 1996. Extreme safety measures were implemented to protect the athletes, visitors, reporters, and dignitaries. During the games, a devastating act of terrorism took place. Using the Gale Forensic Science e-Collection at www.cengage.com/school/forensicscienceadv, read about the bombing of Olympic Park. Use information from the article to write a brief essay describing the events. Include answers to the following questions in your essay: (1) Explain the fears security officials had regarding the venue and potential security briefs. What forensic evidence was recovered at the scene? (2) Do you agree with the decision to continue with the Olympics? Explain.

Explosive	Modified Griess Test	Alcoholic KOH	Diphenylamine
Nitrate	No reaction	No reaction	Blue to blue-black
Nitrocellulose	Orange	No reaction	Blue-black
Nitroglycerine	Orange	No reaction	Blue to blue-black
Trinitrotoluene	No reaction	Red to violet	No reaction
RDX	Orange	No reaction	No reaction
PETN	Orange	No reaction	Blue

Figure 5-13. *Common explosive and the reagent reactions.*

CHROMATOGRAPHY

Crime lab scientists also use thin layer chromatography (TLC) as a presumptive test for explosives. As you learned in Chapter 3, mixtures of substances are separated using various chromatography methods. Scientists use the R_f values calculated from TLC for tentative identification of the components of the mixture. When screening for explosives, scientists use cellulose or silica gel plates as the stationary phase. The mixture of unknown explosives is placed on the plates and rinsed with a solvent. Scientists then compare the resulting chromatogram to chromatograms of known explosives. Chromatograms showing components with similar R_f values suggest that the unknown mixture contains the same explosives as the known. High performance liquid chromatography (HPLC) is also used to separate mixtures of explosive residues (see Figure 5-14).

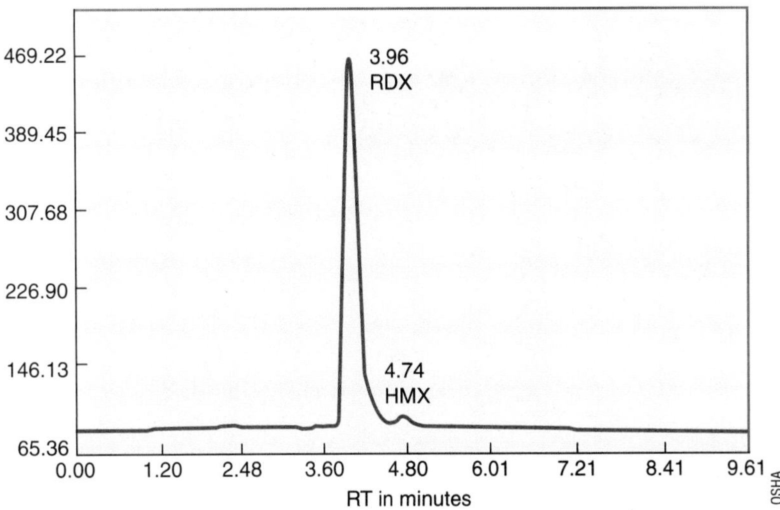

Figure 5-14. *Comparing chromatograms.*

CONFIRMATORY TESTS

Once forensic scientists have a preliminary identification of the explosive material or a short list of possible identities, they perform a confirmatory test. Infrared spectroscopy or gas chromatography–mass spectroscopy is used to confirm the identity of organic explosives, such as TNT and many military explosives. In infrared spectroscopy, a beam of infrared light is passed through a sample. The spectrometer measures how much energy was absorbed by the sample. Each substance absorbs a specific amount of energy during infrared spectroscopy. Therefore, the scientist uses this value to identify the components of the sample. Gas chromatography–mass spectroscopy works in two steps. First, the mixture undergoes gas chromatography. This step separates the compounds in the mixture. The compounds are then sent through a mass spectrometer. In the mass spectrometer, the compounds are broken into smaller charged particles. These particles are separated based on their mass and charge. Scientists use the data to determine which explosives were used. They also learn how much of each kind of explosive was in the mixture.

EXAMINATION OF PHYSICAL EVIDENCE

Obj. 5.5

Analysis of the explosive is most important for the clues it gives investigators about the perpetrator. Investigators gather as much physical evidence at the scene as possible. Pieces of wire, pipe, or devices used to detonate

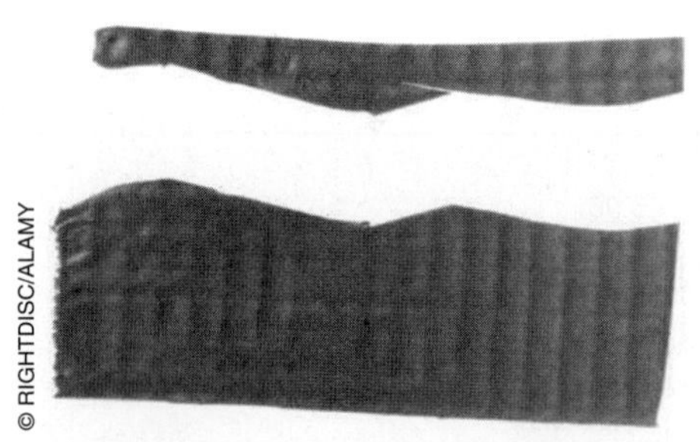

Figure 5-15. *These two pieces of duct tape are a physical match.*

an explosive can provide valuable information about the explosion and the perpetrator. For example, fingerprints on a receipt and explosive residue on his clothes and earplugs connected Timothy McVeigh to the bombing of the Murrah building in Oklahoma City. People who set off bombs often use watches to make timing devices. The serial number on the watch may help investigators identify who made the bomb. Tool marks on pipes and even torn edges of duct tape (see Figure 5-15) are important clues as to how the explosive device was built and maybe even to the identity of the bomber. For example, if one piece of duct tape was found on the remains of the explosive device at the scene and the other was found among the suspect's things, investigators have good evidence that the suspect was involved.

Obj. 5.6 TERRORISM

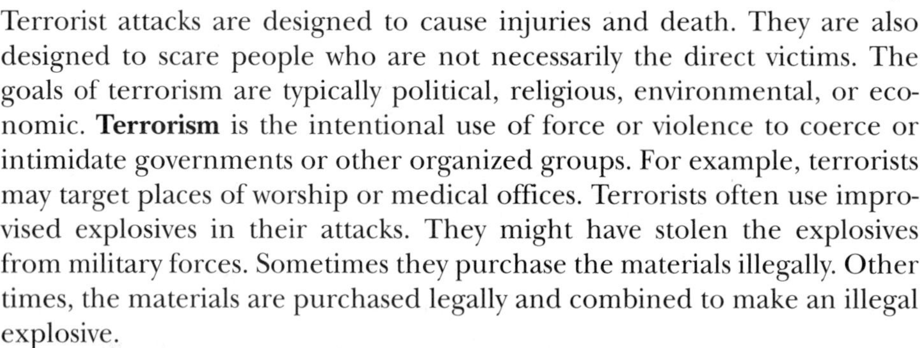

Terrorist attacks are designed to cause injuries and death. They are also designed to scare people who are not necessarily the direct victims. The goals of terrorism are typically political, religious, environmental, or economic. **Terrorism** is the intentional use of force or violence to coerce or intimidate governments or other organized groups. For example, terrorists may target places of worship or medical offices. Terrorists often use improvised explosives in their attacks. They might have stolen the explosives from military forces. Sometimes they purchase the materials illegally. Other times, the materials are purchased legally and combined to make an illegal explosive.

Some acts of terrorism are isolated. The target is an individual or a very small group. Sometimes, however, acts of terrorism kill or hurt a large group of people, capturing the attention of many others. In such cases,

Digging Deeper

with Forensic Science e-Collection

In the United States, the Department of Homeland Security focuses primarily on the prevention of terrorism. Terrorists typically rely on explosives to destroy their targets. Use the Gale Forensic Science e-Collection at www.cengage.com/school/forensicscienceadv, to learn how law-enforcement agencies prevent and respond to the threat of liquid explosives. Then write an essay that addresses the following: (1) Discuss the August 10, 2006, plot to bomb airplanes traveling from the United Kingdom to the United States. Had any similar attempts ever occurred? (2) Discuss the various types of liquid explosives. (3) Explain the issues with detecting liquid explosives. (4) Why are liquid explosives especially dangerous? (5) Describe some problems law-enforcement personnel are likely to have in detecting explosives in the future.

people may also be affected emotionally or economically. The target of the attack might have been a well-known building or other structure. For example, on September 11, 2001, terrorists attacked the World Trade Center (see Figure 5-16) in New York City and the Pentagon in Washington, DC. Another plane heading for either the White House or the Pentagon crashed in Pennsylvania. People all over the United States and the world were indirectly affected. People in the United States were affected later that year by an anthrax scare. In response to these attacks, many people became angry and afraid. As a result, the economy slowed. Several airlines filed for bankruptcy. September 11 has cost the country directly and indirectly hundreds of billions of dollars.

SEAN ADAIR/REUTERS /LANDOV

Figure 5-16. *September 11, 2001, attack on the World Trade Center.*

Domestic terrorists commit terrorism against their own country or against groups of people in their own country. These people are sometimes called "homegrown" terrorists. Timothy McVeigh was a domestic terrorist. Foreign terrorists commit terrorism against other countries or against groups in other countries. For example, foreign terrorists carried out the attacks on September 11, 2001. Members of al Qaeda, an international terrorist network trained and financially backed by Osama bin Laden, organized these attacks.

According to the Bureau of Alcohol, Tobacco, Firearms, and Explosives, the number of explosives incidents declined from 3,790 in 2004 to 3,445 in 2006. An explosives incident is defined as a bombing, attempted bombing, incendiary bombing, or theft of explosives. In 2006, 745 of those incidents were referred for prosecution.

CHAPTER SUMMARY

- An explosion is an oxidation or decomposition reaction that releases a lot of gas and energy very quickly. Even a small explosion can release a large amount of heat.
- The behavior of gases can be explained and predicted based on assumptions of the kinetic molecular theory. Knowledge of the theory is important in the investigation of explosions.
- Low explosives burn rapidly but explode only when confined. The burning of a low explosive is called deflagration.
- High explosives can explode even when they are not confined. They produce a violent disruption to the surroundings.
- The most dangerous high explosives are primary high explosives. They are extremely sensitive to heat, pressure, and movement. Nitroglycerin is a primary high explosive.
- Secondary high explosives are not as sensitive to heat or shock as primary high explosives, but they are capable of violent explosions. Secondary high explosives are typically used by the military or commercially in mining or construction.

- Investigators collect soil samples and debris from an explosion's point of origin. They thoroughly search the blast area for physical evidence such as explosive residue, pieces of the explosive device, or detonating devices.
- In the crime lab, forensic scientists examine evidence under a stereomicroscope. For further examination, the debris is washed with acetone to dissolve the explosives. The solution is then analyzed to identify the type of explosive and to discover other clues.
- Thin layer chromatography is a presumptive test that can be used to provide a preliminary identification. Forensic scientists then perform confirmatory tests, such as infrared spectroscopy and gas chromatography–mass spectroscopy.
- Explosions are often caused by acts of terrorism. Terrorism may be politically, environmentally, or religiously motivated.

CASE STUDIES

John Graham (1955)

On November 1, 1955, United Airlines Flight 629 took off from Stapleton Airport in Denver, Colorado. The plane was carrying 44 passengers and crew to Seattle, Washington. Shortly after takeoff, the pilot heard a loud bang from underneath the airplane. The aircraft erupted into a huge explosion. The fuel tanks were filled to capacity, which fed the explosion. When the wreckage settled, it covered a two-mile radius near the Wyoming border (see Figure 5-17). There were no survivors, and identification of the charred bodies was difficult. The National Guard was called in to secure the crash site. The Civil Aeronautics Board examined the wreckage. Investigators determined that the explosion was not caused by a problem with the aircraft itself. The FBI was called in to complete a criminal investigation. Investigators identified the explosion's point of origin as the rear luggage compartment. Luggage recovered from the

AP PHOTO/EDWARD O. EISENHAND

Figure 5-17. *The wreckage of Flight 629.*

scene had a strong gunpowder odor. Engineers found four small pieces of sheet metal that contained a gray residue from the explosion. Further tests concluded that luggage in that area contained traces of sodium carbonate, nitrates, and sulfates—all chemicals found in dynamite. The FBI concluded that a bomb and a timing device had been placed on the plane.

The FBI investigated the backgrounds of all passengers. Daisy King, a wealthy businesswoman who had died in the explosion, had a large insurance policy. Her son, John Graham, was the only beneficiary. (He would receive all of the insurance money.) He would also inherit a successful restaurant King had owned. Graham was interviewed several times, and his home and car were searched. Among other pieces of evidence, investigators found wire that matched wire found at the crash site.

Graham confessed to building a bomb and placing it in his mother's suitcase. On November 14, 1955, he was arrested and charged with 44 counts of murder. He was convicted of first-degree murder and sentenced to death. Even up to the last moments before his death, he showed no remorse for his actions, and he never tried to explain his enormous hatred toward his mother.

Think Critically **1. What is the significance of Daisy King's insurance policy?**

2. Why was John Graham not accused of domestic terrorism?

USS Cole (2000)

On October 12, 2000, the United States Navy destroyer *USS Cole* (see Figure 5-18) was refueling in the Port of Aden on the Arabian Peninsula. The Port of Aden is in Yemen. A small boat approached the *USS Cole*. This boat was carrying almost 45 kg of explosives, including RDX and TNT, which detonated as the boat approached the *USS Cole*. The explosion left a 12.2 m × 18.3 m hole near the *USS Cole*'s waterline. The occupants on the small boat and 17 American sailors were killed. Investigators from the United States Department of Justice, the U.S. Navy, and the Yemeni government worked together for two years to determine exactly what had happened. Investigators learned that the bombing had been carried out by two suicide bombers, Ibrahim al-Thawr and Abdullah al-Misawa. Al Qaeda claimed responsibility for this attack. On September 29, 2004, a Yemeni court convicted two al Qaeda operatives for planning and organizing the attack. Abd al-Rahim al-Nashiri and Jamal al-Badawi were sentenced to death. Four other men were also convicted for their role in the attacks and sentenced to prison terms ranging from 5 years to 10 years. These other men had given Nashiri and Badawi money, had attempted to videotape the attack, and had falsified identification documents for those involved.

ALADIN ABDEL NABY/REUTERS/CORBIS

Figure 5-18. *The USS Cole.*

Think Critically 1. **What lab techniques described in the chapter may have been used to determine the identity of the explosives on the small boat?**

2. **The attack on the *USS Cole* took place in waters off the coast of Yemen. Citizens of Yemen carried out the attack. Was this an act of foreign or domestic terrorism? Justify your answer.**

Bibliography

Books and Journals

Gaensslen, R. E., Howard A. Harris, and Henry Lee. (2008). *Introduction to Forensic Science & Criminalistics*. New York: McGraw-Hill.

Lerner, K. Lee, and Brenda Wilmoth Lerner, eds. (2006). *The World of Forensic Science*. New York: Gale Group.

Nickell, Joe, and John F. Fischer. (1999). *Crime Scene: Methods of Forensic Detection*. Lexington, KY: The University Press of Kentucky.

Petraco, Nicholas, and Hal Sherman. (2006). *Illustrated Guide to Crime Scene Investigation*. Boca Raton, FL: CRC Press.

Websites

www.atf.gov/publivations/factsheets
www.cengage.com/school/forensicscienceadv
www.dhs.gov
www.ojp.usdoj.gov/nij/training/firearms-training/module12/fir_m12_t05_03_a.htm
www.tsa.gov
www.twgfex.org

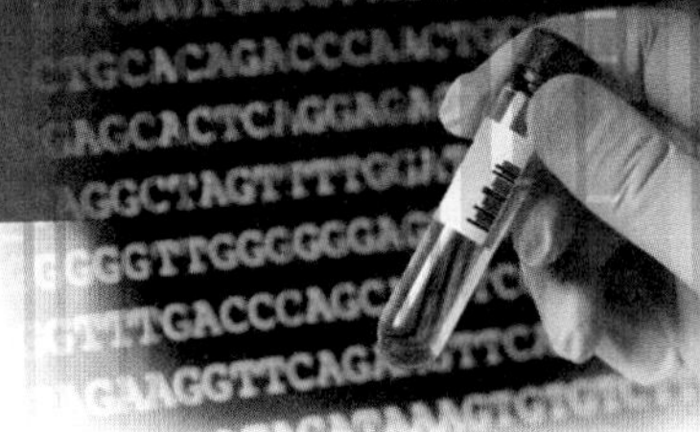

Explosives Technician

A bomb squad technician dismantles explosive devices before they detonate. Sometimes the devices are sitting in a remote area. Sometimes they are in the middle of a crowd. And sometimes, these devices are strapped to the body of a hostage or suicide bomber. The technician must assess the situation and approach the device precisely and carefully to ensure the safety of everyone in the vicinity.

Bomb squad technicians work for local and state police and fire departments, the FBI Explosives Unit, and in the U.S. military. In the military, bomb squad technicians are often faced with roadside bombs, mines, and improvised explosive devises. Bomb squad technicians working for the military are called explosive ordinance disposal specialists or improvised explosive device specialists. All bomb squad technicians approach a suspected explosive device wearing protective gear (see Figure 5-19). Their primary objective is to keep themselves and the public safe. They inspect the bomb and disable or destroy it before it detonates. If a bomb detonates before the bomb squad arrives, bomb squad technicians are trained to evaluate evidence at the scene to learn valuable information about the explosion—to begin to piece together clues about what exploded and why. They are also trained in decontamination and disposal as well as in fragment analysis.

All bomb squad technicians in the United States are trained at the Federal Bureau of Investigation's Hazardous Devices School located at Redstone Army Arsenal in Huntsville, Alabama. In order to qualify for this training, applicants must be working as firefighters, police officers, FBI agents, or other federal investigators and must be assigned to a bomb squad. The Hazardous Devices School is a 300-acre facility that houses many "mock" scenes, such as bus and airline terminals, homes and apartments, a church, and a gas pipeline. Students at the school are trained how to respond to bombs and bomb threats in these real-life environments. They are trained to use the latest equipment, including bomb-disabling robots (see Figure 5-20) and practice many counterterrorism techniques.

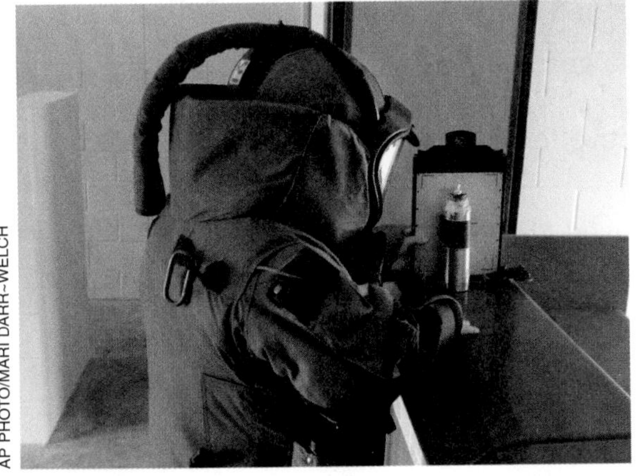

Figure 5-19. *An explosives technician works to disable an explosive device.*

Figure 5-20. *The Boise (Idaho) police department bomb squad uses a bomb-disabling robot.*

Learn More About It

To learn more about a career in explosion investigation, go to www.cengage.com/school/forensicscienceadv.

CHAPTER 5 REVIEW

True/False

Read the following sentences. Decide whether each sentence is true or false. If a sentence is false, change the underlined word or phrase to form a true statement.

1. The detonation of a low explosive produces a pressure wave that travels at less than 340 m/s. *Obj. 5.2*
2. Debris from the site of an explosion is packaged in metal containers or glass jars. *Obj. 5.1*
3. If the volume of a contained gas increases, the pressure increases. *Obj. 5.1*
4. If the temperature of a contained gas decreases, the pressure decreases. *Obj. 5.2*

Multiple Choice

5. A low explosive will explode only ________. *Obj. 5.1*
   a. when present in large amounts
   b. when in the solid state
   c. if exposed to a high explosive
   d. if contained
6. According to the kinetic molecular theory, which of the following statements best describes the particles that make up a gas? *Obj. 5.3*
   a. Gas particles move more quickly than particles in a liquid of the same substance.
   b. The kinetic energy of a gas decreases every time the particles within it collide.
   c. As the temperature of a gas increases, the kinetic energy of its particles decreases.
   d. There are strong attractive forces holding the particles of a gas together.
7. What characteristics are used to categorize explosives? *Obj. 5.3*
   a. how quickly they react
   b. how common they are
   c. how they are used
   d. a & c
8. ________ are used as primers. *Obj. 5.4*
   a. Low explosives
   b. Primary high explosives
   c. Secondary high explosives
   d. All of the above

Short Answer

9. What types of evidence may be revealed with a microscopic examination of explosive evidence? *Obj. 5.2, 5.5*

10. Explain the proper procedure for packaging explosion evidence. *Obj. 5.6*

11. What makes black powder a good choice to make safety fuses? *Obj. 5.2*

12. Define *terrorism* in your own words. Give two examples of terrorist acts that were not described in this chapter. Label your examples as acts of domestic or foreign terrorism. *Obj. 5.1*

13. Describe Alfred Nobel's contribution to the mining and construction industries. *Obj. 5.1*

14. If a gas is held at constant pressure and the temperature is increased, what happens to the volume? *Obj. 5.3*

15. Imagine that you have a helium-filled balloon on a warm day. The temperature is 360 K. The volume of the balloon is 23 L. The pressure is 101.3 kPa. Now, imagine that you carry the balloon into your air-conditioned home. The temperature changes to 293 K and the pressure remains constant. What is the new volume of the balloon? *Obj. 5.3*

____

____

____

16. A gas is contained in a volume of 20 L at 280 K and a pressure of 98 kPa. If the pressure increases to 105 kPa and the volume decreases to 10 L, what is the new temperature of the gas? *Obj. 5.3*

____

____

____

Think Critically

17. An explosive residue is subjected to several tests to determine its identity. If the Modified Griess Test reaction is inconclusive, what other test could be done to identify the explosive? Explain your answer. ***Obj. 5.1***

18. Explain how the characteristics of gases are related to the explosion that takes place when a balloon pops. ***Obj. 5.1***

19. Compare and contrast the components of various types of explosives. ***Obj. 5.3***

20. Explain the procedures for determining the identity of an explosive. ***Obj. 5.5***

"Fireworks" at the Fall Festival

By: **Cody Johnson**

Tavares High School
Tavares, Florida

It was a bright, sunny day in the park—a perfect setting for the Annual Fall Festival. All of the youngsters were getting excited about starting the scavenger hunt. The goal was to collect small metallic boxes that held special prizes (see Figure 5-21). John Acorn fired the blank shot to begin the hunt. Two hours later, the children were gathered around the tent to see who had collected the most boxes.

As volunteers traded the prize boxes for bikes and toys, the metallic—and reusable—boxes were thrown into a large container. Within minutes, the small packages began exploding! Hundreds of explosions sent small, sharp pieces of metal into the crowd of parents, children, and volunteers. Luckily, there were only minor injuries. However, the tent and some of the remaining prizes were not so lucky.

Figure 5-21. *These small metallic boxes held the scavenger hunt prizes.*

The paramedics treated the injured at the scene and only one—a volunteer—needed to be taken to the hospital. Police secured the area and forensic investigators moved in. After photographing the entire scene, CSI Mary Joe began the task of collecting the charred bits of colored metal. While Mary was looking for evidence, police officers were interviewing the witnesses. Only two of the volunteers had handled the boxes prior to the hunt. Don Rutabaga, dressed as a large dinosaur, had been in charge of hiding all of the boxes earlier that morning. Jimmy Buffay had been in charge of bringing the boxes to the park.

Detectives interviewed Buffay and Rutabaga. Buffay was uncooperative because he was supposed to be at a meeting of the local sharpshooters' club. Tests for gunshot residue on his hands and clothing were positive, but Buffay insisted he had been at the rifle range earlier. Other witnesses confirmed that. They also told investigators that Buffay was not too happy about being responsible for bringing the boxes.

Rutabaga, on the other hand, was eager to cooperate. Detectives asked him several questions before noticing some black residue on his dinosaur costume. Tests revealed it was gunpowder. Rutabaga was puzzled—he had no idea how the gunpowder got on his costume. He said he doesn't even own a gun. But Rutabaga freely admits he was the last person to have possession of the boxes.

After reviewing the witness statements, detectives discovered some other interesting facts. Buffay is a loner who got stuck with the responsibility of the boxes because all of the other club members had already volunteered for other duties. He was the last to volunteer, so he got the leftover duty. Rutabaga was a family man who had spent time in the military working as a demolitions expert. He didn't like to talk about his experiences much, but it was general knowledge that he would have made a career of it had he not gotten injured.

In the laboratory, microscopic examination of the pieces of metal revealed unburned grains of gunpowder. After rinsing the debris with acetone, a spot test revealed that nitroglycerin had also been inside some of the boxes. The boxes had been made of aluminum, a soft metal, and then painted to look like the prize packages. The scientists were able to put together a probable scenario. Someone had filled several of the aluminum boxes with gunpowder and then painted them to blend in with the "real" prize packages. One of the packages contained a small amount of nitroglycerin. When it was tossed into the large container, it had acted as a detonator for the gunpowder packages. But who had done it? Who would have had access to all that gunpowder as well as nitroglycerin? And why target kids?

Activity:

Answer the following questions based on information in the Crime Scene S.P.O.T.

1. Who do you think made the bombs? What was the motive?
2. Why didn't the gunpowder explode when the boxes were tossed into the container?
3. What role did the nitroglycerin play in the event?

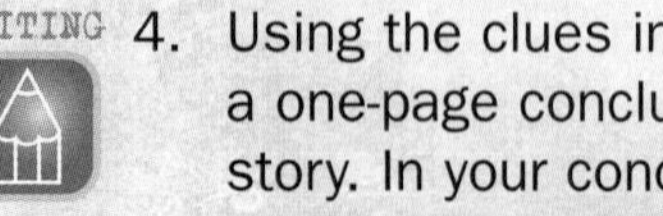

4. Using the clues in the story, write a one-page conclusion to the story. In your conclusion, describe the perpetrator's motive and explain the role each ingredient played in the bomb. Use proper terminology.

Preventing Adolescent Crime Together™ (P.A.C.T)

Introduction:

Every year, children and adults are injured when using fireworks in celebrations. In this activity, you will develop a public awareness brochure to educate others about the dangers and safe handling procedures of explosives, particularly fireworks.

Procedure:

Part 1

1. Research the following information regarding fireworks safety:
   - Current statistics on fireworks injuries/deaths
   - Tips for preventing accidents
   - Laws and regulations regarding the use of fireworks in your area. Be sure to look for local as well as statewide regulations.
2. After completing the research, your group will design an informational brochure for other students. The brochure must be easy to read, informative, creative, and well organized.

Part 2

1. Find out who the legislators are for your area.
2. Using the information from your research, write a persuasive letter to a legislator to:
   a. Request more specific regulations for use of explosive fireworks.
   b. Request stricter or more lenient legislation regarding use of explosive fireworks.
   c. Suggest a safety campaign or contest for students across the country. The winning entry could be presented before the next national celebration involving fireworks.

VAKHRUSHEV PAVEL/GETTY IMAGES

Figure 5-22. *Fireworks.*

ACTIVITY 5-1
3-2-1 LIFTOFF!

Objective:

By the end of this activity, you will be able to:
Describe the behavior of gases in an enclosed system.

Materials:

(per group of two students)
permanent marker
clear film canister or other small, plastic container with a lid
metric ruler
250 mL beaker
hot water
2 effervescent cold or indigestion tablets
stopwatch

Safety Precautions:

Safety goggles must be worn to protect eyes.
Be sure that the film canister is not directed toward anyone during the explosion step.

Procedure:

Before going outside:

1. Use the permanent marker to make a line 1 cm from the bottom of the film canister. Make sure the lid fits tightly.
2. Break each tablet into four equal pieces and set aside.
3. Fill a beaker about half full with hot water.

Once outside:

1. Fill your film canister with warm water up to the 1 cm line.
2. Drop a piece of tablet into the canister and quickly replace the cap tightly. Make sure your lab partner starts the stopwatch as soon as you drop in the tablet.
3. Place the canister upside down on a flat surface, such as a sidewalk or parking lot—away from windows, cars, and other people.
4. Move back and observe. Stop timing as soon as the container "pops." Record the length of time in Data Table 1.

5. Repeat with the other seven pieces of the tablets.
6. Calculate the average time before the explosion.

Data Table 1. Time.

Trial	Time in seconds
1	
2	
3	
4	
5	
6	
7	
8	
Average	

Questions:

1. What happened inside the canister to cause the explosion?
2. Would you have gotten the same result if the lid had been off or if it had fit poorly? Explain.
3. Were the times the same in each of the trials? What might account for the differences?
4. How could you improve the performance of your rocket?

ACTIVITY 5-2
GUNSHOT RESIDUE

Objectives:

By the end of this activity, you will be able to:

1. Distinguish between presumptive and confirmatory tests for gunshot residue.
2. Identify the necessity for a control sample.

Introduction:

Firing a weapon often leaves trace amounts of gunshot residue (GSR) on a person's hands. Therefore, GSR tests can be very helpful to crime-scene investigators. A positive test for GSR on clothing or on the skin of a suspect can indicate that the person has recently fired a weapon. GSR tests, however, may be ineffective if the person has washed his or her hands or clothing prior to the test. Also, the tests may give false positives. Tobacco, fertilizers, and even some cosmetics can yield positive GSR test results. Therefore, the tests are presumptive, and any positive results must be verified by more confirmatory tests.

In this activity, you will use two different tests to detect the presence of trace amounts of chemicals found in gunpowder. The first test detects nitrates using diphenylamine. A color change to blue indicates the presence of nitrates. Nitrates are found in GSR, but also in many common materials. The second test, the sodium rhodizonate test, detects certain metals—primarily barium and lead—found in GSR. A color change to deep red indicates the presence of these metals. False positives from environmental contaminants with this test are possible as well. When both tests are positive, you can be fairly certain GSR is present. However, more tests to provide confirmation are usually performed.

Materials:

Evidence packets for suspects 1, 2, 3, and 4 (each packet contains a piece of cloth from the suspect's shirt)
a control piece of fabric
5 alcohol swabs
5 Petri dishes
diphenylamine solution
sodium rhodizonate solution

Safety Precautions:

Wear aprons, gloves, and safety goggles when handling these chemicals! The diphenylamine solution contains sulfuric acid, which may cause burns. Handle carefully.
The alcohol swabs are flammable. Sodium rhodizonate may cause skin irritation.
If you are allergic to latex, alert your teacher so that you may use alternative gloves.
Wash your hands carefully at the end of this activity.

Procedure:

1. Carefully examine each piece of evidence. Record your observations in Data Table 1.
2. Clean each piece of evidence thoroughly with an alcohol swab. Use a different swab to clean each piece of fabric.
3. Place each swab on a clean Petri dish. Add one drop of the sodium rhodizonate solution to one end of each swab. Note any color changes. Record your results.
4. Add one drop of diphenylamine solution to the other end of each swab. Note any color change. Record your results. (Note: This solution contains sulfuric acid, which will cause the swab to turn black very quickly. The results must be read immediately.)

Data Table 1. Observations.

Sample	Visual Description	Reaction with Sodium Rhodizonate	Reaction with Diphenylamine
Control			
Suspect 1			
Suspect 2			
Suspect 3			
Suspect 4			

Questions:

1. What does it mean to say that a test is *presumptive*?
2. What is the purpose of the control sample?
3. Did any of the suspects test positive for GSR?
4. Did any test positive for one test but not the other? If yes, which one(s)?
5. Why would it be possible to test positive for GSR if no gunpowder were present?

ACTIVITY 5-3
MICROSCOPIC EXAMINATION

Objectives:

By the end of this activity, you will be able to:

1. Separate various components of "explosion debris" based on physical properties.
2. Identify the components of the explosion debris.

Materials:

(per two or three students)
Petri dish
triple beam balance or electronic scale
explosion debris sample
stereomicroscope
white unlined paper
forceps
paper towels

Safety Precautions:

Wear gloves and goggles when completing this lab.
If you are allergic to latex, alert your teacher so that you may use alternative gloves.
Wash your hands thoroughly after performing this activity.

Procedure:

1. Determine the mass of the Petri dish.
2. Place the entire debris sample into the Petri dish and find the mass.
3. Subtract the mass of the Petri dish from the total mass of the sample in the Petri dish. The solution is the mass of the debris sample alone.
4. Place the Petri dish under the stereomicroscope and observe.
5. Cut a white sheet of paper into six squares. Find the mass of each square.
6. Use the forceps to separate the explosion debris into its components. Place each component onto a separate square of paper. Label the squares according to the types of components you find. For example, some of your squares may be labeled "Organic Matter," "Metal," or "Wire." If your explosion debris falls into more than six categories, cut more squares of paper.

7. Determine the mass of the individual samples. Remember to subtract the mass of the appropriate square from the total mass of the sample.
8. Use the following formula to calculate the percentage of the total mass that is made up by each component.

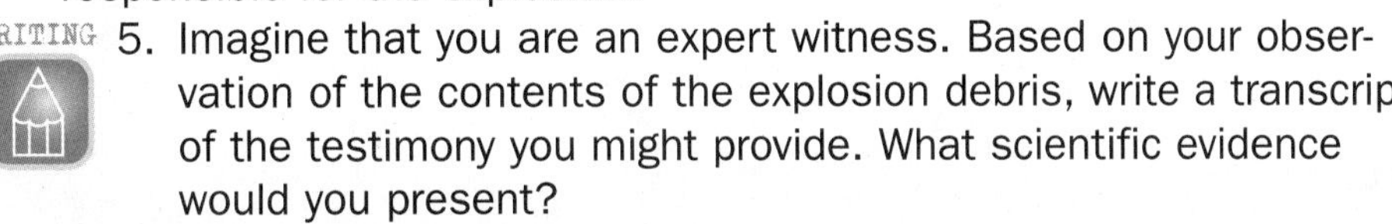

$$\frac{\text{Mass of component}}{\text{Total mass of explosion debris}} \times 100 = \text{percent by mass}$$

9. Design a data table that accurately represents your results.

Questions:

1. What components did you find in the explosion debris? What percentage of the total mass did each component represent?
2. What is the likely source of each component found in the explosion debris?
3. What was the most difficult part of separating the debris mixture?
4. How can evaluating microscopic debris aid in finding the person responsible for the explosion?

WRITING

5. Imagine that you are an expert witness. Based on your observation of the contents of the explosion debris, write a transcript of the testimony you might provide. What scientific evidence would you present?

Extension:

WRITING

Research cases of terrorist bombings around the world from the last year. Make a map showing areas of low, moderate, and high rates of terrorist attacks. Choose one case to study in detail. Write a newspaper or magazine article describing the bombing and the investigation. Be sure to describe the evidence investigators were able to find.

CHAPTER 6

Body Systems

LEANN FLETCHER

AP PHOTO/VAUGHN GURGANIAN, POOL

Figure 6-1. *Michael Fletcher*

Michael Fletcher was in the bathroom when he heard the gunshot. Fletcher ran out of the bathroom and found his wife, Leann, face down in a pool of blood on their bedroom floor. Michael called 911 right away. When police and EMS arrived at the scene, Fletcher led them to the bedroom. A 45-caliber Smith and Wesson automatic pistol was lying next to Leann's right hand. During an initial assessment of the scene and the victim, it appeared that she suffered from a gunshot wound to the head. The cerebral artery is the primary source of oxygen-rich blood to the brain. Even after the artery had been severed, the heart had continued to pump blood to the brain for several seconds. Because of this blood flow, very dark, thick blood surrounded the victim. Upon arrival, emergency medical personnel determined that the victim showed no sign of life.

The crime-scene unit arrived at the scene. After photographing the body, the medical examiner covered the victim's hands with bags to preserve trace evidence. Investigators collected and bagged the gun. The latent fingerprint expert dusted for fingerprints, and crime-scene investigators collected blood from the carpet.

Gravity caused the blood to pool in the parts of Leann's body that were touching the floor. Her thighs and face appeared bruised and bloated due to the pooling blood. Investigators estimated her time since death at just over an hour.

Michael Fletcher was the only suspect in the case. Although he told the same story several times to several investigators, the investigators were "unsettled" by his statement. He said that he and his wife had just returned from the shooting range when he went into the bathroom. He had given the bullets to his wife to reload his gun. He said he heard the shot, ran out of the bathroom, and found his wife on the floor in a pool of blood.

During the autopsy, the medical examiner performed a complete internal and external examination. Characteristics of the gunshot wound were consistent with being shot from 18 inches away. It would have been impossible for the victim to shoot herself from this distance.

Michael Fletcher was arrested and charged with murder. A jury found him guilty of second-degree murder. He was sentenced to life in prison. He is not eligible for parole until 2017.

NEWSCOM

Figure 6-2. Leann's parents, Gloria and Jack Misener, hold up a picture of Leann

OBJECTIVES

By the end of this chapter, you will be able to:

6.1 Discuss the structure and function of the circulatory system.

6.2 Evaluate the forensic implications of the circulatory system.

6.3 Discuss the structure and function of the respiratory system.

6.4 Evaluate the forensic implications of the respiratory system.

6.5 Discuss the structure and function of the muscular system.

6.6 Evaluate the forensic implications of the muscular system.

6.7 Identify body systems and discuss their forensic implications.

VOCABULARY

asphyxiation - a condition in which the amount of oxygen available to the lungs decreases sharply while the level of other gases, especially carbon monoxide, increases

erythrocyte - red blood cell

homeostasis - an organism's relatively stable internal conditions

leukocyte - white blood cell

lividity - pooling of blood in the lowest portion of the body

platelets - cell fragments that help form blood clots at wound sites; also called *thrombocytes*

rigor mortis - the stiffening of the skeletal muscles after death

suffocation - condition in which the amount of oxygen available to the lungs is quickly diminished

INTRODUCTION

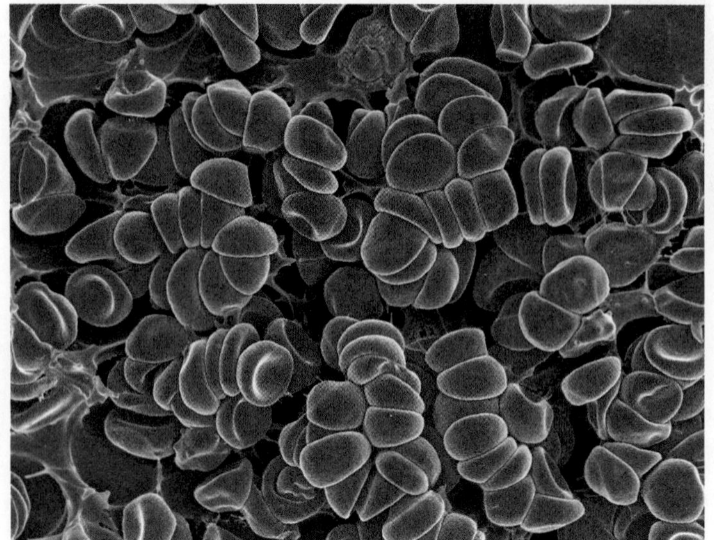

Figure 6-3. *Red blood cells.*

The study of the structure of the human body is called *anatomy.* The study of the function of the body systems is *physiology.* Anatomy and physiology are often studied together because the structure of a cell or organ supports its function. For example, red blood cells carry oxygen to all the cells of the body and return carbon dioxide to the lungs. The large surface area of a red blood cell (see Figure 6-3) allows the cell to absorb and release oxygen and carbon dioxide. The shape also allows the cells to be flexible as they move through the blood vessels, preventing clotting.

An understanding of anatomy and physiology is extremely important in forensic science. This knowledge allows scientists to interpret information about the crime from physical and biological evidence found at the crime scene.

Your body systems work together to keep your body temperature, blood pH, and all other physical and chemical conditions relatively constant. This stable internal environment is called **homeostasis.**

Obj. 6.1, 6.2

THE CIRCULATORY SYSTEM

The circulatory system pumps blood throughout the body, delivering oxygen and nutrients to cells. Blood also carries carbon dioxide and other cellular wastes away from cells. Because blood is so vital to sustaining life, the circulatory system is very important in a death investigation. The major structures of the circulatory system (see Figure 6-4) are the heart, blood, arteries, and veins. The system is divided into two pathways—systemic circulation and pulmonary circulation. The heart pumps oxygenated blood through arteries to the cells in *systemic circulation.* The cells absorb the oxygen from the blood and release carbon dioxide and other waste products into it. Some of these waste products are released in the kidneys and other organs for elimination. The systemic circulation pathway then carries the blood back to the heart. The heart pumps the deoxygenated blood, carbon dioxide, and other wastes to the lungs in *pulmonary circulation.* In the lungs, the blood releases carbon dioxide and other wastes and picks up oxygen. The oxygenated blood then flows back to the heart to enter systemic circulation.

PROPERTIES OF BLOOD

Blood serves many vital functions, including distributing heat and transporting oxygen to cells and wastes from them. Blood is composed of red blood cells, white blood cells, platelets, and plasma. Figure 6-5 shows a red blood cell, a white blood cell, and a platelet.

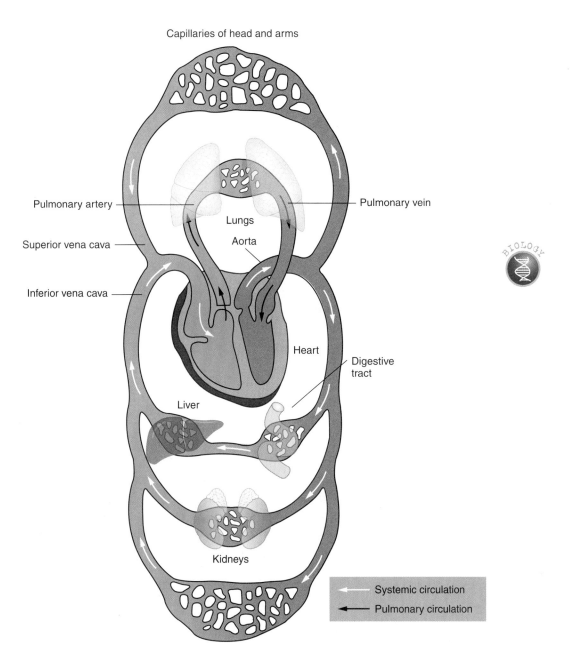

Capillaries of head and arms

Pulmonary artery

Pulmonary vein

Lungs

Superior vena cava

Aorta

Inferior vena cava

Heart

Digestive tract

Liver

Kidneys

Systemic circulation

Pulmonary circulation

Capillaries of abdominal organs and legs

Figure 6-4. *The circulatory system.*

BIOLOGY

Red Blood Cells

Red blood cells, called **erythrocytes,** carry oxygen throughout the body. The concave shape of red blood cells gives them a large surface area, which allows for efficient gas exchange. *Hemoglobin* is a protein in red blood cells. The hemoglobin binds to oxygen and transports it throughout the body. Red blood cells form in the bone marrow and live about 120 days. The production of red blood cells is controlled by the amount of oxygen in your body. If oxygen levels fall, the body produces more red blood cells. Oxygen levels fall when a person travels to higher elevations, where there is less oxygen. People who suffer from emphysema and other lung diseases may also have low oxygen levels. When oxygen levels return to normal, the body slows production of red blood cells.

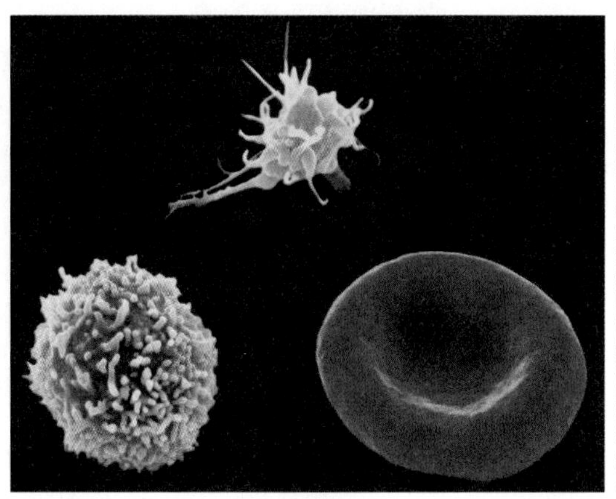

Figure 6-5. *Left to right: White blood cell, platelet, and red blood cell.*

CHAPTER 1 2 3 4 5 6 7 8 9 10 11 12 13 14 15

White Blood Cells

White blood cells are also called **leukocytes.** White blood cells protect the body against infection and fight viruses and bacteria. Typically, a healthy person has between 5,000 and 10,000 white blood cells in each microliter of blood. However, the body produces more white blood cells when it encounters a virus or bacteria. Therefore, doctors sometimes analyze white blood cell counts to check for infection. A white blood cell count of more than 10,500 may be an indication that the body is fighting an infection, such as tonsillitis or appendicitis. If the white blood cell count is below 3,500, the person's immune system is suppressed. Low white blood cell count can be caused by a variety of infections, including HIV, influenza, or malaria.

Platelets

Certain bone marrow cells divide to produce thousands of cell fragments. Each fragment is enclosed in a cell membrane. When these fragments leave the bone marrow and enter the blood, they are called **platelets.** Platelets are also called *thrombocytes.* When a blood vessel rips or tears, platelets help form blood clots. The platelets stick together at the wound site to form a plug. The platelets release chemicals called *clotting factors.* The clotting factors cause a series of reactions, including the production of thin, sticky filaments at the site of the wound. The filaments prevent more blood from escaping through the wound (see Figure 6-6). After the clot hardens, it becomes a scab. Patients with a relatively rare genetic condition called *hemophilia* have very low platelet counts. Without enough platelets, the body is not as able to form clots. Therefore, simple cuts can lead to severe bruising and internal bleeding. The loss of blood can cause death.

In addition to disease, certain medications and medical conditions affect the body's normal blood-clotting mechanism. Several other factors, including smoking, bed rest, and childbirth, increase a patient's risk of deep-vein thrombosis (DVT). DVT is a condition in which a blood clot forms in a deep vein, such as a vein in a leg. If this clot is carried to the heart or lungs, it can cause heart attack, stroke, or a pulmonary embolism. DVT can be caused by damage to a blood vessel wall or decreased blood flow. Physical trauma can cause either of these factors. Additionally, people who have recently

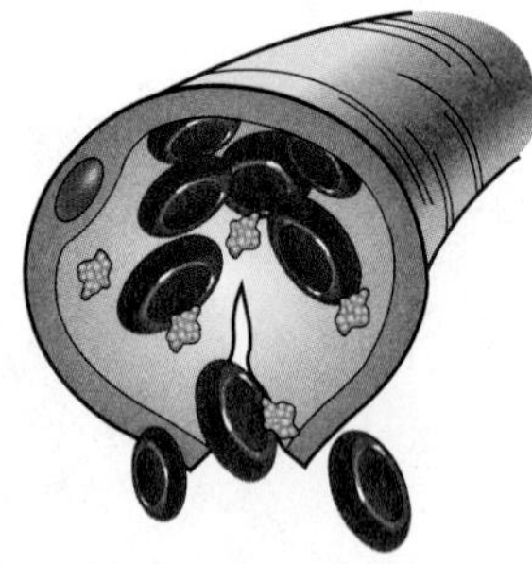

1 Cut or Scrape
A cut or scrape causes damage to the wall of a blood vessel.

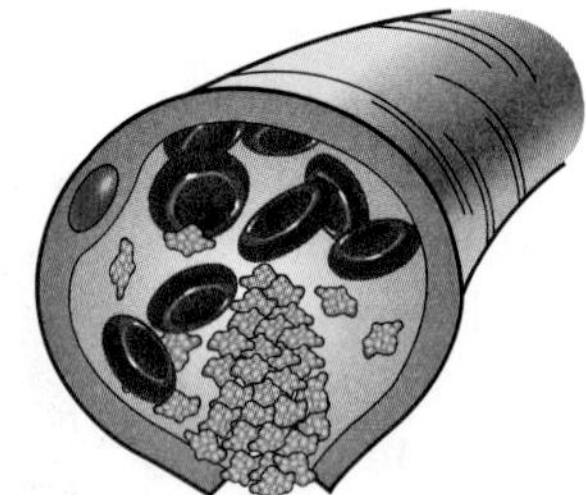

2 Platelets Arrive
Platelets from the blood enter the site and stick together. They release a clotting factor, causing a series of reactions.

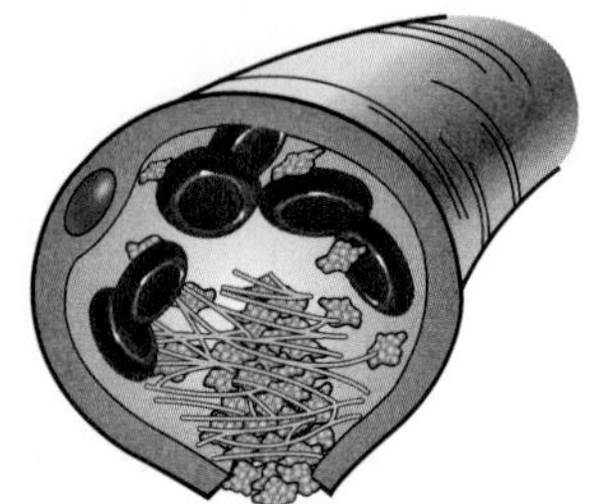

3 Filaments Form a Clot
The clotting factors cause the body to produce thin, sticky filaments at the site of the wound. These filaments form the clot, preventing more blood from escaping.

Figure 6-6. *Platelets are involved in the formation of blood clots.*

undergone surgery and are not mobile are at higher risk of DVT because their blood circulation has been affected. Certain prescription and over-the-counter medications, including aspirin, can reduce clotting. A doctor must be fully aware of a patient's medical history before prescribing or recommending medications.

Plasma

Blood *plasma* is the liquid portion of blood. Blood cells and platelets are suspended in the plasma. Plasma is 92 percent water. The rest of the plasma is made up of salts, nutrients, enzymes, proteins, dissolved gases, and other important compounds. Plasma is responsible for transporting nutrients, vitamins, and gases. It also regulates pH, maintains fluid balance, and controls body temperature.

PATH OF BLOOD THROUGH THE HEART

The human heart is divided into four chambers. The upper chambers are called *atria* and the lower chambers are called *ventricles.* The right atrium receives the deoxygenated blood returning back to the heart. The blood moves into the right ventricle through the tricuspid valve (see Figure 6-7) and is pumped into the pulmonary artery to the lungs. Capillaries—the smallest blood vessels—surround tiny air sacs in the lungs. The air sacs are called *alveoli.* Carbon dioxide diffuses from the blood in the capillaries into the alveoli. Oxygen diffuses from the alveoli into the blood in the capillaries. This process is called *gas exchange.* The oxygenated blood then returns to the left atrium via the pulmonary vein. From the left atrium, blood flows into the left ventricle through the bicuspid valve. After the blood moves into the left ventricle, it travels to the aorta and then to the rest of the body. Blood travels away from the heart toward the rest of the body through arteries. Blood flows back to the heart through veins.

The human heart is about the size of your clenched fist. It beats approximately 30 million times per year and pumps blood through more than 60,000 miles of blood vessels.

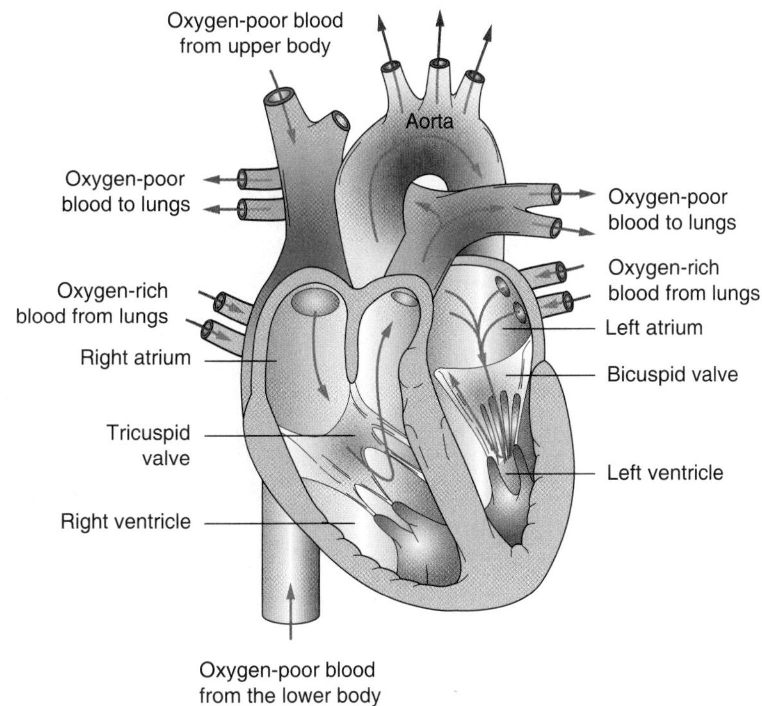

Figure 6-7. *Diagram of the human heart.*

FORENSIC IMPLICATIONS OF THE CIRCULATORY SYSTEM

The circulatory system is important to forensic scientists for many reasons. Blood found at a crime scene can be used to identify the victim or the perpetrator. Blood type is a class characteristic. For example, if the victim had O blood, finding AB blood at the crime scene suggests that the perpetrator had that blood type. Many people may have that blood type, but suspects with any other blood type can be excluded. Nuclear DNA found in the blood can be used as individual evidence. Red blood cells do not have a nucleus or DNA. However, white blood cells have a nucleus and nuclear DNA. Except for identical twins, each person's nuclear DNA is unique. Therefore, DNA can be used to identify the victim or the perpetrator.

BIOLOGY

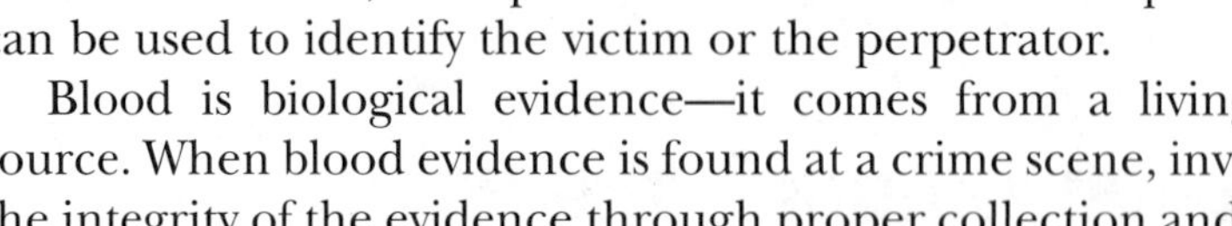

Blood is biological evidence—it comes from a living or once-living source. When blood evidence is found at a crime scene, investigators ensure the integrity of the evidence through proper collection and evaluation techniques. As you may have learned in previous forensic science coursework, when a red stain is found, investigators must ask three basic questions:

- Is it blood?
- Is it human blood?
- If it is human blood, can the blood be traced to a single person?

JIM VARNEY/PHOTO RESEARCHERS, INC.

Figure 6-8. *The reagent on these strips reacts with hemoglobin in a presumptive test for blood.*

Hemastix® provides a presumptive test used to indicate whether the red stain might be blood (see Figure 6-8). Hemastix are plastic strips that have been treated with a special blood reagent. The strips are moistened with distilled water. When crime-scene investigators find a red stain at a crime scene, they touch the stain with the tip of the moistened strip. If the stain is blood, the reagent on the test strip turns green. The iron in hemoglobin acts as a catalyst in this reaction. A *catalyst* is any substance that speeds up a chemical reaction. If investigators need to test a large area for the possible presence of blood evidence, they can use a different presumptive test—luminol. Luminol is mixed with hydrogen peroxide. During the reaction with hydrogen peroxide, the luminol is oxidized. The electrons in the oxygen atoms are energized—they move to a higher orbital. As these electrons release the additional energy and return to the lower orbital, they emit light (see Figure 6-9). Therefore, the luminol test is viewed in a darkened area. The leucomalachite green presumptive test for blood is based on this same reaction. In the presence of iron, the leucomalachite green turns blue-green. A solution of phenolphthalein is used in the presumptive Kastle-Meyer test. This solution turns pink in the presence of traces of blood.

CHEMISTRY

PHOTO BY JOHN KIRK-ANDERSON/GETTY IMAGES

Figure 6-9. *This attorney is holding a photograph of the results of an luminol test. The test shows traces of blood on a footprint.*

After investigators have fully photographed and documented the scene, they collect samples of possible blood evidence (see Figure 6-10) to process at the lab. In order to avoid cross-contamination of the evidence, each item of evidence is packaged separately in a paper bag. If bloody hairs and fibers are found, investigators make sure that the evidence is evaluated by all labs—serology, DNA, and trace evidence. If the blood evidence is found on a small object, such as a pencil or soda can, the object is packaged and the blood evidence is removed at the lab. If the blood evidence is found on a large object, such as a door or wall, only the blood evidence is collected. Investigators collect dried blood in a variety of ways. They often use a wet swab to remove the blood. Alternatively, they might place fingerprint tape over the blood and lift the stain, or they might use a sharp instrument to scrape blood into a paper bag. Investigators use similar methods to collect and package wet blood. However, the item will first need to be air dried. Drying the blood prevents mold or other microorganisms from forming. Microorganisms can destroy the evidence.

Figure 6-10. *Blood evidence.*

When a suspect is apprehended, evidence is collected and identified. This evidence, such as blood samples, acts as a *control* or *known sample.* Comparative tests will be done on the samples from the crime scene and the control samples to determine whether they are consistent with each other, possibly connecting the suspect to the crime scene.

Identification

At the crime lab, processing blood evidence to produce a DNA profile requires sophisticated equipment and highly skilled forensic experts. DNA evidence is considered one of the most compelling pieces of evidence at trial and can be important in the process of convicting or exonerating a suspect. The techniques used to develop a DNA profile will be discussed in depth in Chapter 10.

Lividity

When the heart stops pumping blood through the body, blood will travel in the direction of gravity. As the blood collects, the body changes color, especially in parts that are touching the ground. The pooling of blood in the direction of gravity is called **lividity.** The postmortem change in color caused by lividity is called *livor mortis* (see Figure 6-11). Lividity begins to appear about 30 minutes to 2 hours after death and continues to become more apparent and darker for up to 12 hours. For the first few hours after death, livor mortis is not fixed. If the area of lividity is pressed, the pressure will push the pooled blood away from the spot. For a moment, the skin will lighten again. When the pressure is removed, the blood quickly pools again. After approximately 12 hours, however, the lividity becomes fixed—irreversible and permanent. As the body cools, the layer of fat that surrounds the capillaries hardens. At this point, pressing on the area of

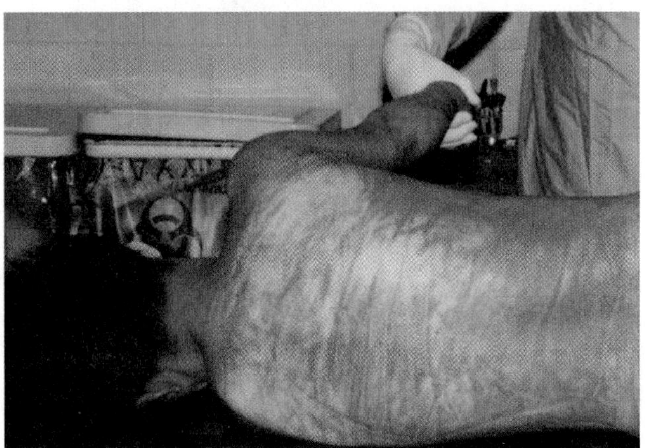

Figure 6-11. *The lighter areas of livor mortis indicate that this person's shoulders were touching a hard surface.*

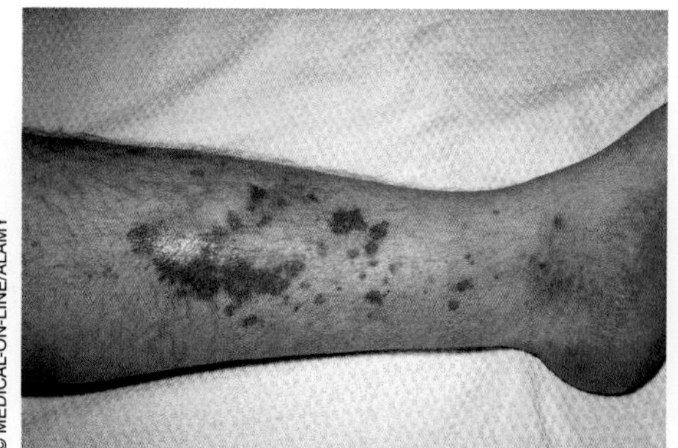

Figure 6-12. *Petechiae.*

lividity no longer moves the pooled blood or changes the color of the skin. Because lividity becomes fixed, forensic investigators can use it to determine whether the body was moved after the victim died. Within the first hours after death, patches of lividity in different areas of the body indicate the body has been moved. After lividity has become fixed, lividity that is not consistent with the position in which the body was found is an indication that the body was probably moved.

If an arm or leg is in a hanging position, such as over a bed or chair, when a person dies, petechiae may appear. *Petechiae*, small red dots underneath the surface of the skin, occur as capillaries near the skin rupture (see Figure 6-12).

Livor mortis is normally a bluish-purple or reddish-purple color on parts of the body closest to the ground or floor. The color becomes darker as time passes because after death, oxygen begins to separate from hemoglobin. This change produces a purple pigment—deoxyhemoglobin—in the red blood cells. Areas away from the ground and at the edge of the lividity tend to be pink. However, variations in the color of the lividity can provide clues to the cause of death. For example, a victim of carbon monoxide poisoning is likely to show bright red lividity. Because cold temperatures slow the formation of deoxyhemoglobin, lividity in a victim of hypothermia will be bright pink. A body refrigerated shortly after death and the body of a victim of cyanide poisoning also exhibit bright pink lividity. Dark brown lividity indicates exposure to lethal doses of nitrates, aniline, and potassium chlorate. These chemicals are used in the manufacture of various materials including herbicides. In the presence of these chemicals, the body forms meth-hemoglobin rather than deoxyhemoglobin. This pigment causes the brown coloring. To make a final assessment of the cause of various livor mortis, investigators rely on lab analysis of the blood and information from the death scene and autopsy reports.

Crime-scene investigators sometimes misinterpret lividity as bruising or signs of trauma. In such cases, an autopsy easily confirms the discolorations as livor mortis. Bruising is the result of blood leaking into extracellular spaces—the spaces between cells. If the discoloration is indeed lividity, the blood will be confined to the blood vessels.

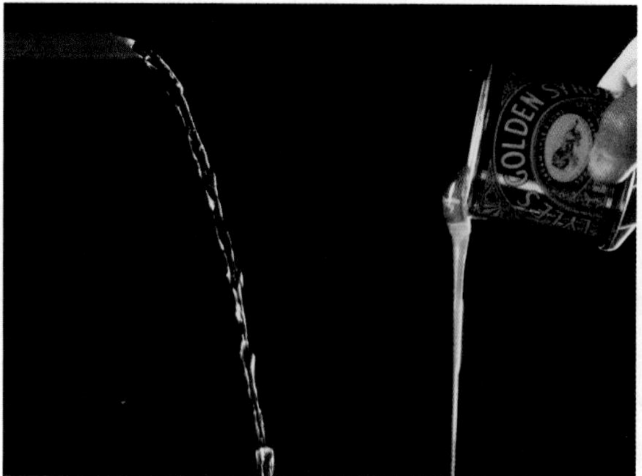

Figure 6-13. *Viscosity of various liquids. Notice that syrup has a high resistance to flow, a high viscosity. The viscosity of water is much lower.*

Blood Spatter

If a victim is violently attacked, the blood-spatter evidence can give the investigators clues about the attack. The spatter evidence might help investigators determine the position of the victim and the perpetrator. Some criminal investigation units have blood-spatter experts who provide analysis of the blood evidence. These experts use the laws of physics to interpret the blood spatter. Blood is viscous. Viscosity describes a liquid's resistance to flow (see Figure 6-13). Water, for example, has low viscosity. On the other hand, blood has higher viscosity, and thick syrup has very high viscosity.

Blood has high surface tension. Surface tension allows the blood spatter to retain much of its shape when it comes in contact with another object. Once blood leaves the body and travels through the air, gravity pulls the blood downward. The *velocity*, or speed and direction, that the blood is traveling as it exits the body combines with gravity to produce a certain path. If a person continues to move after he or she begins losing blood from a stab wound, the blood is likely to fall straight down, producing large droplets of blood on the ground. This blood spatter is said to be low-velocity spatter (see Figure 6-14). This type of spatter will likely form a pool of blood. Medium-velocity blood spatter (see Figure 6-15) can be caused by blunt-force trauma, such as injuries caused by a baseball bat. This type of trauma causes blood to spurt out of the body. A third pattern, known as arterial spray, is based on the heartbeat pattern. Gunshot wounds produce high-velocity blood spatter. This type of spatter looks like tiny droplets, similar to a fine spray (see Figure 6-16). Blood-spatter evidence is useful in helping investigators determine the position of the victim at the time of the crime; the type of weapon or tool used; how many times the victim was hit, shot, or stabbed; and whether the victim moved after the assault. After evaluating the blood spatter, investigators may be able to determine the following:

- What events transpired during the crime
- The sequence of events
- Who was or was not present

With this information, criminal investigators can compare witness statements with the findings and look for misleading or contradictory information.

ANDY CRAWFORD/GETTY IMAGES

Figure 6-14. *Low-velocity blood spatter.*

ANYKA/SHUTTERSTOCK

Figure 6-15. *Medium-velocity blood spatter.*

THE RESPIRATORY SYSTEM

Obj. 6.3, 6.4

The primary functions of the respiratory system are bringing oxygen into the body and removing carbon dioxide and other gaseous wastes from it. The process of gas exchange is called *respiration*. The respiratory system interacts directly with the circulatory system. The respiratory system brings air into the lungs. The oxygen gas (O_2) moves across the alveoli and capillary membranes into the bloodstream (the circulatory system). Carbon dioxide (CO_2) moves into the lungs from the blood across the same membranes.

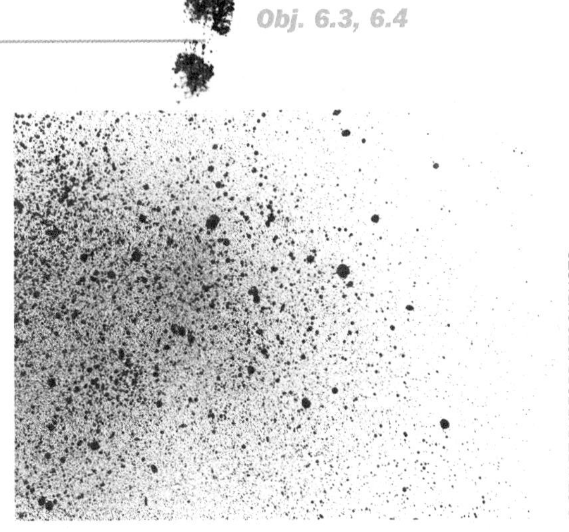

© STUART BURFORD/ISTOCKPHOTO

Figure 6-16. *High-velocity blood spatter.*

ORGANS OF THE RESPIRATORY SYSTEM

The organs of the respiratory system (see Figure 6-17 on page 172) are divided into two tracts—the upper respiratory system or tract and the lower

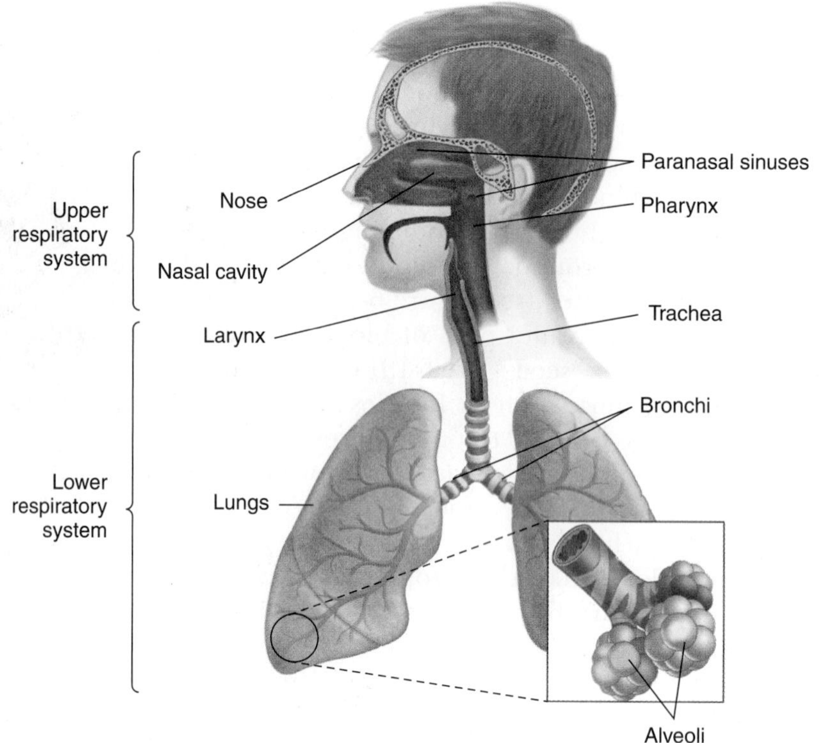

Figure 6-17. *The respiratory system.*

Did You Know?

The choking game is blamed for up to 1,000 adolescent deaths each year in the United States. The victims, typically 9 to 16 years of age, restrict blood flow to the brain with a belt, rope, or even their bare hands. When the pressure is released, blood rushes into the brain, producing a feeling of euphoria. If the person passes out before the pressure is released, blood flow is not restored and the person dies. Playing the game in any form causes death of brain cells and may cause seizures, strokes, or permanent brain damage.

respiratory system or tract. The upper respiratory tract includes the nose, nasal cavity, paranasal sinuses, and pharynx. The lower respiratory tract includes the larynx, trachea, bronchi, and lungs.

The nose is made up of bone and cartilage. The two openings through which air can enter and exit are called *nostrils.* Tiny hairs inside the nostrils prevent large particles from entering the body. Some dangerous particles are able to enter the body through the nose even with the protection of the tiny hairs in the nose. For example, anthrax is a bacterium that is only 0.5 μm wide. People coat the bacterial spores with a powder to produce an airborne bio-weapon. Theses spores can get through the hairs in the nose and be inhaled into the lungs where the deadly bacteria are released. Behind the nose is a hollow area called the *nasal cavity.* The nasal septum divides the nasal cavity into two distinct sections.

The pharynx, also known as the *throat,* is the bridge between the mouth and the esophagus. The pharynx also supports the production of sounds and speech.

The *larynx* is the large area slightly below the pharynx. The larynx is composed of muscle and tissue. It allows air to move into and out of the trachea and reduces the risk of toxic substances entering the lungs. The vocal cords are found on the larynx. Inside the larynx are two pairs of horizontal vocal folds. The upper folds contain the false vocal cords. These do not produce sound. Instead, the muscles in the false vocal cords close the airway when a person swallows. The lower folds contain the true vocal cords. As air reaches these cords, the cords vibrate. The vibration generates sound waves. The *glottis* and the *epiglottis* are additional structures within the larynx. These structures allow food and liquids to pass through the larynx without entering the trachea and air passages. The *trachea* is a long, narrow tube that travels from the larynx to the bronchi. Mucous-coated cilia, or short, hair-like projections, line the walls of the trachea.

Upper Respiratory System		Lower Respiratory System	
Structure	**Function**	**Structure**	**Function**
Nose	Contains nostrils with tiny hairs to prevent large particles in the air from invading the body	Larynx	Moves air into and out of the trachea and helps ensure that toxic or foreign substances do not enter the body; where vocal cords are located
Nasal cavity	Contains mucous membranes to increase surface area; assists in adjusting the temperature of the inhaled air to the internal body temperature	Trachea	Long tube that connects the larynx and bronchi; cilia filter air and move foreign particles into the pharynx and coat the particles with mucus to be swallowed
Paranasal sinuses	Reduce the weight of the skull; affect the overall quality of the voice	Bronchi	Two large tubes that bring air from the trachea into the lungs
Pharynx	Moves food to the esophagus; aids in producing sound and speech	Lungs	Contain bronchi; in the lungs, bronchi divide into smaller and smaller tubes until they reach alveoli, which are surrounded by capillaries; gas exchange occurs between the alveoli and the capillaries

Figure 6-18. *Structures and functions of the respiratory system.*

The mucosa traps unwanted particles and moves them up and into the pharynx. Once in the pharynx, the unwanted material in the mucus can be swallowed.

Air moves from the trachea into two large tubes, called *bronchi* (singular, *bronchus*). Each bronchus leads into a lung. The bronchi branch into two distinct parts—the left bronchus and the right bronchus. These bronchi continue to branch into smaller and smaller tubes and structures inside the lungs. Capillaries line tiny air sacs, called *alveoli,* at the end of the bronchi. Gas exchange takes place between the alveoli and the capillaries. Oxygen diffuses across cell membranes from the alveoli into the blood in the capillaries. Carbon dioxide diffuses from the blood into the alveoli.

The right lung is divided into three lobes and is larger than the left lung. The left lung is divided into only two lobes. The lungs contain air passageways, alveoli, blood vessels, and nerves of the lower respiratory tract. Figure 6-18 summarizes the structures and functions of the respiratory system.

FORENSIC IMPLICATIONS OF THE RESPIRATORY SYSTEM

The process of using oxygen and producing carbon dioxide on a cellular level is called *cellular respiration.* Cellular respiration involves many individual reactions. However, the overall process can be summarized by the following equation:

$$6O_2 + C_6H_{12}O_6 \longrightarrow 6CO_2 + 6H_2O + \text{Energy}$$

$C_6H_{12}O_6$ represents glucose, which the body gets from food. The oxygen gas from the air you breathe combines with glucose to provide your cells with energy they need to carry out all life processes. The chemical energy released through cellular respiration is stored in a molecule called adenosine triphosphate, or ATP. ATP is made up of adenine, ribose (a five-carbon sugar), and three phosphate groups (see Figure 6-19). ATP's energy is stored in the chemical bond between the second and third phosphate groups. When the bond is broken, energy is released and can be used by the body.

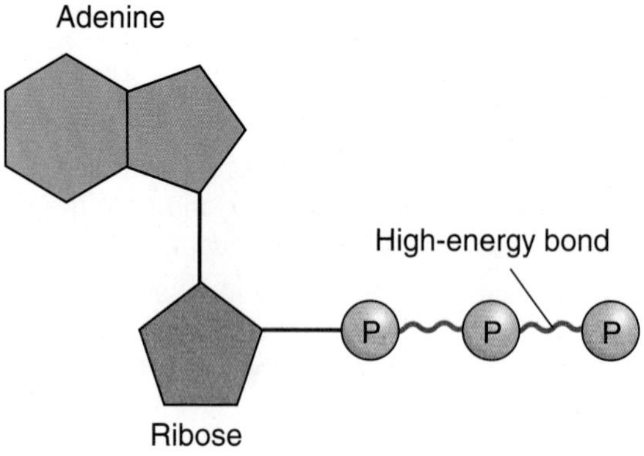

Adenine
High-energy bond
P P P
Ribose
ATP

Figure 6-19. *Molecule of ATP.*

Without oxygen, a person will eventually die. For example, in a fire, a person may die of **asphyxiation,** a condition in which the amount of available oxygen decreases sharply while the level of toxic gases (especially carbon monoxide) increases.

As the carbon dioxide levels in a person's blood increase, the brain stimulates the body to inhale. The higher the carbon dioxide levels, the stronger the signals from the brain become. Eventually, the person cannot keep from breathing. Therefore, smoke from a fire can be lethal if the victim is not rescued quickly.

Suffocation can occur when the amount of available oxygen is stopped or slowed quickly. Sealing a plastic bag over a person's head blocks available oxygen. The person will soon pass out. Without immediate medical assistance, the victim will likely die. If a person is found in water, a medical examiner might be able to determine whether the victim drowned or was placed in the water after death. If the medical examiner finds water in the lungs that is consistent with the water where the victim was found, the victim likely drowned. If, however, the medical examiner determines that there is no water in the lungs, the victim might not have drowned. Instead, the body might have been placed in the water after death.

Digging Deeper

with Forensic Science e-Collection

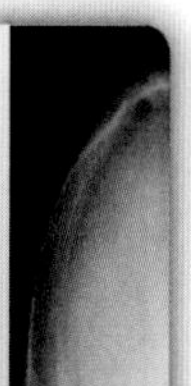

One law-enforcement technique used to control violent suspects is physically restraining them so they cannot move. The feet and hands are bound together behind their backs, and suspects are placed face down. Unfortunately, there have been some deaths reported from positional asphyxia when this type of restraint is used. Using the Gale Forensic Science e-Collection at www.cengage.com/school/forensicscienceadv, research positional asphyxia. Then answer the following questions: (1) What changes in procedure have been made since the deaths have occurred? (2) What other contributing factors are implicated in the deaths? (3) How do medical examiners make proper judgment as to cause of death in such cases?

THE MUSCULAR SYSTEM

Obj. 6.5, 6.6

Muscles provide the most basic conscious and unconscious movements. Voluntary movements are movements you intend to make and include walking, waving, and many facial movements. Involuntary movements are movements, such as breathing and digestion, which you do not control completely.

Did You Know?

The tongue is the only muscle in the body attached at only one end.

TYPES OF MUSCLES

There are three types of muscle—skeletal, smooth, and cardiac (see Figures 6-20 and 6-21). Skeletal muscle is responsible for voluntary movement and is attached to bone. Two kinds of protein—actin and myosin—form fibers within muscles. These materials are called *myofibrils.* The organization of the myofibrils in skeletal muscles produces light and dark bands, or striations. The myosin and actin work together to allow a muscle to contract and relax. Smooth muscle is found in internal organs, such as the stomach or bladder. Smooth muscles have no striations and are not under voluntary control. Cardiac muscle is found only in the heart. It is striated and very complex. Like smooth muscle, cardiac muscle is not under voluntary control.

FORENSIC IMPORTANCE OF THE MUSCULAR SYSTEM

When a muscle is relaxed, the actin and myosin are not connected. However, when a muscle contracts, the myosin attaches to the actin, forming a bridge.

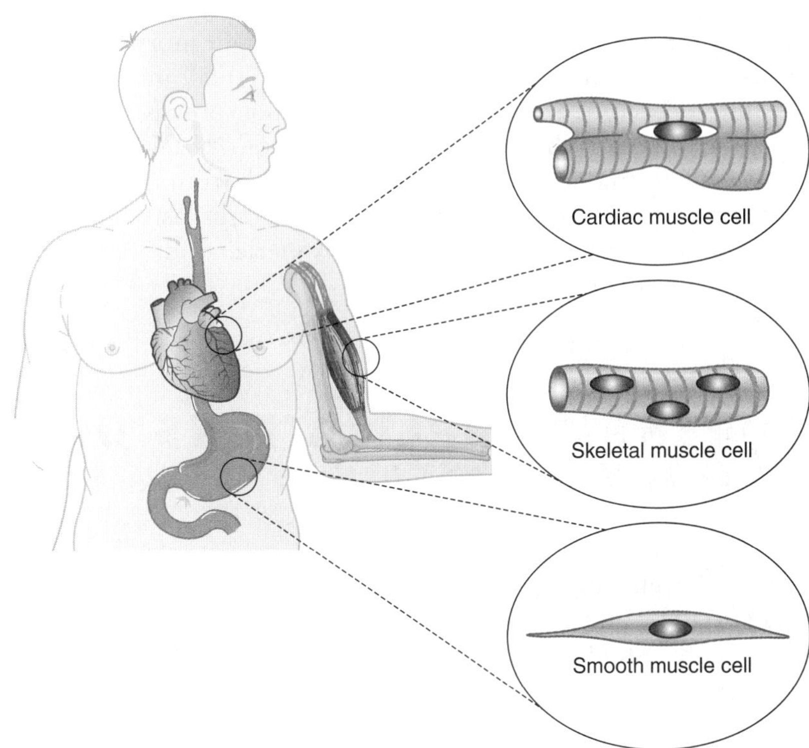

Figure 6-20. *Cardiac, skeletal, and smooth muscles.*

A.

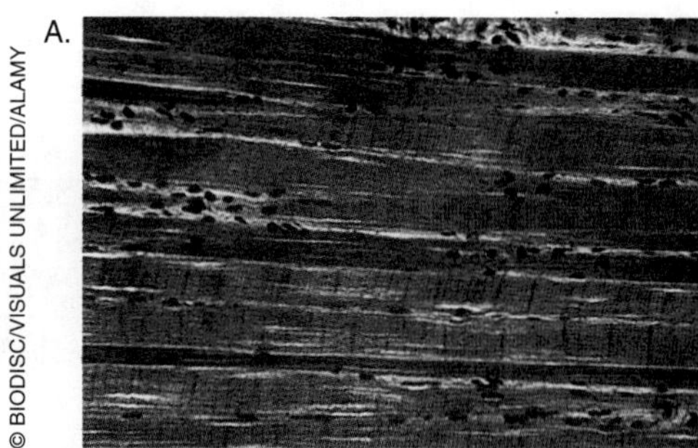

B.

C.

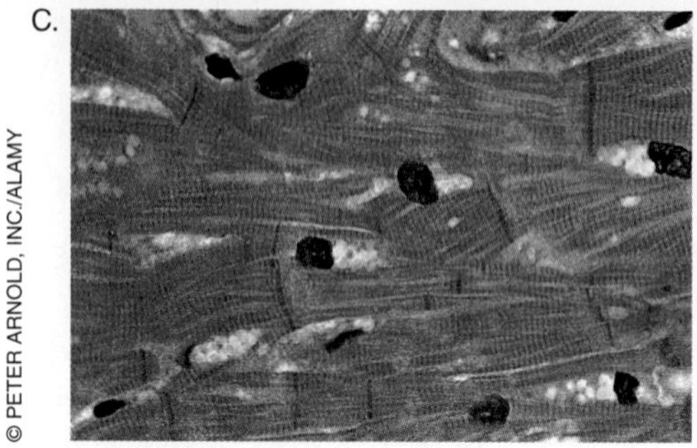

Figure 6-21. *Photomicrographs of skeletal muscle at 80× (A), smooth muscle at 140× (B), and cardiac muscle at 250× (C).*

Rigor can set in more quickly in victims of drowning. The victim likely struggled vigorously to survive. The struggle used up the muscles' energy source—adenosine triphosphate, or ATP. Likewise, when a victim is chased prior to death, his or her leg muscles might show evidence of rigor more quickly than other muscles.

The bridges pull the actin, causing the muscle to get shorter and shorter, or to contract. The cell uses energy in the form of adenosine triphosphate (ATP) to break and rebuild the bridges.

Because oxygen is no longer present after death, no additional ATP is produced in the cells. Actin and myosin are no longer able to relax to extend the skeletal muscles, and the muscles become stiff. The process in which the muscles of a body begin to stiffen is called **rigor mortis**. This process begins about two hours after someone dies. The small muscles in the face are the first to become stiff, and rigor eventually spreads throughout the entire body in about 12 hours. Rigor can last from 24 to 48 hours. As the body tissues begin to decay, digestive enzymes leak from the lysosomes in the cells and the muscles relax. The length of time rigor lasts depends on the condition of the body at death and on atmospheric conditions, such as temperature and humidity.

THE FORENSIC IMPLICATIONS OF OTHER BODY SYSTEMS

Obj. 6.7

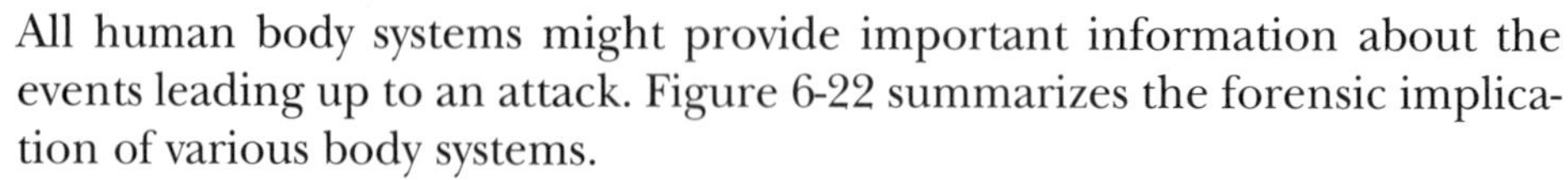

All human body systems might provide important information about the events leading up to an attack. Figure 6-22 summarizes the forensic implication of various body systems.

Body System	Function	Organs	Forensic Connection
Integumentary	Protects the body; prevents water loss	Skin, hair, nails, glands	• Bruising is often the first visual sign of trauma. • Elder abuse can be caused by a patient not being turned enough, resulting in bedsores. • A diaper left on an infant/elderly person too long can lead to ulcers. • HaXir is often found at a crime scene and can offer a link to the crime.
Skeletal	Protects internal organs; supports the body	Bones, cartilage, ligaments, tendons	• Fused epiphysis may indicate that a skeleton is that of an adult. • A pelvic girdle can help a forensic anthropologist estimate sex. • A broken hyoid indicates that a victim might have been strangled. • A crushed skull can indicate blunt-force trauma.
Nervous	Receives stimuli; directs the body via nerve impulses	Nerves, sensory organs, brain, spinal cord	• Some drugs or toxins can slow down nerve impulses while others can stimulate them. When the toxin is unknown, the victim's symptoms might help narrow down possible causes of illness or death. • Driving while intoxicated slows response times and often leads to accidents.
Endocrine	Regulates body functions using hormones	Pituitary gland, adrenal glands, thyroid	• Victims of violent crime can recover physically but might suffer long-lasting psychological effects, which could lead to post-traumatic stress disorder.
Digestive	Processes and absorbs food	Mouth, esophagus, stomach, intestines, liver, pancreas	• Stomach contents and undigested food can aid investigators in building a timeline of events leading up to the crime • Teeth can serve as a method of identifying an individual.
Urinary	Removes metabolic waste materials from the blood	Kidney, bladder	• Toxicological analysis for poisons, drugs, and alcohol can assist the medical examiner in determining cause of death. • Random drug testing is utilized in various professions. • In cases of sexual assault, a screening for various date rape drugs can be completed.
Reproductive	Carries out reproduction	Testes, ovaries, and associated organs	• Semen found at a crime scene indicates that a sexual assault may have occurred. It may be possible to retrieve DNA and make an individual identification.

Figure 6-22. *Systems of the human body.*

CHAPTER 1 2 3 4 5 6 7 8 9 10 11 12 13 14 15

Digging Deeper
with Forensic Science e-Collection

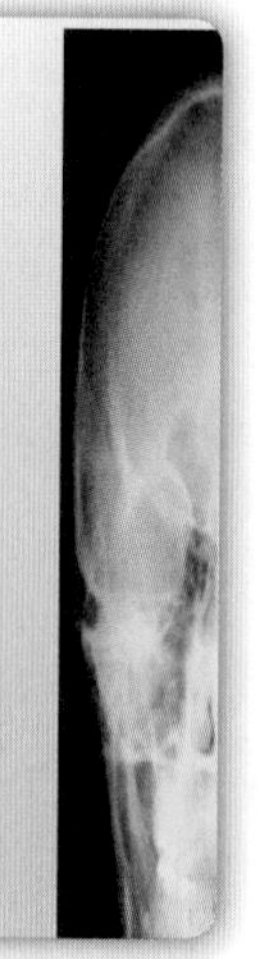

Hair is one of the unique characteristics that separate mammals from other animals. Although the structure of hair in all mammals is the same—consisting of a medulla, cortex, and cuticle—there are differences between human hair and hair of other mammals. Using the Gale Forensic Science e-Collection at www.cengage.com/school/forensicscienceadv, research articles about hair as forensic evidence. Then answer the following questions: (1) Why are results from hair composition analysis controversial? (2) What can hair found at the crime scene tell investigators? Write a one-page scenario in which hair is used as evidence.

CHAPTER SUMMARY

- Each body system can offer clues regarding the events that led up to or occurred during the crime.
- The circulatory system is divided into two pathways—systemic circulation and pulmonary circulation. Primary structures include the heart, blood, arteries, and veins.
- The primary role of blood is to transport oxygen to the cells and wastes from them. Blood is composed of red blood cells, white blood cells, platelets, and plasma.
- Gas exchange occurs in the alveoli of the lungs. Oxygen diffuses into the capillaries to be transported to the body's cells and carbon dioxide diffuses out of the capillaries into the lungs to be exhaled.
- Lividity occurs when blood pools in the lowest part of the body after a person dies. Lividity provides clues about time since death and the position of the body at the time of death.
- The primary function of the respiratory system is to remove gases and other wastes from the blood and to transport oxygen.
- The respiratory tract is divided into the upper respiratory tract and the lower respiratory tract.
- Asphyxia is a condition that occurs when the amount of available oxygen decreases while toxic gases increase. Suffocation is a form of asphyxia that occurs when the amount of available oxygen decreases.
- Rigor mortis is the process where the muscles of the body begin to stiffen after death. Rigor begins within two hours after death and can last from 24 to 48 hours.

CASE STUDIES

Mark Unger (1998)

AP PHOTO/CHARLES V. TINES, POOL

Figure 6-23. *Mark Unger*

Mark and Florence Unger appeared to have the perfect marriage. They lived in an upscale neighborhood in a Detroit suburb. Florence was a stay-at-home mother for their two boys. However, in 1998, things began to fall apart. Mark hurt his back and became addicted to the painkiller Vicodin. Soon, he was hooked on alcohol and gambling. When Mark stopped working, Florence returned to work to pay their mounting debt. In October 2003, the Unger family took a trip to Watervale, a lake resort in northern Michigan.

One evening during the trip, the boys were watching a movie in the cabin while Mark and Florence went down to the boat dock. Mark left Florence at the dock to put the boys to bed. When he returned 15 minutes later, Florence was gone. He assumed she was at the neighbor's cottage, so he went back inside and went to bed. When Florence was still not back by morning, Mark called the neighbors, Linn and Maggie Duncan, to help him search. Maggie and Linn found Florence face down in the water and called 911.

In May 2004, Mark Unger (see Figure 6-23) was arrested and charged with premeditated murder. Mark passed a polygraph, but prosecutors were still suspicious. Most of the evidence against him was circumstantial. For example, he had not been told where the body was, but he went straight to the spot where Florence was found. He packed the car and left the resort before his wife's body was pulled from the water. He sobbed but did not shed any tears. Florence had a broken hip, internal injuries, and a fractured skull. These injuries indicated that she had fallen onto the concrete before she entered the water. Blood on the concrete several feet from the water confirmed that she had fallen. The autopsy also indicated that the head wound would have knocked her unconscious. Therefore, investigators suspected that Florence had been dragged into the water, where she ultimately drowned. Prosecutors were certain that Mark put her in the water after he pushed her over the railing onto the concrete.

Mark Unger was convicted and sentenced to life in prison without the possibility of parole.

Think Critically

1. **What kinds of tests might investigators have performed on the blood found on the concrete? How might these tests indicate that Florence had fallen before entering the water?**
2. **What information did forensic scientists likely gather from examining Florence Unger's lungs? What conclusions might detectives draw from this evidence?**

Anthrax Scare (2001)

In the fall of 2001, several Americans were diagnosed with anthrax infection. Some of these victims suffered from inhalation anthrax. Victims breathe the bacteria into the lungs. Fluid builds up in the lungs and in the sac around the heart, making breathing difficult and leading to organ failure. Even with aggressive antibiotic treatments, 20 percent of victims of inhalation anthrax die. After a string of news-affiliate employees were diagnosed with anthrax, postal workers began to exhibit symptoms of the bacteria. Although most postal workers made a full recovery, a few fell victim to the bacterial toxin. In all, five people died after inhaling the anthrax spores. There was very little physical evidence, no fingerprints, no suspects, and no motive.

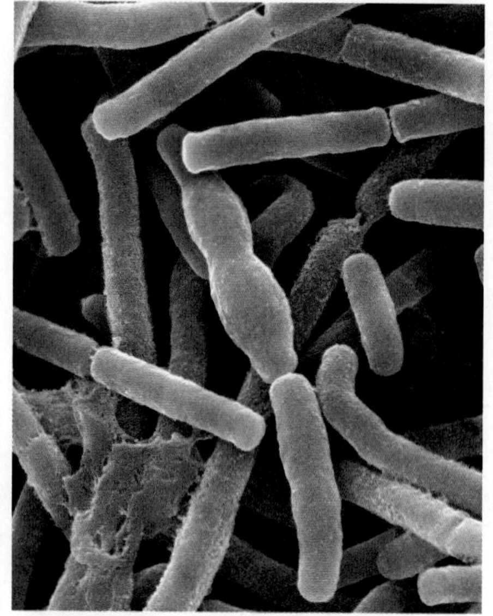

Figure 6-24. *Anthrax under a microscope.*

Epidemiologists from the U.S. Centers for Disease Control examined the spores and found that the spores used in the poisoning had certain genetic mutations in common with samples found in a government lab (see Figure 6-24).

The FBI employed the help of a behavioral and linguistic expert. After a complete analysis of the evidence, it was determined that the letters all came from the same person, who was likely a male with a scientific background. Circumstantial evidence connected Bruce Ivins, a U.S. government scientist who specialized in anthrax research, to the spores collected at the crime scenes. Just before he was to be arrested, Ivins committed suicide.

Think Critically

1. **What is the significance of the fact that several U.S. postal workers exhibited signs of anthrax poisoning?**
2. **How would examining the specific strain of bacteria help investigators narrow down a possible suspect? Be sure to use the concept of class and individual evidence in your discussion.**
3. **Various forms of anthrax exist. Examine each type of anthrax and discuss the body's physiological response. Draw or use visuals in your discussion.**

Bibliography

Books and Journals

Calaluce, Robert, and Jay Dix. (1998). *Guide to Forensic Pathology*. Boca Raton, FL: CRC Press.

DiMaio, Vincent J., and Dominick DiMaio. (2001). *Forensic Pathology* (2nd ed.). Boca Raton, FL: CRC Press.

DiMaio, Vincent J. M., and Suzanna E. Dana. (2007). *Handbook of Forensic Pathology* (2nd ed.). Boca Raton, FL: CRC Press.

Dix, Jay, and Michael Graham. (2000). *Time of Death, Decomposition and Identification: An Atlas* (Causes of Death Atlas Series). Boca Raton, FL: CRC Press.

Gaensslen, R. E., Howard A. Harris, and Henry Lee. (2008). *Introduction to Forensic Science & Criminalistics*. New York: McGraw-Hill.

Tortora, Gerard J., and Bryan Derrickson. (2006). *Principles of Anatomy and Physiology*. Hoboken, NJ: John Wiley & Sons, Inc.

Websites

www.biologycorner.com

www.cengage.com/school/forensicscienceadv

CAREERS IN FORENSICS

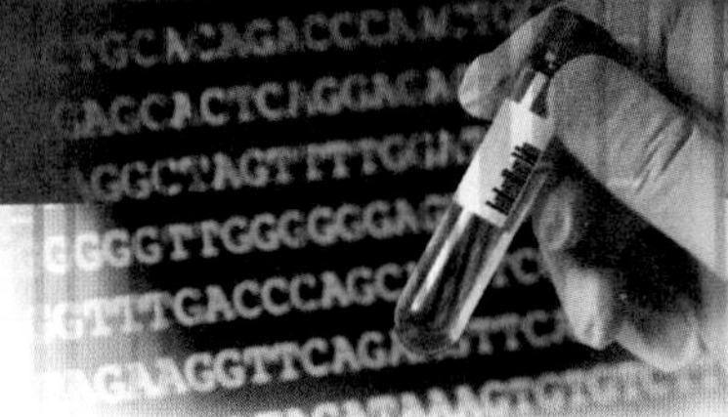

Science, Technology, Engineering & Mathematics

Dr. Kathy Reichs: Forensic Anthropologist

Dr. Kathy Reichs (see Figure 6-25) serves as a forensic anthropologist for the North Carolina Office of the Chief Medical Examiner and for the Laboratoire des Sciences Judiciaires et de Médecine Légale in Quebec, Canada. She is certified by the American Board of Forensic Anthropology and has served on the board of directors of the American Academy of Forensic Sciences. Dr. Reichs is also a professor of anthropology at The University of North Carolina at Charlotte. She divides her time between Charlotte and Montreal and frequently serves as an expert witness in criminal trials.

Figure 6-25. *Dr. Kathy Reichs*

Dr. Reichs has traveled to Rwanda to testify at the UN Tribunal on Genocide, and she has helped identify individuals from mass graves in Guatemala. In New York, she worked at Ground Zero after the 9-11 terrorist attacks on the World Trade Center. During a typical day, Reichs investigates cases often unrelated to homicides. For example, she has worked cases in which someone wandered off and died in the woods. The body may have been found years after the person went missing. When the body is discovered, Reichs estimates the age, sex, and stature of the person. With this information, investigators search missing person databases to find someone that fits the profile. Medical records and dental records also aid in making an identification of the missing individual. Approximately 30 to 40 percent of the cases Reichs works on involve the analysis of trauma, such as examining fracture patterns, entrance and exit wounds, or stab wounds to help determine the cause of death. She collaborates on cases only when requested by the pathologist, in order to help with identification. Dr. Reichs has also used her experience as a forensic anthropologist to write a series of novels. Her main character, Temperance Brennan, uses skills and tools important in the work of a forensic anthropologist. Dr. Reichs's novels have been adapted for television in the series *Bones*.

Learn More About It

To learn more about a career in forensic pathology, go to www.cengage.com/school/forensicscienceadv.

CHAPTER 6 REVIEW

True/False

Determine whether each of the following statements is true. If the statement is false, correct the underlined term to correct the statement.

1. In pulmonary circulation, <u>oxygenated blood is pumped from the heart to the arteries and to the cells.</u> *Obj. 6.1*

2. <u>Leukocytes</u> are important in helping to form clots at the site of a wound. *Obj. 6.1*

3. The upper respiratory tract consists of the nose, nasal cavity, <u>paranasal sinuses</u>, and pharynx. *Obj. 6.3*

Multiple Choice

4. Rigor mortis _____. *Obj. 6.6*
   a. begins about two hours after death
   b. is useful in estimating time of death during the first 36 to 48 hours after death
   c. is permanent
   d. both a & b

5. Which of the following is *not* part of the circulatory system? *Obj. 6.1*
   a. heart
   b. blood
   c. trachea
   d. capillaries

6. What role does an erythrocyte play? *Obj. 6.1*
   a. Erythrocytes help the body fight off disease.
   b. Erythrocytes help form blood clots.
   c. Erythrocytes transport gases through the body.
   d. Erythrocytes regulate acidity levels.
7. Based on information in the chapter, what kind of blood spatter is a gunshot wound most likely to produce? *Obj. 6.2*
   a. high-velocity blood spatter
   b. medium-velocity blood spatter
   c. low-velocity blood spatter
   d. none of the above
8. Which of the following medical conditions would *not* be indicated by a low white blood cell count? *Obj. 6.1*
   a. HIV
   b. influenza
   c. chicken pox
   d. tonsillitis

Short Answer

9. Describe the path of blood through the heart and the rest of the body. *Obj. 6.1*

____

____

____

____

10. What information can lividity and blood spatter provide about a victim at a crime scene? *Obj. 6.2*

____

____

____

____

11. Explain the similarities and differences between asphyxia and suffocation. Provide an example of each. *Obj. 6.2, 6.4*

____

____

____

____

12. Explain how the respiratory and circulatory systems work together to sustain basic life functions. *Obj. 6.1, 6.3*

13. In your own words, describe the process of blood clotting. *Obj. 6.1*

14. Compare and contrast the three types of muscles. *Obj. 6.5*

15. Design a graphic organizer that portrays all body systems, their similarities, differences, and how they can overlap in forensic investigations. *Obj. 6.1–6.7*

Think Critically 16. A young patient comes into the hospital with a broken arm. An emergency room doctor suspects the patient has been physically abused. The doctor also discovers several small, circular burn marks on the patient's arm. What body systems provided the doctor with clues about this patient's well-being? *Obj. 6.7*

17. If a doctor suspects that a victim has been poisoned, what body system would the doctor most likely test to confirm the suspicion? *Obj. 6.6*

18. Jane Davis was 72 years old. She was discovered dead, lying on her back in her living room. Rigor mortis was just beginning to set in. A large amount of blood covered her face. There were red and blue contusions (bruises) to her chest. Upon examination of the victim's right side, it was noted she had a lacerated (torn) spleen and liver. There was a patterned bruise on her left cheek. The pattern appeared to be the imprint from a heel of either a shoe or a boot. No head trauma was noted. The next day a suspect was arrested. The suspect was wearing boots, and investigators collected them for processing. Fingerprint powder was dusted on the soles of the boots, and a transparency print of the sole was made. This print was then placed over the wound on the face. This print was consistent with the wound on the victim's face. The suspect was indicted, tried, and convicted of second-degree murder.

    No other evidence against the suspect was found in this case. Were the interpretations made by the crime-scene investigator scientifically valid? As an expert witness in this case, justify your position. *Obj. 6.2, 6.7*

19. Randall Blyth was a 46-year-old homeless man. He was discovered in the middle of the road under a bridge. Blood was seeping out of his left ear and he had lacerations (tears) on the left side of his scalp. Blood draining from the ear was caused by a fracture of the left middle fossae (a bone at the base of the skull). The victim had cerebral edema (fluid accumulating on the brain). He also had multiple brain contusions (bruises) on the left side of the brain. An additional laceration was noted on the back right side of the scalp. This laceration pierced through to the bone. Three ribs and the left scapula were fractured. A laceration and several contusions of the left lung were noted. Fifty mililiters of blood was found in the left chest cavity.

    When police and investigators examined the scene, they determined that the victim fell to his death by jumping or falling off the bridge. Do you agree with police and investigators? Explain, in detail, how you would justify your opinion in court. *Obj. 6.2, 6.7*

CRIME SCENE

Student-Prepared Original Titles

Love's Murder
By: Reginald Wright and Amber Farner

Tavares High School
Tavares, Florida

"Hey, I thought I was meeting you. What are you doing here?" Sara was applying cherry-red lipstick to her pale pink lips. As she reached for her comb, a deep, raspy voice said, "Yeah, babe. I know we were supposed to meet at the restaurant, but I couldn't wait to see you!" Now turning away from the mirror, Sara saw her date. His blue eyes, bloodshot, pierced through the darkness of the brown hoodie, which was drenched in sweat. Startled by his demeanor, she uttered, "Sam? What's wrong— ?" Before she could finish her sentence, he took out some old rope and wrapped it around her neck. Sara conjured up enough strength to yell "Stop!" but before she could finish, she fell to the ground. She looked her murderer in the eye, and asked, "Why . . .?" He then took out a knife and unmercifully stabbed her. A pool of blood surrounded her. The hooded killer then cleaned the knife on the victim's blouse and walked toward the door. Before exiting, the killer looked at Sara's dead body and said, "You know what, sweetie? I really did love you. Why couldn't you feel the same?" The murderer then closed the door to the one-bedroom apartment and fled.

Figure 6-26. *The crime scene.*

JIM VARNEY/PHOTO RESEARCHERS, INC.

Four hours passed and detectives, along with CSIs, arrived at the scene (see Figure 6-26). "I know" was written in the victim's blood on the bathroom wall. The detectives and CSIs noticed that the frame of the door and other points of entry had no scrapes, indicating no forced entry. The victim must have let the murderer in or else the murderer had a key. Everything in the apartment was still intact, except for the blood spatter on the bathroom walls and an overturned love seat. They also noticed bruises on Sara's arms. The detectives thought they looked like carpet burns. What appeared to be brown lint was found on the victim's shoulder. Some rope strands were found as well. A marriage certificate in a silver frame was found next to the sofa. The glass had been smashed. Under the certificate was a wedding picture.

"Mr. Newson, you are Sara's husband, correct?" Detective Jones asked. With tears in his eyes, he stated that, yes, he was her husband. The detective asked where he was this evening. Mr. Newsome explained he was at Bonsir and Stroven, waiting for Sara. It was their anniversary. They were to meet there at 7:00 P.M, but she had called around 7:15 P.M. She was running late from work and Cindy Love needed a ride home. Also, she was going to run to the apartment to "freshen up." She never showed up.

Detective Jones inquired about Cindy Love. Mr. Newson told the detective that Cindy was once a friend of Sara's. However, he was surprised to hear his wife was giving her a ride, because three months ago, Sara had been promoted, and Cindy had applied for the same position. As time progressed, their friendship had become strained.

Detective Jones seemed finished with his interview of Mr. Newson. "Oh, wait . . . before you leave," said the detective, "did anyone else have access to your apartment besides you and your wife?" At first, Mr. Newson said no, but then quickly remembered that his brother, Rob, had a spare key. He continued to explain that Rob works at a clothing store. The store had just received new inventory, and Rob had been working nearly around the clock. The CSIs made sure that the crime scene was properly secured and then everyone left the apartment.

In the patrol car, Detective Jones called Bonsir and Stroven. The headwaiter confirmed that Mr. Newson had been there the entire evening, from 7:00 P.M. until 10:00 P.M. Phone records indicated he had, indeed, talked to the victim an hour before she was murdered.

The detectives arrived at Cindy Love's home the next morning and asked if they could talk to her about Sara Newson. Detective Jones explained what had happened. He then proceeded to ask her how she knew the victim. "She was my coworker," she replied. Jones asked her where she had been last night, and she explained that she was at the hair salon. When they left Miss Love's home, Jones called the salon to confirm her alibi. The stylist, Ms. Patty, confirmed that Cindy Love had been at the salon. The surveillance camera showed her sitting in the chair and talking for hours.

Activity:

Answer the following questions based on information in the Crime Scene S.P.O.T.

1. Sara was brutally attacked and suffered multiple injuries before dying. Discuss the body systems and organs affected by this attack. Be sure to use proper terminology whenever possible.
2. What could have caused the bruises on Sara's arms?
3. Who is the most likely suspect? Why?

WRITING

4. Write the ending to the story. Be sure to connect the victim to the murderer and to provide a motive. Be creative, but be forensically accurate. Your ending to the story should be 500 words or less.

Preventing Adolescent Crime Together™ (P.A.C.T)

Introduction:

Health and wellness are ongoing concerns in every community. Screening for potential health problems provides a way to detect early-warning signs of many common illnesses. Knowing your risks allows you to take measures to correct or prevent disease. Making healthy lifestyle decisions is one way people can be proactive in maintaining good health.

Procedure

1. Select a date for the health fair. Secure the facility you plan to use—the school gymnasium or cafeteria works well.
2. At least six weeks before the health fair, invite local health-care professionals to participate in a health fair to be held at your school. Consider including several of the following services: blood-pressure screening, cholesterol screening, diabetes evaluation, eye tests, nutrition information, and exercise physiology. Some businesses might be willing to donate refreshments in return for publicity as a sponsor of the event.
3. Design a brochure to promote the event. Include the professionals who are sponsoring or participating in the event. List services that will be provided and whether there are any costs involved for the service.
4. Create posters to market the health fair. Posters may be placed around the school and some of the businesses in the area. Be sure to check with business managers and school administrators before hanging your posters.
5. The day before the health fair, set up tables and allow participants to set up their exhibits for the event.
6. After the event, send thank-you notes to all participants.

Figure 6-27. *During the health fair, members of your class may want to display information about healthy habits.*

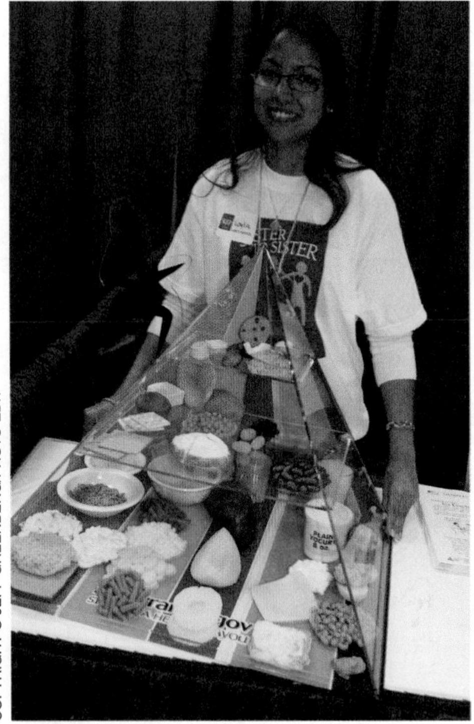

ACTIVITY 6-1 *Ch. Obj: 6.5, 6.7*
ANATOMY OF A CHICKEN WING

Objectives:

By the end of this activity, you will be able to:

1. Identify the relationship between structure and function of bones and muscles.
2. Compare and contrast a chicken wing to the arm of a human.

Background:

Clearly, there are many differences between the human body and a chicken. However, many of the bones and muscles in the wing of a chicken and the arm of a human are similar in structure. These structures are called *homologous structures*. In this lab, you will study the bones and muscles of the chicken wing.

Materials:

(per group of two students)

raw chicken wing
water
paper towels
dissecting tray
scissors or scalpel
blunt probe
forceps

Safety Precautions:

Safety goggles, apron, and protective gloves must be worn. Raw chicken might be contaminated with the bacteria *Salmonella*. Do not put your hands near your face or mouth during this dissection.

Be cautious when using dissection instruments.

Wash your hands thoroughly with soap and water when finished.

Procedure:

1. Rinse the chicken wing well with cool water. Pat it dry with paper towels. Place the wing in a dissecting tray.
2. Stretch the wing out to examine the separate parts of the wing. Sketch the chicken wing (see Data Table 1 on page 190).

Data Table 1. Observations.

External Chicken Wing
Muscles and Tendons of the Chicken Wing
Bones and Ligaments of the Chicken Wing

3. Examine the cut end of the wing. What human structure compares to this? On your sketch, label the shoulder and elbow joints.
4. Using a pair of sharp scissors or a scalpel, cut the skin down the middle of the wing, as shown in Figure 1. Be very careful not to cut through the muscles underneath the skin. Continue cutting until the cut extends all the way to the shoulder joint.

Figure 1. Cut the skin so that you can examine the tissue underneath.

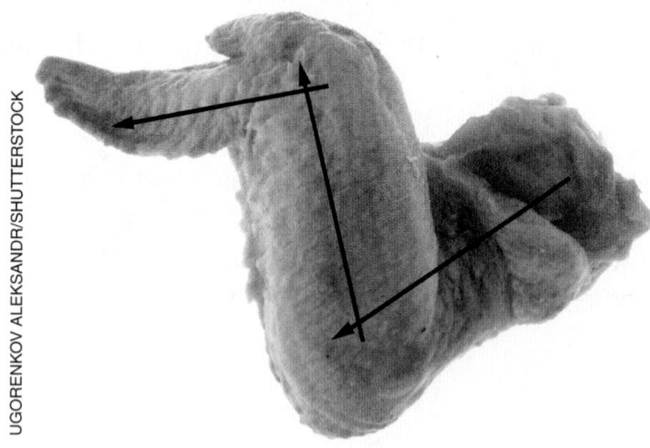

UGORENKOV ALEKSANDR/SHUTTERSTOCK

5. Remove the skin.
6. Find the yellowish tissue beneath the skin. This is fat tissue. Carefully remove as much of the fat tissue as you can.
7. Examine the muscles and bones. Sketch the muscles in the wing (Data Table 1).
8. Find the individual muscles of the upper wing. Hold the shoulder down and pull one of the muscles. Repeat with the other muscle. Record your observations in Data Table 2.

Data Table 2. Observations.

Observations of Upper Wing	Observations of Lower Wing

9. Locate the muscles in the lower part of the wing. Hold down the wing at the elbow, and pull on each muscle group one at a time. Record your observations in Data Table 2.
10. Find the shiny white tissue at the ends of the muscles. These are tendons. Tendons connect muscles to bones. Label the tendons on your sketch.
11. Carefully remove the muscles and tendons. Locate the ribbon-like structures between bones. These are called *ligaments*. Ligaments connect bone to bone. Sketch the bones and ligaments of the chicken wing (Data Table 1).
12. Dispose of the chicken wing, skin, and fat tissue according to your teacher's instructions. Clean your work area thoroughly. Wash the dissecting tray and all instruments and set them aside to dry. Wash your hands with soap and water.

Questions:

1. What are homologous structures?
2. What is the function of muscle tissue?
3. What causes muscle fibers to contract?
4. Describe the ligaments and tendons. What is the function of ligaments? Tendons?
5. Why is it important for a forensic pathologist to understand the relationship between muscles and bones?

ACTIVITY 6-2 *Ch. Obj: 6.5, 6.7*
DEM BONES

Objectives:

By the end of this activity, you will be able to:

1. Determine sex by examining features of various bones.
2. Estimate the height of an individual based on the length of long bones of the body.
3. Compare the relationship between height and bone length in males and females.

Materials:

(per student)

butcher paper (2 m)
pencil
meter stick
chart of skeletal system with bones labeled
various bones

Procedure:

Part 1 *(in pairs)*

1. Working with a partner, take turns lying on the butcher paper while your partner traces an outline of your body.
2. After the outlines are finished, draw in the major bones of the body to scale. Refer to a picture or chart of the skeletal system.
3. Label the bones.

Part 2 *(as a class)*

1. On a chart similar to Data Table 1, record each person's name, sex, and height.
2. Measure the length of each person's femur and ulna. Record the measurements in centimeters.

Data Table 1. Student information.

Student's Name	Sex	Height (cm)	Length of Femur (cm)	Length of Ulna (cm)

Part 3 *(in groups of three or four students)*

In this portion of the activity, your group will move from station to station around the classroom, observing various bones and drawing conclusions about them.

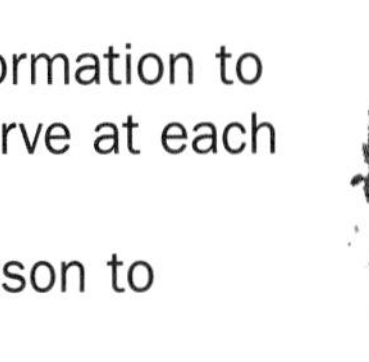

1. At each station, identify the type of bone. Record the information in Data Table 2.
2. Measure and record the length of the long bones.
3. Look back at the information in Data Table 1. Use that information to estimate the height of the individual whose bone you observe at each station. Record the information.
4. Examine all of the bones and determine the sex of the person to whom the bone belongs. Record your conclusions.

Data Table 2. Bone observations

Name of Bone	Length of Bone (cm)	Estimated Height of the Individual	Sex of the Individual

Questions:

1. Is there a correlation between a person's height and the length of the person's ulna and femur? If so, what is the correlation and is it the same for males and females? Explain your answer.
2. How do you think this information is valuable to a forensic scientist?
3. If available, compare your calculations/correlations with a bone length chart. Do your answers fall within the range shown on the chart? If not, why do you think your numbers are different?
4. Other than height and sex, what other information do you think a forensic scientist might make from bones?

CHAPTER 7

Physical Trauma

THE SIREN AND THE SLUGGER

Just after midnight on Sunday, February 9, 2009, police were called to the scene of a dispute in Los Angeles, California. At the scene, officers encountered a rented Lamborghini sports car and a young woman who had visible injuries from a physical assault. The young woman was later identified as Grammy Award-winner Rihanna.

Rihanna had been romantically involved with fellow recording artist Chris Brown for more than a year. That night, the pair was traveling in the rented sports car through the Hancock Park area of Los Angeles. At that time, a verbal disagreement began about a text message Brown had received from a woman with whom he had had a previous relationship. Brown tried to halt the argument by pulling the vehicle over and attempting to force the female singer from the car.

When Rihanna refused to exit the vehicle, Brown shoved Rihanna against the window with sufficient force to cause a raised circular contusion to form on her forehead. Then, he began to punch her in the face. Rihanna attempted to protect herself by covering her face and head with her arms. The barrage of blows resulted in numerous contusions on her left hand and a large contusion to her left triceps.

Brown placed Rihanna in a headlock and bit her left ear. Brown stopped the car and began to choke Rihanna until she attempted to gouge his eyes. He bit her left ring and middle fingers and then released her. She turned to face him, placed her feet on his chest, and pushed him away. He continued to punch her legs and feet, causing several more contusions. When Rihanna began screaming for help, Brown got out of the vehicle and walked away from the scene. A resident in the neighborhood heard Rihanna's cries and called 911. When the police arrived, they processed the scene, took a statement from Rihanna, and sent the singer to the hospital for treatment. At the hospital, photographs were taken of her injuries.

At about 7:00 P.M. that evening, Chris Brown turned himself in at a Los Angeles police station. In June 2009, Brown pled guilty to a charge of felony assault and received a sentence of five years' probation, six months of community service, and one year of domestic violence counseling.

AP PHOTO/LORI SHEPLER, POOL

Figure 7-1. Chris Brown at his sentencing hearing.

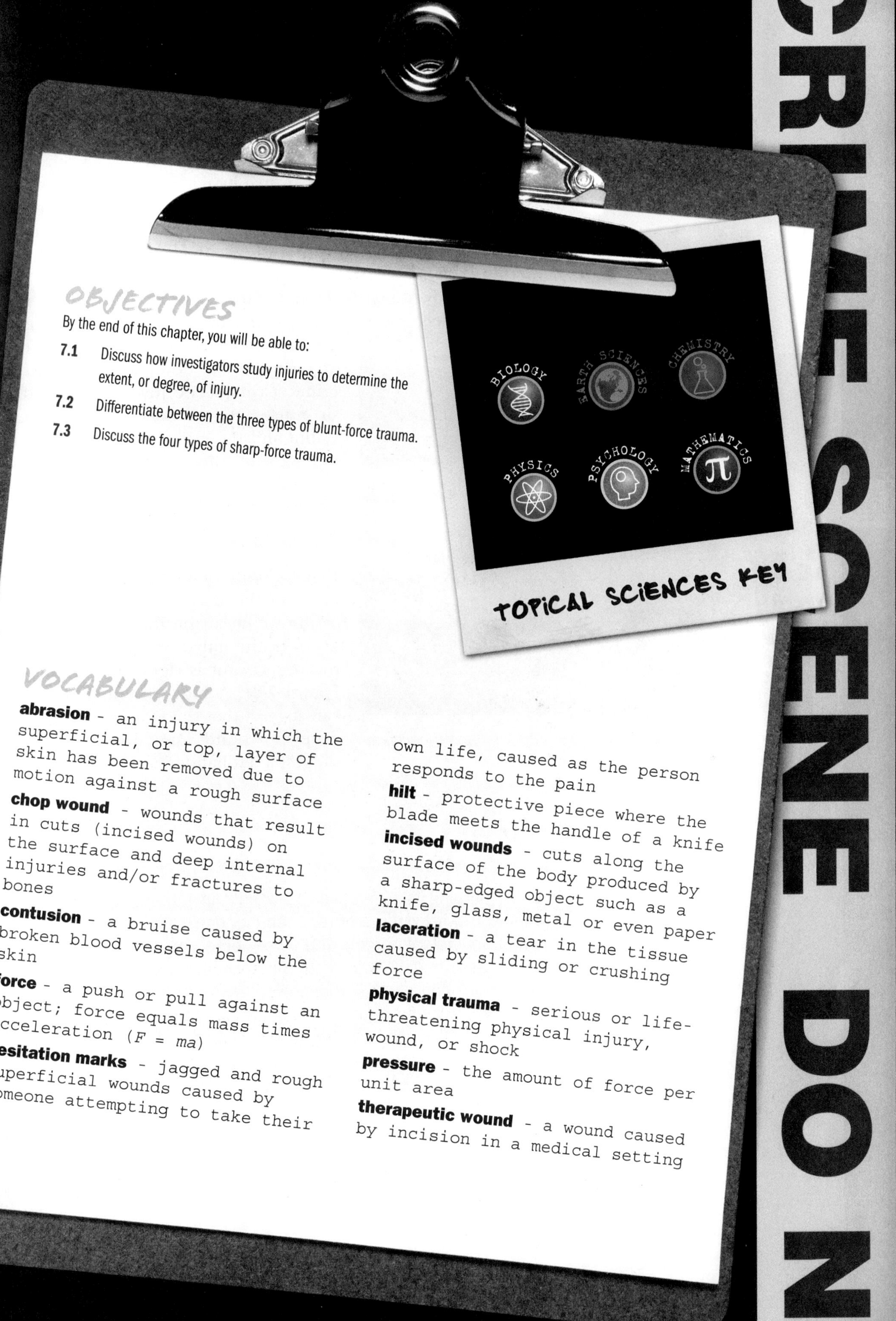

OBJECTIVES

By the end of this chapter, you will be able to:

7.1 Discuss how investigators study injuries to determine the extent, or degree, of injury.

7.2 Differentiate between the three types of blunt-force trauma.

7.3 Discuss the four types of sharp-force trauma.

VOCABULARY

abrasion - an injury in which the superficial, or top, layer of skin has been removed due to motion against a rough surface

chop wound - wounds that result in cuts (incised wounds) on the surface and deep internal injuries and/or fractures to bones

contusion - a bruise caused by broken blood vessels below the skin

force - a push or pull against an object; force equals mass times acceleration ($F = ma$)

hesitation marks - jagged and rough superficial wounds caused by someone attempting to take their own life, caused as the person responds to the pain

hilt - protective piece where the blade meets the handle of a knife

incised wounds - cuts along the surface of the body produced by a sharp-edged object such as a knife, glass, metal or even paper

laceration - a tear in the tissue caused by sliding or crushing force

physical trauma - serious or life-threatening physical injury, wound, or shock

pressure - the amount of force per unit area

therapeutic wound - a wound caused by incision in a medical setting

INTRODUCTION

According to the FBI, more than 11 million crimes were reported in the United States in 2008. About 1.3 million of those were classified as violent, and about 16,000 resulted in death. Many of the other victims of these violent crimes suffered from physical trauma. **Physical trauma** is any serious or life-threatening physical injury, wound, or shock. Blunt-force trauma is caused when the victim hits or is hit by a hard object. Sharp-force trauma is caused when the victim is poked, cut, or stabbed by something sharp.

Crime-scene investigators, forensic nurses, and other trained forensic professionals collect and process evidence in an effort to link a suspect to a crime. Detectives interview the victim and any witnesses to develop an account of the crime. Crime-scene investigators and forensic nurses collect trace evidence from the victim and the crime scene. Forensic nurses evaluate the physical trauma and interpret the injuries to look for clues regarding the events that lead up to the crime. Also, some injuries can provide a profile of the suspect. The location of the trauma on the victim's body helps investigators determine how tall the perpetrator is or the position of the victim at the time of the attack. The angle of the impact might indicate whether the perpetrator is right- or left-handed (see Figure 7-2). A blow that angles to the left can suggest that the attacker is right-handed while a wound angled to the right can suggest a left-handed attacker.

Figure 7-2. *If this right-handed perpetrator stabs the victim, the wound will likely angle to the left.*

Digging Deeper
with Forensic Science e-Collection

Forensic nursing is a relatively new field in forensics. Forensic nurses work with victims of rape, abuse, domestic violence, and trauma from violent crimes. The career of forensic nursing offers opportunities to work with both crime victims and perpetrators, to collect or photograph evidence for law-enforcement agencies, and to assist during death investigations. Forensic nurses may also counsel schoolchildren who use weapons and provide care in the correctional system as well as act as legal nurse consultants and expert witnesses in court. Using the Gale Forensic Science e-Collection at www.cengage.com/school/forensicscienceadv, research forensic nursing and answer the following questions: (1) What are some of the specialties within the forensic nursing field? (2) What special training is involved in becoming a forensic nurse? (3) Describe the duties of a forensic nurse.

EVIDENCE OF PHYSICAL TRAUMA

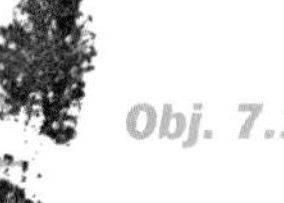

Obj. 7.1

Medical personnel—often an emergency room doctor or a forensic nurse—will need to look closely at the victim's injuries. The appearance and extent of the injuries will depend on the following:

- The amount of force applied to the body
- The weapon's surface area and mass
- The part of the body affected

FORCE

Force is an important factor in the extent of physical trauma. **Force** is a push or a pull applied to an object. Force describes how hard the weapon hit the victim. Force can be calculated using the following equation:

$$\text{force} = \text{mass} \times \text{acceleration}$$

or

$$F = ma$$

Acceleration is the change in velocity over a period of time. When a car is traveling at 45 miles per hour, its velocity is 45 miles per hour, or 20 meters per second (m/s). If that car then stops suddenly, the change in velocity is 20 m/s. If it takes two seconds for the vehicle to come to a complete stop, the acceleration is 10 m/s/s or 10 m/s^2 (20 m/s divided by 2 s). If the car crashed into a stationary object, such as a brick wall, the force of the impact is the mass of the car multiplied by the car's acceleration. If the car has a mass of 1,000 kg, the force would be 1,000 kg × 10 m/s^2. The force would be 10,000 $kg(m/s^2)$ or 10,000 Newtons. If the car were more massive, the force would be greater (see Figure 7-3). In the case of a physical attack, the force on the victim depends on the mass and the acceleration of the weapon.

Mass = 1,000 kg
Acceleration = 10 m/s^2

Force = 1,000 kg × 10 m/s^2 = 10,000 N

A

Mass = 1,500 kg
Acceleration = 10 m/s^2

Force = 1,500 kg × 10 m/s^2 = 15,000 N

B

Figure 7-3. *Force is equal to mass times acceleration.*

Because of the relationship between mass, acceleration, and force, a more massive weapon or a weapon that is moving more quickly upon impact will cause a more serious injury.

SURFACE AREA AND PRESSURE

Surface area is another factor in the severity of physical trauma. If the same amount of force is exerted over a larger surface area, the injury will be less severe. This is because **pressure** is the amount of force per unit area.

$$\text{pressure} = \frac{\text{force}}{\text{surface area}}$$

$$P = \frac{F}{A}$$

Imagine that you push against a wall with your palm (see Figure 7-4). If you apply a force of 44.5 Newtons (10 pounds) and your hand is 129 cm^2 (20 in^2), you are exerting 0.34 N/cm^2 of pressure. If, however, you exert the same amount of force with a closed fist, the pressure will be much higher. Assuming that your closed fist has a surface area of 38.7 cm^2 (6 in^2), you are exerting pressure of 1.15 N/cm^2.

Pressure also explains why women wearing high-heeled shoes often "tiptoe" across a soft surface. Normally, most of the force is applied to the heel of these shoes. The surface area of the heel of high-heeled shoes is often much smaller than that of regular shoes. When the force (the person's weight) is placed on the smaller heel, the pressure can puncture soft flooring or muddy ground.

Because of the relationship between pressure, force, and surface area, a weapon with a larger surface area will inflict a less severe injury. For example, assuming that the same amount of force is applied, a flat wooden board will cause a less severe injury than a narrow metal rod. Additionally, rounded areas of the body, such as the skull, are more prone to severe trauma than flatter areas, such as the back. Because the back will have more contact with the weapon, the pressure will be lower and the injury will be less severe. Another way to think about this is that on a larger area, the effect of the force is more spread out and, therefore, less severe.

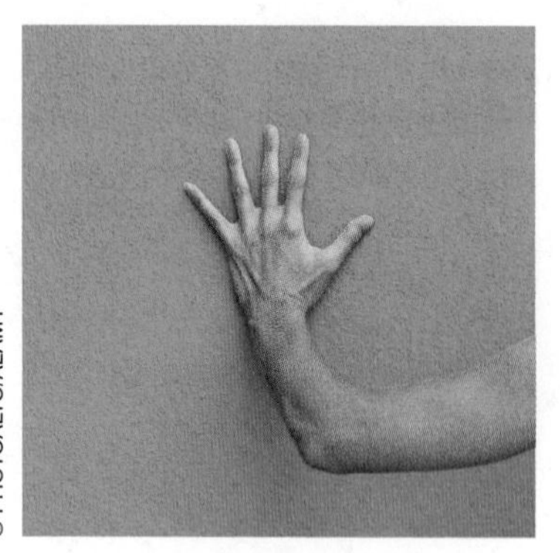

© PHOTOALTO/ALAMY

Force = 44.5 N
Surface area = 129 cm^2

$$\text{Pressure} = \frac{44.5\text{ N}}{129\text{ cm}^2} = 0.34\text{ N/cm}^2$$

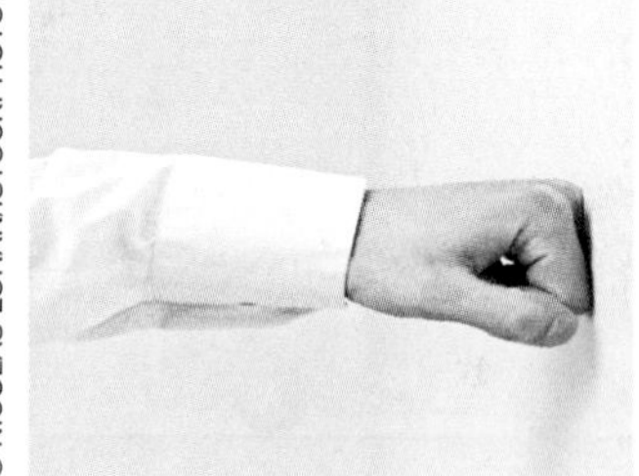

© NICOLAS LORAN/ISTOCKPHOTO

Force = 44.5 N
Surface area = 38.7 cm^2

$$\text{Pressure} = \frac{44.5\text{ N}}{38.7\text{cm}^2} = 1.15\text{ N/cm}^2$$

Figure 7-4. *Pressure is equal to force divided by surface area.*

Digging Deeper
with Forensic Science e-Collection

First responders to accidents and crime scenes suffer tremendous stress. Many of these first responders are able to cope with the stress. However, after years of this stress, many exhibit some of the symptoms of post-traumatic stress disorder. Others become depressed or even contemplate suicide. Using the Gale Forensic Science e-Collection at www.cengage.com/school/forensicscienceadv, research critical incident stress and answer the following questions: (1) What professions are considered first responders? (2) Explain the warning signs that a first responder is experiencing critical-incident stress. (3) What might affect how a first responder reacts to a critical incident?

TYPES OF TRAUMA

Injuries to the body are usually classified as either *blunt-force* or *sharp-force trauma.* This classification is based on the type of wound produced and the weapon used. Several types of injuries fall into each of the classifications.

Obj. 7.2

BLUNT-FORCE TRAUMA

Blunt-force trauma is caused when the victim is hit by something hard. Blunt-force trauma can also be caused when the victim falls or is pushed into a hard object. Blunt-force trauma falls into one or more of the following categories:

- Abrasions
- Contusions
- Lacerations

Blunt-force trauma to the abdomen is usually the result of automobile accidents, assaults, recreational accidents, or falls. Organs most commonly injured in these cases include the spleen, liver, small bowel, kidneys, bladder, diaphragm, and pancreas.

Abrasions

The skin is the major organ of the integumentary system. Skin has many functions, including temperature regulation and excretion. The skin contains receptors to detect pain, touch, pressure, and heat. However, the most important functions of the skin are to protect the body against pathogens, including bacteria, and to prevent water loss. The skin is made up of three layers—the epidermis, dermis, and hypodermis (see Figure 7-5). The *epidermis* is the outermost layer of the skin. The cells that make up the epidermis are continually being replaced. The *dermis*, just below the epidermis, is denser and contains the blood vessels and nerves. The *hypodermis* is a layer of fat and connective tissue beneath the dermis.

If part of the body rubs against a rough surface, a portion of the epidermis may be removed. The injury that results is called an **abrasion**. Skinned knees and elbows are common abrasions. Also, if someone is dragged down a sidewalk, the victim's legs will probably have abrasions. Abrasions are not usually very deep, and they heal relatively quickly. When a living person gets an abrasion, the trauma appears reddish-brown in color. However, if the abrasion occurs postmortem (after death), the abrasion will appear yellow, almost transparent. Occasionally, abrasions are the only external sign of trauma. There are three basic forms of abrasions—brush abrasions, impact abrasions, and patterned abrasions.

Brush Abrasions When force is applied parallel to the skin, a brush, or scrape, abrasion forms (see Figure 7-6 on page 200). For example, if the victim is dragged along a rough surface, then he or she will sustain brush abrasions. In these cases, the skin is scraped off in layers. More fluid, including blood, will be produced by brush abrasions that affect the dermis than by those that affect only the epidermis. When this fluid dries, a scab forms. Forensic scientists are often able to determine the direction of the force by observing the scrape under a magnifying lens. The skin is damaged in the direction of the force. Additionally, the scientist might find debris, gravel, or glass in the abrasion. These materials can provide information about the location of the incident and the cause of the abrasion. A noose or tie around ankles, wrists, or neck can produce a brush abrasion and might leave fibers in the wound.

Figure 7-5. *Structure of the skin.*

Impact Abrasions When force is applied perpendicular to the skin, the skin may be crushed. The resulting abrasion is called an *impact abrasion*. An impact abrasion is most commonly seen over bony regions of the body, such

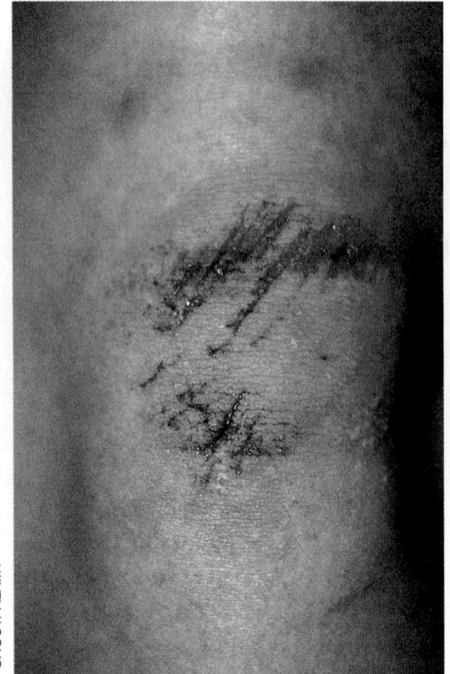

© SHOUT/ ALAMY

Figure 7-6. Brush abrasion.

as an eyebrow or cheekbone. An impact abrasion may form on the side of the nose if an unconscious victim falls to the ground.

Patterned Abrasions Sometimes, the object that causes an impact or brush abrasion leaves an imprint on the victim's skin. This kind of abrasion is called a *patterned abrasion* and is caused by the intensity of the impact and the crushing effect of the blunt object on the skin. The pattern left on the victim may be consistent with a particular surface. In these cases, the patterned abrasion can provide important evidence about the crime, such as clues about the weapon used.

Contusions

Blunt force often causes contusions. A **contusion**, also known as a *bruise*, is trauma caused by broken blood vessels below the surface of the skin (see Figure 7-7). Contusions can also occur on internal organs, such as the lungs, heart, or brain. There is a clear distinction between a contusion and livor mortis. In a contusion, the blood vessels have been broken. In livor mortis, however, the blood is contained within blood vessels. Before livor mortis has become fixed, the blood can be squeezed away from the area.

Sometimes, contusions are large enough that the collected blood causes swelling. This kind of contusion is called a *hematoma*. The outer surface of the brain is called the *dura*. When a head injury causes bleeding under the dura, blood collects and can cause the brain tissue to be compressed. This kind of injury is called a *subdural hematoma* and can cause death very quickly. Figure 7-8 compares a healthy brain with an injured brain. The image on the left shows a healthy brain. The image on the right shows a brain with a subdural hematoma. Notice that the hematoma is below the dura.

The amount of force and the type of tissue affected determine how severe the contusion is. For example, a contusion is easily identifiable where skin is thin and loose. The bruise may even look similar to the object that produced it. This kind of bruise is called a *patterned contusion*. These patterns sometimes provide important information about the crime, including information about the weapon used. A pattern that is consistent with a suspected weapon can be an important piece of evidence.

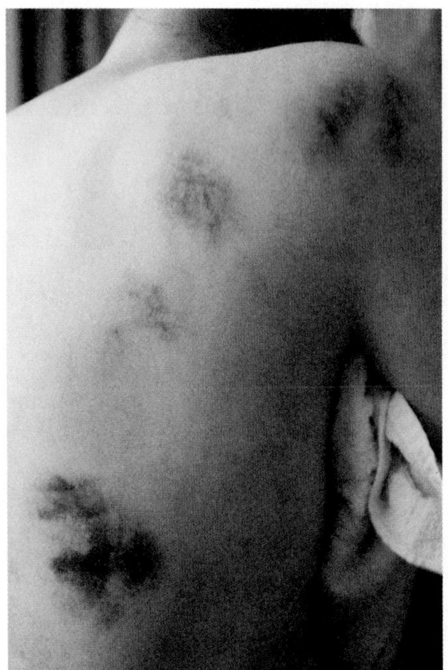

© DESIGN PICS INC./ALAMY

Figure 7-7. Contusions.

Contusions on internal organs are not visible externally. Doctors can perform computerized tomography (CT) scans to check for contusions in deep tissue and internal organs. CT scans use X-rays to produce three-dimensional images of the organs and internal tissues, showing the injury.

A victim's skin pigmentation can make detecting contusions more difficult. The darker the skin, the harder it is to identify a contusion. The two groups most vulnerable to bruising are children and the elderly. Children have looser, more delicate skin than adults, while the underlying layers of an elderly person's skin have less supportive tissue than the skin of younger adults. Additionally, obese individuals bruise more easily than people with athletic builds. Firm muscle supports and protects the blood vessels, making the tissue less prone to bruising. An alcoholic who has cirrhosis of the liver or individuals on aspirin therapy may also suffer from bleeding disorders. Bleeding disorders increase the likelihood of bruising.

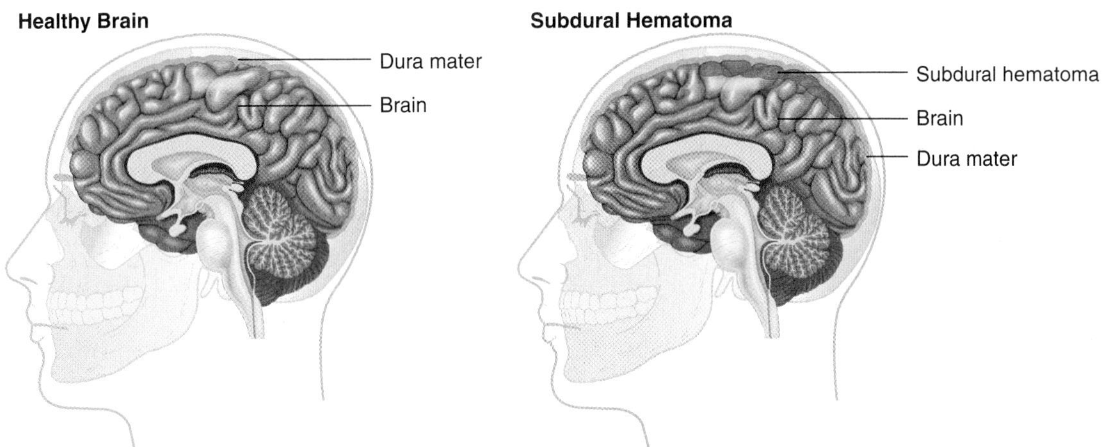

Figure 7-8. *Compare the image of a healthy brain (left) to the image of a brain with a subdural hematoma (right).*

A single contusion usually does not cause death. However, if a victim sustains many contusions at once, he or she might suffer from extensive blood loss. When this happens, the body is unable to provide the organs and tissues with enough oxygen. This condition is called *shock* and can cause death. Shock is common in young children who are severely physically abused.

Lacerations

A **laceration** is a tear in the tissue caused by a sliding or crushing force (see Figure 7-9). Lacerations are caused by extreme force produced by blunt objects, a fall, or vehicle impacts. The marks produced have irregular edges. Sometimes, the object that caused the trauma leaves a patterned mark. However, this is not always true because the skin tears around the object. Linear lacerations are produced by objects, such as pool cues, that are long and narrow. An object that has a flat surface will produce an uneven or Y-shaped laceration. Lacerations may occur internally or externally.

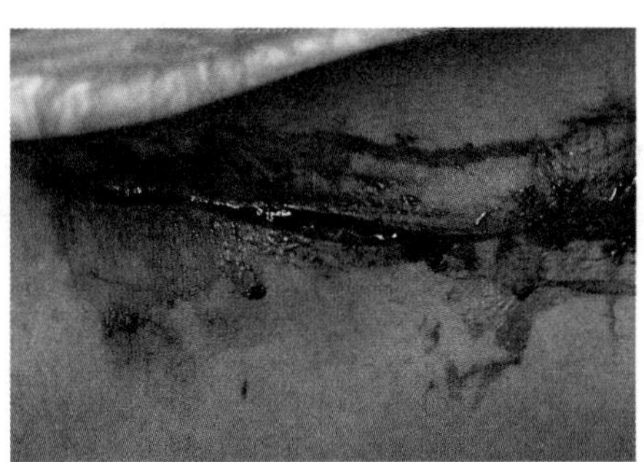

© HERCULES ROBINSON/ALAMY

Figure 7-9. *Laceration.*

Defensive Wounds Caused by Blunt Force

Defensive wounds are usually abrasions and contusions on the hands, wrists, and arms. Defensive lacerations often provide important clues to investigators. These wounds sometimes hold trace evidence from the weapon. Injuries to the perpetrator caused by the victim are also called defensive wounds. If the perpetrator injures himself or herself during the struggle, the wounds are called *offensive wounds.* The appearances of offensive and defensive wounds are useful in substantiating claims by the victim and perpetrator about the assault.

Did You Know?

A traumatic brain injury can occur when the head receives a blow, jolt, or penetrating injury that disrupts normal brain function. The leading cause of traumatic brain injury to 14- to 19-year-olds is automobile accidents.

SHARP-FORCE TRAUMA

Obj. 7.3

Sharp weapons, such as knives, cause sharp-force trauma. Sharp-force trauma usually falls into one of four categories:

- Stab wounds
- Incised wounds
- Chop wounds
- Therapeutic wounds

CHAPTER 1 2 3 4 5 6 7 8 9 10 11 12 13 14 15

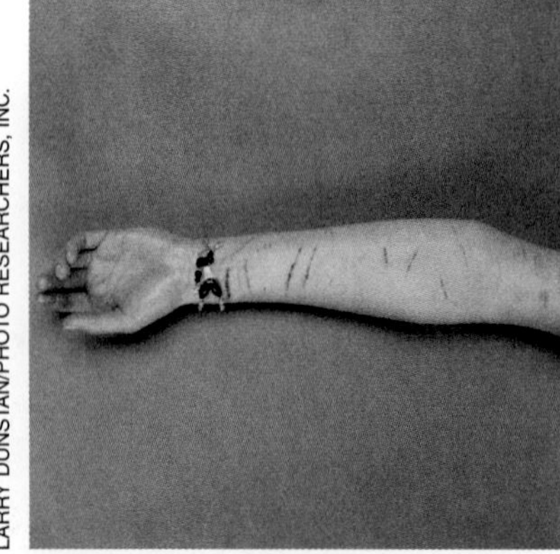

Figure 7-10. *Hesitation marks suggest that the wounds are self-inflicted.*

Stab Wounds

Most stab wounds are caused by single-edged kitchen, pocket, or folding knives. The blade is usually four to five inches long. Ice picks, scissors, forks, pens, or broken glass can also cause stab wounds. The weapon goes deeper into the body than the layers of skin. The weapon may damage muscles, fat, and organs. Because blades tend to be smooth and sharp, the edges of the wound are typically clean and smooth. However, wounds found on suicide victims are often less smooth. People attempting suicide often cut themselves several times before they cut all of the way through the skin. These surface wounds are called **hesitation marks** (see Figure 7-10). In an attack, on the other hand, irregular wounds can be caused when the victim struggles with the attacker. As the victim moves away, the blade may twist.

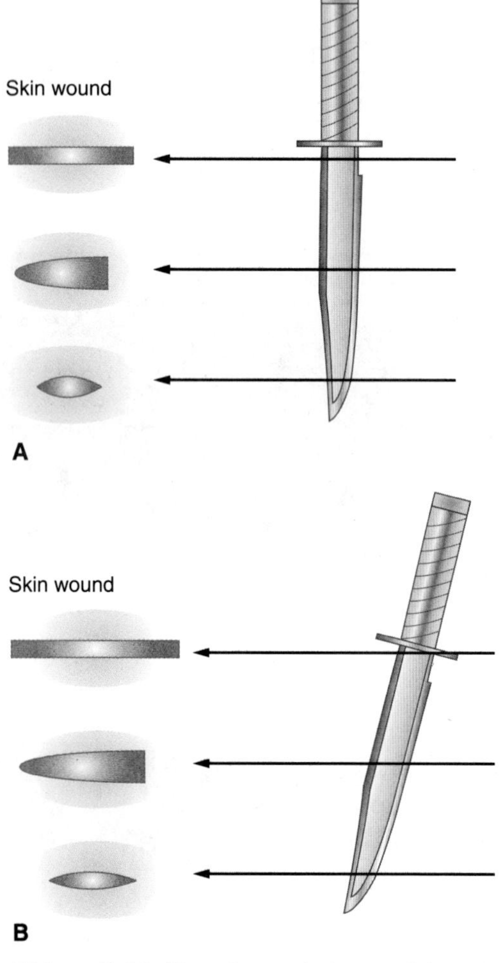

Figure 7-11. *The size and shape of the wound are affected by the angle and depth of entry.*

A very sharp knife will move through the body with very little force. Just as with blunt-force trauma, pressure affects the injury. With a very sharp knife, all of the force is placed on a very small amount of surface area. Therefore, little force produces a lot of pressure. As long as the weapon does not come into contact with bone, it will even pass through most organs with very little force. Wounds that result in a punctured organ are called *penetrating wounds,* and wounds that puncture an organ and come out the other side of the organ are called *perforating wounds.* If a knife enters the body with a great deal of force, the protective piece, also called a **hilt**, at the bottom of the knife's handle might leave a patterned abrasion around the wound. If the knife goes straight into the body, the hilt will be aligned with the knife. If the knife entered the body in a downward direction, the hilt will be distinctly above the stab wound. Figure 7-11 illustrates how the size and shape of the wound is affected by the way the knife entered the body.

A forensic nurse or pathologist will examine the location, depth, and path of the wound. X-rays provide information about the wound inside the body. The pathologist uses information from this examination to estimate the length and width of the weapon and the angle of penetration. The characteristics of the wound provide clues to investigators about the weapon. Ice picks leave very distinctive wounds. A knife is usually at least an inch wide at its widest point. However, an ice pick is very narrow and straight. An ice pick will leave small, round wounds that can easily be missed or even confused with .22 caliber bullets. A barbecue fork produces clusters of two or three equidistant wounds. If the victim has been stabbed multiple times, these evenly spaced markings help the investigator determine how many times the victim was stabbed. If a victim has been stabbed with scissors, the appearance of the wound will depend on whether the scissors were opened or closed. If the scissors are closed, they will split the skin instead of cutting the skin. The edges of the wound will be rough. If the blades of the scissors are apart, two wounds will be produced each time the victim is stabbed.

Examination of the wound may also reveal broken tips of the weapon, especially if the weapon hit bone. The broken tips may be compared to a recovered weapon or provide further clues about the type of weapon that was used. Further, forensic scientists are sometimes able to show that tool marks left on the bone are consistent with a particular weapon.

Incised Wounds

Incised wounds, or cuts, are produced by sharp-edged objects such as a knife, glass, metal, or even paper (see Figure 7-12). A stabbing wound is deeper than it is wide. This happens because the force is applied toward the body. An incised wound, on the other hand, is generally longer than it is deep. The deepest point of the incised wound is usually in the center of the wound. Generally, the force is applied along the body, parallel to the skin. Incised wounds do not offer clues about the weapon. For example, a 4-inch incised wound could have been produced by an 8-inch blade, a 2-inch blade, or a piece of glass. These types of wounds have clean edges. They are not rough and ragged like the edges of a laceration caused by blunt force. Incised wounds are not typically fatal. Instead, the victim might need to get stitches to close the wound so that it can heal properly.

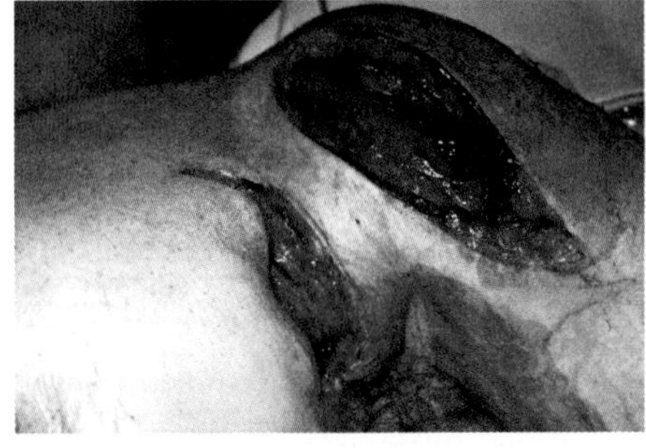

Figure 7-12. *Incised wounds to the shoulder.*

Fatal incised wounds most often appear on suicide victims. The victims cut an easily accessible part of the body—for example, the wrists or neck. Incised wounds can also be evidence of defensive wounds. These wounds are typically found on the palms of the hands (see Figure 7-13). As the victim tried to grab the weapon, the weapon cut the victim's hands.

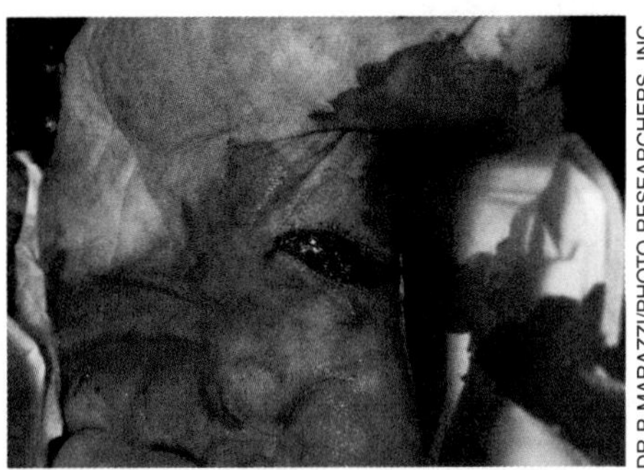

DR P. MARAZZI/PHOTO RESEARCHERS, INC.

Figure 7-13. *Defensive wounds.*

Chop Wounds

Heavy tools such as axes, machetes, and meat cleavers produce **chop wounds**. These tools have a cutting edge and produce incised wounds on the surface of the skin and deep internal injuries, including bone fractures (see Figure 7-14). The characteristics of the wound help forensic investigators narrow down the type of weapon used. For example, a cleaver will form a clean, thin wound. A cleaver is unlikely to break bone. An axe will often crush tissue and bone. A *forensic pathologist* is a medical doctor trained to identify why and how someone died and to present those findings in court. A forensic pathologist might observe the victim's bones under a microscope. Under a microscope, it is apparent that cleavers produce thin, sharp lines on the bone. Machetes produce more obvious lines that are coarse and less distinct. An axe is more likely to crush bone and does not tend to produce lines that are visible under a microscope.

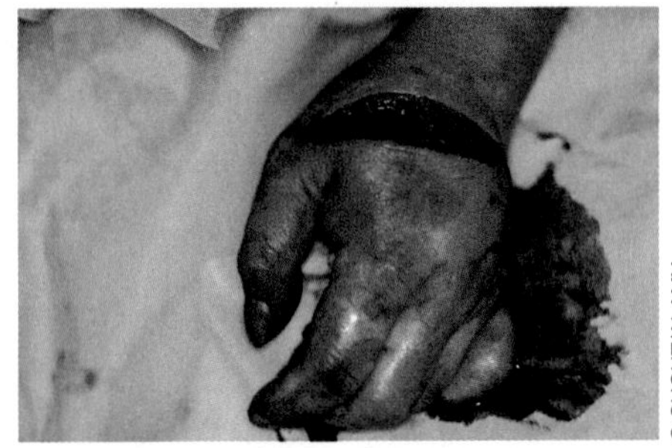

© SHOUT/ALAMY

Figure 7-14. *Chop wounds.*

Therapeutic Wounds

Surgery produces a **therapeutic wound**. For example, if a patient arrives at the hospital and is unable to breathe, a doctor may perform a tracheotomy. In this procedure, the doctor cuts the superficial layer of skin at the throat to expose the trachea. This cut allows the medical staff to insert a tube into the trachea, opening the airway and allowing the patient to breathe. Some therapeutic wounds can be mistaken for criminal wounds. For example, if a patient had a chest tube inserted to drain fluids, this wound might be interpreted as a criminal stab wound. Conversely, an old stab wound could be used for a chest drain. If the patient dies, the wound could be

CHAPTER 1 2 3 4 5 6 7 8 9 10 11 12 13 14 15

Did You Know?

Evidence of surgeries can be helpful to investigators in making identification of an unknown victim. Surgical scars, clips and clamps, artificial joints, plates, nails, screws, and wires confirm that a surgical procedure was done and may also provide model or serial numbers. The numbers on a surgical appliance can often be traced to a specific hospital and even to a patient.

misinterpreted as being therapeutic. To avoid confusion, tubes for therapeutic purposes are left in place for the medical examiner. The medical examiner completes a thorough evaluation of the victim's chart. If there are additional questions, the pathologist will consult the physician that had been treating the patient.

CHAPTER SUMMARY

- The extent of the physical trauma depends on the amount of force applied to the body, the part of the body affected, how much of the body's surface was affected, and the type of weapon used.
- Force is equal to mass times acceleration. The more massive the weapon, the greater the force exerted on the body. Additionally, the more the weapon is accelerating upon impact, the greater the force exerted on the body.
- Pressure is equal to force divided by surface area. Assuming that the force is equal, a weapon with a smaller surface area will exert greater pressure than a weapon with a larger surface area.
- Abrasions are blunt-force trauma caused when the top layers of skin are damaged.
- Contusions are broken blood vessels caused by blunt force and are commonly called bruises.
- Lacerations are tears in tissue caused by sliding or crushing blunt force.
- Stab wounds are an example of sharp-force trauma. Stab wounds are deeper than they are wide. The force is applied toward the body.
- Incised wounds, or cuts, are sharp-force trauma caused when a sharp object is forced along the body. These wounds tend to be wider than they are deep.
- Chop wounds are sharp-force trauma caused by sharp, heavy objects. These wounds involve incised wounds on the surface and deep internal injuries, often including broken bones.
- Therapeutic wounds are a result of surgery.

CASE STUDIES

Jesse L. Pitts (2007)

At 8:18 A.M. on September 20, 2007, police received a 911 call reporting the body of a girl in the river in Edgewater Park in Anderson, Indiana. The caller identified the girl as 14-year-old Amanda Brinker. Upon arriving at the park, the police discovered the girl's body lying face down in shallow water near the river's edge. The back of the girl's head was bloody, and a large gash was visible. Autopsy later revealed that Amanda had received seven or eight strikes to the head with a blunt object, causing multiple deep lacerations, a skull fracture, and bleeding and swelling of the brain.

Upon further investigation of the surrounding area, police identified an area where a large amount of blood covered the grass. They were able to determine that Amanda had been killed in the grass, and then her body had been moved to the water.

Police interviewed the man who had called 911 to report the body. Twenty-year-old Jesse Pitts told several different stories about the events of that morning. Originally, he claimed simply to have found the body while walking in the park with his girlfriend, Barbara Howard. Later, Pitts admitted to being in the park with Amanda when another man attacked Amanda. Then, he admitted to a physical altercation with Amanda in which he knocked her down and she struck her head. Finally, he admitted to swinging a metal pipe he had found on the trail and accidentally striking Amanda when she got too close.

Figure 7-15. *Jesse Pitts was convicted of Amanda Brinker's murder in 2007.*

A search of Pitts's car and residence revealed blood-spattered clothing and a jack handle covered with blood, hair, and fibers. The shape and size of the jack handle were consistent with the wounds on Amanda's head, and DNA analysis confirmed that the blood and hairs on both the clothing and the jack handle matched Amanda's. Pitts was arrested (see Figure 7-15) and tried for the murder of Amanda Brinker. He was convicted and sentenced to serve 65 years in prison.

Think Critically **1. Describe how Locard's exchange principle relates to this case.**

2. Were Amanda Brinker's injuries sharp-force trauma or blunt-force trauma?

Amanda Knox (2007)

Meredith Kercher, of London, England, was found murdered in her room in Perugia, Italy, on November 2, 2007. One of her roommates—Amanda Knox, of Seattle, Washington—would soon be one of the main suspects in the case.

Perugia police had been called to identify and return two cell phones found discarded in a yard on Via Sperandio. The phones were registered to Romanelli Filomena, who resided nearby on Via Della Pergola. When police arrived at Romanelli's house, they encountered two young people: Amanda Knox and her Italian boyfriend, Raffaele Sollecito. Knox and Sollecito claimed that they had called the police to report a broken window and suspected burglary. When Romanelli arrived and showed police into the house, nothing appeared to be missing. However, Romanelli thought it odd that the door to Meredith Kercher's room was locked and that her phones were off. When police entered the bedroom, they found the body of Meredith Kercher on the bed. Several wounds that had been caused by a sharp, pointed weapon could be seen on her neck.

Strange behavior and frequently changing stories led police to suspect Knox and Sollecito, as well as a third man, Rudy Guédé, who admitted to having been at the house for a date with Kercher. A bloody handprint in Meredith's room was sufficient evidence to confirm his presence at the crime scene and to cause his conviction for murder.

Originally, investigators believed the murder weapon had been a penknife. A small "flick" knife, 3.5 inches long and 1 inch wide, was deemed consistent with the wounds and was confiscated from Sollecito as a potential murder weapon. However, that weapon contained no DNA to

CHAPTER 1 2 3 4 5 6 7 8 9 10 11 12 13 14 15

© ALESSANDRO BIANCHI/REUTERS/CORBIS

OLI SCARFF/GETTY IMAGES

Figure 7-16. *Raffaele Sollecito (left), Amanda Knox (right), and Rudy Guédé have been convicted of killing Meredith Kercher.*

connect it to the crime. Instead, a kitchen knife with a 6.5-inch blade was presented as a murder weapon at the trial. The knife was found at Sollecito's home. The knife had Amanda Knox's DNA on the handle. There is some controversy as to whether the knife also had Meredith Kercher's DNA on the blade.

The coroner, Dr. Francesco Introna, claimed that at least one of the wounds on Meredith's neck had been made by a knife with a 3- to 3.5-inch blade. Independent forensic examiner Mariano Cingolani also contended that one of the three cuts on Meredith's neck should have been larger, based on its depth, if the kitchen knife had been the murder weapon. However, the angle of the victim's neck and the elasticity of her tissues could have affected the shape of the knife wounds.

Despite the many contradictions in the case, Amanda Knox and Raffaele Sollecito were found guilty on December 3, 2009, of killing Meredith Kercher and were sentenced to life sentences for the crime (see Figure 7-16). At the time of this publication, all three defendants were appealing the verdicts based on several types of evidence.

Think Critically

1. **Were Meredith Kercher's injuries sharp-force trauma or blunt-force trauma?**
2. **Mariano Cingolani suggested that, at the same depth, the kitchen knife would have produced a larger wound than one of the wounds found on Meredith Kercher's neck. Explain the relationship between the size and depth of the wound and the size of the knife blade.**

Bibliography

Books and Journals

Calaluce, Robert, and Jay Dix. (1998). *Guide to Forensic Pathology.* Boca Raton, FL: CRC Press.

DiMaio, Vincent J., and Dominick DiMaio. (2001). *Forensic Pathology* (2nd ed.). Boca Raton, FL: CRC Press.

DiMaio, Vincent J. M., and Suzanna E. Dana. (2007). *Handbook of Forensic Pathology* (2nd ed.). Boca Raton, FL: CRC Press.

Dix, Jay, and Michael Graham. (2000). *Time of Death, Decomposition and Identification: An Atlas* (Causes of Death Atlas Series). Boca Raton, FL: CRC Press.

Hanzlick, Randy, M.D. (2007). *Death Investigation: Systems and Procedures.* Boca Raton, FL: CRC Press.

Websites

http://anthropology.si.edu/writteninbone/making_ids.html
www.cdc.gov/ncipc/factsheets/tbi.htm
www.disastercenter.com/crime/uscrime.htm
http://emedicine.medscape.com/article/433404-overview
www.forensic-evidence.com/site/ID/toolmark_id.html

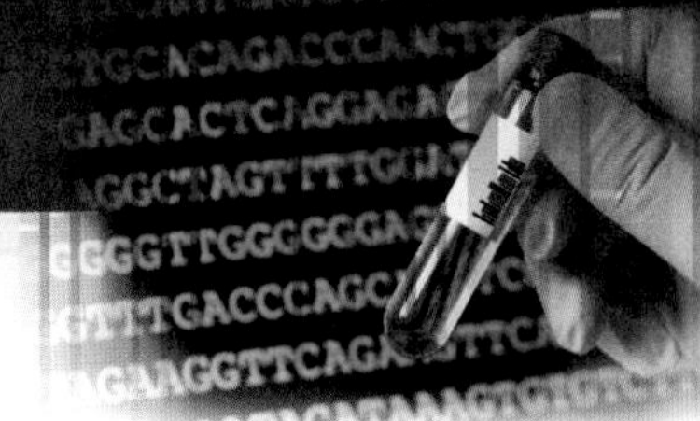

CAREERS IN FORENSICS

M. Margaret Knudson, M.D. F.A.C.S.: Trauma Surgeon

You are not likely to go to Dr. Margaret Knudson if you need care for the flu or a broken arm. However, many critically injured U.S. military troops who arrived at the Landstuhl Regional Medical Center in Germany after being injured in Iraq in 2006 needed her services. (Most of the troops there had suffered burn, blast, and blunt-force trauma.) At home, Dr. Knudson is a Professor of Surgery at the University of California at San Francisco.

Trauma surgeons complete a one- or two-year training internship in surgical critical/trauma care. The training qualifies them to treat most injuries to the neck, chest, abdomen, and extremities (other than bones). The trauma surgeon must be familiar with a wide variety of general surgical procedures and must be comfortable making quick, definite decisions, often without the benefit of having a patient's medical history at hand.

Perhaps the most important responsibility of a trauma surgeon is quick and efficient "triage" of a trauma victim—the surgeon assesses the victim and prioritizes the most life-threatening injuries. The surgeon then develops an overall plan for the patient's care, even as urgent care has begun.

Most trauma surgeons work in emergency rooms. However, like Dr. Knudson, many trauma surgeons are invaluable in war zones. Trauma surgeons are also needed at the sites of natural disasters. Dr. Knudson was one of thousands who volunteered her time and expertise to aid the most desperately injured in Haiti after the devastating earthquake on January 12, 2010.

Dr. Knudson's experience as a military trauma surgeon served her well in Haiti. She spent two weeks working on a retired U.S. Navy oil tanker, *Comfort*, which has been restored as a floating hospital with eleven operating rooms (see Figure 7-17). Over the course of just one month, Dr. Knudson and other medical personnel performed more than one thousand surgeries on Haitians who had orthopedic and traumatic injuries that were life-threatening or not healing.

After that experience, Dr. Knudson was able to take what she learned and share it with her colleagues and students. It is in this way that medical advances are made and that medical treatment improves.

U.S. NAVY PHOTO BY MASS COMMUNICATION SPECIALIST 2ND CLASS CHELSEA KENNEDY

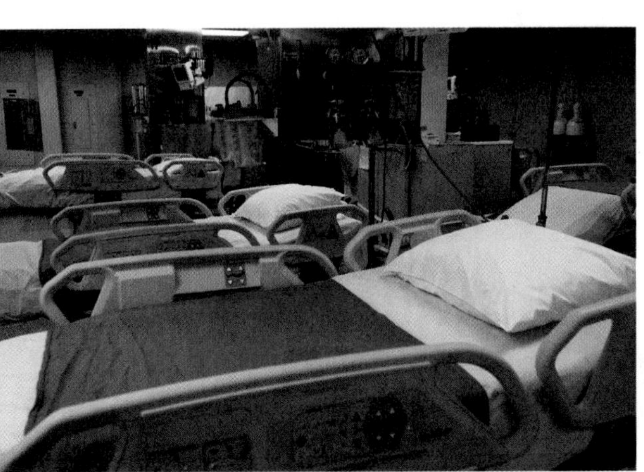

ALEX WONG/GETTY IMAGES

Figure 7-17. *Dr. Knudson and many other volunteers worked in the floating hospital on the* USS Comfort *after the January 2010 earthquake in Haiti.*

Learn More About It
To learn more about a career in trauma surgery, go to www.cengage.com/school/forensicscienceadv.

CHAPTER 1 2 3 4 5 6 7 8 9 10 11 12 13 14 15

CHAPTER 7 REVIEW

Multiple Choice

1. Which of the following is a type of blunt-force trauma? *Obj. 7.2*
   a. therapeutic wound
   b. hesitation marks
   c. impact abrasion
   d. incised wound
2. The outermost layer of skin is called the ________. *Obj. 7.1*
   a. hypodermis
   b. epidermis
   c. dermis
   d. endodermis
3. A five-year-old child is playing on the sidewalk with a next-door neighbor. While playing tag, the little boy starts running down the sidewalk and falls. The child scrapes his knee. What type of injury did he sustain? *Obj. 7.2*
   a. abrasion
   b. contusion
   c. laceration
   d. hematoma
4. Assuming that each weapon is accelerating at the same rate, which of the following would exert the most force? *Obj. 7.2*
   a. 10 kg brick
   b. 2 kg block
   c. There is not enough information to determine the answer.

Short Answer

5. An attacker swings a board with a mass of 2 kg toward a victim. The acceleration is 10 m/s^2 when the board hits the victim. How much force did the board exert on the victim? *Obj. 7.2*

6. Using the force you calculated in question 5, determine the pressure exerted on the victim. (The board impacted 232 cm^2 of the victim's back.) *Obj. 7.2*

7. Discuss how a therapeutic wound could be misinterpreted as a wound caused by a violent crime or assault. *Obj. 7.3*

8. Differentiate between blunt-force wounds and sharp-force wounds. *Obj. 7.2, 7.3*

9. Compare and contrast the three types of abrasions. Give examples of each. *Obj. 7.3*

10. A body is discovered, and what appears to be bruising is observed on the back. The body is not yet in rigor. How would investigators determine whether the discoloration is due to lividity or a contusion? *Obj. 7.1*

CHAPTER 1 2 3 4 5 6 7 8 9 10 11 12 13 14 15

11. Describe the factors that affect the appearance and severity of a victim's injury. *Obj. 7.1*

____

____

____

____

12. Examine the role that force and pressure have in blunt-force trauma. *Obj. 7.2*

____

____

____

____

Think Critically

13. A death investigator arrives at an apartment where a woman has been brutally stabbed with a barbeque fork. Initially, he is able to determine that the victim was stabbed multiple times. How will the investigator study the marks on the body to determine how many times she was stabbed? *Obj. 7.3*

14. A man is discovered in his dormitory room. He has been stabbed with a knife multiple times. By looking at the victim, how could investigators develop a profile of who may have committed this crime? *Obj. 7.3*

15. Using the force, mass, and acceleration formulas, design a scenario in which a victim sustains physical trauma. Two of the three components must be provided to determine the third. Include the solution to your scenario. *Obj. 7.2*

Best of Friends
By: Pavan Patel

Tavares High School
Tavares, Florida

Bill, John, Jane, Lily, Jessie, Carlos, and Shayna met in a class in college. Even though they all came from very different environments and home situations, they became fast friends.

Always the prankster, Bill was the clown of the group. He had grown up in a single-parent household and was more of a friend than a son to his mother. His mother strongly suggested that he join the military or go to school. Little did he know, the "safe" choice wasn't so safe.

Jessie was hard-working and very studious. She got along with the stepfather who had helped to raise her. Jessie wanted to make a difference in the world and was studying criminology.

The athlete of the group was Jane. She had been a shy, pudgy child but had blossomed into a beautiful young woman. She aspired to be a nutritionist and personal trainer.

John had grown up poor with an abusive stepfather and a mother who didn't care. He wanted to be a social worker and psychologist. John thought this would put him in a position to protect kids who were in situations like his.

Carlos wanted nothing more than to make his father proud of him. So far, nothing had worked. Men in his family had always been hard workers, yet none had pursued education beyond high school. Carlos had been the first to go to college, and he hoped that would help him earn his father's respect.

Shayna grew up in a house with two supportive parents. She had always been interested in people and wanted to do something that would make a difference. Shayna's history of blackouts might have prevented her from being accepted into nursing school, so she didn't tell anyone except her best friends.

And then there was Lily. Lily's only interests were makeup, clothes, and jewelry. She was in college to socialize.

The group of friends decided to rent a house together to help cut expenses and everything was working well—for a while, at least. The problems began when Shayna, Jessie, and Lily all began competing for the affections of Carlos. On one afternoon, Jane and Lily left to go shopping while the others stayed home. Carlos and Shayna had argued earlier in the day but seemed to have ironed out their differences. By early evening, everyone had drifted off to different parts of the house. All was quiet until a scream echoed down the hallway. Bill came running out of his room and met Shayna coming out of the bathroom. Jessie appeared seemingly out of nowhere as John stumbled groggily out of his room after a nap.

They all saw it at the same time. Carlos was lying on the sofa with a pillow over his face as blood began to appear. Shayna cautiously removed the pillow to reveal a dead Carlos. His throat had been slit, and several lacerations were visible on his chest. Jessie and Bill stood in stunned silence, but John ran for the bathroom and threw up. Shayna was the one who sprang into action and called the police.

Lily and Jane rushed back home when they heard what had happened. Since the investigation was ongoing and they could not be at the house while evidence was being gathered, they went out for pizza. After they returned and finished giving their statements, they all went to their rooms.

A scream at 1:00 A.M. woke them. Shayna was the first one in the room. Lily's eyes had been gouged out and she had been stabbed several times. Shayna screamed again, and the others came running into the room, The scene was gruesome. Jessie began yelling and accusing Shayna of the crime. Shayna claimed she had blacked out and couldn't remember anything. Once more the police were called.

Investigators collected hair found at the scene. One of the hairs was under the body. There also appeared to be some blood under Lily's fingernails. Bloody clothes found in Shayna's room seemed to point to Shayna, so she was taken into custody.

The hair was inconclusive, but was consistent with that of Jessie. Lab analysis confirmed the DNA under the fingernails belonged to a female, but matched nothing in the database. The skin cells and sweat on the clothing provided DNA evidence that was a match to that found under the fingernails. None of the DNA matched the sample collected from Shayna.

Police returned later the next day to arrest the real murderer.

Figure 7-18. *Police arrest the real murderer. Who is it?*

Activity:

Answer the following questions based on information in the Crime Scene S.P.O.T..

1. Who killed Carlos and Lily? Explain how Locard's exchange principle helped police find the murderer.
2. Why did the police suspect Shayna?
3. Describe the wounds found on the victims. What type(s) of trauma are the wounds?

WRITING

4. After reading the story, write a one-page paper that describes the roles of the crime-scene investigator and various lab personnel in this case.

Preventing Adolescent Crime Together™ (P.A.C.T)

Introduction:

Many elderly people live in assisted-living facilities for various reasons. The most common reason is that neither the person nor his or her family is able to provide the care the person needs. At other times, an older person may be sick and wish to be surrounded by people while coping with and managing the illness. Unfortunately, some elderly people are neglected or physically abused. Young people who take time to volunteer at assisted-living facilities build character traits, including empathy, interpersonal skills, and self-confidence.

Procedure:

1. Contact a local assisted-living facility.
2. Set up an appointment to meet with the director.
3. Develop a questionnaire for residents—likes, dislikes, birthday, etc.
4. Take the questionnaire to the appointment with the facility director.
5. Ask the director if you can "adopt a grandparent" and that you are particularly interested in an individual who receives few or no visitors.
6. After a person is selected, ask if you can meet your adopted grandparent and introduce yourself. This would be a good time to give your questionnaire to your grandparent.
7. If possible, have your adopted grandparent fill out the questionnaire while you are there, and take it with you. You may need to help him or her fill out the form.
8. Make arrangements with the director for when and how often you will visit. Plan to visit at least once a week. Try to bring something with you each time you visit. Handmade gifts and cards are always well appreciated.
9. On special occasions make an effort to visit, even if it is for just a few minutes.
10. Reflect on your experience and discuss with the class.

Figure 7-19. *Many older people move to retirement communities before they need medical support.*

ACTIVITY 7-1 *Ch. Obj: 7.1, 7.3*
STAB WOUNDS

Objectives:

By the end of this activity, you will be able to:

1. Distinguish between the wounds made by different knives.
2. Identify marks on "bone" made by stab wounds.
3. Explain the difference between perforating and penetrating wounds.

Materials:

(per group of four students)
1 gelatin "body"
foam tray
4 plastic knives: A, B, C, and D
metric ruler
4 toothpick flags (labeled A, B, C, and D)
scalpel

Safety Precautions:

Wear safety goggles to protect your eyes and an apron to protect clothing. Gloves may also be worn.
Make sure long hair is pulled back and dangling jewelry is removed.
Use caution when making the stab wounds and when using the scalpel.

Procedure:

1. Obtain the gelatin "body" from your teacher. Place the body on the foam tray.
2. Examine each of the knives. Note anything unusual or distinct about each one. Record your observations in Data Table 1.
3. Measure each knife. Record the length, width, and thickness of the blade in Data Table 1.

Data Table 1. Knives.

	Length of Blade	Width of Blade	Thickness of Blade	Distinguishing Characteristics
Knife A				
Knife B				
Knife C				
Knife D				

4. Using knife A, make a stab wound through the body. Use a toothpick flag to mark this wound A.
5. Repeat step 4 with knives B, C, and D.
6. Carefully examine each wound. Measure the width and thickness of each entry wound. Record the measurements in Data Table 2.

7. Using the scalpel, make a cut that is perpendicular to the wound closest to the outside edge of the body. This cross-section cut will allow you to measure the depth of the wounds. Observe the shape and texture of each wound. Record the measurements and qualitative observations in Data Table 2.

Data Table 2. Wounds.

	Width of Wound	Thickness of Wound	Depth of Wound	Distinguishing Characteristics
Wound A				
Wound B				
Wound C				
Wound D				
Unknown				

8. Observe the unknown provided by your teacher. Repeat step 7 for each wound on the unknown.
9. After all measurements and comparisons are made, carefully remove the "bones" and "organs" from the body.
10. Examine each body part for damage from the stabbings. Look for penetrating and perforating wounds. Note any distinguishing characteristics. Record your observations in Data Table 3.

Data Table 3. Internal injuries.

Organ/Bone	Type of Damage	Observations

Questions:

1. Which knife was most likely used to produce the wounds on the unknown? What evidence do you have to support your answer?
2. Compare and contrast perforating and penetrating wounds.
3. Without the flag labels, would you have been able to distinguish between markings on the "bone" made by the various knives? Why or why not?

4. Write one or two paragraphs describing how this activity modeled the investigation of sharp-force trauma. Evaluate the strengths and weaknesses of this model.

ACTIVITY 7-2 *Ch. Obj: 7.2, 7.3*
BONE TRAUMA

Objectives:

By the end of this activity, you will be able to:

1. Develop a plan for testing the effects of different forms of trauma on "bones."
2. Describe the effects of various forms of trauma on bones.

Background:

Degree of decomposition is one significant factor that medical examiners evaluate when they are completing a death investigation. Forensic anthropologist Dr. William Bass has studied in great detail the decomposition process of human remains. Bass established the Anthropology Research Facility at the University of Tennessee. At this facility, which is often called the "Body Farm," Bass and his team place donated bodies in various environments to study the body's response. For example, he and his teams study how the stages of decomposition relate to entomological (insect) evidence. Medical examiners, crime-scene investigators, and other experts in forensic science look to the Body Farm for cutting-edge information about estimating time of death.

Introduction:

The Body Farm is currently interested in expanding the department to include skeletal trauma. Your team, an independent forensic anthropology firm, has been hired to conduct experiments on various skeletal remains. These tests must demonstrate a wide range of trauma that could be inflicted on bones. Ultimately, the Body Farm wants to develop a database of as many examples of skeletal trauma as possible with pictures and explanations to be used by law-enforcement personnel around the country. Your team will complete this task and write a report of your findings. Your team will be competing for the opportunity to be long-term consultants for the Body Farm, an incredibly prestigious opportunity.

Materials:

(per group of three or four students)
modeling clay
resealable plastic bags (15)
permanent marker

(for the entire class)
various tools, such as a hammer, keys, ice pick, screwdriver, stapler, forks, crowbar, tire iron, fireplace poker, knives, meat mallet, or bat

Safety Precautions:

Wear gloves, aprons, and goggles throughout the activity.
Be careful when working with tools.

Procedure:

Part 1

1. Using bone models and/or bone charts, fashion 15 bones out of the clay for your group. Make the bones to scale and as close in shape as possible. (Note: The bones do not have to be all different, but make more than one type and at least three of each.)
2. Package your bones to be used later in Part 2.

Part 2

1. Plan three to four experiments to determine the effects on bones of various types of trauma. In order to obtain valid results, you must inflict only one trauma on each bone, and you must repeat each trial at least three times. Be sure to include the following in your plan:
    - The five types of trauma you will inflict
    - The type of bone you will test
    - Your hypothesis
    - Your procedures
    - A list of materials you will need
2. Give your written plan to your teacher for approval. If a camera is available, your plan should include taking photographs of the results. Your teacher will let you know if any of the materials you need are not available or if your procedure is not feasible. In either case, you will need to revise your procedure.
3. Once your plan is approved, follow your procedures. Take careful notes and prepare any necessary data tables or graphs.
4. After each experiment, place each injured bone in a plastic bag. Seal the bag. Label it with your team name and the type of trauma.
5. Write a thorough and systematic report of your procedures, data, analysis, and overall findings.

Questions:

1. What tools did you use to cause trauma to your set of "bones"? Why did you choose those tools?
2. Did you identify a pattern of trauma with any of the tools you used? Were any traumas caused by one tool similar to traumas caused by another tool? Explain why or why not.
3. When testing bones with the same tool multiple times, did you get the same outcome each time? If not, what are some considerations for the variation you experienced in your results?
4. Discuss the importance of truthful reporting of experimental data.
5. How could the development of a database on various skeletal traumas aid law-enforcement officials in solving crimes?
6. Evaluate how well this activity modeled the work of a forensic anthropologist. What are some strengths and weaknesses of this activity as a model?

CHAPTER 8

Autopsy

MADISON RUTHERFORD

Madison Rutherford was a 34-year-old financial advisor traveling to Mexico in July 1998. One evening, Rutherford drove his rental car off the side of the road. The car caught on fire, and Rutherford died at the scene. Rutherford had life insurance policies worth $7 million. His wife was the sole beneficiary—she would get all of the money. Before Kemper Life Insurance Company agreed to pay the claim, however, they investigated the death.

Kemper Life hired Dr. Bill Bass, a world-renowned forensic anthropologist, to consult on the case. Bass had founded the Forensic Anthropology Center, also known as the "Body Farm," at the University of Tennessee. Scientists at the Body Farm have researched extensively the effect of fire on the human body. Dr. Bass went to Mexico to examine Rutherford's remains. Because the fire had been so intensely hot, the body had burned very quickly, leaving behind mostly bone fragments. In a fire, the body undergoes very distinct changes. According to Bass, the extremities burn first and then the skull fills with fluid. The fluid puts large amounts of pressure on the skull, and the skull bursts. Typically, the pelvic region is the last part of the body affected. Rutherford's skull had exploded during the fire. However, fragments of the skull were found in an unlikely position on the floor of the vehicle.

AP PHOTO/NORTHERN VIRGINIA DAILY, RICH COOLEY

Figure 8-1. Dr. Bill Bass

Bass was able to determine from fracture patterns that the body was one that was freshly decomposing. The bones showed signs of arthritis, not typical of a man in his 30s. Bass took the teeth from the scene and compared them to Rutherford's dental records. Bass knew almost immediately that the teeth were from an elderly man of Native American descent. Comparison with Rutherford's dental records confirmed that these were not Madison Rutherford's teeth.

Based on Bass's findings, the insurance company decided to complete a more formal and thorough investigation. A private investigator eventually tracked Rutherford down in Boston. To fake his death, Rutherford had stolen a body from a Mexican mausoleum and placed it in his rental car. He then lit the car on fire. The insurance company notified the FBI. Rutherford pled guilty to wire fraud. Wire fraud is intentional misrepresentation of the truth using an electronic communication system, such as a telephone line or money wiring service. Rutherford was sentenced to five years in federal prison.

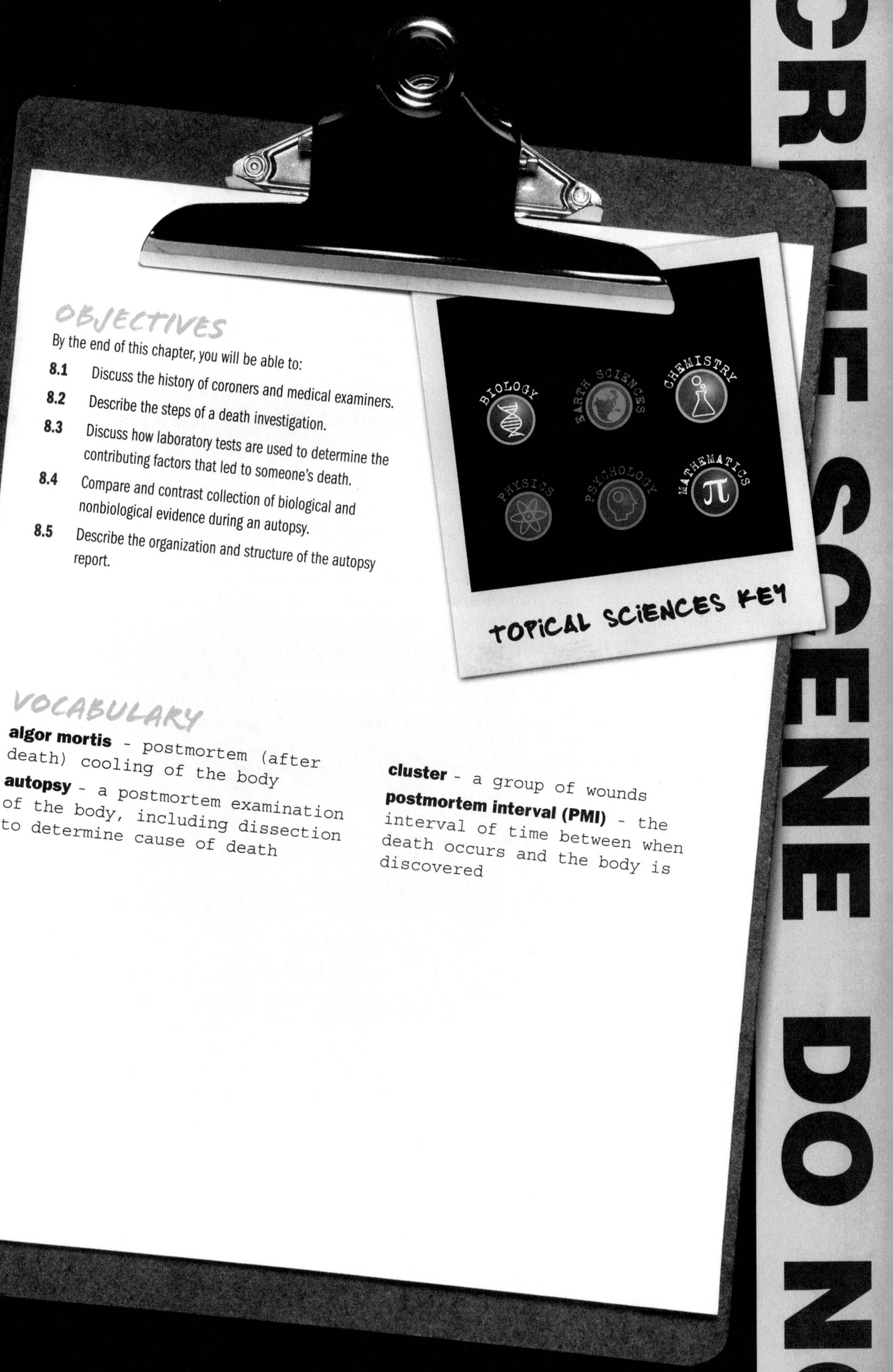

OBJECTIVES

By the end of this chapter, you will be able to:

8.1 Discuss the history of coroners and medical examiners.

8.2 Describe the steps of a death investigation.

8.3 Discuss how laboratory tests are used to determine the contributing factors that led to someone's death.

8.4 Compare and contrast collection of biological and nonbiological evidence during an autopsy.

8.5 Describe the organization and structure of the autopsy report.

VOCABULARY

algor mortis - postmortem (after death) cooling of the body

autopsy - a postmortem examination of the body, including dissection to determine cause of death

cluster - a group of wounds

postmortem interval (PMI) - the interval of time between when death occurs and the body is discovered

INTRODUCTION

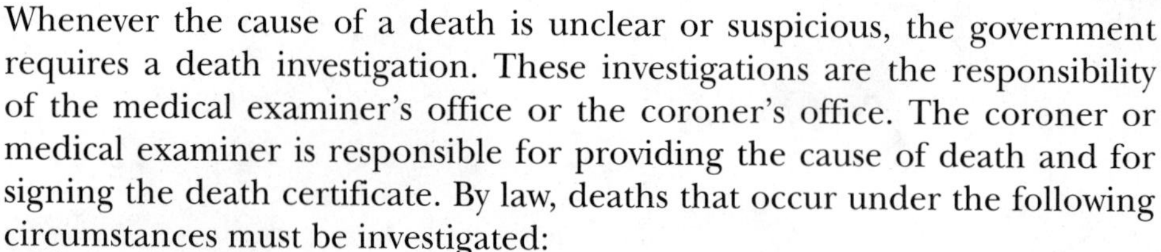

Whenever the cause of a death is unclear or suspicious, the government requires a death investigation. These investigations are the responsibility of the medical examiner's office or the coroner's office. The coroner or medical examiner is responsible for providing the cause of death and for signing the death certificate. By law, deaths that occur under the following circumstances must be investigated:

- Violent crime, suicide, or accident
- Within 24 hours of entering a hospital or as a result of surgery
- A natural death when a doctor is not present or the patient is not under the care of a medical facility
- Occurs in police custody or in a correctional facility
- Results from a communicable disease that may pose a threat to public health

When possible, before the body is removed from the scene, the death investigator completes an initial assessment of the body. The death investigator acts as a liaison between law enforcement and the medical examiner's office. Ultimately, the medical examiner will be responsible for estimating the time of death and determining the manner of death. Recall from previous coursework that the *manner of death* is the way a person dies—natural death, accidental death, suicidal death, homicidal death, or undetermined (see Figure 8-2). Additional preliminary assessments at the scene may include identifying signs of trauma, sexual assault, or a struggle. Upon arrival at the medical examiner's office or the coroner's office, extensive chemical and physical tests will be completed to determine how the victim died.

Did You Know?

Death certificates must be complete and accurate because information gathered from a death certificate is used in statistical reporting of diseases and conditions. State and national death statistics are also utilized to determine research and development funding, set public health goals, and measure health status at the local, state, national, and international levels.

Manner of Death	Description
Natural	Most common manner of death; body function failure as a result of age, illness, or disease
Accident	Unintentional
Suicide	Victim intentionally takes his or her own life; the cause of death is usually a gunshot, hanging, or poisoning
Homicide	One individual takes the life of another intentionally or through a negligent or reckless act; up to the court to determine whether the death is a murder; in certain circumstances, a person is justified in killing someone else in self-defense
Undetermined	The pathologist is not able to determine the manner of death, even after all internal and external examinations are completed and toxicological tests are evaluated

Figure 8-2. *There are five possible manners of death.*

HISTORY OF CORONERS AND MEDICAL EXAMINERS

Obj. 8.1

Two systems of death investigation exist in the United States—the coroner system and the medical examiner system. As of 2006, counties in 29 states had some form of coroner system, and counties in 40 states had a medical examiner system. A coroner is often an elected official. Some jurisdictions require that the coroner be a medical doctor. In these areas, the coroner is required to perform an autopsy. An **autopsy** is a postmortem (after death) examination of the body, including dissection of the corpse. Some jurisdictions do not require that the coroner be a medical doctor. In this case, the coroner must hire a forensic pathologist to complete autopsies. A *forensic pathologist* is a medical doctor who is trained to identify why and how someone died and to present those findings in court. Medical examiners are medical doctors; some are also forensic pathologists. Medical examiners perform autopsies. Coroners and medical examiners provide official reports on the cause and manner of death as well as an estimate of the time of death.

Did You Know?

The term *autopsy* has been in use since 1678, when it referred to determining the cause of death by examining a body. The term *autopsy* means literally "to see for one's self."

The coroner system dates back to twelfth century England. At the time, coroners were called *crowners* and were elected officials who protected the people from "less than honest" sheriffs. They also were responsible for overseeing financial matters. Eventually, coroners began to investigate deaths. They were able to arrest suspects (including the sheriff) and were in charge of gathering a group of individuals who would decide the fate of the suspect. Coroners soon recognized the need for an autopsy. The autopsy helped the group determine the cause of death and, in many cases, the guilt or innocence of the suspect. The original settlers to the United States brought this system with them, and it remained intact until the 1870s.

The first medical examiner was appointed in Massachusetts in the late 1870s. However, the first true medical examiner system was not formed until 1918 in New York City. Today, a fully functional medical examiner's office serves the public and judicial needs of the community or the state. The primary goal of a medical examiner's office is to determine cause and manner of death. The chief medical examiner is required to be an experienced physician with advanced training in *pathology*—the study of the causes and processes of disease. The medical examiner is able to perform autopsies and is an integral part of legal investigations.

THREE STEPS OF A DEATH INVESTIGATION

Obj. 8.2, 8.3, 8.4

When there is a suspicious death, investigators conduct a death investigation. The purpose of the investigation is to determine the cause, mechanism, and manner of death. If the death was a homicide, investigators try to determine who caused the victim's death. The investigation can be divided into three major phases:

- First, a preliminary investigation is conducted at the death scene.
- Then, the body is transported to the morgue, where the medical examiner examines the body and performs an autopsy.
- Finally, the medical examiner orders lab tests on biological evidence collected during the autopsy.

AT THE DEATH SCENE

A death investigator is employed by the coroner's office or the medical examiner's office and is responsible for the initial assessment of the body and the death scene. Crime-scene and death investigators take photographs and make sketches of the body at the scene (see Figure 8-3). Photographs document the position of the body. Death investigators take photographs of the face for identification purposes. They also photograph the underside of the body and the surface beneath the body to make a visual record of lividity, blood and trace evidence, and any other characteristics that could provide clues to the medical examiner. Death investigators document any signs of trauma. Additionally, they make observations that will help the medical examiner estimate the time of death. It is not possible to determine an exact time of death. However, the approximate time of death can help investigators identify suspects and confirm alibis. The time between the death and the discovery of the body is called the **postmortem interval (PMI)**. The estimation of time of death is less accurate the longer the postmortem interval. At the scene, death investigators look for signs of livor mortis and rigor mortis. Recall from Chapter 6 that livor mortis is the body's change in color as the blood pools due to gravity. Livor mortis is first apparent 30 minutes to 2 hours after death. It becomes fixed 8 to 12 hours after death. Recall also that rigor mortis is the stiffening of the body due to the lack of ATP in muscle cells. Rigor mortis begins to appear 2 to 4 hours after death and usually subsides within 36 hours. All of this information helps investigators estimate time of death.

Figure 8-3. *Investigators take photographs around the crime scene.*

Digging Deeper
with Forensic Science e-Collection

Investigators use a number of methods to help determine a victim's time of death. One method utilizes the fluid behind the eyes. This fluid is called *vitreous humor.* The concentrations of chemicals found in the vitreous humor change in a predictable manner after a person dies. Using the Gale Forensic Science e-Collection at www.cengage.com/school/forensicscienceadv, research the postmortem changes in the vitreous humor, and then answer the following questions: (1) What chemical changes occur in the vitreous humor after a person dies? (2) How are the changes in concentrations used to determine PMI? (3) Why is determining the time of death or the PMI just an estimate at best?

The degree of decomposition also helps death investigators estimate the time of death. At the scene, death investigators look for obvious signs of decomposition. They also look for evidence, called *scene markers,* at the scene. A scene marker is nonbiological evidence that provides clues about the time of death. For example, is unopened mail or a newspaper near the body? Are dated sales receipts in the victim's pockets? Scene markers are especially helpful if the body is severely decomposed.

PHILIPPE PSAILA/PHOTO RESEARCHERS, INC.

Figure 8-4. *Proper handling of the body to protect trace evidence.*

At the conclusion of the initial assessment, evidence is collected, a chain of custody is established, and the evidence is sent to the lab. As soon as the victim is identified, the victim's name is added to the chain of custody. This may happen at the scene or at the morgue. Paper bags are placed over the body's hands and feet to protect trace evidence (see Figure 8-4). The body is then wrapped in a clean sheet or bag and transported to the morgue. Placing bags over the victim's hands and feet and wrapping the body in a sheet or bag protects the trace evidence, preventing a cross transfer of material. Look back at Chapter 1 if you need to review Locard's exchange principle.

While the body is being processed and transferred, death investigators begin to interview witnesses. Once the victim has been identified, they also interview the victim's family. These interviews help investigators gather information about the victim's normal routine. Additionally, the investigator might review the victim's medical history and police reports. This information is used to determine which tests the medical examiner orders or what evidence he or she might expect to find. For example, checking for trace evidence during an external and internal exam will be different depending upon whether the victim was shot or died in a car accident.

MEDICAL EXAMINATION

As soon as the body arrives at the morgue, the medical examiner assigns a case number. The chain of custody is updated to include the date and time of arrival, who transported the body, and who received the body. The medical examiner then performs a thorough examination of the body. The purpose of the autopsy is to determine the mechanism, manner, and cause of death. *Cause of death* is the immediate reason for a person's death. For example, the victim might have died from a fall. The *mechanism of death* is the body's physiological response that caused the cessation of life. For example, the victim of a fall might have bled to death or the victim's brain function might have stopped.

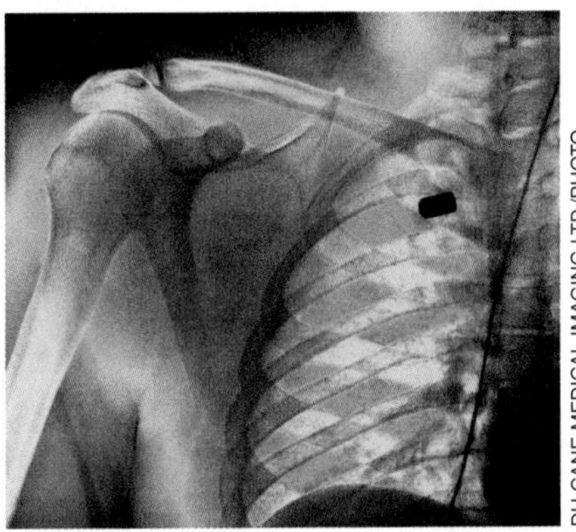

DU CANE MEDICAL IMAGING LTD./PHOTO RESEARCHERS, INC.

Figure 8-5. *This X-ray shows a bullet lodged in the victim's chest.*

External Examination

After all of the preliminary paperwork has been completed, the medical examiner looks at the victim's clothing, completes an external examination of the body, and orders X-rays (see Figure 8-5). The medical examiner then examines the body for any biological and nonbiological evidence. *Biological evidence* is material from living

Digging Deeper
with Forensic Science e-Collection

A *pathologist* is a medical doctor who specializes in the investigations of death and disease. Using the Gale Forensic Science e-Collection at www.cengage.com/school/forensicscienceadv, research the various subspecialties in pathology. Write a brief description of each.

or once-living sources. Hairs, blood, and plant debris are common types of biological evidence. *Nonbiological evidence* is material from nonliving sources. For example, the medical examiner might find glass, soil, and artificial fibers on the victim. He or she collects, preserves, and processes any trace evidence found on the body. For instance, if a victim attempted to fight off the attacker, the victim may now carry several pieces of trace evidence, including skin cells under fingernails, fibers from clothing, and blood (see Figure 8-6). Each piece of evidence is given an individual chain of custody and sent to the crime lab.

Autopsy

As part of the preliminary examination, the medical examiner determines body temperature. A healthy human body temperature is approximately 37°C (98.6°F). Upon death, the body temperature begins to fall at a constant rate. This postmortem cooling of the body is called **algor mortis**. The most accurate measurement of core body temperature is taken by inserting a

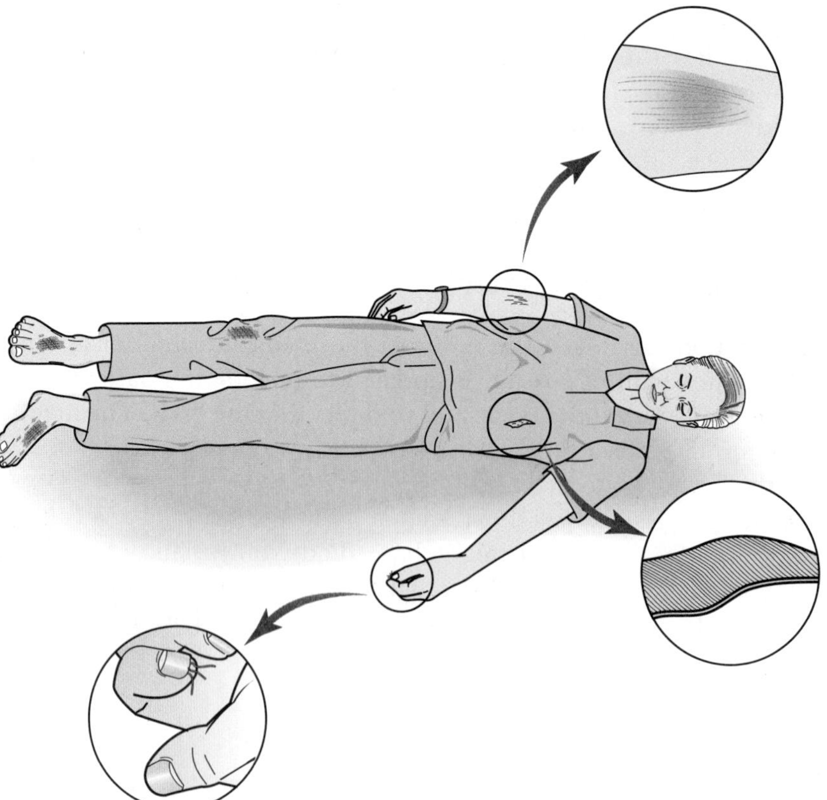

Figure 8-6. *During the external examination, biological and nonbiological materials are often collected from the victim. For example, the medical examiner may collect blood, fibers, skin cells from under the fingernails, grass, and soil.*

thermometer into the liver. However, at the death scene, death investigators sometimes measure body temperature rectally. This method is less reliable but significantly more convenient. Investigators must be careful not to disturb or contaminate any evidence around the site where the temperature is being taken. On average, the temperature tends to fall by about 1.5°F every hour until it reaches the *ambient temperature,* which is the temperature of the area surrounding the body. Assuming that the body temperature was normal at the time of death, the following formulas can be used to estimate the time of death within the first 12 hours:

$$\text{Time since death (in hours)} = \frac{37^\circ\text{C} - \text{Current body temperature}}{0.78}$$

$$\text{Time since death (in hours)} = \frac{\text{Normal body temperature (F)} - \text{Current body temperature}}{1.4}$$

If the body is found more than 12 hours after death, investigators look for specific clues to determine that the PMI is greater than 12 hours. In the first 12 hours, the head, arms, and legs cool rapidly. After 12 hours, most of the heat has escaped from the extremities. The remaining heat dissipates from the torso more slowly. In the first 12 hours the body temperature will drop 9.4°C. If the body temperature is below 27.6°C, then the victim has been dead for more than 12 hours. The following formula would be used if the victim died more than 12 hours before the body was found:

$$\frac{27.64^\circ\text{C} - \text{Current temperature}}{0.39} + 12$$

The exact rate of algor mortis is affected by several factors, including the temperature in the area where the person died. Because so many variables affect the rate at which a body cools, algor mortis can offer investigators only an estimation of the time of death. Therefore, investigators rely on multiple clues to narrow down the time of death.

During the autopsy, the medical examiner removes and examines the following structures: brain, larynx, hyoid bone, heart, and lungs. The hyoid bone is in the neck between the lower jaw and the larynx. It provides support for the tongue and is attached to various muscles that aid in swallowing. If a victim has been strangled, the hyoid bone may be broken. The medical examiner also collects samples of body fluids, such as blood, urine, and bile. He or she then removes the internal organs and records the mass of each organ. The medical examiner analyzes the stomach contents. Stomach contents provide clues about the time of death.

If the body is in advanced decomposition, the medical examiner collects tissue from various organs. These tissues can be used for DNA analysis and toxicology screening. Proper handling, collection, and preservation of biological evidence maintain the integrity of the evidence. Each item of evidence is placed in an appropriate container (see Figure 8-7) and labeled with the following information:

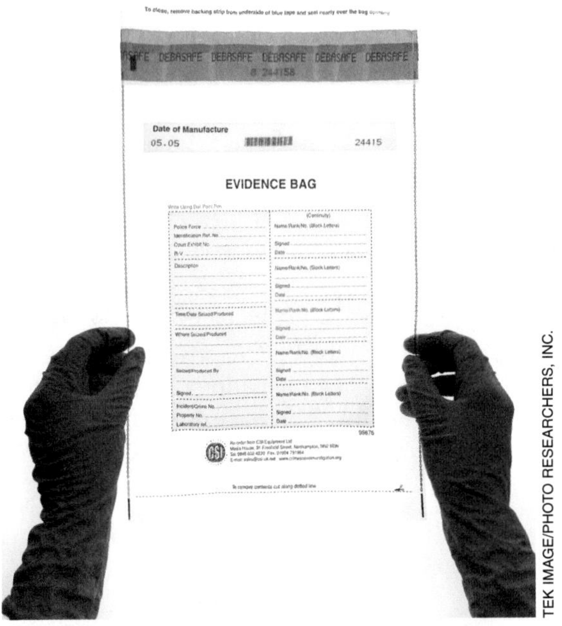

TEK IMAGE/PHOTO RESEARCHERS, INC.

Figure 8-7. *Evidence bag.*

- Name of victim
- Case number
- Date of exam
- Where item was recovered
- What the item is (or is assumed to be)
- Initials of medical examiner who collected it

Although each local and state agency has its own procedures for collecting biological evidence, Figure 8-8 provides a general list of guidelines.

Nonbiological evidence is also collected during an internal exam. When bullets are retrieved from the body, the investigator will use a gloved hand or a tool with a rubber tip to prevent marking the bullet. The medical examiner is careful to keep the bullet intact. The bullet is placed in a labeled envelope. The investigator identifies the area of the body from which the bullet was taken.

Biological Evidence	Proper Method of Collecting Evidence
Blood for DNA	If the victim is unidentified, the medical examiner collects blood from an artery, such as the femoral artery in the thigh, and places it on a blood card. This card is sent to the DNA lab for identification. If dried blood is found on the victim's body, it might have come from the perpetrator. Using a scalpel, the examiner removes the dried blood from the skin and places it in a test tube. If there is too little dried blood on the skin to use a scalpel, the examiner uses a swab moistened with saline solution. The swab must be air dried before the examiner places it in a sterile test tube. The techniques used to analyze DNA are covered in detail in Chapter 10.
Semen	If the medical examiner suspects that the victim was sexually assaulted, he or she places cotton swabs inside the vagina and rectum. The examiner removes the swabs and makes slides of the evidence. The slides and swabs are air dried. The examiner then packages and labels the slides and swabs. The examiner also swabs the mouth around the gums and teeth. Moist, sterile gauze is used to collect any additional suspicious material in the mouth. The gauze is then air dried and placed in a test tube, envelope, or bag.
Hair	The examiner collects hair from the victim's body and places it in an envelope. The origin of the hair is unknown—some may belong to the perpetrator; some may belong to the victim. The examiner documents where the hair was collected. He or she also collects control hairs from multiple areas of the body. The control hairs must be pulled from the root to get the follicular tag so that DNA analysis can be performed. These control hairs are compared to the hairs of unknown origin to determine whether the unknown hairs are consistent with the victim's hair. In rape cases, the pubic hair is combed for hairs not belonging to the victim.
Tissue	The examiner places each tissue sample in a separate sterile container. Tissue samples are stored in a cool area, such as a refrigerator, to await further examination.
Fingernails	If the examiner finds loose pieces of fingernail on the body, he or she collects them and places them in an envelope. The examiner then cuts the victim's fingernails and places each cut fingernail into its own clean, sterile envelope. The envelopes are labeled with the location of the fingernail, such as "right index finger."
Bite Marks	A *forensic odontologist* is a doctor trained to examine dental records and bite mark patterns during a criminal investigation. If one is available, a forensic odontologist will examine any bite marks prior to the autopsy, before the body is cleaned. If an odontologist is not available, the medical examiner will wipe the bite mark with a cotton swab moistened with saline and then with a dry swab. Both are air dried and placed in test tubes. This technique collects saliva around the wound. The saliva may contain DNA. The medical examiner will also take pictures of the wound with a ruler in the field of view. If the materials are available, the examiner will make a cast of the bite mark. Forensic odontology is covered in detail in Chapter 11.

Figure 8-8. *Processes for collecting biological evidence.*

Other nonbiological evidence includes any surgical equipment, such as artificial limbs or rods inserted after serious fractures. This equipment often includes a serial number that can be traced to the hospital or even the surgeon who performed the surgery. This information can then be used to identify the victim.

Did You Know?

In August 2009, a woman's body was discovered in a suitcase in a dumpster in Santa Cruz, California. Her fingers and teeth had been removed, making identification particularly difficult. Investigators were able to identify the body of Jasmine Fiore from the serial numbers found on her breast implants.

LABORATORY ANALYSIS

The medical examiner sends evidence to the lab for analysis. He or she might order toxicology, histology, neuropathology, microbiology, and serology tests on the biological evidence. These tests can help determine the cause or manner of death. The outcome of lab tests can also help identify or exclude any contributing factors that may have caused death. The medical examiner often also requests DNA testing.

The medical examiner also collects fingerprints and palm prints during an autopsy. The prints help identify the victim.

Toxicology

Toxicology is the science related to the detection of drugs, alcohol, and poisons. Information from a toxicology report might suggest that the victim was poisoned. A combination of alcohol and other drugs might also explain a motor vehicle accident.

Alcohol is the most widely abused drug in the United States. Toxicologists usually test for the presence of alcohol first. They use gas chromatography to test the concentration of alcohol in the victim's blood.

Toxicologists might then test urine for the presence of barbiturates or other depressants by performing an immunoassay on the urine (see Figure 8-9). The immunoassay is a presumptive screening tool for the presence of a specific chemical. If this test is positive, then a second test is performed. Confirmatory tests utilize gas chromatography–mass spectrometry and provide quantitative results. Immunoassay and other toxicology tests will be discussed in greater detail in Chapter 9.

Figure 8-9. *Forensic scientists perform immunoassays to screen for drugs and other chemicals.*

As a third screening, the toxicologist might test for the presence of tranquilizers, narcotics, and antidepressants along with hundreds of other drugs. This general screening requires the use of gas chromatography. Only mass spectrometry can provide confirmatory, quantitative results.

Finally, the toxicologist might use immunoassay to test urine for narcotics. Narcotics are opium; drugs made from opium, such as morphine and heroine; and synthetic substitutes for these drugs. This test screens for the presence of opiates, cocaine, and methadone. Methadone is used in drug rehabilitation facilities to help patients end their addiction to opiates. If the immunoassay is positive, the toxicologist uses gas chromatography–mass spectrometry to perform further confirmatory tests.

Histology

Histology is the study of tissues. The histologist slices the tissue into very thin sheets and stains the tissue samples for viewing under the microscope. Different stains are used to make different structures or abnormalities in the tissue visible. For example, eosin is an acidic stain used to stain basic structures. With it, the cytoplasm appears pink. Red blood cells appear bright red. Eosin stain is often coupled with a hematoxylin stain. This stain turns certain structures blue. Generally, hematoxylin binds to structures, such as ribosomes and the nucleus. These structures contain nucleic acids.

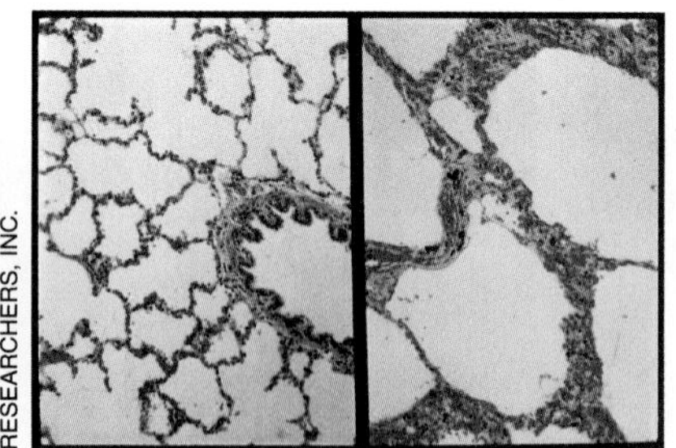

BIOPHOTO ASSOCIATES/PHOTO RESEARCHERS, INC.

Figure 8-10. *Comparing healthy lung tissue (left) with lung tissue of an emphysema patient (right). Emphysema breaks down the alveolar walls, making gas exchange less efficient.*

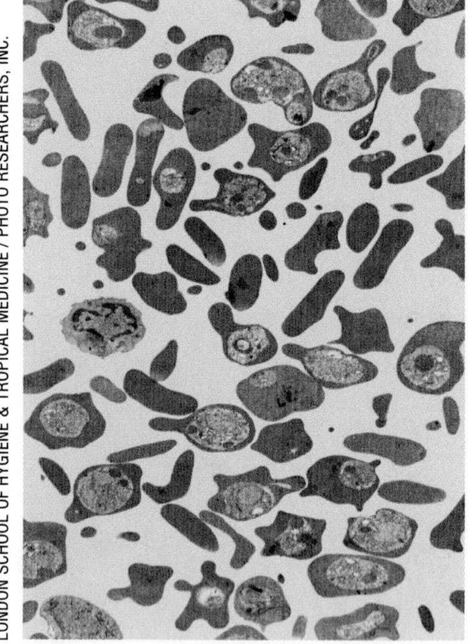

LONDON SCHOOL OF HYGIENE & TROPICAL MEDICINE / PHOTO RESEARCHERS, INC.

Figure 8-11. *Micrograph (6000×) green showing malaria parasites (green clusters) in the victim's blood.*

Together, eosin and hematoxylin can be used to detect cancers (see Figure 8-10).

If the medical examiner suspects that the victim suffered from a bacterial infection, the histologist might use the crystal violet stain. In solution, the crystal violet stain forms a positive and a negative ion. The positive ion reacts with the negatively charged parts of the bacterial cell. Under a microscope, the bacterial cell appears purple.

Neuropathology

Neuropathology is the study of disease and trauma associated with the nervous system, including the brain, spinal cord, and peripheral nerves. More than half of all deaths encountered by medical examiners are associated with the nervous system—most often the brain. In many cases, the neuropathologist is able to determine the cause of death. A forensic neuropathologist may be asked to consult on a case to determine whether the trauma the victim experienced was due to a fall or blunt force. In cases of suspected child abuse, the neuropathologist examines the brain trauma. The findings may confirm or disprove the suspicion. Finally, a neuropathologist studies tissue samples of various types of diseased tissue and tumors such as Alzheimer's disease and brain cancer. Although a forensic neuropathologist often collaborates with the medical examiner, it is the responsibility of the medical examiner to make the final determination of the cause of death for the death certificate.

Serology

Forensic serology is the study of blood, semen, and other body fluids with reference to legal matters. A serologist might find that the victim died from infection by a blood-borne pathogen (see Figure 8-11). A serologist might also help the medical examiner identify the victim or perpetrator. As you may recall from Chapter 1, Karl Landsteiner discovered the ABO blood-typing system. Blood type is a class characteristic. Knowing the blood type can narrow the list of possible suspects. Further advancements in technology have made it possible to develop DNA profiles from body fluids. This DNA profile is individual evidence, narrowing the suspect down to just one. DNA is considered the most compelling piece of physical evidence provided at trial. If DNA from a suspect is consistent with DNA found at the crime scene, a link between the suspect and the crime scene is made. However, DNA evidence is not always available. Therefore, many cases rely on other pieces of physical and circumstantial evidence.

Obj. 8.5

THE AUTOPSY REPORT

After the medical examiner completes the autopsy, he or she must prepare the autopsy report. Careful and truthful reporting is the cornerstone of the investigation. A medical examiner is responsible for presenting findings from the autopsy in court. The autopsy report must be accurate, detailed, and free from bias. Every local, county, and state medical examiner's office has its own standard form for documenting results of the autopsy and all laboratory tests. However, all autopsy reports contain the same kind of information. Figure 8-12 shows a portion of a basic autopsy report.

Autopsy/Specimen Description Sheet

Date:	Place barcode sticker here:
ME Number:	
Name:	
Body Intact? Y / N	DNA? Yes No
Presumptive ID? Y / N	Tox? Yes No
Confirmed Finger/Foot? Y / N	Weight:
Confirmed Dental? Y / N	Dimensions:
Manner:	Interval:
COD:	
Description:	

Mark out missing body parts or circle applicable fragment.

Comments:

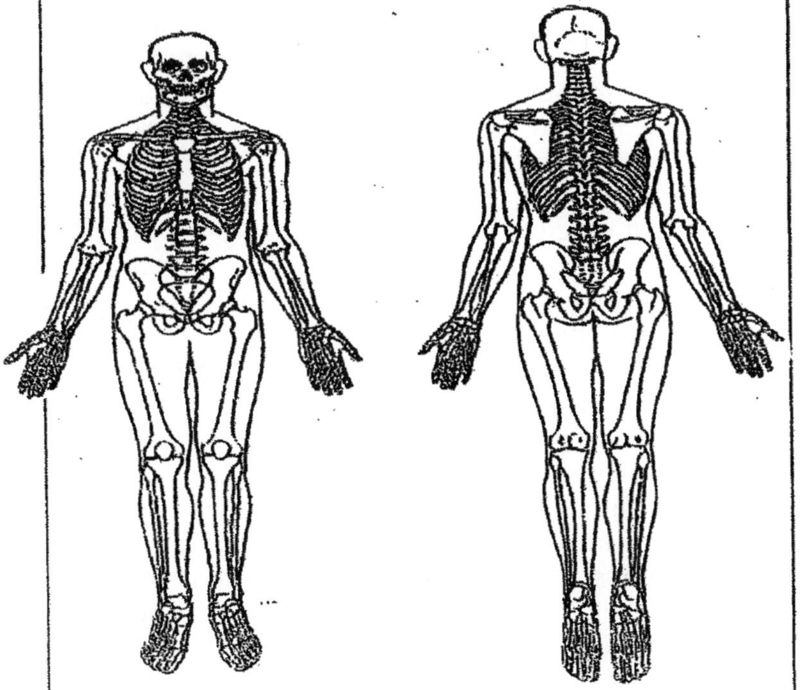

Pathologist________________________________

ARMED FORCES INSTITUTE OF PATHOLOGY

Figure 8-12. Sample autopsy report.

AUTOPSY REPORT: HEADING

Each autopsy report includes a heading. The heading contains the name, age, and gender of the victim. The medical examiner documents the case number as well as the date and time of death. Additionally, he or she records the date, time, and location of the exam and the names of all personnel present for the exam.

AUTOPSY REPORT: EXTERNAL EXAMINATION

The medical examiner records all findings from the external examination. This information is usually located in the second section of the autopsy report. This section includes the overall description of the deceased, including his or her build, height, and weight. The medical examiner also notes any congenital diseases, disorders, or malformations and includes a thorough description of the victim's clothing. This section of the report concludes with a full description of the body including, but not limited to, the degree of rigor mortis and lividity, hair, eyes (including shape and color), acne, missing teeth, scars, tattoos, evidence of disease, and old injuries unrelated to the death.

AUTOPSY REPORT: EVIDENCE OF INJURY

The medical examiner examines any recent internal or external injuries and documents the evidence in the third section of the autopsy report. He or she describes evidence of any injuries—even apparently minor injuries. In the case of a gunshot or stab wound, the examiner must track the movement of the bullet or knife through the body. The medical examiner assigns the entrance wound a number and records the overall dimensions of the entrance wound. Figure 8-13 is a photo of an entrance wound. Notice that the medical examiner included a ruler in the photograph to show scale.

The medical examiner records each organ and tissue that came into contact with the bullet. If the bullet exited the body, the autopsy report describes the exit wound. If a bullet is removed from the body, the medical examiner records information about where the bullet was found, the condition of the bullet, and the caliber, if possible. Stab wounds are handled in a very similar manner. If a victim was stabbed or shot several times, the medical examiner studies the wounds in groups called **clusters**.

The medical examiner might include several diagrams and photographs in this section of the report to help people not affiliated with the medical community understand the report.

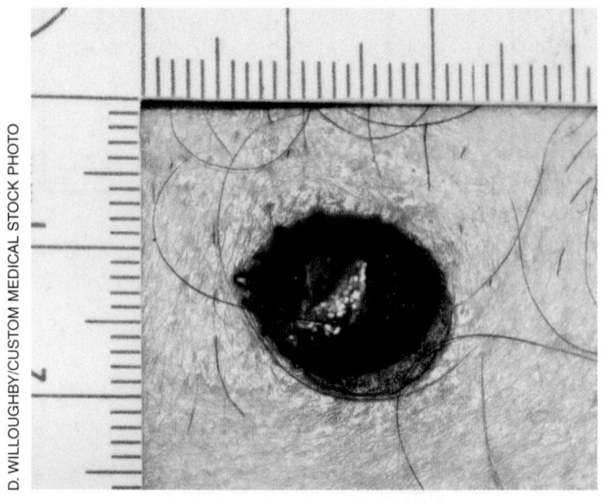
D. WILLOUGHBY/CUSTOM MEDICAL STOCK PHOTO

Figure 8-13. *Bullet entrance wound.*

AUTOPSY REPORT: INTERNAL EXAMINATION

In the fourth section of the autopsy report, the medical examiner records the findings from the internal examination. This part of the report illustrates the thorough and systematic approach used in studying all the major organ systems during the autopsy. The medical examiner weighs and describes each major organ and then notes any abnormalities. All of this information is indicated on the report. This section also contains the findings from the microscopic examination, the toxicology screenings, and any other tests run. Finally, the examiner records the analysis methods in this section of the report.

AUTOPSY REPORT: MEDICAL EXAMINER'S FINDINGS AND OPINION

In the last section of the autopsy report, the medical examiner summarizes the findings and states his or her opinion. The opinion is a brief description of the cause and manner of death. The findings include all results and outcomes of various tests and examinations listed in order of importance. This section is written for people not affiliated with the medical community and provides an explanation of the events that surrounded the death (see Figure 8-14). The opinion provides closure to

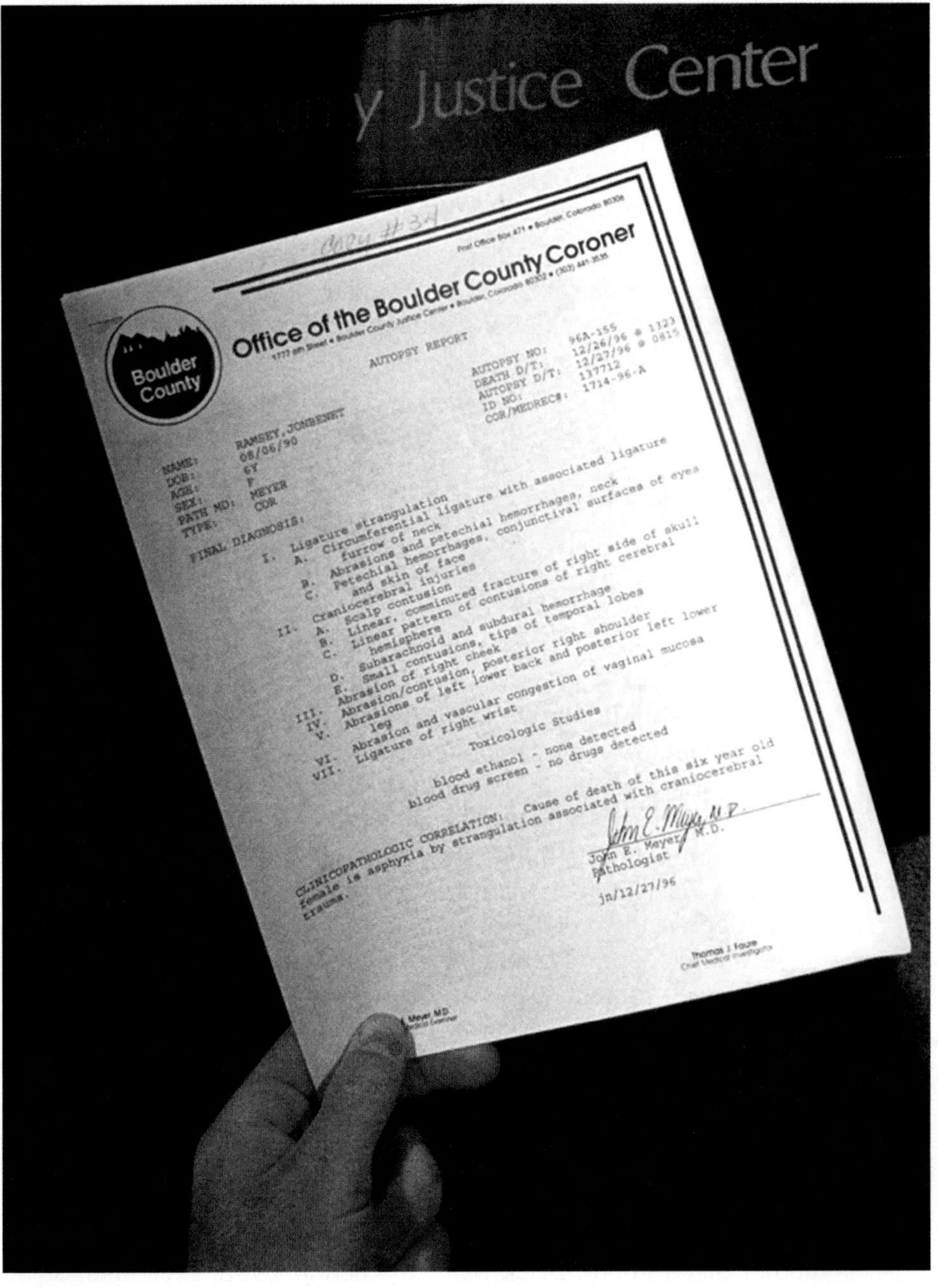

AP PHOTO/JOE MAHONEY

Figure 8-14. *Medical examiner's findings and opinion.*

family as well as valuable information to a jury. This section is often important in helping a jury decide whether someone should be held responsible for the victim's death.

CHAPTER SUMMARY

- In the United States, either the coroner or the medical examiner leads the death investigation when the cause of death is unclear or suspicious.
- Manner of death is classified as natural, accidental, suicide, homicide, or undetermined.
- The three steps of a death investigation include determination of the events that surrounded and led to the death, internal and external examination, and laboratory analysis.
- An autopsy is performed if the cause of death is not known, to document injuries, to exclude other causes of death, and to determine factors that contributed to death.
- Biological evidence is material from sources that are living or were once living. Nonbiological evidence is material from nonliving sources.
- An autopsy report typically includes a heading, information about the internal and external examinations, description of any evidence of injury, and the Medical Examiner's findings and opinions.

CASE STUDIES

Dr. Stephen Scher (1976)

On June 2, 1976, Dr. Stephen Scher and his friend, Martin Dillon, went hunting. Later in the day, Scher returned, stating that Dillon was dead. Scher said that he and Dillon were headed to get cigarettes when Dillon saw a porcupine. Dillon grabbed the shotgun and ran after the animal. A moment later, Scher heard a shot and ran to find his friend. Dillon was lying on top of the gun. It appeared that Dillon had tripped and shot himself. The local doctor performed an autopsy the day after the shooting, and the death was ruled accidental. There was no further investigation and the case was closed. As time passed, it was rumored that Scher was involved romantically with Dillon's wife. Two years later, they were married and moved away.

NEWSCOM

Figure 8-15. *Stephen Scher*

In 1992, Martin Dillon's father, Lawrence Dillon, hired his own private investigators. Blood spatter on Scher's clothes and a small cut in his pants indicated that Scher had been standing closer to Dillon than he had admitted. The evidence was brought to the local medical examiner, who agreed to reexamine the case. The medical examiner reviewed photos and evidence and ordered an exhumation of Dillon's body. Dillon was exhumed in 1995. Experiments conducted using similar weapons did not support Scher's original statement. Scher told police that Dillon was 250 feet away when the accident happened. The boots that Scher was wearing were reexamined by the FBI. Based on the blood spatter on the boots and wound evidence, it was determined that Dillion had been kneeling with a clay pigeon in his hand. He had been getting the skeet machine ready when Scher shot him. The size of the round in the weapon was for hunting, not skeeting. This suggested premeditation. Scher was convicted of first-degree murder. Figure 8-15 shows Dr. Scher being escorted into the courthouse for his trial in 1997.

Think Critically **What methods might the medical examiner have utilized to establish the cause of death in this cold case?**

Chris Benoit (2007)

Figure 8-16. *Chris and Nancy Benoit.*

Chris Benoit was a professional wrestler. He and his wife, Nancy (see Figure 8-16), lived in Fayetteville, Georgia with their seven-year-old son, Daniel. After Benoit missed several scheduled public appearances, police were called to his home for a "welfare check." On June 25, 2007, police discovered Benoit, his wife, and their son dead inside their home. It was apparent that Chris Benoit had murdered his wife and son and then committed suicide. Medical examiners performed an autopsy on each of the three victims. Nancy had bruising on her back and stomach. Her hyoid bone was broken, and there was blood on the back of her head. This information, along with the position in which Nancy was found, indicated that Benoit had held her down with his knee as he tied a cord around her body and neck. The blood at the back of her head indicated that she struggled to get away before the cord strangled her. Toxicology reports showed that Daniel had been given a sedative. The autopsy revealed internal injuries, but no external bruising. Benoit strangled himself with a cord from a weight machine. He formed a noose on the end of the weight cord. When he released the 240 pounds of weights, he died of strangulation.

Although the motive for this murder-suicide is unknown, investigators suggest several possibilities. Benoit was taking testosterone, a steroid, for therapeutic purposes, and toxicology reports corroborate therapeutic levels in the body. Potential side effects of steroids include uncontrollable anger and paranoia. However, evidence at the scene was not consistent with rage but with a more deliberate plot to kill his family. A second, and more scientifically supported, motive was the number of concussions he suffered as a result of his career in professional wrestling. Tests completed on Benoit's brain indicated that he suffered from chronic traumatic brain injury. All four lobes of his brain showed signs of injury. According to Julian Bailes at the Sports Legacy Institute, who performed the tests on Benoit's brain, his brain was so severely damaged that it looked like the brain of an 85-year-old Alzheimer's patient. Bailes suggests that this brain trauma might have been a leading cause in this murder-suicide.

Think Critically **1. What evidence might have supported this crime being caused by rage and not a plot to kill his family?**

2. Had Benoit not killed himself, how would the evidence have been compiled against him? What do you think the medical examiner's testimony would have included?

Bibliography

Books and Journals

Calaluce, Robert, and Jay Dix. (1998). *Guide to Forensic Pathology.* Boca Raton, FL: CRC Press.

DiMaio, Vincent J., and Dominick DiMaio. (2001). *Forensic Pathology* (2nd ed.). Boca Raton, FL: CRC Press.

DiMaio, Vincent J. M., and Suzanna E. Dana. (2007). *Handbook of Forensic Pathology* (2nd ed.). Boca Raton, FL: CRC Press.

Dix, Jay, and Michael Graham. (2000). *Time of Death, Decomposition and Identification: An Atlas* (Causes of Death Atlas Series). Boca Raton, FL: CRC Press.

Echaore-McDavid, Susan, and Richard A. McDavid. (2008). *Career Opportunities in Forensic Science.* New York: Checkmark Books.

Hanzlick, Randy, M.D. (2007). *Death Investigation: Systems and Procedures.* Boca Raton, FL: CRC Press.

Tortora, Gerard J., and Bryan Derrickson. (2006). *Principles of Anatomy and Physiology.* Hoboken, NJ: John Wiley & Sons.

Websites

http://anthropology.si.edu/writteninbone/making_ids.html
www.biologycorner.com
www.cdc.gov/nchs
www.cdc.gov/ncipc/factsheets/tbi.htm
www.disastercenter.com/crime/uscrime.htm
http://emedicine.medscape.com/article/433404-overview
www.msnbc.msn.com
www.ncjrs.gov
www.nytimes.com/1996/07/28/us/doctor-faces-murder-charge-in-76-killing.html
www.trutv.com

CAREERS IN FORENSICS

Science, Technology, Engineering & Mathematics

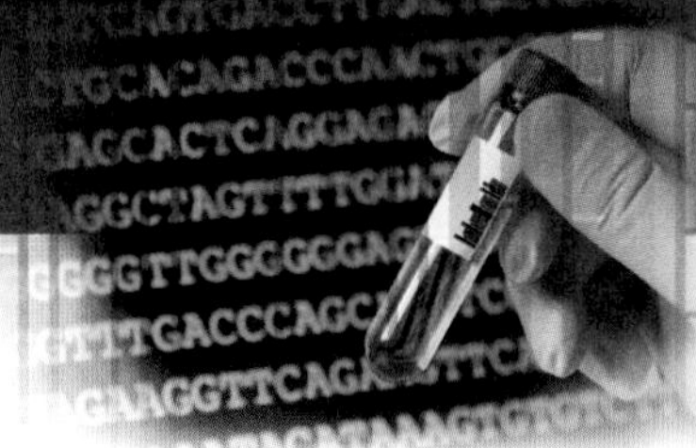

Dr. Jan Garavaglia: Medical Examiner

Jan C. Garavaglia, M.D., has been the chief medical examiner for the Orange-Osceola County Medical Examiner's Office in Florida since 2004 (see Figure 8-17). Prior to joining the office in Florida, Dr. Garavaglia was a medical examiner at the Bexar County Forensic Science Center in San Antonio, Texas, for 10 years. She also served at the University of Texas Health Science Center as the clinical assistant professor for the department of pathology and as a member of the Graduate Faculty Council for the Graduate School of Biomedical Science. She has also worked as a medical examiner in metropolitan Atlanta and in Jacksonville, Florida.

AP PHOTO/PETER COSGROVE

Figure 8-17. *Dr. Jan Garavaglia*

Dr. Garavaglia starts most days by reviewing cases, looking at microscope slides, and meeting with families. As her day progresses, she completes autopsies and autopsy reports. Her goal is to complete an autopsy and have a determination of cause of death within 24 hours. Dr. Garavaglia emphasizes that her assessments of the cause and manner of death are opinions, based on fact and the preponderance of evidence. She catalogs photographs and biopsy and tissue samples in case questions arise later. She takes fingerprints during each autopsy, and she runs the fingerprints through the AFIS database. She also keeps the fingerprints as a permanent record in case there is ever a question of identity. Dr. Garavaglia completes the death certificate and helps interpret findings from pictures and medical records.

Dr. Garavaglia is probably best known as "Dr. G," the host of the *Dr. G.: Medical Examiner* television series on Discovery Health. In each episode, a real case is followed from discovery of the victim through the investigation, including the autopsy. She likes to select cases that do not produce the expected answers. She partners with the show's producers to select the most intriguing and informative cases that will most educate her viewing audience.

Dr. Garavaglia first entered medicine because she wanted to help people. Her decision to specialize in forensic pathology allows her to fulfill her dreams of helping society. According to Dr. Garavaglia, when families get answers about how and why a loved one has died, it sometimes helps them cope with the loss. "Some families are angry and some are thankful, but the bottom line is that they want honest answers."

Learn More About It

To learn more about a career as a medical examiner, go to www.cengage.com/school/forensicscienceadv.

CHAPTER 8 REVIEW

Multiple Choice

For items 1-5, read each brief scenario and decide which body system or systems will be examined when trying to determine cause of death.

1. Upon arrival at a scene, the medical examiner notices the victim is in full rigor and exhibits signs of livor mortis on the lower-middle part of the back. *Obj. 8.2*
   a. muscular
   b. integumentary
   c. circulatory
   d. a & b
   e. a & c
2. During the autopsy, the medical examiner discovers the victim had not eaten for at least 8 hours. *Obj. 8.2, 8.3*
   a. endocrine
   b. digestive
   c. integumentary
   d. a & b
   e. all of the above
3. At the scene, the victim is found face down in a pool. At the autopsy the victim has water in the lungs, and the toxicology screening shows the presence of alcohol. *Obj. 8.3*
   a. circulatory
   b. respiratory
   c. urinary
   d. a & b
   e. b & c
4. At a rape case, due to a broken hyoid bone, the examiner determines that the victim has been strangled. *Obj. 8.2*
   a. integumentary
   b. reproductive
   c. skeletal
   d. b & c
   e. none of the above
5. At a double murder-suicide, the blood-spatter evidence offers clues to investigators regarding the chain of events that occurred during the attack. *Obj. 8.2, 8.4*
   a. circulatory
   b. respiratory
   c. integumentary
   d. a & b
   e. b & c

6. Which of the following is *not* found in an autopsy report? *Obj. 8.5*
   a. full description of the body
   b. notation of disorders or malformations
   c. recent internal or external injuries
   d. a detailed family medical history
7. The role of the forensic pathologist is to ________. *Obj. 8.3*
   a. examine a corpse to determine how death occurred
   b. produce evidence to present in court
   c. study how medicine relates to the law
   d. all of the above
8. Which of the following is *not* true about estimating time of death? *Obj. 8.2*
   a. It is an exact science.
   b. It often relies on witness statements.
   c. It uses body temperature.
   d. Insect infestation may provide relevant information.
9. How quickly does a body lose heat after death? *Obj. 8.2*
   a. 1.5°C/hr
   b. 1.0°C/hr
   c. 0.8°C/hr
   d. 37°C/hr

MATHEMATICS π

10. Rigor mortis ________. *Obj. 8.2*
    a. begins about two hours after death
    b. is useful in estimating time of death during the first 36–48 hours after death
    c. is permanent
    d. a & b
11. Which of the following is *not* part of a death investigator's preliminary investigation? *Obj. 8.2*
    a. talking to witnesses and family of the deceased
    b. reviewing medical history and police reports
    c. doing an internal examination
    d. talking to law enforcement about recommended laboratory analysis
12. An example of trace evidence found on a victim would be ________. *Obj. 8.3*
    a. clothing
    b. a bullet
    c. jewelry
    d. a fiber

Short Answer

13. Distinguish between a coroner and a medical examiner. *Obj. 8.1*

14. Compare and contrast cause of death, mechanism of death, and manner of death. Give an example of each. *Obj. 8.5*

15. Describe factors to be considered when an investigator estimates the time of death. *Obj. 8.5*

16. A body is discovered in an alley. The body is still in rigor. Explain the significance of rigor mortis in the determination of the time of death. *Obj. 8.2, 8.5*

17. The body of a 30-year-old woman is found in her home. The police think that her husband poisoned her for insurance money. What tests could be done to prove whether the victim was poisoned? *Obj. 8.3*

18. Explain the proper methods of collecting biological evidence. *Obj. 8.4*

Think Critically 19. A healthy 22-year-old woman was shot in the abdomen. She was taken to the emergency room, where she underwent emergency surgery to repair the wound. All life-threatening injuries were repaired. During her postoperative care, she developed peritonitis (when the bowels leak into the abdomen). This condition caused her to develop an infection. She died a week later. The autopsy showed a severe infection in the abdominal region. The medical examiner determined that the cause of death was a gunshot wound. The manner of death was ruled a homicide. During the suspect's trial, the defense argued that the woman died of an infection in the abdomen. The prosecution argued that she died of a gunshot wound. Which side should the jury believe? Should the doctor be held accountable? Explain. *Obj. 8.5*

20. It was 98°F when the body of a 65-year-old man was found lying in a ditch off a winding country road on August 12. The victim was fully dressed. He had on overalls, a short-sleeved shirt, and socks. He was missing his shoes. From the abdomen down, he was covered with a blanket. He had maggot infestation in the head and neck regions. He had a four-inch cut to the scalp but no brain or bone injury. A small towel was wrapped tightly around his neck. The knot was at the back of his neck. His stomach contents revealed bacon, brown liquid, and yellow food particles. How could family members assist in determining the time of death? How could the environmental factors affect estimating the time of death in this case? An entomologist was not called in to collaborate on this case. Explain why an entomologist was not necessary. What was the cause of death, mechanism of death, and manner of death? *Obj. 8.2*

CRIME SCENE S.P.O.T. Student-Prepared Original Titles

Crime Scene S.P.O.T. Expert Testimony

By: **Lyndsie Pickren**
East Ridge High School
Clermont, Florida

Dr. Marks coolly eyed the crowded courtroom; he'd been through all of this before. Dr. Marks was the defense team's expert witness specializing in evaluating suspicious homicides. He was also the county's medical examiner. Part of his job over the years has been to provide expert testimony at trials just like this one.

He sighed deeply and began. "After extensive research into this case, I believe the accused, 15-year-old Marie Johnson, is innocent in the death of her mother, Gabriella Johnson," said Dr. Marks. Mrs. Johnson suffered from an intracerebral hemorrhage—bleeding in the brain (see Figure 8-18).

Figure 8-18. Intracerebral hemorrhage—the orange region shows a hematoma caused by blood leaking out of the blood vessels into the brain.

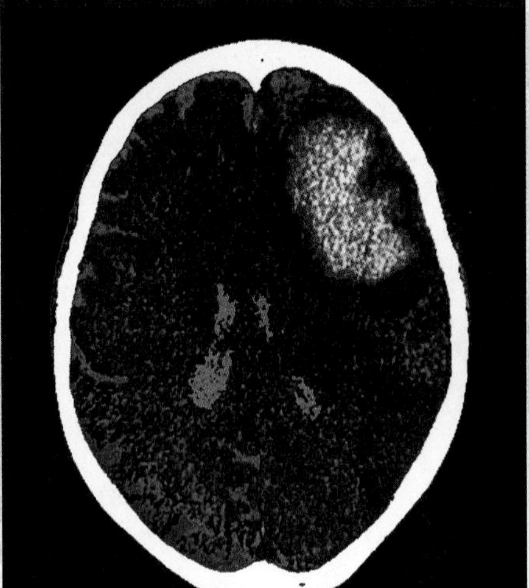

ZEPHYR/PHOTO RESEARCHERS, INC.

"And you intend to prove that Marie Johnson did not brutally beat and strangle her own mother to death?" the district attorney asked.

"Yes, sir, I do. Gabriella Johnson was diagnosed several years ago with severe epilepsy. As the court may or may not know, epilepsy, which is manageable with medication, causes the victim to seize when encountered by certain stimuli. The stimuli are unique to each victim. Fluctuations or strobing of light, sound, or even temperature may cause a seizure. Mrs. Johnson had frequent and often violent seizures when she was not taking her medication. It is noted in her medical records that Mrs. Johnson had been admitted into the emergency room repeatedly for bleeding wounds and broken bones sustained during epileptic episodes."

Dr. Marks consulted his notes, and continued: "According to the toxicology report, Mrs. Johnson had not been taking her medications. Additionally, the autopsy revealed that she was three months pregnant. According to her physician, Gabriella Johnson was concerned that her medication would affect her baby and requested a lower dosage." Dr. Marks looked around the room, analyzing his audience's faces.

"Do you think that a single seizure, regardless of its violence, was enough to cause that kind of trauma to Mrs. Johnson? What about the fact that it is very obvious she was strangled? What about the trauma done to her skull?" retorted the district attorney.

Dr. Marks glanced at the judge and back at the courtroom. "It is not at all obvious that Mrs. Johnson was strangled" he stated. "Often in violent seizures, the victim thrashes because muscles in the body contract. The victim has no cognitive reactions. Gabriella Johnson's neck muscles contracted so violently that there was actual bruising, but the hyoid bone located over the trachea in the neck was not broken. It likely would have been if strangulation had occurred. The blow to Mrs. Johnson's head was inflicted by a hard, round object, such as the leg of the dinner table. It is physically impossible for Marie Johnson to have inflicted the damage, unless the leg was actually broken off, which it wasn't. The blow to the head was in such a position that Mrs. Johnson clearly was lying on the floor when her head hit the table leg." Silence greeted this news.

Dr. Marks continued: "It is my firm professional belief that the cause of death was head trauma brought on by an epileptic seizure. The mechanism of death was intracerebral hemorrhaging—bleeding in the brain. I have concluded in my investigation that Gabriella Johnson died a natural death, and that her daughter, Marie Johnson, is innocent of homicide."

Activity:

Answer the following questions based on information in the Crime Scene S.P.O.T.

1. The victim had extensive bruising around the neck. However, the expert witness argued that she had not been strangled. What is the most significant evidence to support the assertion that the victim was not strangled?
2. Based on the information provided, what was the cause and mechanism of death?
3. Discuss what aspects of this story would not happen in a real courtroom. Discuss the implications of the behavior of the ME on the stand.

WRITING

4. After reading the story, write an ending to the story that describes the final testimony of Dr. Marks, the expert witness for the defense. Then, write one or two paragraphs describing the testimony of the expert witness for the prosecution. Be sure to use correct terminology and logical, well-reasoned arguments.

Preventing Adolescent Crime Together™ (P.A.C.T.)

Background:

Childhelp® is a nonprofit organization that provides help and support to victims of child abuse. According to Childhelp, approximately five children die every day in the United States from injuries caused by physical abuse. Seventy-five percent of these victims are younger than four years old. More than 3 million reports of child abuse are made every year. Many of these children receive treatment for their injuries; blunt-force trauma is the cause of many of the injuries.

According to a report for the U.S. Department of Justice, 95 percent of all domestic violence victims are women. Teen dating violence, however, affects boys and girls equally. Many victims seek emergency medical treatment for their injuries.

Neglect is the most common form of elderly abuse. The majority of elder abuse victims are females, and the majority of abusers are males. Adult children are the most frequent abusers of the elderly.

Due to the ongoing need for support and shelter from abuse, various organizations provide protection, counseling, and guidance to victims as they transition to healthier and safer situations.

Introduction:

Research local nonprofit organizations that provide support or shelter for victims of abuse. Determine what volunteer services each organization provides. Volunteer your services to the organization of your choice. For example, you may contact battered women's shelters, advocates for victims of child abuse or domestic violence, homes for babies born addicted to drugs, or any similar organization in your area. Also, volunteer to provide literature and information about the organization to the public.

Procedure:

1. In groups of four to six students, research local nonprofit organizations that help victims of abuse. The victims can be any age or either gender. Each group should choose a different organization.
2. Call the charity and offer to volunteer. Ask whether the organization would like other support, such as a donation drive. (The charity may need supplies, such as clothing, office supplies, and food.)
3. If possible, make an appointment to meet with the director of the organization to discuss additional needs and literature they may like you to develop. Take notes and compile a list of the organization's needs. Tell the director that you plan to develop a school-wide marketing campaign to raise awareness of the organization. Determine whether the director would like to approve your campaign before you implement it.
4. Once you have approval from the organization and from your teacher, implement your marketing plan.

5. If possible, your literature should describe the representative physical trauma inflicted on the victim. Demonstrate caution when discussing abuse. Be sensitive to your audience and to the victims.
6. Make another appointment with the director to bring donated items collected and, if possible, schedule a time when your group can volunteer at the organization.

PURESTOCK/JUPITERIMAGES

Figure 8-19. *Many shelters and other children's advocacy organizations need volunteers to read and spend time with kids affected by violence.*

ACTIVITY 8-1 *Ch. Obj: 8.2*
FETAL PIG DISSECTION

Objectives:

By the end of this activity, you will be able to:

1. Perform a whole-body dissection of a vertebrate.
2. Identify the major anatomical features of the vertebrate body in a dissected specimen.
3. Explain how organs of a system function together for performance of a specific process.

Materials:

(per group of four students)

fetal pig
dissecting tray
paper towels
metric ruler
dissection kit (scissors, scalpel, blunt probe, needle probe, forceps)
apron
latex gloves
string

Safety Precautions:

Wear proper lab safety attire—gloves, aprons, safety goggles.

Do not place your hands near your mouth or eyes while handling preserved specimens. Most preservatives in use today are nontoxic to the skin, but may cause minor skin irritations. If skin comes in contact with a preservative, wash area of skin thoroughly with soap and warm water. If the preservative splashes into the eyes, rinse thoroughly using the safety eyewash.

Procedure:

Part 1: External Anatomy

1. Before completing the dissection, review the following terms (see Figure 1):
   anterior—top or head end
   posterior—bottom or hind end
   dorsal—back
   ventral—front
   medial—toward middle of the body, the midline
   lateral—toward the outside of the body
   proximal—close to a point of reference
   distal—farther from a point of reference

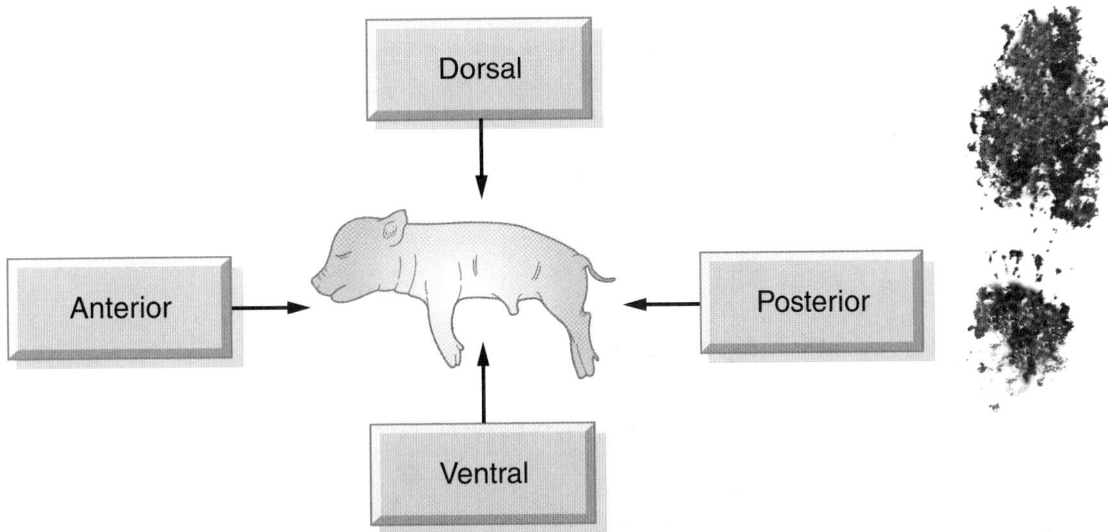

Figure 1. *It is important to understand scientific terms for regions of the body.*

2. Measure the length of your fetal pig from the end of the snout to the beginning of the tail. Use Table 1 to estimate the age of your pig (normal gestation for the fetal pig is between 112 and 115 days).

Table 1. Average length over time.

Length in Cm	Gestational Days
1.1	21
1.7	35
2.8	49
4	56
22	100
30	birth

3. Look for the urogenital opening on the pig. This opening is located near the anus on females; on males, near the umbilical cord. Be aware that both males and females have rows of nipples. Is your pig male or female?
4. Observe the feet of the pig. How many toes are on each foot?
5. Look at the eyes of the pig. Carefully remove the eyelid to view the eye underneath. Does it appear well developed?
6. Place your tongue on the roof of your mouth. Locate the hard and soft palates inside your mouth. Probe inside the pig's mouth. Find the hard and soft palates on the roof of the mouth. Observe the taste buds on the side of the tongue. Feel the edge of the mouth. Are there any teeth?

7. Find the esophagus at the back of the mouth. Locate the epiglottis—a cone-shaped structure at the back of the mouth. Find the pharynx, the cavity in the back of the mouth.
8. Place the pig on one side in the dissecting pan. Carefully cut away the skin from the side of the face and upper neck to expose the lymph nodes, the salivary glands, and the muscle that works the jaw. Be careful not to cut too deeply—you risk damaging the glands. The salivary glands look a little like chewing gum.

Part 2: Internal Anatomy

1. Place your fetal pig in the dissecting pan with the ventral side up. Use string to tie the legs of the pig to get them out of your way.
2. Look at Figure 2 to determine where you will make your first cuts. Use scissors to cut through the skin and muscles. Do not cut through the umbilical cord at this time.

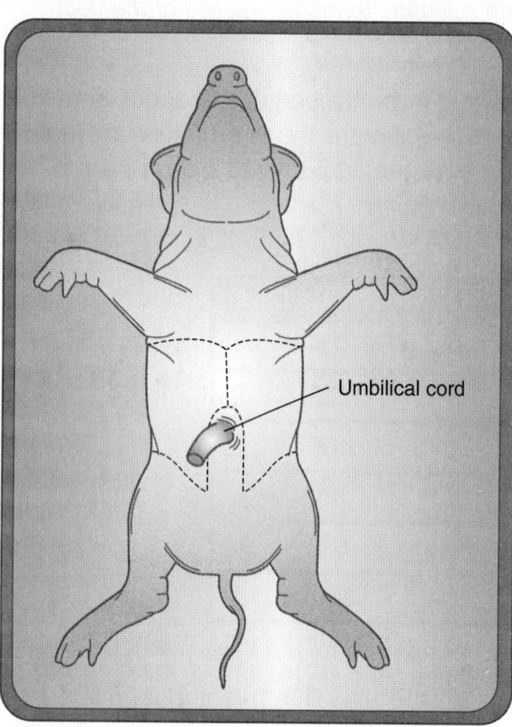

Figure 2. *Use scissors to cut through the skin and muscles.*

3. Find the umbilical vein that leads from the umbilical cord to the liver. Cut the vein. This allows you to open the abdominal cavity. If your pig is filled with water and/or preservative, drain it over the sink and rinse the organs.
4. Locate each of the following organs. Use Figure 3 to help you.
   a. **diaphragm**—a muscle that divides the thoracic and abdominal cavity, located near the rib cage to aid in breathing
   b. **liver**—a lobed structure, the largest internal organ in the body, responsible for making bile for digestion
   c. **gall bladder**—a greenish organ located underneath the liver; stores bile and sends it to the duodenum (through the bile duct)
   d. **bile duct**—attaches the gall bladder to the duodenum

e. **stomach**—a pouch-shaped organ just underneath the liver and slightly to the pig's left; responsible for churning and breaking down food
f. **esophagus**—located at the top of the stomach
g. **small intestine**—leads from the stomach; composed of the **duodenum** (straight portion just after the stomach) and the **ileum** (folded part)
h. **pancreas**—a bumpy organ along the underside of the stomach; makes insulin (necessary for the proper regulation of sugars in the blood); a pancreatic duct leads to the duodenum
i. **spleen**—a flattened organ that lies across the stomach and toward the extreme left side; stores blood; is not part of the digestive system
j. **large intestine**—located at the end of the ileum, where it widens; reabsorbs water from the digested food
k. **cecum**—a seemingly dead end of the large intestine; helps the pig digest plant material
l. **rectum**—located at the end of the large intestine, toward the back of the pig, and will not be moveable; any undigested food is stored in the rectum as feces; opens to the **anus**
m. **kidneys**—two bean-shaped organs on either side of the spine; responsible for removing harmful substances from the blood; these substances are excreted as urine
n. **urinary bladder**—a flattened sac between two **umbilical vessels**

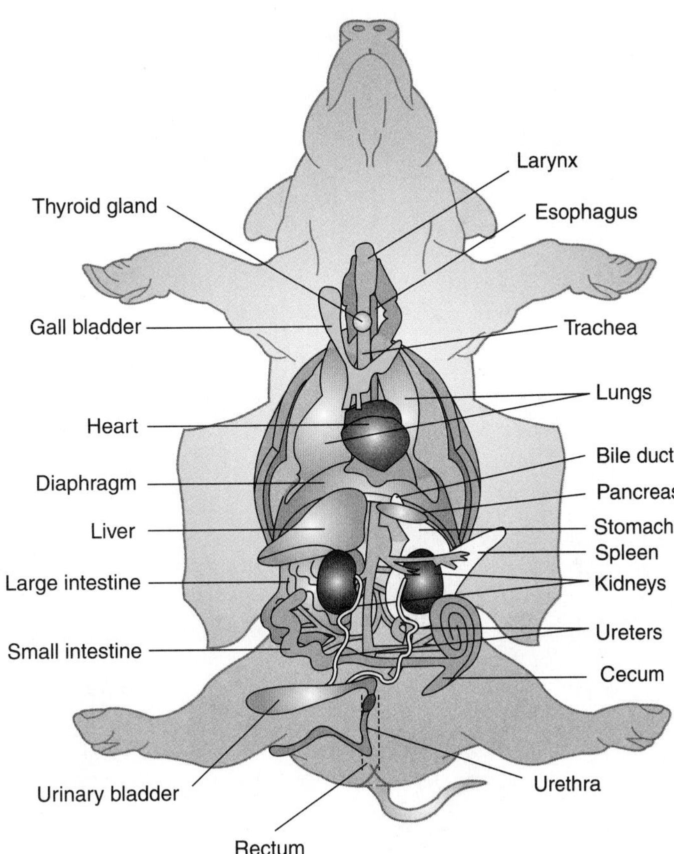

Figure 3. *The internal organs of a fetal pig.*

o. **ureters**—the tubes leading from the kidneys; carry urine to the urinary bladder
p. **urethra**—under the bladder, a tube that carries urine out of the body
q. **heart**—located above the diaphragm in the center of chest; note the veins and arteries entering and leaving the heart
r. **lungs**—two spongy structures located to the left and right sides
s. **trachea**—in the chin area above the heart, easy to identify because of the rings of cartilage that keep it from collapsing
t. **thyroid gland**—a pinkish-brown, V-shaped structure atop the trachea; secretes hormones that control growth and metabolism
u. **larynx**—located at the anterior trachea; a hard, light-colored structure; produces sounds

5. If your pig is a male: (see Figure 4)
   a. **scrotal sacs**—posterior end of the pig
   b. **testes**—inside the scrotal sac (you will need to cut through the sac to see them); produces sperm
   c. **epididymis**—coiled structure that lead into a tube (vas derferens), allowing sperm to leave the testes and travel into the urethra
   d. **penis**—tube-like structure containing the urethra

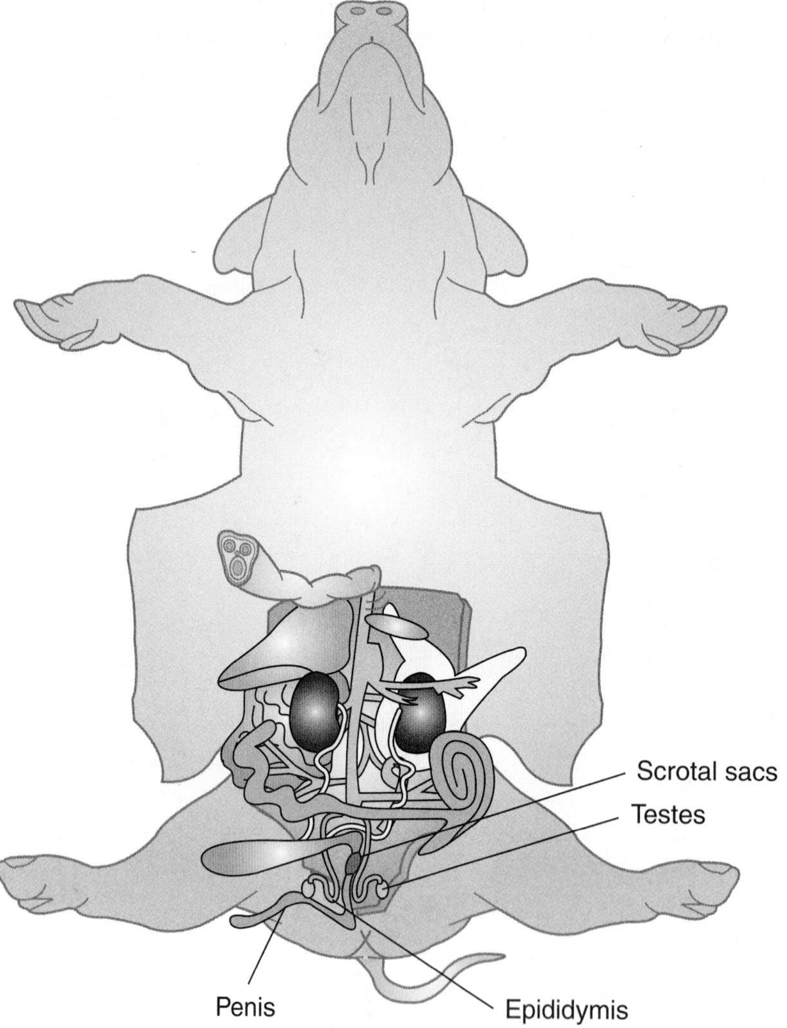

Figure 4. *Male.*

6. If your pig is a female: (see Figure 5)
   a. **ovaries**—two bean-shaped structures located just posterior to the kidneys
   b. **oviducts**—curly structures connected to the ovaries
   c. **uterus**—can be found by tracing the oviducts to the posterior of the pig
   d. **vagina**—appears as a continuation of the uterus

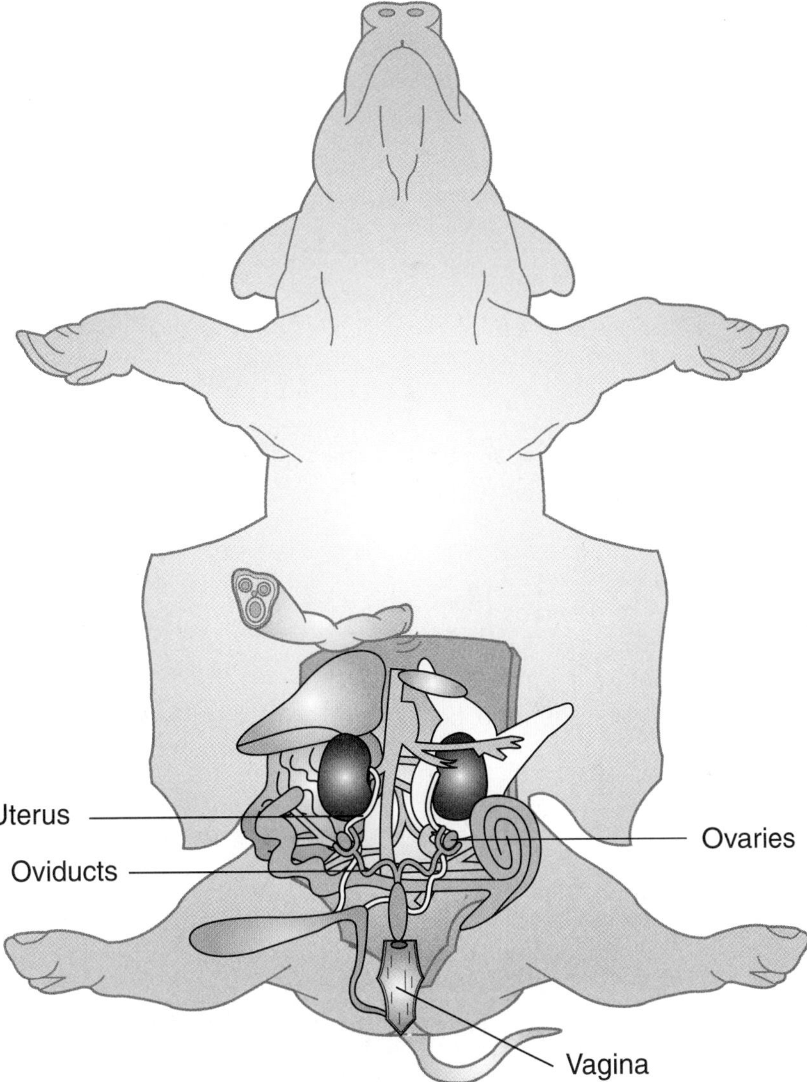

Figure 5. *Female.*

Questions:

1. What is the gestational age of your fetal pig?
2. Compared to the pig, the intestines of humans are much shorter. Why?
3. Do you think pigs are born with their eyes open or closed? Why do you think so?

ACTIVITY 8-2 *Ch. Obj: 8.2*
DEATH SCENARIOS

Objectives:

By the end of this activity, you will be able to:

1. Determine cause of death, mechanism of death, and manner of death from descriptions of events.
2. Create a scenario to illustrate cause of death, mechanism of death, and manner of death.

Procedure:

Read the following scenarios and determine the cause of death, the mechanism of death, and the manner of death in each.

SCENARIO 1

An elderly woman in her mid-70s was walking to the local drugstore. On her way, a man shoved her as he walked by and stole her purse. She was not injured in the altercation but was clearly shaken up. About 10 minutes later, she began to experience shortness of breath and chest pain. She was taken to the emergency room. While she was being examined, she developed an irregular heartbeat and died.

Questions:

1. What is the cause of death?
2. What is the mechanism of death?
3. What is the manner of death?
4. Should anyone in this scenario be charged with murder? Why or why not?

SCENARIO 2

Britta was babysitting for her neighbor's two-year-old and had just put little Justin down for a nap. As she was sitting in the living room texting her friends about the movie they had planned to see later in the evening, she heard the sound of breaking glass. Britta grabbed a baseball bat that was propped up in the corner and went to investigate. As she rounded the corner to the kitchen, she noticed the broken glass from the door. When Britta turned around, she saw the intruder coming toward her, wielding a very large knife. To his surprise, she raised the bat and crushed his skull. Britta called 911, and police and paramedics responded within minutes. Unfortunately for the intruder, it wasn't fast enough and he died before they arrived.

Questions:

1. What is the cause of death?
2. What is the mechanism of death?
3. What is the manner of death?
4. Should anyone in this scenario be charged with murder? Why or why not?

SCENARIO 3

Devin (age 17) and Carlos (age 16) were known around the suburban Maryland neighborhood as "future felons" because of their frequent run-ins with the law. One afternoon, they decided they wanted to run the local drug dealer out of business and take over his clientele. In order to accomplish this, the boys needed money. So they came up with a plan. The next evening, Devin and Carlos put on dark clothing and ski masks and broke into the corner drugstore about an hour after closing. Neither of them was aware of a night watchman, so when Mr. Franklin surprised them, Devin, who was carrying a gun, accidentally shot and killed Mr. Franklin. Frightened, both fled the scene. When the pharmacist arrived the next morning, she found Mr. Franklin lying in a pool of blood in front of the prescription counter.

Questions:

1. What is the cause of death?
2. What is the mechanism of death?
3. What is the manner of death?
4. Should anyone in this scenario be charged with murder? Why or why not?

SCENARIO 4

WRITING

Create a scenario. Identify the manner, mechanism, and cause of death.

CHAPTER 9

Physiology of Alcohol and Poisons

AN ACCIDENTAL HOMICIDE

PHOTO BY MJ KIM/GETTY IMAGES

Figure 9-1. Michael Jackson died at his home after being administered several drugs, including valium, propofol, lorazepam, and midazolam.

Two months after his death on June 25, 2009, singer Michael Jackson's death was ruled a homicide. *Homicide* means that a person died at the hands of another, but in Jackson's case, the homicide appeared to be accidental. Forensic tests indicated that Jackson had died from a deadly drug interaction between propofol—a powerful anesthetic—and two other sedatives, lorazepam and midazolam. All three drugs had been prescribed and administered to Jackson by his physician, Dr. Conrad Murray.

Murray wrote an affidavit for the police, explaining that he had been treating Jackson for insomnia, or the inability to sleep, for more than a month and a half. Initially, Murray treated the insomnia with propofol alone. However, Murray became concerned that Jackson was developing an addiction to the strong drug, and Murray says he decided to decrease the amount of propofol that he administered to Jackson every day. Murray started giving Jackson only a low dose of propofol with a dose of lorazepam and midazolam each.

In addition to the propofol, lorazepam, and midazolam, investigators also discovered that Jackson had taken a 10 mg tablet of valium, another sedative, approximately 9 hours before his death. Murray reported that the final dose of propofol that he gave Jackson was only 25 mg, a relatively small amount. However, even in small amounts, drugs can interact with deadly consequences. The coroner determined that the interaction of anesthetics and sedatives in Jackson's body caused Jackson to stop breathing. The lack of oxygen caused Jackson's heart to stop beating, leading to his sudden death. The death was ruled a homicide because the drug cocktail administered to Jackson was atypical for the treatment of insomnia.

At the time of this publication, Dr. Murray had been charged with involuntary manslaughter, but the trial had not yet begun.

AP PHOTO/ISAAC BREKKEN, POOL

Figure 9-2. Dr. Conrad Murray was using a cocktail of various drugs to treat Michael Jackson for insomnia.

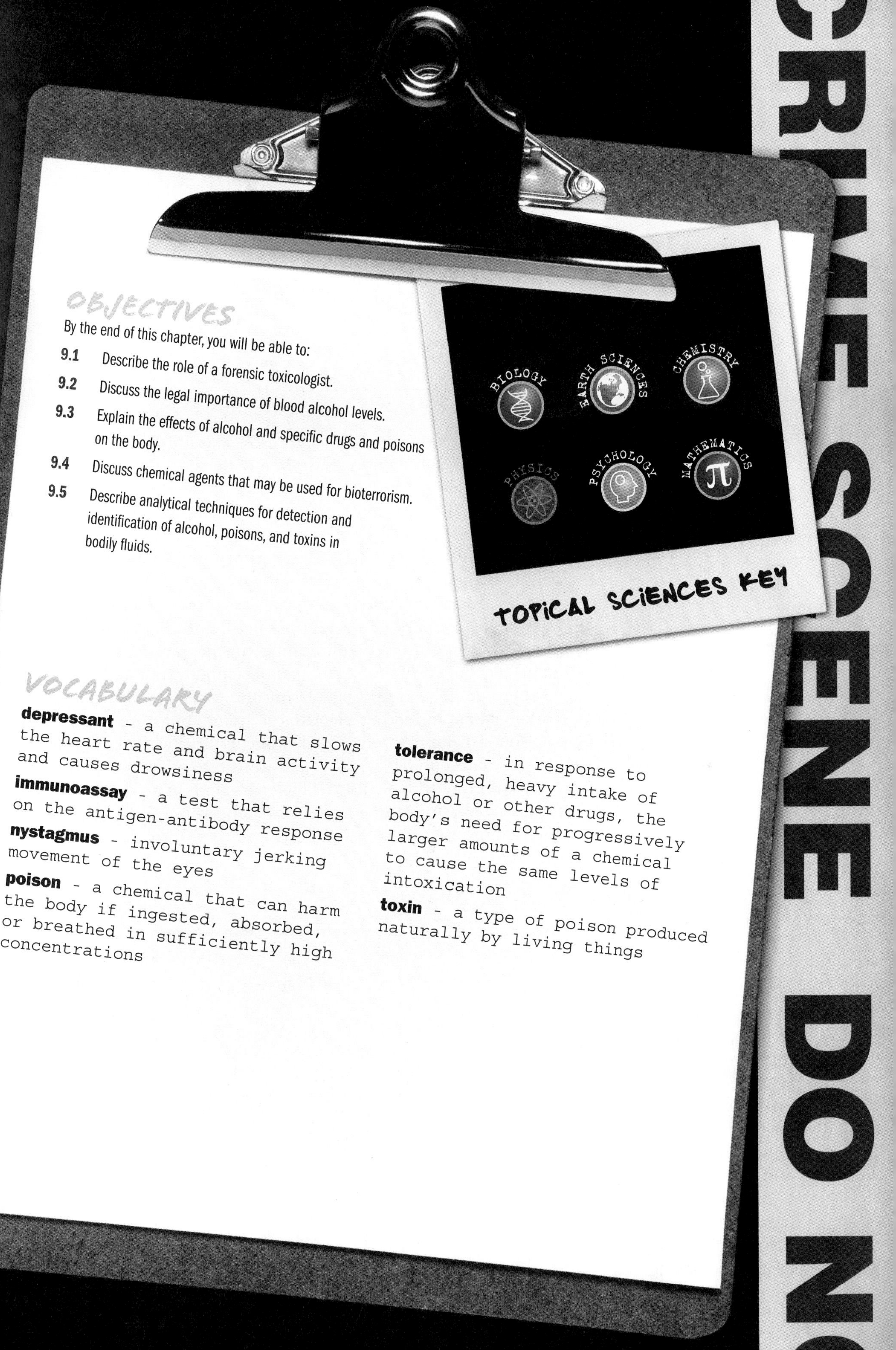

OBJECTIVES

By the end of this chapter, you will be able to:

9.1 Describe the role of a forensic toxicologist.

9.2 Discuss the legal importance of blood alcohol levels.

9.3 Explain the effects of alcohol and specific drugs and poisons on the body.

9.4 Discuss chemical agents that may be used for bioterrorism.

9.5 Describe analytical techniques for detection and identification of alcohol, poisons, and toxins in bodily fluids.

VOCABULARY

depressant - a chemical that slows the heart rate and brain activity and causes drowsiness

immunoassay - a test that relies on the antigen-antibody response

nystagmus - involuntary jerking movement of the eyes

poison - a chemical that can harm the body if ingested, absorbed, or breathed in sufficiently high concentrations

tolerance - in response to prolonged, heavy intake of alcohol or other drugs, the body's need for progressively larger amounts of a chemical to cause the same levels of intoxication

toxin - a type of poison produced naturally by living things

INTRODUCTION

Murder by poisoning has been a common theme in books, television programs, and movies. Even mythology refers to the use of poisons to kill. **Poisons** are chemicals that can harm the body if ingested, absorbed, or inhaled in sufficiently high concentrations. For a long time, arsenic, odorless and tasteless, was often used because it was undetectable in the body and the symptoms resembled those of disease or illness. Today, however, chemists have discovered techniques to detect and measure small amounts of arsenic, as well as many other poisons, in human tissue.

HISTORY OF TOXICOLOGY

Obj. 9.1

The first use of chemistry to detect poisons was in the early 1700s. Hermann Boerhaave, a Dutch chemist, discovered that arsenic had a distinct odor when heated. However, his method was not effective in detecting arsenic in a human body. Many other scientists developed processes for detecting poisons in the victim's body. Each scientist expanded on the research of previous scientists. By the early 19th century, toxicology had become a new scientific discipline.

The first use of toxicology in a legal case was the Lafarge murder trial in France in 1840. Charles Lafarge became ill and died. His wife Marie (see Figure 9-3) was accused of killing him with rat poison. A few years earlier, an English chemist, James Marsh, had developed a test to detect arsenic, the major ingredient in rat poison. Using the Marsh test, investigators found traces of arsenic in food at the Lafarge home, but were not able to detect the poison in Lafarge's body. The leading authority in medical toxicology at that time, Mathieu Orfila, was called in to retest Lafarge's exhumed body. Orfila found traces of arsenic in Lafarge's body. The soil surrounding the body was also tested and no traces of arsenic were found, disproving claims by the defense that poison had been absorbed into the body after burial. Marie Lafarge was convicted of murder.

In 1821 the emperor of France, Napoleon Bonaparte, died—reportedly of cancer. However, in 2001, an analysis of locks of his hair indicated that he had elevated levels of arsenic in his body and may have actually died from arsenic poisoning.

Figure 9-3. *Marie Lafarge.*

Today, a toxicologist studies the harmful effects of many types of chemicals on the human body. Forensic toxicologists are concerned with the legal and medical aspects of alcohol, drugs, poisons, and toxins in bodily fluids. Because many autopsies involve analysis of body fluids and tissues for the detection of poisons or drugs, toxicologists often work with pathologists. Other duties include analyzing evidence to determine whether alcohol or drugs were contributing factors in criminal cases, such as vehicular homicide or sexual assault. Forensic toxicologists may present expert testimony in court, offering an opinion or explaining a procedure.

ALCOHOL

Obj. 9.2, 9.3

The term *alcohol* is used in everyday speech to mean beer, wine, or liquor. In chemistry, the term is used to refer to a group of organic compounds with a hydroxyl (–OH) group. Some alcohols are used in industry as solvents and can be found in antifreeze, perfumes, fuels, hairspray, and medications. In sufficiently high concentrations, all alcohols are *toxic*, or poisonous, to humans. Methanol is so toxic that consuming just 30 milliliters may cause death. When ingested, propanol, or rubbing alcohol, causes decreased heart rate, dizziness, and internal hemorrhage. Even ethanol, the alcohol used in alcoholic beverages, can be fatal if consumed in large quantities. Figure 9-4 shows the chemical structure of these three common alcohols.

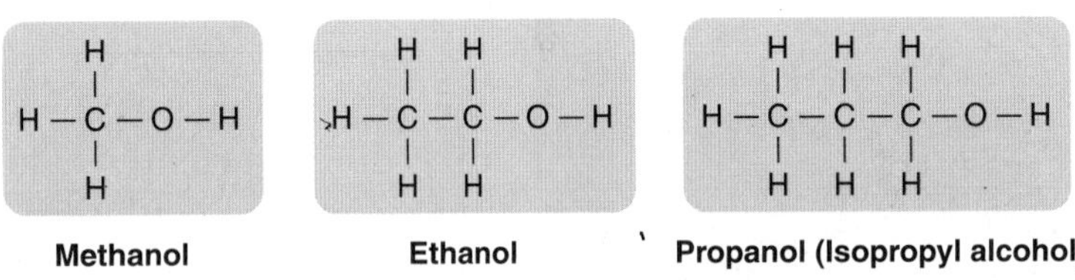

Figure 9-4. *Methanol, ethanol, and propanol (isopropyl alcohol).*

ALCOHOL IN THE BODY

Ethanol, also called *ethyl alcohol*, is a colorless liquid obtained from fermented grains or fruits. People often mistakenly think that ethyl alcohol is a stimulant because it may initially produce feelings of euphoria. However, it is actually a **depressant**, a chemical that slows the heart rate and brain activity and causes drowsiness.

Absorption of alcohol in the body begins as the chemical diffuses into the bloodstream through the walls of the stomach and small intestine. The blood distributes the alcohol throughout the body. Because alcohol is a depressant, as the concentration in the body increases, the person's ability to respond to stimuli decreases. At first, the person may feel less inhibited and feel very happy and talkative. The initial, and seemingly stimulated, response to alcohol is actually caused when the part of the brain that controls inhibitions is depressed. As more alcohol is absorbed, the person begins to lose coordination and to become confused. The part of the brain responsible for learning and memory becomes affected.

In the liver, alcohol is metabolized, or broken down, by the enzyme alcohol dehydrogenase (ADH). The resulting product is acetaldehyde. The acetaldehyde is eventually broken down by more enzymes into carbon dioxide and water. The liver can normally metabolize one or two alcoholic drinks, about 15 to 30 milliliters of alcohol, in an hour (see Figure 9-5). About 90 percent of the ingested alcohol is converted to carbon dioxide and water. The remaining 10 percent is excreted (eliminated) from the body in the breath, perspiration, and urine. When a

Prior to 1984, the legal drinking age varied from state to state. In some states, the age was set as low as age 18. The National Minimum Drinking Age Act of 1984, also called the Uniform Drinking Age Act, mandated that no one under the age of 21 be allowed to purchase alcohol within the United States.

© ART-4-ART/ISTOCKPHOTO

Figure 9-5. *This photo shows a typical serving size of wine, beer, and whiskey. Each serving contains about 15 mL of ethanol.*

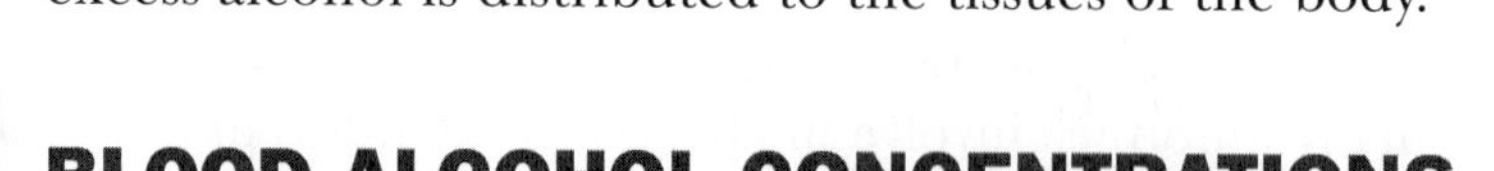

person drinks more alcohol than the liver can metabolize, the excess alcohol is distributed to the tissues of the body.

BLOOD ALCOHOL CONCENTRATIONS AND STAGES OF INTOXICATION

Determining the amount of alcohol in a person's blood is a good indication of the level of intoxication the person is experiencing. A blood alcohol concentration of 0.08 means there are 8 g of alcohol to every 10,000 mL of blood. If a person consumes alcohol faster than the body can eliminate it, the blood alcohol concentration increases. As the concentration of alcohol increases, the level of intoxication also increases.

Alcohol affects different people differently. However, the effects within certain ranges have been determined. At a blood alcohol concentration of 0.01 to 0.05, most people will behave normally. Somewhere between 0.03 and 0.12, euphoria begins. The person will become very talkative, less inhibited, and more confident. He or she will also begin to experience a loss of motor skills. As the concentration increases from 0.09 up to 0.25, vision is blurred and the person experiences loss of balance and coordination. Sleepiness sets in at this stage of intoxication (see Figure 9-6). By the time the blood alcohol concentration reaches 0.18 to 0.30, the person's speech becomes slurred and he or she may become dizzy and disoriented. This is the stage where many drinkers get very emotional and confused. At the upper levels in this range, there may be an increased tolerance for pain and the respiratory system may be affected. Somewhere between 0.25 and 0.40, standing and walking is very difficult. The person may begin vomiting and lose control of the bladder. By the time blood alcohol concentrations reach 0.35 to 0.50, both the respiratory and circulatory systems are impaired. Breathing may become difficult and the body temperature will drop. The drinker may slip into a coma and possibly even die. Any concentration over 0.45 is potentially fatal due to respiratory arrest.

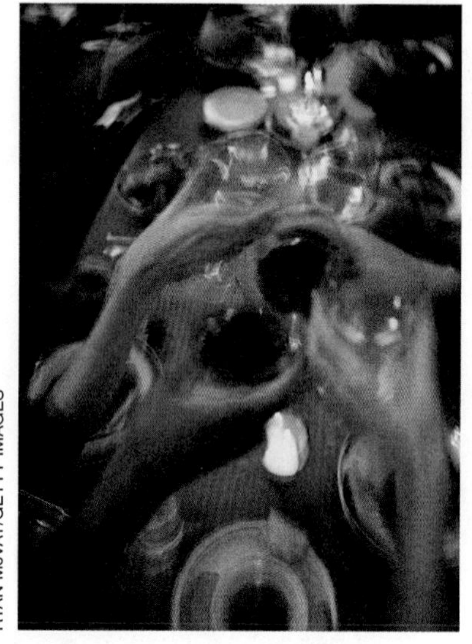

RYAN McVAY/GETTY IMAGES

Figure 9-6. *After reaching a blood alcohol concentration of only 0.09, a drinker may become sleepy and experience blurred vision and loss of coordination.*

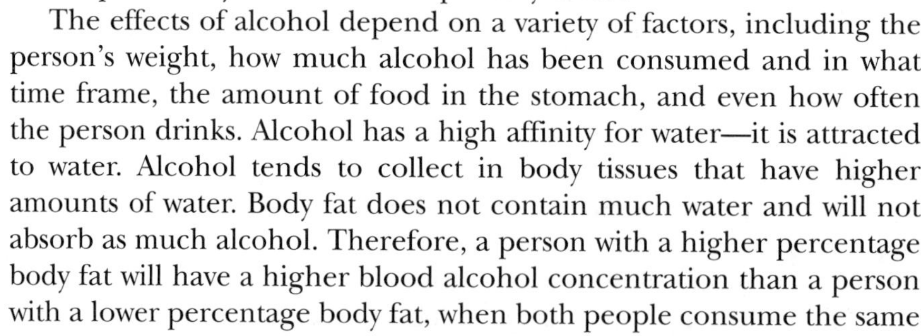

The effects of alcohol depend on a variety of factors, including the person's weight, how much alcohol has been consumed and in what time frame, the amount of food in the stomach, and even how often the person drinks. Alcohol has a high affinity for water—it is attracted to water. Alcohol tends to collect in body tissues that have higher amounts of water. Body fat does not contain much water and will not absorb as much alcohol. Therefore, a person with a higher percentage body fat will have a higher blood alcohol concentration than a person with a lower percentage body fat, when both people consume the same amount of alcohol. Because the body fat is not absorbing much alcohol, more alcohol is left in the bloodstream, increasing blood alcohol levels.

Food in the stomach slows alcohol absorption as well. When there is food or fluid in the stomach, the valve at the bottom of the stomach, called the *pyloric sphincter*, holds the contents in the stomach longer for digestion. Even water or fruit juice will slow the absorption of the alcohol. On the other hand, carbonation will speed up the absorption of alcohol. Only about 20 percent of the alcohol is absorbed directly into the bloodstream through the walls of the stomach. The majority of the alcohol—about 80 percent—is absorbed through the walls of the small intestine (see Figure 9-7).

Many medications affect how alcohol is metabolized in the body. Even popular energy drinks and herbal and over-the-counter medications can

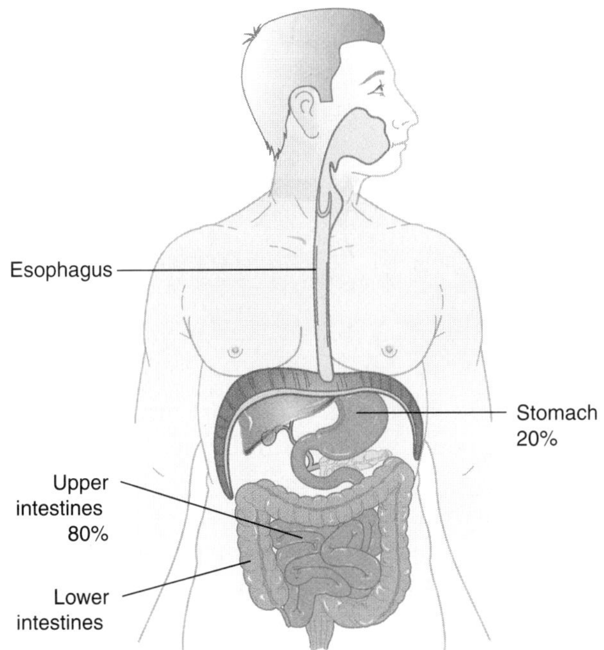

Figure 9-7. *About 80 percent of the alcohol a person drinks is absorbed through the walls of the intestines.*

interact with alcohol. Any medication that has a sedative effect will enhance the effect of the alcohol. Because alcohol is also a depressant, the effects of both the alcohol and the sedative will be increased. Alternatively, the alcohol might interfere with or block the action of the medication, decreasing the medicine's effectiveness.

Digging Deeper
with Forensic Science e-Collection

In 1984, the legal age for consumption of alcohol in the United States was set at 21. Now, however, there is a debate about this topic. Many college presidents are joining forces to get the drinking age lowered to 18 to help curb the incidents of binge drinking. Using the Gale Forensic Science e-Collection at www.cengage.com/school/forensicscienceadv, research the arguments for lowering the drinking age and those for keeping the drinking age at its current 21. After reviewing the pros and cons of each, write a one-page position paper regarding your views. Make sure to state your position clearly. Defend your viewpoint with the reasons you have chosen this position.

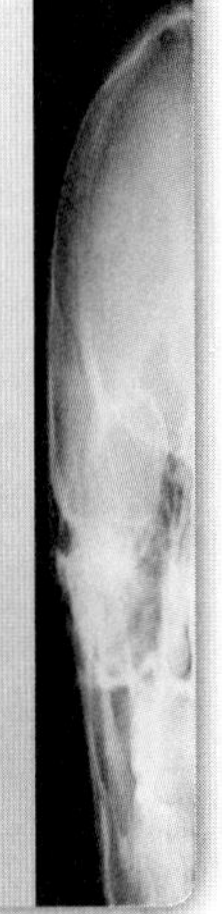

TOLERANCE

Over time, a person who drinks often may develop alcohol tolerance. **Tolerance** is the body's need for progressively larger amounts of a chemical to cause the same levels of intoxication. There are two kinds of tolerance—metabolic tolerance and functional tolerance. In response to prolonged heavy exposure to alcohol, the liver produces larger amounts of the enzyme alcohol dehydrogenase. This enzyme is important in the process of metabolizing alcohol. An increase in alcohol dehydrogenase speeds up the elimination process and lowers the person's blood alcohol concentration. This kind of tolerance is called *metabolic tolerance* and can lead to serious liver damage. *Functional tolerance* causes a person

CHAPTER 1 2 3 4 5 6 7 8 9 10 11 12 13 14 15

Figure 9-8. *Results of an alcohol-related car crash.*

to display fewer symptoms of intoxication. However, functional tolerance has no effect on the person's blood alcohol concentration or impairment.

ALCOHOL AND THE LAW

Thousands of people lose their lives in alcohol-related traffic accidents each year (see Figure 9-8). In most states, driving with a blood alcohol level above 0.08 is considered driving under the influence and is illegal. In the United States, all states have implied consent laws concerning alcohol. These laws prevent impaired motorists from refusing to submit to a breath test or blood test by invoking their Fifth Amendment rights. According to implied consent, there is an agreement that you will not operate a motor vehicle if you are under the influence of alcohol. Anyone who is suspected of driving while under the influence of alcohol may be requested to submit a breath or blood sample for analysis. If the driver refuses, he or she will lose his or her driver's license for a specified time. In most states the length of time is six months to a year.

In the 1966 court case of *Schmerber v. California*, Anoncenda Schmerber sued police because a blood sample was taken and tested for alcohol without his consent. Schmerber was being treated in the hospital emergency room for injuries he suffered in a traffic accident. The results of the blood alcohol test indicated that he had been driving while intoxicated and he was arrested. He argued that taking the sample of blood violated his Fifth Amendment right to refuse to testify against himself. The case eventually reached the United States Supreme Court, which ruled that the Fifth Amendment privilege applied only to testimonial evidence, not to physical evidence such as fingerprints or blood samples.

Obj. 9.5

DETERMINING ALCOHOL LEVELS

According to the National Survey on Drug Use and Health (NSDUH), 86.1 percent of the 20-year-olds surveyed admitted to using alcohol. More than 62 percent claimed to have used alcohol before their eighteenth birthdays. Alcohol is used by more young adults than tobacco or illicit drugs.

Testing the blood for alcohol is best done as quickly as possible because the alcohol will continue to metabolize in the suspect's body. Urine may also be tested; however, urine is a less reliable indicator of blood alcohol level because the alcohol concentrates in the urine for 24 hours or more after the person has stopped drinking.

In most cases, a driver pulled over because of suspicion of operating a motor vehicle while under the influence of alcohol is subjected to sobriety tests. Because actual blood testing must be done in a laboratory, presumptive field tests are often performed to determine whether further testing is required.

FIELD SOBRIETY TESTS

One obvious clue that a driver has been drinking is the odor of alcohol. When a police officer smells alcohol, he or she may perform a horizontal gaze nystagmus (HGN) test. A human's eyes make involuntary jerking motions, called **nystagmus**. When a person is intoxicated, nystagmus is more pronounced. To perform the HGN test, a small flashlight or penlight is held at eye level and moved slowly from side to side. The subject is asked to follow the movement of the light with his or her eyes, not moving the head. Police officers are trained to recognize the eye movements that are consistent with intoxication (see Figure 9-9). A driver's inability to "pass" these tests is an indication that more tests should be completed.

Tests that require concentration on more than one task at a time, called *divided-attention tests*, can also be used to judge alcohol impairment. For example, a person who is asked to count or recite the alphabet backwards would have to concentrate on that task. If an additional task, such as standing on one foot, is added, the intoxicated person would have difficulty performing the tasks simultaneously. The familiar "walk the straight line, heel to toe" test (see Figure 9-10) measures the person's ability to concentrate on two tasks at the same time.

JOE RAEDLE/GETTY IMAGES

Figure 9-9. *Horizontal gaze nystagmus testing.*

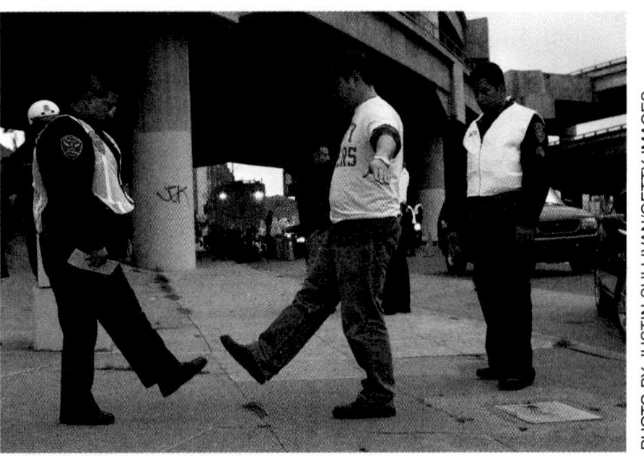

PHOTO BY JUSTIN SULLIVAN/GETTY IMAGES

Figure 9-10. *Divided-attention testing.*

BREATH TESTS

Alcohol is excreted from the body through the urine, perspiration, and breath. When a person drinks, the alcohol diffuses from the blood into the alveoli in the lungs. The air exhaled from the lungs of a person who has been drinking contains alcohol. There is a direct correlation between the amount of alcohol in the breath and the blood alcohol level. The amount of alcohol in the breath is $\frac{1}{2{,}100}$ of the amount of alcohol in the blood at any particular time. In other words, 2,100 mL of exhaled air contains the same amount of alcohol as 1 mL of blood.

A breath test, such as a breathalyzer (see Figure 9-11), measures the amount of alcohol in exhaled air. A person breathes into the breathalyzer and the air bubbles through a vial containing a mixture of sulfuric acid, potassium dichromate, silver nitrate, and water. The silver nitrate is a catalyst for the reaction and keeps the chemicals in solution. Ethyl alcohol is removed from the exhaled air by the sulfuric acid in the mixture. The alcohol, now in liquid form, reacts with the reddish orange potassium dichromate to form chromium sulfate, potassium sulfate, acetic acid, and water. The chromium sulfate is green. The intensity of the green color is directly proportional to the amount of alcohol in the breath. The vial containing the reacted breath sample is then compared to a second vial containing the same, but unreacted, mixture. Using a meter to measure the difference, the device automatically calculates the amount of alcohol in the blood based on the amount of alcohol given off in the exhaled air. Figure 9-12 on page 260 shows the chemical equation for the breathalyzer test.

BLOOD AND URINE TESTS

Because blood tests provide a direct measure of the alcohol concentration within the blood, they are more accurate than urine and breath tests. Toxicology labs use gas chromatography/mass spectrometry (GC/MS)

© CULTURE-IMAGES GMBH/ALAMY

Figure 9-11. *Breath test.*

Did You Know?

People often think that rinsing out their mouth or brushing their teeth eliminates the alcohol, at least for a breathalyzer. Because the test is designed to test the alcohol in the exhaled air, not the odor from the mouth, brushing, rinsing, or chewing gum will not alter the results of the breath test significantly.

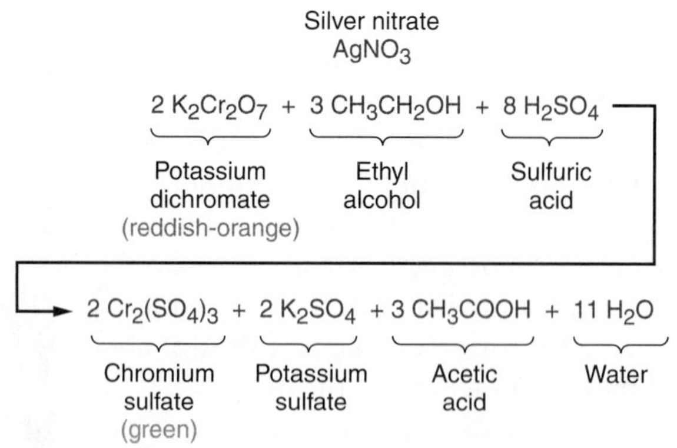

Figure 9-12. *The chemical equation of the breathalyzer test.*

to measure alcohol levels. Gas chromatography separates the sample into individual components. Each component is recorded as a separate peak on the chromatogram. The width of the peaks provides a quantitative measurement of the components. Review more information about GC/MS in Chapters 3 and 4.

Obj. 9.3, 9.4

POISONS

Many plants and animals produce **toxins**—poisons produced naturally by organisms—as protection against predators. Venomous snakes, such as the rattlesnake in Figure 9-13, and spiders produce venoms that cause nausea and vomiting; others may even cause paralysis, convulsions, or death. Some plants also produce toxins. If eaten, fruit or leaves from the nightshade plant (see Figure 9-14) or bulbs of daffodils can be fatal.

Insects, including wasps and bees, may also inject toxins into the human body. These toxins usually cause symptoms, such as mild swelling and itching. However, people who are allergic to the toxin may have much stronger responses, such as dizziness, weakness, confusion, and even unconsciousness and death.

PHYSIOLOGY OF POISONS

Poisons enter and affect the body in several different ways. Some of the poisons are *ingested* (eaten). Others can be inhaled, injected, or absorbed

PONTUS EDENBERG/SHUTTERSTOCK

Figure 9-13. *Rattlesnake.*

MARILYN BARBONE/SHUTTERSTOCK

Figure 9-14. *Nightshade.*

through the skin. Some poisons affect the body in different ways depending on how the substance enters the body.

Metal compounds are poisonous and may enter the body through ingestion, inhalation, or absorption through the skin. The metals are stored in soft tissues. Organs throughout the body are damaged when the metals bind to organic molecules in cells, preventing the molecule from functioning properly. The metal's interference with the function of organic molecules ultimately leads to cell death. Heavy metals such as arsenic, lead, mercury, antimony, and thallium are some of the most serious. The sources of contamination range from pesticides used in agriculture to industry pollution to contaminated food or water.

Ingested Poisons

The vast majority—90 percent—of all poisonings occur at home and involve young children swallowing household products (see Figure 9-15), including cleaning products, cosmetics, and over-the-counter medications such as *analgesics*, or painkillers. When swallowed, the poisons are absorbed through the lining of the intestinal tract. Depending on the poison ingested, the target may be a specific organ, a region of the body, or the nervous system.

IMAGE SOURCE/JUPITERIMAGES

Figure 9-15. *Children account for the majority of accidental poisonings.*

Inhaled Poisons

Some chemicals are dangerous when they are inhaled. For example, when carbon monoxide gas is inhaled, it combines with the hemoglobin in red blood cells, replacing oxygen and causing asphyxia. Low concentrations of the gas may cause symptoms such as dizziness, shortness of breath, or loss of consciousness. In higher concentrations, death occurs. Blood gas analysis and the color of lividity help the medical examiner and toxicologist identify carbon monoxide poisoning.

Other poisonous gases, such as the nerve gas sarin, interfere with nerve impulses when inhaled. Sarin causes convulsions, respiratory failure, and even death when inhaled in large doses. Symptoms of exposure appear within seconds after inhaling the nerve toxins. If exposure is mild and the victim is removed from the area of contamination and treated with proper medications, recovery is possible. Sarin is also easily absorbed through the skin when it is mixed with organic solvents. Sarin has been ruled a weapon of mass destruction. In 1995, Japanese domestic terrorists released sarin gas in the Tokyo subway. Twelve people died and approximately 6,000 were injured in the attack.

Injected Poisons

Some poisons are injected. For example, heroin abusers inject themselves with heroin. Heroin is a poison that produces euphoria almost instantly, followed by drowsiness and a sense of well-being. In the brain, heroin is converted into morphine. If the dose of the drug is high enough, the respiratory centers in the brain become depressed. This respiratory system depression can lead to respiratory arrest very quickly.

Absorbed Poisons

Some poisons may be absorbed into the body through the skin, eyes, or mucous membranes. Sarin can be absorbed as well as inhaled. Symptoms

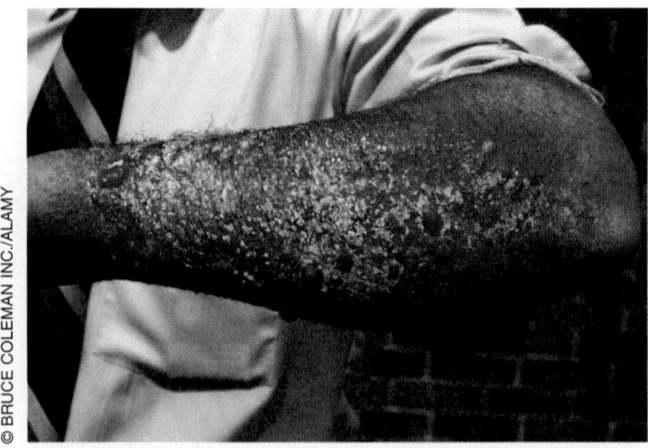

Figure 9-16. *Blisters from contact with poison sumac.*

of the absorbed poison include redness and irritation, blisters, or even death. Poisons like the sticky sap of poison sumac produce a blistering rash (see Figure 9-16). Others, such as yperite, also known as *mustard gas,* are much more dangerous. In low concentrations, mustard gas causes blistering of the skin. When inhaled, mustard gas strips away the mucosal lining of the bronchial tubes, causing difficulty in breathing. The gas also attacks the cornea in the eyes, leading to blindness. Mustard gas was first used as chemical warfare in 1917 during World War I.

In recent years, the fear of bioterrorism has prompted preparation and planning against attacks using various toxins such as ricin and anthrax.

Ricin is a very poisonous chemical that can gain entry into the body by injection, inhalation, or ingestion. Ricin is a by-product of processing the castor bean plant. In the body, ricin inhibits protein synthesis by inactivating the ribosomes. Without functioning ribosomes, cells die. On the skin, ricin poisoning is not likely to be fatal unless it is mixed with an absorbable solvent. If ricin is inhaled, it produces coughing, tightness in the chest, and fever within a few hours. Cells in the respiratory tract eventually die and respiratory failure occurs. When ingested, ricin causes lesions in the digestive tract. These lesions cause abdominal pain, nausea, vomiting, and bloody diarrhea. Death from ricin poisoning usually occurs from three to five days after exposure.

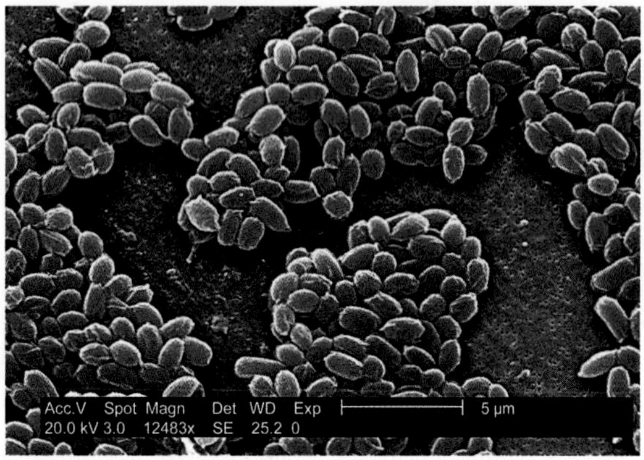

Figure 9-17. *Scanning electron micrograph of spores of the bacterium that causes anthrax:* Bacillus anthracis.

Anthrax has been in the news recently because it has been used in several attacks. Like ricin, it can enter the body in several ways. Anthrax is an infection caused by the spores of the bacterium *Bacillus anthracis* (see Figure 9-17). Ingestion of anthrax is rare but can be caused from eating undercooked or raw meat from an infected animal. It causes inflammation of the intestinal tract, leading to symptoms similar to food poisoning: abdominal pain, vomiting, and diarrhea. Anthrax on the skin—cutaneous anthrax—causes blisters and is usually nonfatal. The most common method of acquiring the skin infection is from contact with contaminated soil, animals, or animal products, such as hides and wool. Inhalation anthrax is the deadliest form of anthrax. This kind of poisoning occurs when dried bacterial spores become airborne and are inhaled. Symptoms may not appear for as long as a month or two after exposure. Following the latent period, fever and chills will develop, along with a cough, shortness of breath, and headache. The victim may experience chest and abdominal pain. Symptoms worsen over several days as the lungs fill with fluid, and the victim may go into shock. In more than 80 percent of inhalation anthrax cases, death occurs within three to seven days after the first symptoms appear.

The Centers for Disease Control and Prevention and the Department of Homeland Security have guidelines in place for the prevention of and response to such bioterrorism attacks. Currently, no cure exists for ricin poisoning, although measures are underway to develop a vaccine. A vaccine for anthrax is available to individuals who are at high risk of exposure, mostly military personnel and those who work in laboratories with the bacterium spores.

DRUGS

Obj. 9.3, 9.5

Crimes associated with drug use and abuse include possession, sale and use, and drug-related violent crimes such as robberies and homicides.

One key task for forensic toxicologists is detecting or identifying drugs in the body or a body fluid. Additionally, many businesses hire toxicologists to perform drug testing before a potential employee may be hired. Athletes are tested to make sure they are not using performance-enhancing drugs. Crime victims may be screened for the presence of specific club or "date rape" drugs. The chart in Figure 9-18 summarizes the various types of drugs that are commonly encountered. Toxicologists use gas chromatography and mass spectrometry to detect and identify various chemicals within the blood and other tissues in the body.

Class of Drug	Description/Physiology	Common Examples
Hallucinogen	Produce changes in perceptions, thought processes, and mood by interfering with neural pathways and sensory perception	Marijuana, MDMA, PCP (phencyclidine), LSD (lysergic acid diethylamide),
Narcotic	Opium derivatives; induce sleep and relieve pain by suppressing the central nervous system's ability to send pain messages to the brain	Codeine, oxycodone, morphine, heroin
Stimulant	Increase alertness, provide a sense of well-being and confidence by speeding up functions of the central nervous system	Cocaine, crack cocaine, methamphetamine
Depressant	Produce mild euphoria followed by drowsiness or reduce anxiety by slowing down the central nervous system, including brain activity	Alcohol, barbiturates, tranquilizers
Anabolic steroid	Promote cell division and tissue growth, leading to muscle growth, but have severe side effects (depression, anger, destructive behavior)	Synthetic testosterone

Figure 9-18. *Comparison of various classes of drugs.*

TOXICOLOGICAL TESTING AND PROCEDURES

Obj. 9.5

Clues at the scene of a crime or death may guide the toxicologist in determining the type of analysis needed (see Figure 9-19). For example, syringes found in a park next to a body may suggest a drug overdose or empty bottles in the backseat of a car involved in a deadly car crash might suggest alcohol as a contributing factor. If there is no evidence to assist investigators, the toxicologists conduct tests to screen for possible poisons and drugs.

Thin layer chromatography, gas chromatography, and immunoassay techniques may be used to detect trace amounts of drugs in the blood, urine, and body tissues. Once the possibilities are narrowed, confirmation is achieved using GC/MS. You may want to refer to Chapter 3 to review each of these analytical techniques.

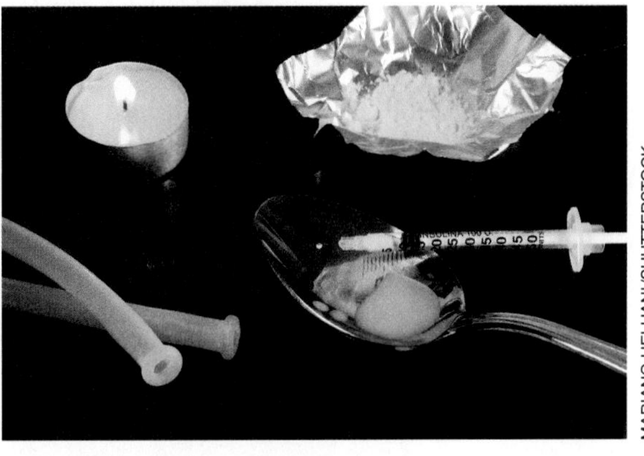

MARIANO HELUANI/SHUTTERSTOCK

Figure 9-19. *Drug paraphernalia found at a crime scene suggest that the victim should be tested for a drug overdose.*

CHAPTER 1 2 3 4 5 6 7 8 9 10 11 12 13 14 15

Digging Deeper

with Forensic Science e-Collection

Forensic toxicologists commonly rely on blood and urine to test for the presence of various drugs and toxins postmortem. Most recently, the process of examining and analyzing hair for the presence of drugs and toxins also has become a generally accepted practice. Using the Gale Forensic Science e-Collection at www.cengage.com/school/forensicscienceadv, discuss the advantages of using hair rather than blood or urine to test for the presence of drugs. Explain how hair absorbs the drugs and toxins and how samples are collected for evaluation. Write a short essay discussing the process of analyzing hair. What are additional applications for examining hair? Be sure to cite the sources you used in the research for your essay.

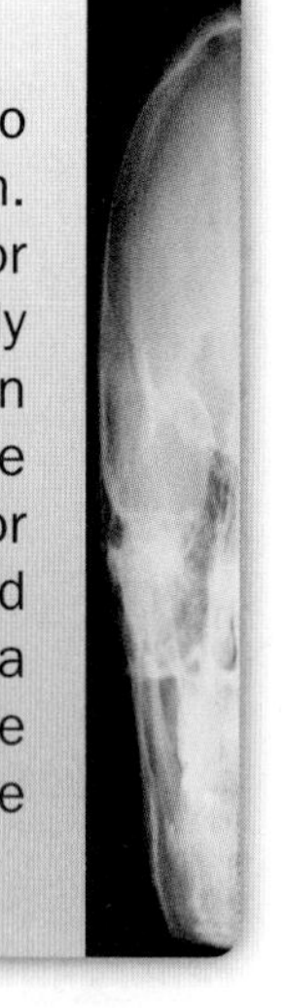

The Reinsch test is used when heavy-metal poisoning is suspected. In a solution of hydrochloric acid, ions of metals such as bismuth, arsenic, mercury, and thallium will produce a silver-colored coating on copper dipped into the solution. If the test is positive for heavy metals, the exact metal can be identified and confirmed through emission spectroscopy or atomic absorption spectrophotometry.

Medical examiners often order blood-gas screenings in victims found at a fire scene or bodies with cherry red lividity. Carbon monoxide in the blood is detectable using spectrophotometry or gas chromatography. An absence of carbon monoxide in the fire victim suggests that the victim did not inhale any of the gases produced in the fire and likely died before the fire was set. In these cases, investigators suspect that the fire was set to destroy evidence of a murder.

An **immunoassay** is a test that relies on the antigen-antibody response. An antigen is any substance that enters the human body and produces an immune response. In response to an antigen, the immune system produces more immune cells. Some of these cells produce antibodies. Antibodies are proteins that attach to specific antigens. When poisons, including illegal drugs, enter the body, they act as antigens and cause the immune system to produce antibodies. The immunoassay detects or measures the amount of antibodies found in the blood or urine. A toxicologist adds specific drug antigens to a sample of the victim's or suspect's blood or urine. If antibodies to the drug are present, the antigens and antibodies will bond. Enzyme-multiplied immunoassay technique (EMIT) is an immunoassay that can rapidly detect the presence of several drugs and their metabolites. Cocaine, marijuana, heroin, amphetamines, barbiturates, and many other drugs can be detected by EMIT.

CHAPTER SUMMARY

- Poisons are chemicals that harm the body if ingested, inhaled, or absorbed through the skin. Nearly anything can be poisonous in sufficiently high concentrations.

- Forensic toxicologists investigate the legal and medical aspects of alcohol, drugs, poisons, and toxins in the body. Their job may also include testifying as expert witnesses, collaborating with the medical examiner, and analyzing evidence.
- The role of forensic toxicologist has expanded to include not only detection of drugs in bodies, but also testing employees for drug use, screening athletes for drug use, and testing sexual assault victims for the presence of date rape drugs.
- In everyday speech, *alcohol* usually refers to the ethanol used in alcoholic beverages. In chemistry, alcohols are a group of substances with a hydroxyl group.
- Alcohol is a depressant. Initially, consumption of alcohol causes feelings of euphoria. Depending on the amount of alcohol consumed, motor skills become impaired, brain function is depressed, and the person may even lapse into a coma and die.
- The liver can metabolize only about one or two alcoholic beverages (15 to 30 mL of alcohol) in an hour. If more alcohol is consumed, a person becomes intoxicated.
- If a police officer suspects that a driver is intoxicated, he or she will order several field sobriety tests. Horizontal gaze nystagmus and divided-attention tests are often used as presumptive tests of intoxication. Breath tests can measure the actual amount of alcohol in the breath to determine blood alcohol levels.
- How a poison affects the body differs depending on how the poison enters the body. For example, inhaled poisons can cause asphyxia whereas ingested poisons often target specific organs.
- Determining the types of toxins or other types of poisons in the body requires many different testing procedures. Chromatography techniques, mass spectrometry, and immunoassay can be used to detect various poisons and drugs in urine and blood. The Reinsch test detects heavy-metal poisoning.

CASE STUDIES

Jane Toppan (1901)

Jane Toppan (see Figure 9-20) attended nursing school at a hospital in Cambridge, Massachusetts, in the early 1880s. She was dismissed before completing the program because so many of her patients died in her care. She found a nursing job, but she was fired when the hospital discovered that she had not finished nursing school. Toppan forged her nursing certificate and became a private-duty nurse working for prominent families in New England.

For many years, Toppan was considered a kind, sensitive, and competent caregiver. All of that changed in 1901. Over a period of only six weeks, several members of a Cape Cod family died. The husband of Mary Gibbs, one of the victims, was suspicious and demanded an investigation. An autopsy on Gibbs and several of the exhumed bodies confirmed that all had been poisoned. Nurse Toppan was arrested.

Figure 9-20. *Jane Toppan, the "Angel of Death."*

During the trial, Toppan confessed to poisoning her patients with a mixture of morphine and atropine. Morphine causes the pupils to constrict; pinpoint-size pupils is a characteristic effect of morphine poisoning. Atropine causes pupils to dilate, so it was administered to mask the signs of morphine. She said she began with very small amounts and gradually increased the dosage of their "medicine" until her patients died. Toppan said she enjoyed watching people die. After serving 36 years of her life sentence, Jane Toppan died of natural causes in the Taunton State Asylum for the Criminally Insane in Massachusetts.

Think Critically

1. **If a similar case occurred today, what type(s) of tests would be done to determine whether the victims died from natural causes or poisoning?**
2. **Toppan enjoyed watching people die, so why did she administer such small doses of the drugs and gradually increase the amounts?**

Elvis Presley (1977)

Elvis Presley (see Figure 9-21) was known as the King of Rock 'n' Roll. On August 16, 1977, Presley died in his bathroom. He was only 42 years old, and although he was overweight, he had no known medical problems that would explain his sudden death. Rumors swirled that drugs were involved, but devout fans hoped his death was from some natural cause.

Initially, the Memphis medical examiner stated that drugs were in no way involved in Presley's death. The medical examiner stated that Presley's heart stopped as a result of cardiovascular disease. For several weeks after his death, however, the medical examiner refused to release the autopsy and toxicology reports, stating that they were not part of any official investigation. In October, a Baptist Hospital pathologist claimed Presley's death was due to drugs. The pathologist stated that the cause of death was a combined drug effect. Each drug found in Presley's system worked to depress the central nervous system; the interaction of all the drugs depressed his heart and lungs. The pathologist ruled the death accidental but suggested that the fact that several doctors were prescribing medication to Presley without communicating with each other was a leading factor in his death. The pathologist's report continued by dismissing the opinion of the Memphis medical examiner, stating there was no evidence of cardiovascular disease, no blood clots, and no hardened arteries. Presley did have a slightly enlarged heart, but the pathologist claimed that was not a factor in his death.

Figure 9-21. *Elvis Presley.*

Think Critically **Drug interactions from medications prescribed by more than one doctor can be very dangerous. As of January 2010, 34 states have a prescription drug–monitoring program (PDMP) in place to alert doctors, pharmacists, and law enforcement when prescription drugs are being overprescribed or when prescriptions of the same drug are being written by multiple doctors for the same patient.**

1. **What are the benefits of having such a system?**
2. **Explain how such a system, had it been in place in 1977, might have had an impact on the health of Presley.**

Kathy Augustine (2006)

On the morning of July 8, 2006, Chaz Higgs made a frantic call to 911. He reported that he had come home to find his wife, Kathy Augustine (see Figure 9-22), passed out and not breathing. Higgs was a nurse, so he said he had tried CPR to revive her, but it did not work. Augustine was rushed to a hospital, but she never regained consciousness and died three days later of an apparent heart attack. Augustine was a Nevada state politician and had been working hard on her next campaign. Higgs was quick to give this as the likely reason behind her heart attack. No evidence of heart damage indicative of a heart attack was found during the autopsy. Additionally, the autopsy found an unexplained mark on her left hip that could have come from an injection. These findings turned Augustine's death into a possible homicide. Investigators searched Higgs's house, where they found a vial of succinylcholine, a powerful muscle relaxant that paralyzes muscles used for breathing, causing a person to suffocate. Toxicologists do not typically screen victims for succinylcholine poisoning, which is likely why Higgs chose it as his murder weapon. He would have had easy access to succinylcholine as a nurse. Higgs was arrested and eventually convicted of his wife's murder.

FILE PHOTO TAKEN 12/19/05/NEWSCOM

Figure 9-22. *Kathy Augustine.*

Think Critically **1. What two findings during the autopsy indicated that Kathy Augustine did not die of natural causes?**

2. What is a possible reason for Chaz Higgs choosing succinylcholine as his murder weapon?

Bibliography

Books and Journals

Barile, Frank A. (2004). *Clinical Toxicology: Principles and Mechanisms.* Boca Raton, FL: CRC Press.

Conklin, Barbara, Robert Gardner, and Dennis Shortelle. (2002). *The Encyclopedia of Forensic Science.* Westport, CT: Onyx Press.

Garriott, James C., ed. (1998). *Medical Aspects of Alcohol Determination in Biological Specimens.* Littleton, MA: PSG Publishing Company, Inc.

Lerner, K. Lee, and Brenda Wilmoth Lerner, eds. (2006). *The World of Forensic Science.* New York: Gale Group.

Newton, Michael. (2008). *The Encyclopedia of Crime Scene Investigation.* New York: Info Base Publishing, pp. 125–126.

Ricciuti, Edward. (2007). *Science 101: Forensics.* New York: HarperCollins.

Websites

www.bxscience.edu/publications/forensics/articles/toxicology/r-toxi01.htm
www.cc.utah.edu/~sa11170/EMT/DOTRefresh/Poison/poisonbot.htm
www.cdc.gov/HealthyYouth/alcoholdrug/index.htm
http://forensicscience.suite101.com/article.cfm/forensic_toxicology_drug_analysis_takes_time
www.grsproadsafety.org/themes/default/pdfs/Drinking%20Age%20Limits.pdf
www.intox.com/physiology.asp
www.madd.org
www.nlm.nih.gov/visibleproofs/galleries/cases/orfila.html
www.oas.samhsa.gov/2k9/138/138AlcBefore21stBday.htm
www.oregoncounseling.org/ArticlesPapers/Documents/ETOHBIOFx.htm
www.rochester.edu/uhs//healthtopics/Alcohol/tolerance.html
https://webapps.ou.edu/alcohol/docs/13EtohandMedicationInteractions40-54.pdf

CAREERS IN FORENSICS

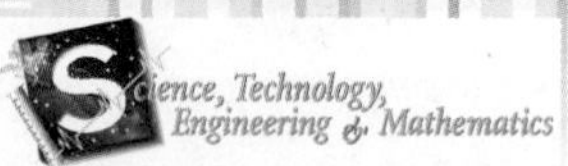

Forensic Toxicologist

A forensic toxicologist (see Figure 9-23) takes part in the legal investigation of a death. According to the American Chemistry Society, there are about 21 million registered compounds and there are tests for more than 3,000 of those compounds.

With such overwhelming possibilities, where does the toxicologist start? The crime scene. There is the physical evidence to consider: pill bottles, trace residue, needles, food, and chemicals of any kind and in any form. Using these items as a starting point, the forensic toxicologist must determine the effect of the various substances on the human body. These calculations are complicated by the fact that no two people metabolize a substance in the same way. Height, weight, body fat, and other factors will impact how a substance is absorbed into the blood. Every bit of evidence is critical in narrowing the list of possible substances for the toxicologist to test.

For example, in 1962, when actress Marilyn Monroe was found dead in her bedroom, toxicologists immediately tested her body for Nembutal and chloral hydrate, because empty prescription bottles were found on her nightstand. She did indeed die from an overdose of these drugs. These findings prompted a psychological analysis to determine the likelihood of accidental death versus suicide. That analysis painted a picture of a woman who was prone to frequent suicidal depressive episodes. However, at the time of her death, Monroe was experiencing a very positive, up-beat time in her life. Because Monroe was such a famous person, the media followed the story closely. There were charges of possible conspiracy, cover-up, and incompetence. In the end, however, her death was ruled a suicide.

A toxicologist can obtain a sample to test for toxins in many different ways. If the subject is living, the toxicologist can sample blood, urine, breath, or even saliva. If the subject is no longer living, samples are taken during an autopsy. Hair samples can give evidence of long-term drug use—the drugs affect hair as it grows and leave a "record" of drug intake. Stomach contents can reveal when and possibly where a victim last ate, and the liver harbors toxins as it cleans the body, so the forensic toxicologist tests liver samples for toxic substances. The brain, spleen, fibrous layer of the eyeball, and the eye socket also sometimes reveal clues.

Even with the most advanced technology and a team at work on a case, a full toxicology report can take weeks or more. The correct answers are often not the most obvious ones. Toxicology is a profession that requires dedication and a great deal of perseverance.

KLAUS TIEDGE/JUPITERIMAGES

Figure 9-23. *Forensic toxicologist.*

Learn More About It

To learn more about a career in forensic toxicology, go to www.cengage.com/school/forensicscienceadv.

CHAPTER 9 REVIEW

Matching

Match the following poisons with their method of entry into the body. Some answers may be used more than once and some review items may have more than one answer.

1. carbon monoxide *Obj. 9.3, 9.4*
2. ricin *Obj. 9.3, 9.4*
3. anthrax *Obj. 9.3, 9.4*
4. snake venom *Obj. 9.3, 9.4*

a. ingestion
b. inhalation
c. injection
d. absorption

Multiple Choice

5. Alcohol is eliminated from the body through excretion in the ________. *Obj. 9.3*
   a. breath
   b. urine
   c. perspiration
   d. all of the above
6. In what organ is alcohol metabolized? *Obj. 9.3*
   a. liver
   b. stomach
   c. lungs
   d. small intestine
7. Currently, what is the legal threshold for driving while intoxicated in most states? *Obj. 9.2*
   a. 0.08 percent
   b. 0.10 percent
   c. 0.15 percent
   d. 0.80 percent
8. The presence of high levels of carbon monoxide in the blood of a victim at the scene of a fire suggests that the victim ________. *Obj. 9.5*
   a. died somewhere else and was brought to the scene postmortem
   b. perished after the fire started
   c. was already dead when the fire started
   d. was a cigarette smoker
9. Most accidental poisonings occur as a result of ________. *Obj. 9.3*
   a. inhaling carbon monoxide
   b. children consuming household products
   c. contact with a biological poison
   d. none of the above

10. Which of the following types of alcohol is found in alcoholic beverages, such as beer and wine? *Obj. 9.5*
    a. ethanol
    b. methanol
    c. isopropanol
    d. any of these

Short Answer

11. Distinguish between a stimulant and a depressant. Give an example of each. *Obj. 9.3*

12. Distinguish between a toxin and a poison. *Obj. 9.3*

13. Explain the significance of the court case *Schmerber v. California*. *Obj. 9.2*

14. What are implied consent laws? Why were they enacted? *Obj. 9.2*

15. Describe the following field sobriety tests: *Obj. 9.2*
    a. Horizontal gaze nystagmus

    b. Divided-attention test

16. Describe how immunoassay detects drugs in the body. *Obj. 9.5*

17. Summarize the stages of intoxication. *Obj. 9.3*

18. Explain the difference between metabolic and functional tolerance. *Obj. 9.3*

19. Describe the role of a forensic toxicologist. *Obj. 9.1*

20. Explain the absorption, distribution, and elimination of alcohol through the body. *Obj. 9.3*

21. If a blood test reveals that an adult has 0.2 g of alcohol in 1 L of blood, what is the person's blood alcohol concentration? What symptoms of intoxication or physical impairment is the person likely to be experiencing? Is the person within the legal limits to drive? *Obj. 9.2*

22. If a blood test reveals that an adult driver has 1 g of alcohol in 1,000 mL of blood, what is the person's blood alcohol concentration? What symptoms of intoxication or physical impairment is the person likely to be experiencing? Is the person within the legal limits to drive? *Obj. 9.2*

____

____

____

____

____

Think Critically

23. Why can a breath test be used to determine the amount of alcohol in the blood? Explain how the metabolism of alcohol in the body can be used to determine the amount of alcohol a person drank even when the test is administered hours later. ***Obj. 9.5***

24. Two people of equal weight drink the same amount of alcohol over a period of two hours. One of them appears intoxicated after only three drinks while the other seems barely affected. Discuss possible factors that make this possible. ***Obj. 9.3***

CRIME SCENE S.P.O.T. Student-Prepared Original Titles

The Case of the Sleepy Driver

By: **Kyle Banas, Kelsey Janos, and Melissa Pena**

East Ridge High School
Clermont, Florida

The phone rang at 3:33 A.M., waking Gill suddenly. Gill was the crime-scene investigator in Blue Springs. "Hello?" said Gill sleepily, as he answered.

"Hey, Gill. It's Rick. We got one for you. We're at Joe's Bar."

Gill forced himself to his feet, threw on his uniform, grabbed his wallet and keys, and headed for the car. On the ride over to Joe's, he called Sarah, his partner. She answered on the third ring, and Gill told her what he knew. They agreed to meet at the bar as soon they could both get there.

Upon arrival at the scene, Gill spoke to paramedics while Sarah and Rick began their investigation. Rick ran the license plate and found that the car belonged to Jason Schwartz, the driver. They walked around to the front of the car. A man was pinned between the car and the outside wall of the bar.

"Looks like the driver fell asleep at the wheel. We don't have an ID on the second vic. He was apparently at the wrong place at the wrong time." Rick's assessment was as good as any, for now.

Gill slipped on his state-issued gloves in order to avoid contaminating evidence before he began his examination of the car. The first thing he noticed was the driver's clothing. Schwartz had either gotten dressed in a hurry or in the dark. His hair was disheveled. There was a cell phone lying on the ground next to the car. After Sarah finished the photographs, Gill picked up the phone and examined it. He hit the redial button. After a few rings, a woman answered.

"Logan! Are you on your way home yet?"

Gill answered, "This is CSI Gill Barkin. To whom am I speaking?"

The woman identified herself as Logan's girlfriend, Taylor. After a few questions, Gill was able to determine that the other victim (who was pinned to the wall) was most likely Logan Mefford. He also discovered that the two dead men worked together.

Witnesses at Joe's agreed that Jason and Logan had been having drinks with another man, their boss Ryan Little. Little also happened to be Taylor's brother. According to rumors, the boss was not happy that both of the men had been dating his baby sister, and he had been trying to convince them to leave her alone. The glasses the three had been drinking from were still on the table.

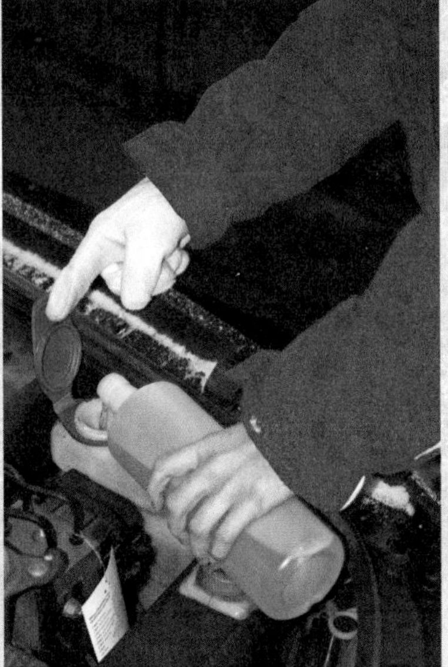

© MAURITIUS IMAGES GMBH/ALAMY

Figure 9-24. *Antifreeze is used in car radiators to prevent the water from freezing. When ingested by animals and humans, it is highly toxic.*

Continuing their investigation, the team popped the trunk and searched the rest of the interior of the car. Two gym bags were collected as evidence. One was blue; the other was a black bag with the initials TL on it. By the time all of the evidence had been tagged and packaged, the medical examiner showed up to transport the bodies.

It took a couple of days to get the autopsy report. Schwartz had suffered a broken neck in the crash, but according to the toxicology report, crystals were found in the kidneys—calcium oxalate crystals, which was consistent with antifreeze poisoning (see Figure 9-24). Gill had already spoken to the ME about this. According to the doctor, antifreeze is sweet and can easily be mixed into tea or some other sweetened beverage. The symptoms of antifreeze poisoning include dizziness—a possible cause of the crash.

Sarah asked, "So you are saying he was poisoned?"

"Yes," replied Gill. "It looks as if our accident has just become a homicide."

A search of Schwartz's apartment revealed evidence of a female visitor to the apartment. A glass with lipstick on the rim was collected for possible DNA and fingerprint analysis. Other than that, everything seemed to be in order. Mefford's home revealed a little more. He was not as neat and tidy as Schwartz, and hidden under the kitchen sink was a nearly empty container of antifreeze. The container was bagged and taken to be processed for trace evidence.

Weeks later, when all of the evidence had been processed, the team sat down for their final assessment of the case.

Gill addressed the group. "The antifreeze found in Mefford's home appears to have been planted there. A hair stuck to the inside of the cap was analyzed for mitochondrial DNA. It showed a familial relationship to Taylor. The fingerprint on the glass in Schwartz's apartment belonged to Taylor. We believe Schwartz's murder was supposed to be pinned on Mefford, but, instead, Mefford ended up dead as well. We have an arrest warrant coming, and by this afternoon, the real murderer will be behind bars."

Activity:

Answer the following questions based on information in the Crime Scene S.P.O.T.

1. Based on the information provided, whose name is on the arrest warrant? What evidence supports your theory?
2. If Schwartz died of a broken neck in the crash, why was his death ruled a homicide?

WRITING

3. After reading the story, write a one- or two-paragraph conclusion to the story. Be sure to use correct terminology and logical, well-reasoned arguments.

Introduction:

According to Mothers Against Drunk Driving (MADD):

- In 2007, the average age of initial alcohol use was 16.6 years of age—almost five years younger than the legal drinking age.
- Teenagers who consume alcohol are five times more likely to drop out of high school before graduation.
- Consumption of alcohol among teens leads to more deaths than all illegal drugs combined.

These statistics have prompted many schools across the United States to host drug- and alcohol-free parties, hold reenactments of alcohol-related traffic accidents, and work toward drug-free campuses. In this P.A.C.T. activity, you will develop an awareness program designed to educate students in your school about the dangers of alcohol/drug use.

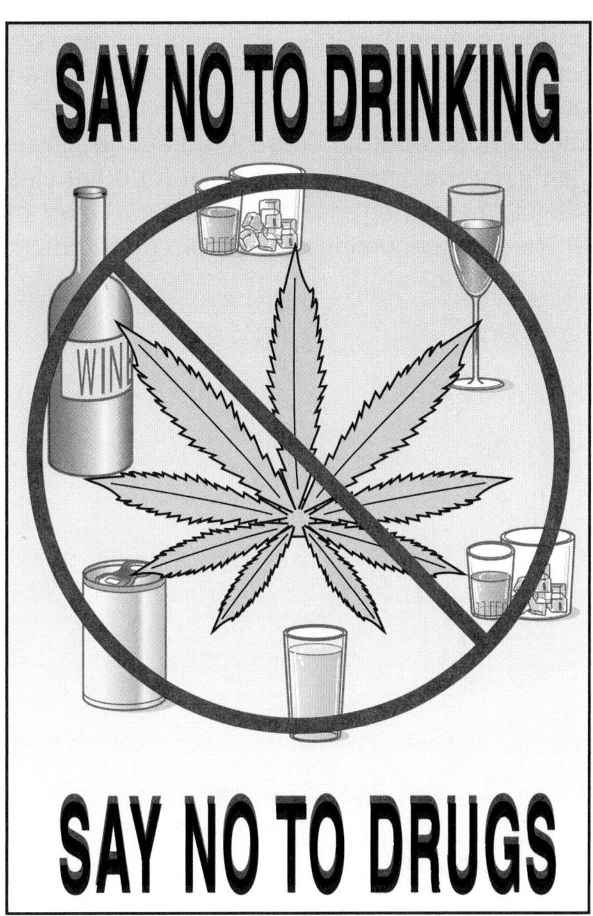

Figure 9-25. *Sample poster.*

Procedure:

Part 1

1. Working with a partner, choose a topic for your campaign. You may choose alcohol, smoking (cigarettes contain many harmful chemicals), or illegal drugs. If you choose to focus on the prevention of illegal drug use, you may focus on a specific type of drug (club drugs, hallucinogens, performance enhancement, etc.) or on illegal drugs in general.
2. Research the statistics for your topic. Look for prevalence of use, symptoms of use, long-term effects of use, and short-term dangers of use.
3. Design a brochure that may be distributed to students in your school or a poster that can be placed in the hall or cafeteria to make students aware of the problem.

Part 2

1. In groups of four or five students, write a public service announcement to increase awareness of alcohol and drug use among teens.
2. Design a commercial or skit and include the public service announcement. Videotape the presentation to be played on your school's morning news or perform the skit for other classes.
3. If timing is right, the skit or commercial could be part of Red Ribbon Week activities or homecoming or pre-prom festivities.

ACTIVITY 9-1 *Ch. Obj: 9.5*
GET TO THE VET

Objectives:

By the end of this activity, you will be able to:

1. Differentiate between physical and chemical reactions.
2. Perform differential testing on four known samples and one unknown sample.
3. Identify the insecticide that poisoned Sherlock.

Materials:

(per group of two or three students)
15-well spot plate
samples of poisons:
- Anti-Ant
- Kritter Kill
- Rid-A-Roach
- The Insectinator
- unknown (from Sherlock)

candle
matches
protective gloves
reagents A, B, and C
3 pipettes or droppers
15 toothpicks
5 small spatulas
5 small squares of aluminum foil
grease pencil
spring type clothespin
hand lens (optional)
black construction paper (optional)

Safety Precautions:

Safety goggles must be worn when working with chemicals and flames.
Aprons should be worn to protect clothing from the reagents.
Vinyl or latex gloves will protect skin from chemicals.
Wash hands thoroughly after handling the chemicals in this lab.

Background:

Your dog, Sherlock, crawled under the fence in the backyard and wandered into the neighbor's open garage. When he came home, Sherlock was having difficulty breathing and collapsed at your feet. There was a white powder residue around his mouth and nose.

The veterinarian is certain that the dog got into a poison and will die if the proper treatment is not started soon. Because the treatment is different for various poisons, Dr. Bassett needs to know which chemicals Sherlock may have inhaled or ingested.

There are four different insecticides stored in your neighbor's garage—all of which are open. The vet's assistant is out sick and Dr. Bassett has asked you to help him.

Procedure:

Part A: Chemical Tests

1. Label the spot plate using the grease pencil. See Figure 1 for proper labeling.

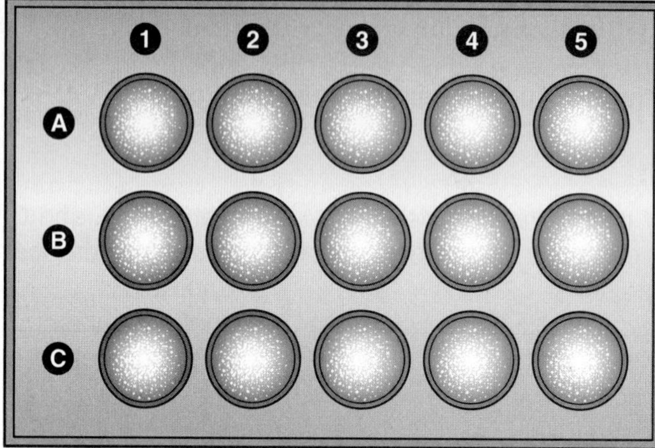

Figure 1. *Spot plate observations.*

2. Using a small clean spatula, place a sample of Anti-Ant in each of the wells numbered 1.
3. Examine the powder. Observe and record the color and texture. Use a clean toothpick to see how it holds together, separates, or clumps. You may use a hand lens for these observations. If the spot plate is transparent, placing it over a piece of black construction paper may help you observe the powders more clearly. Record the physical characteristics in Data Table 1.

Data Table 1. Chemical tests.

Insecticide	Visual Examination	Reaction with Reagent A	Reaction with Reagent B	Reaction with Reagent C
Anti-Ant				
Kritter Kill				
Rid-A-Roach				
The Insectinator				
Sample from Sherlock				

4. Carefully add 3 to 5 drops of reagent A to the wells in row A. If needed, use the toothpick to mix the chemicals. Record the results in Data Table 1.
5. Repeat step 4 using reagents B and C, making sure to put the reagents in the proper wells: reagent B in row B, reagent C in row C.
6. Repeat steps 2 through 5 with the remaining insecticides and unknown. Record your results in Data Table 1.
7. Wipe the spot plate clean with paper towels. Dispose of paper towels properly.
8. Thoroughly wash and dry the spot plate.

Part B: Heat Test

9. Make the foil into a shallow container to hold the powders.
10. Mark the containers 1, 2, 3, 4, and 5.
11. Place a small amount of Anti-Ant in container 1.
12. Light the candle.
13. Use the clothespin to hold the container of poison over the flame for 60 to 90 seconds. Observe any odors, smoke, or color changes. Record your results in Data Table 2.
14. Repeat with the remaining poisons, making sure to place the samples in the proper corresponding container. Record your results in Data Table 2.

Data Table 2. Heat tests.

Insecticide	Heat Test Observations
Anti-Ant	
Kritter Kill	
Rid-A-Roach	
The Insectinator	
Sample from Sherlock	

Questions:

1. Which of the insecticides caused Sherlock to become ill? How do you know?
2. Which of the reactions with the reagents were physical changes? Which ones were chemical changes? Justify your answers.
3. Would you be able to determine whether Sherlock had been poisoned by more than one of the insecticides? Give an example.

ACTIVITY 9-2 *Ch. Obj: 9.5*
IMMUNOASSAY

Objectives:

By the end of this activity, you will be able to:

1. Describe the process of an immunoassay.
2. Perform an immunoassay test.
3. Interpret the results of the test.

Materials:

(per group of two or three students)
blood sample from victim
test reagents for:
    Dioxin
    Special K
    Ecstasy
    Roofies
test tubes (4)
test tube rack
transfer pipette

Safety Precautions:

Safety goggles must be worn when working with chemicals.
Aprons should be worn to protect clothing from the reagents.
Vinyl or latex gloves will protect skin from chemicals.
Wash hands thoroughly after handling the chemicals in this lab.

Background:

There is a local club that is very popular with the high school students. Every third Friday of the month, the club is open exclusively to high-school-age teenagers. The rules are very strict—no alcohol or drugs are permitted. One Friday night, a 16-year-old female was found lying unconscious on the ground behind the club. When police and paramedics arrived on the scene, they at first suspected the girl was intoxicated. There was no odor of alcohol, no indication that she had been drinking, and she was unresponsive. The victim was transported to the hospital.

Investigators were able to determine from witnesses that the girl had been seen earlier in the evening arguing with a male who had just turned 19. Rumors were that he was trying to convince her to try some drugs and she had refused.

At the hospital, the victim was stabilized and blood samples were collected for toxicology screenings. You will perform a series of immunoassay tests to find out whether the victim may have been drugged or poisoned.

Procedure:

1. Label the test tubes, each with the name of a suspect poison or drug.
2. Use a transfer pipette to put approximately 10 mL of blood into each of the test tubes.
3. In the first test tube, add 10 drops of the test reagent for dioxin.
4. Swirl the test tube gently to mix the reagent with the blood sample.
5. Let stand for one minute.
6. Note any changes in the appearance of the sample. Record your results in Data Table 1. (Note: A color change is not an indication of a positive test result. A positive result will produce a precipitate.)
7. Determine whether the test showed a positive or negative result. Record.
8. Repeat steps 3 through 7 for the other drugs or poisons.

Data Table 1. Observations.

Suspected Drug/Poison	Observations	Positive/Negative
Dioxin		
Special K		
Ecstasy		
Roofies		

Questions:

1. Explain how an immunoassay test works.
2. Were any of the drugs or poisons present in the victim's blood? Explain.
3. Why was it important to determine which drugs or poisons were in the victim's body?
4. What other types of tests might be performed on this victim if the immunoassay is inconclusive?
5. How would investigators determine whether the victim was intentionally poisoned or whether she took them voluntarily?

ACTIVITY 9-3 *Ch. Obj: 9.5*
URINALYSIS

Objectives:

By the end of this activity, you will be able to:

1. Perform a urinalysis test.
2. Interpret the results of the test.

Materials:

(per group of two or three students)
urine samples from players
test reagent for drug screen
wax pencil
spot plate
24 transfer pipettes
data sheet

Safety Precautions:

Safety goggles must be worn when working with chemicals.
Aprons should be worn to protect clothing from the reagents.
Vinyl or latex gloves will protect skin from chemicals.
Wash hands thoroughly after handling the chemicals in this lab.

Background:

The football team has made it to the state finals! In order to remain eligible, all players must pass a drug screening. State officials are concerned that some of the players may be using performance-enhancement drugs. You will act as independent toxicologist to perform the urinalysis and determine whether any players are indeed using the drugs.

Procedure:

1. Number the wells of the spot plate.
2. Use transfer pipettes to put 10 drops of urine into each of the wells. Make sure you put the sample from athlete #1 in the #1 well, the #2 sample in the #2 well, and so on. (If you have more samples than you have wells, finish testing one set, wash the spot plate thoroughly, and renumber.)
3. To each well, add 1 drop of the test reagent.
4. Note any changes in appearance of the urine.
5. Determine whether the test showed a positive or negative result. For this test, a color change is an indication of a positive result. Record the results in Data Table 1.
6. If needed, clean the spot plate and repeat steps 1 through 5 to finish the testing for the remaining players.

Data Table 1. Observations.

Player	Observations	+/−	Player	Observations	+/−
1			13		
2			14		
3			15		
4			16		
5			17		
6			18		
7			19		
8			20		
9			21		
10			22		
11			23		
12			24		

Questions:

1. Were any players using performance-enhancement drugs? Explain.
2. If any of the players tested positive, what might be the next step? Explain.
3. What precautions need to be taken to ensure the samples were not mixed up and are accurately correlated to the correct player?
4. Why would the state be concerned that members of a team had been using performance-enhancement drugs?

ACTIVITY 9-4 *Ch. Obj: 9.2, 9.3, 9.5*
BLOOD ALCOHOL CALCULATIONS

Objectives:

By the end of this activity, you will be able to:

1. Determine the correlation between breath alcohol levels and blood alcohol levels.
2. Calculate the blood alcohol level based on breathalyzer results.
3. Calculate the expected breathalyzer results given a specific blood alcohol level.

Materials:

paper
pencil
calculator

Procedure:

Part 1

When a person consumes alcohol, some of the alcohol is eliminated from the body through the air exhaled from the lungs. A breathalyzer traps 52.5 mL of air. Using this information and information from the chapter, determine the correlation between breath alcohol concentration and blood alcohol concentration.

Part 2

Read the following scenarios and make the appropriate calculations.

SCENARIO 1

Jack Delmar was pulled over after police officers saw him weaving in and out of traffic. As soon as Delmar got out of the car, officers noted that the odor of alcohol was overwhelming. Jack explained that someone had spilled a drink on him at dinner. Not convinced, the officers asked Delmar to stand on one foot and recite the alphabet backwards. He could not do it without much hesitation and wobbling. Based on that result, he was given a breath test. The breathalyzer measured 0.00375 g/mL.

Questions:

1. What was his blood alcohol level?
2. Will he be charged with driving under the influence?
3. What type of field sobriety test was administered prior to the breathalyzer?

SCENARIO 2

Ryan Jones drank several mixed drinks while having dinner at a restaurant. As he was leaving, he collapsed. The manager immediately called 911 and the unconscious man was transported to the nearest hospital. Because he had obviously been drinking, alcohol poisoning was suspected. Blood was drawn and the results were a blood alcohol level of 0.35.

Questions:

1. If Jones had been conscious and had been given a breath test instead, what would the reading have been?
2. What other symptoms was Jones most likely exhibiting just prior to his collapse?
3. Was he in any danger of dying?

Extension:

Research other types of breath tests for determining blood alcohol concentrations. Find out how the tests work and the advantages and disadvantages of using the other tests. Present your research in an essay or classroom presentation.

CHAPTER 10

Advanced Concepts in DNA

SOLVING A 200-YEAR-OLD MYSTERY

Figure 10-1. Louis, Prince Royal of France, died on June 8, 1795.

In the 1790s, France was in the midst of a bloody revolution that led to the overthrowing of the royal family. On August 13, 1792, the entire royal family was arrested and imprisoned. A few months later, the king and queen were executed. However, the two royal children—Louis and Marie-Thérèse—remained locked in Paris's Temple prison for another two years.

The young Louis was the heir to his father's throne. Although Louis was only eight years old when his father was executed, supporters of the royal family still considered him to be the new king of France. However, prison life was difficult for the young king, and he soon became very ill.

On June 8, 1795, prison officials reported that young Louis had died from a lung disease. But many people did not believe this story. Rumors almost immediately began circulating that supporters of the royal family had somehow smuggled Louis out of prison. Soon, dozens of people across Europe and the United States claimed that they were the missing French king.

Louis's sister, Marie-Thérèse, was released from prison a few months after Louis's reported death. She did her best to squelch the rumors of Louis's escape, but the stories continued, and books have supported the theory for more than 200 years.

Finally, DNA technology helped scientists solve the mystery. France's custodian of relics, the Duke of Bauffrement, asked scientists to take DNA samples from the remains of the executed king and queen. He also provided a heart that had been removed from the young prince's body before he was buried, according to official records.

In addition, scientists analyzed hair samples from the queen's sisters and blood samples from descendants of those sisters for mitochondrial DNA. The DNA evidence proved that the preserved heart was the heart of the young Louis. The mystery was finally solved. The young prince had indeed died in prison.

Figure 10-2. Both Louis's father and his mother, Marie-Antoinette (above), were executed two years prior to his death.

OBJECTIVES

By the end of this chapter, you will be able to:

10.1 Discuss the structure and function of DNA.

10.2 Explain what causes variation in DNA.

10.3 Differentiate between the various types of DNA analysis.

10.4 Compare and contrast the methods of extracting DNA.

10.5 Explain the proper methods of collecting DNA evidence at a crime scene.

VOCABULARY

DNA extraction - process of removing DNA from a cell; procedures include Chelex extraction and organic extraction

homologous chromosomes - pair of chromosomes in which one chromosome was inherited from the male parent and the other was inherited from the female parent

nucleotide - subunit of DNA and other nucleic acids; made up of a 5-carbon sugar, a phosphate group, and a nitrogenous base

polymerase chain reaction (PCR) - technique used to make billions of copies of a specific segment of DNA; allows very minute samples of DNA to be copied billions of times

primer - small piece of DNA used to begin and end replication during PCR

short tandem repeat (STR) - short segment of DNA in which the same sequence of two to six base pairs is repeated many times; varying numbers of repeats found among individuals

INTRODUCTION

Deoxyribonucleic acid (DNA) is the genetic material that makes each individual unique. Other than identical twins, no two people have the exact same DNA. DNA serves two primary functions. First, it makes copies of itself, or replicates, in order for cells to divide properly. Second, DNA is the blueprint for making proteins that cells use to carry out life processes. DNA can be found in biological evidence, including blood, semen, saliva, and hair. Because DNA evidence is unique and can be attributed to a single source, it is *individual evidence.*

STRUCTURE AND FUNCTION OF DNA

Obj. 10.2

DNA is made up of subunits called **nucleotides.** Each nucleotide has three components: deoxyribose (a 5-carbon sugar), a phosphate group, and a nitrogenous base (see Figure 10-3). The four nitrogenous bases are adenine, guanine, cytosine, and thymine. Each nitrogenous base can form hydrogen bonds with a specific base on a complementary strand of DNA (see Figure 10-4). The adenine on one strand always pairs with thymine on the complementary strand; and guanine always pairs with cytosine. When the two strands align in this way, they form a structure similar to a twisted ladder. The base pairs form the rungs, and the sugar and phosphate groups form the edges of the "ladder." The structure of DNA is often referred to as a *double helix* (see Figure 10-5).

The weak hydrogen bonds between the complementary nitrogenous bases make it possible for the DNA strands to separate in order for each strand to be copied when cells divide and during protein synthesis. The specific sequence of nitrogenous bases carries a lot of information. *Genes* are sections of DNA that code for a protein. These proteins catalyze, regulate, and control chemical reactions in the cell. The proteins play an important role in determining the individual's traits. Different forms of each gene are

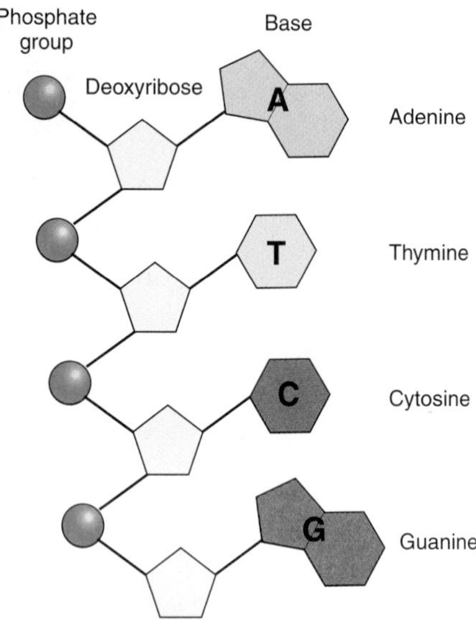

Figure 10-3. *DNA is made up of nucleotides. Each nucleotide is composed of a phosphate group, deoxyribose, and a nitrogenous base.*

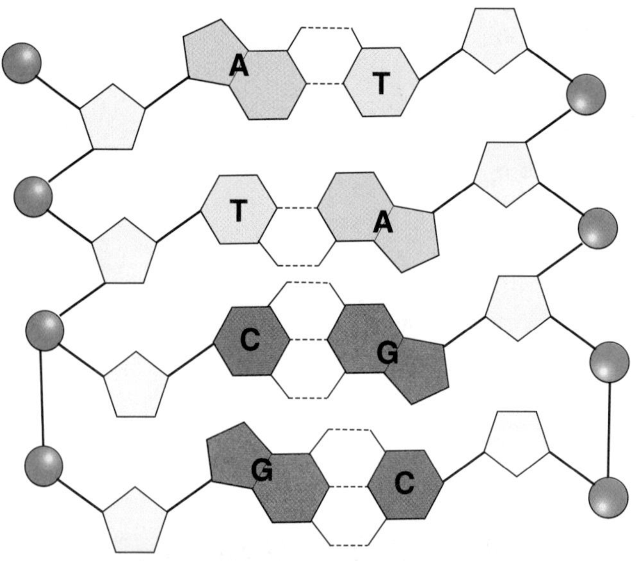

Figure 10-4. *The nitrogenous bases on complementary strands of DNA are held together by weak hydrogen bonds.*

called *alleles.* For example, there is an allele for blue eyes, one for brown eyes, and alleles for many other shades. Each person has 23 pairs of chromosomes. Each chromosome is made up mostly of nuclear DNA, which codes for specific genes on the chromosome. One chromosome of each pair is inherited from the mother and one chromosome is inherited from the father. The chromosomes in each pair are called **homologous chromosomes**.

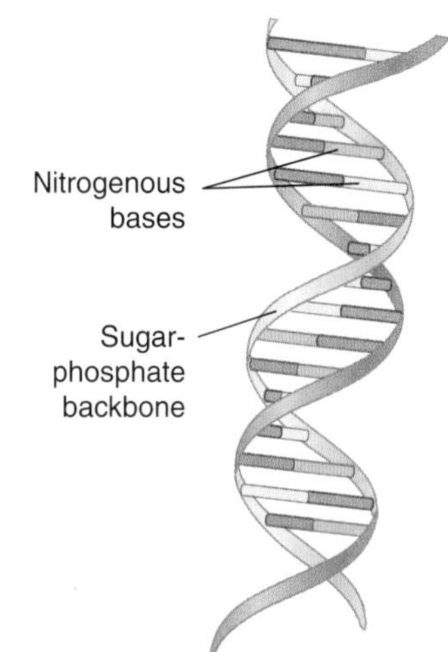

Figure 10-5. *The structure of DNA is similar to a twisted ladder and is often called a double helix.*

In a process called *transcription,* portions of DNA are copied to form complementary ribonucleic acid (RNA) molecules (see Figure 10-6). In eukaryotic cells, including human cells, RNA is produced in the nucleus. The RNA then moves to the cytoplasm, where proteins are assembled based on the sequence of base pairs found on the RNA molecule. This process is called *translation.*

The four nitrogenous bases in RNA form the genetic code. The genetic code is made up of "three-letter words" formed by consecutive bases. Each set of three bases is called a *codon* and codes for a single amino acid. During translation, the amino acids are strung together to form proteins. The proteins, then, play an important role in the organism's traits.

The genetic code is a nearly universal "language." In all living things, the code is read three bases at a time. In some organisms, the codons code for slightly different amino acids; but the basic processes are the same for all living things.

If a mistake is made during transcription, a different codon is produced. A change in the codon could result in coding for an entirely different amino acid or even a stop codon. If the amino acid is different, the protein is different and will likely affect the phenotype of the individual. The *phenotype* is the individual's outward appearance, If a stop codon is coded for, protein synthesis will cease. Depending on where in the protein strand synthesis stops, the mistake could change the phenotype or even be fatal. For example. UAU is the codon for tyrosine. If a mistake is made during replication and the new codon is UAC, there is no change in the amino acid, the protein, or the phenotype. If the new codon is UCU, a different amino acid, serine, results. A change of even a single amino acid changes the protein. If however, the mistake codes for UAA, the result is a stop codon. These mistakes in the genetic code are called *mutations.* See the chart in Figure 10-7.

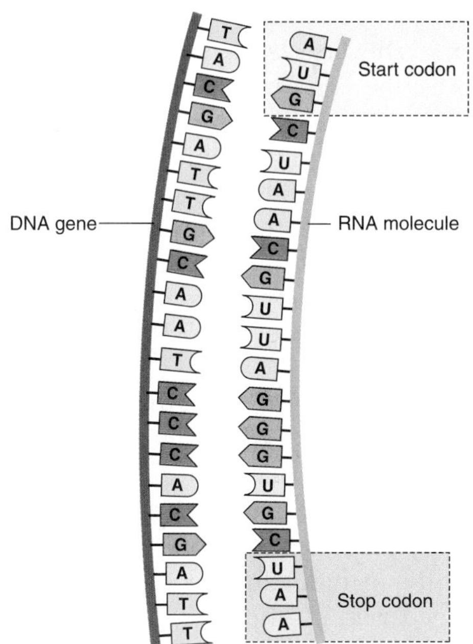

Figure 10-6. *Each gene is a sequence of nucleotides that codes for a specific protein.*

Codons and Amino Acids

2nd base in codon

1st base in codon	U	C	A	G	3rd base in codon
U	Phe	Ser	Tyr	Cys	U
	Phe	Ser	Tyr	Cys	C
	Leu	Ser	STOP	STOP	A
	Leu	Ser	STOP	Trp	G
C	Leu	Pro	His	Arg	U
	Leu	Pro	His	Arg	C
	Leu	Pro	Gln	Arg	A
	Leu	Pro	Gln	Arg	G
A	Ile	Thr	Asn	Ser	U
	Ile	Thr	Asn	Ser	C
	Ile	Thr	Lys	Arg	A
	Met	Thr	Lys	Arg	G
G	Val	Ala	Asp	Gly	U
	Val	Ala	Asp	Gly	C
	Val	Ala	Glu	Gly	A
	Val	Ala	Glu	Gly	G

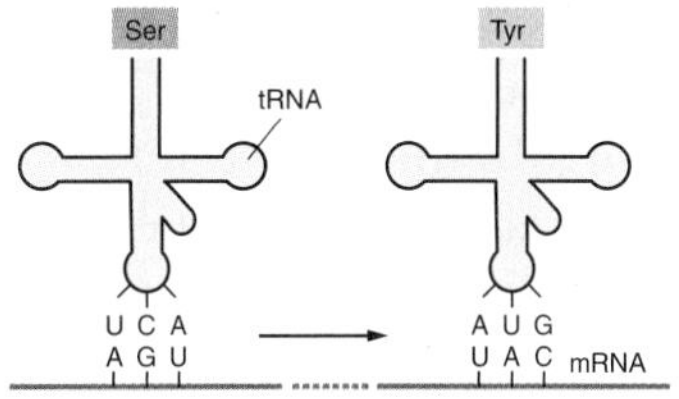

Ser
Tyr
tRNA
UCA
AGU
AUG
UAC
mRNA

Figure 10-7. *The anticodon on the tRNA is complimentary to the codon on the mRNA. Together, they work to get the amino acids to the ribosomes for protein synthesis.*

Did You Know?

The original model that Watson and Crick developed was a triple helix. The sugars and phosphates were on the inside and the nitrogenous bases were on the outside. They shared their information with colleagues who were using X-ray diffraction. Watson and Crick were told their model did not even closely resemble the model produced on the X-ray diffraction.

Digging Deeper

with Forensic Science e-Collection

Research is a fundamental component of the advancement of medical science. When a medical company or pharmaceutical company invests millions of dollars in research, development, and clinical trials, they patent their discoveries and treatments to assist in recouping the massive costs. In 1982 the first gene was patented. The medical community and society alike were shocked that information on the genome could be patented. More than 40,000 patents now exist for genes on human DNA. Using the Gale Forensic Science e-Collection at www.cengage.com/school/forensicscienceadv, research gene patents and decide whether you support the patents or think the discoveries should be open to the general public and other competing medical corporations.

After researching the topic, complete one of the following assignments: (1) Choose a side and write an essay supporting your point of view. (2) Write an essay discussing the merits of each side of the debate. (3) Working in a small group, prepare for a debate on the topic. Your group will then need to debate a group that has chosen the other point of view.

Obj. 10.2 WHAT CAUSES VARIATION IN DNA?

Approximately 99.5 percent of DNA is the same from person to person. It is the sequence of bases in the other 0.5 percent of DNA that makes humans unique. Ten million base pairs are found in that tiny portion of human DNA. If two alleles are different on a homologous chromosome, they are *heterozygous*. If the two alleles are the same, they are *homozygous*. For example, the ability to roll your tongue into a tube is an inherited trait. If the mother and father both pass on the allele for the ability to roll the tongue, the child will be homozygous because both alleles are the same. If, however, the mother passes down the allele for the ability to roll the tongue and the father passes down the allele for the inability to roll the tongue, the child will be heterozygous because the alleles are different. In this case, the allele for the ability to roll the tongue is dominant. Therefore, the child would be able to roll his or her tongue but would carry the recessive allele for the inability to roll the tongue.

Forensic biologists use DNA fingerprinting to compare reference samples (from a victim or suspect) to evidence samples. This comparison will show whether the reference samples match the evidence samples. During the tests, the biologist examines several markers on multiple chromosomes. A *marker* is a sequence of DNA base pairs associated with a specific trait. The degree of variation is used to identify an individual or cell carrying that gene. If the examination reveals any variation, the reference sample and evidence sample are from different people. Conversely, if the examination reveals several matching markers, the two samples are more likely to be from the same or related people. The more matching markers found, the higher the probability that the two samples are from the same person.

DNA ANALYSIS PROCEDURES

Obj. 10.3, 10.4

For DNA to be processed and compared in a laboratory, it must first be isolated, or extracted. **DNA extraction** is the process of removing DNA from the cell. Extracting only the DNA from the cell must be done carefully. The method used to isolate the DNA will vary based on the type of biological evidence being analyzed, the amount of evidence available, and the type of cells present.

CHELEX EXTRACTION

When only a very small amount of biological evidence is available, forensic scientists use Chelex extraction. A serologist places the sample into boiling water with Chelex beads, causing the cell to burst open, exposing the DNA. Chelex beads are coated with an acid that binds to metal ions such as magnesium. Removing the magnesium from the cellular material inactivates the enzymes that destroy DNA. The mixture is spun in a centrifuge, and the DNA solution rises to the top (see Figure 10-8).

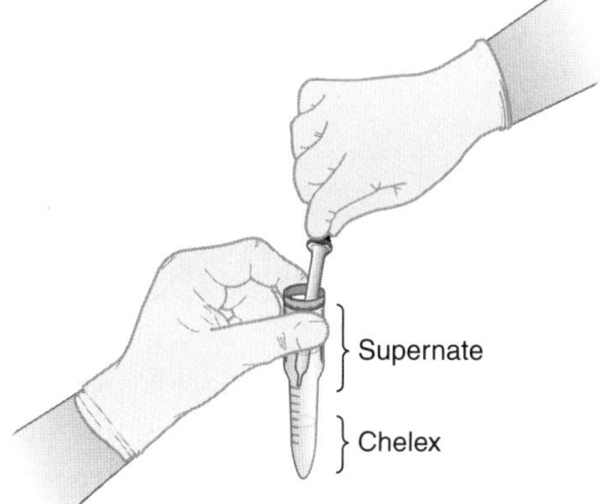

Figure 10-8. *During the Chelex extraction, the mixture is placed in a centrifuge. The desired portion sits at the top of the centrifuge tube.*

ORGANIC EXTRACTION

When more biological evidence is available, forensic scientists typically use organic extraction. This method separates DNA from the sample more thoroughly than the Chelex extraction method.

Simple Organic Extraction

The DNA in blood, saliva, and epithelial cells is relatively easy to isolate. In these cases, a simple organic extraction is completed.

When biological evidence is found on material such as a bedsheet, T-shirt, or shoelace, the forensic biologist first cuts the material into small pieces. Next, the sample is placed into a warm solution, which causes the cells to detach from the material. Additional chemicals and low heat cause the cells to burst open, exposing the DNA. The cellular material that is soluble in the organic solvents is then removed. DNA and other nucleic acids remain because they are not soluble in the solvents. Using filters, the DNA is purified and concentrated. The DNA is now prepared to be copied or analyzed.

In 1987, the DNA profile of Tommy Lee Andrews proved he was responsible for the sexual assault of several women in Florida. This was the first time DNA evidence was used to solve a crime in the United States.

Differential Extraction

In cases of sexual assault, forensic scientists often find a mixture of sperm cells and other cells, including vaginal cells. Vaginal cells are epithelial cells and have very different characteristics than sperm cells. In these cases, the scientists must separate the victim's cells from the perpetrator's cells using a specific organic extraction process called differential extraction. First, the material with the biological evidence is placed into a mild solution of phenol and chloroform. This solution causes only the vaginal cells to burst. The liquid that contains these cells is then moved to a different test tube. The vaginal epithelial DNA is isolated, purified, and concentrated.

Once the vaginal cells have been removed, the remaining sample is called the *sperm fraction.* Because the outer coating of sperm cells has sulfur bonds, it is more difficult to break open the sperm cells, exposing the DNA. Therefore, sperm cells are treated with additional chemicals. Dithiothreitol (DTT) is used to break the sulfur bonds. Then, the sample is placed into a solution of phenol and chloroform, which causes the cells to burst. The sperm DNA is then isolated, purified, and concentrated.

Successful differential extraction separates the victim's DNA from the perpetrator's DNA, allowing for the production of two distinct DNA profiles—one from the victim and one from the perpetrator.

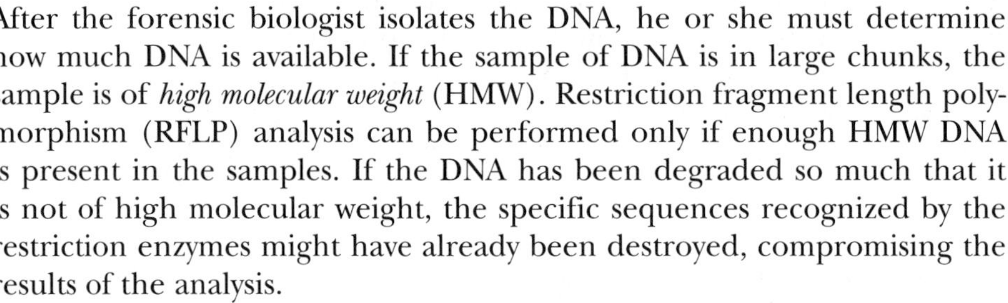

MEASURING QUALITY AND QUANTITY OF DNA

Obj. 10.3

After the forensic biologist isolates the DNA, he or she must determine how much DNA is available. If the sample of DNA is in large chunks, the sample is of *high molecular weight* (HMW). Restriction fragment length polymorphism (RFLP) analysis can be performed only if enough HMW DNA is present in the samples. If the DNA has been degraded so much that it is not of high molecular weight, the specific sequences recognized by the restriction enzymes might have already been destroyed, compromising the results of the analysis.

For the most part, RFLP analysis has been replaced by polymerase chain reaction (PCR). RFLP analysis is not effective if much of the DNA has been degraded or if there is very little DNA available. PCR can be performed on very small amounts of DNA. Therefore, this technique makes it possible to evaluate biological evidence even when most of the DNA has been degraded.

DNA FINGERPRINTING

Sometimes, biological evidence is found on materials that inhibit PCR. For example, blue denim contains dyes that mix with the DNA and restrict the PCR process. New technology is being developed that will remove the dye and allow the biological sample to be collected and analyzed.

The number of times a specific sequence of base pairs repeats is variable from one person to another. The repeating sections of DNA are called *variable number of tandem repeats* (VNTR). The VNTRs are cut from the surrounding DNA by restriction enzymes to produce restriction fragments. The length of these fragments varies from one individual to another. Scientists use the term *polymorphic* to describe the variations in length and sequence. These fragments, therefore, are called *restriction fragment length polymorphisms* (RFLPs). Analysis of RFLPs produces a genetic profile that is unique to each individual. The technique is known as *DNA fingerprinting* or *DNA profiling*. DNA fingerprinting can be completed using a variety of techniques including RFLP, short tandem repeat (STR), PCR, mitochondrial analysis, and Y-chromosome analysis.

DNA fingerprinting can be divided into several steps. First, the forensic biologist isolates and purifies the DNA. Next, the biologist uses restriction enzymes to cut the DNA into fragments (RFLPs). The biologist then performs gel electrophoresis to sort the DNA bands according to size. In gel electrophoresis, a mixture of DNA fragments is placed at one end of a gel. An electric charge is applied, and the negatively charged DNA molecules move toward the positive end of the gel. Smaller fragments move further through the gel.

Through a transfer process called *Southern blotting*, the scientist transfers the DNA bands from the gel to a nylon sheet (see Figure 10-9). The scientist treats the sheet with radioactive probes to target and visualize the desired DNA. Finally, the scientist uses population genetics to determine the probability that another individual could have the same profile.

PCR

Polymerase chain reaction (PCR) makes it possible to analyze DNA even when only very small samples of biological evidence are available at the crime scene. Additionally, PCR is fast, sensitive, and does not require a high-quality sample.

PCR uses an enzyme to separate a region of DNA and then copies that region billions of times. The process of PCR requires samples to be heated and cooled in a very specific three-step cycle. The steps are very similar to the natural process of DNA replication. Each cycle takes about five minutes. PCR will go through approximately 30 cycles in which about 1 billion copies of the specific segment of DNA are made. The three steps of this process are denaturing, annealing, and polymerization.

Denaturing

The first step in PCR is to separate, or denature, the two DNA strands (see Figure 10-10 on page 294). Each strand of DNA provides a template for making a new complementary strand. The weak hydrogen bonds that hold the two strands together break in the presence of high temperatures. Therefore, in the first step of PCR, the DNA is heated to 94°C.

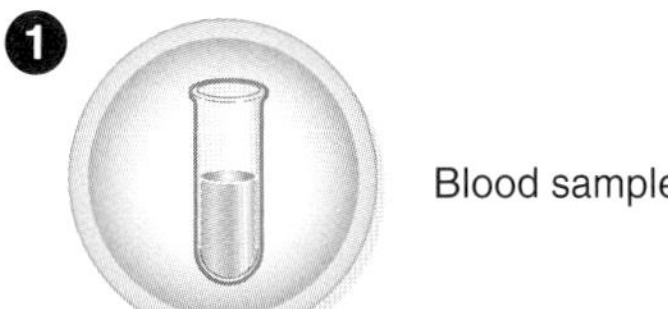
1
Blood sample

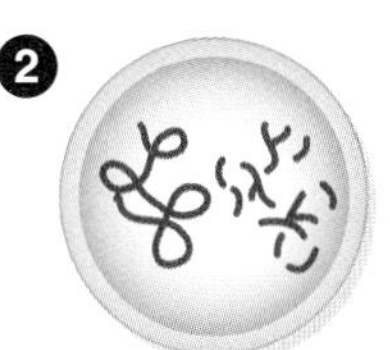
2
Biologist isolates DNA and uses restriction enzymes to cut the DNA into fragments.

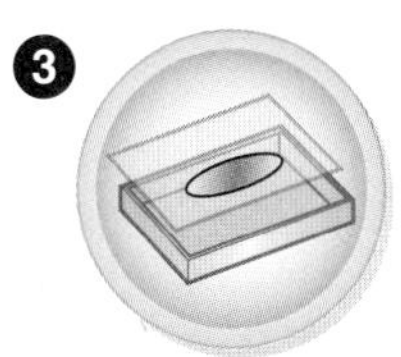
3
Gel electrophoresis sorts the DNA into bands. Biologist transfers the bands to a nylon sheet.

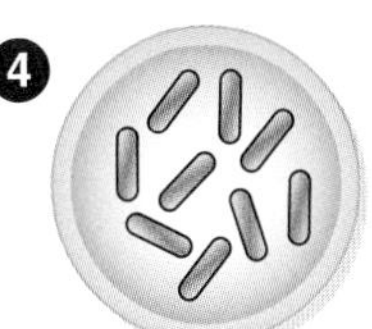
4
Biologist prepares the radiactive probe.

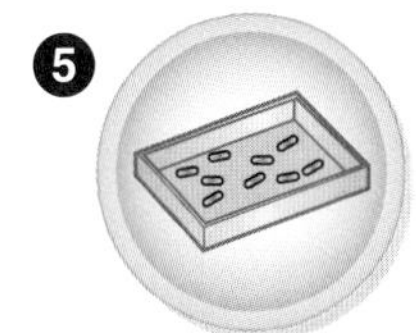
5
The radioactive probe binds to target DNA.

6
Excess probe is washed off.

7
X-ray film detects the radioactive pattern. Development of the film makes the pattern visible.

Figure 10-9. *Together, electrophoresis and Southern blotting are used to make a DNA fingerprint.*

Primer Annealing

Once the DNA has been denatured, the sample is cooled. A targeted sequence of 100 to 600 base pairs is selected on the individual DNA strand for replication. The scientist adds **primers**, 20 to 30 base pairs in length, to the beginnings and ends of the target sequence. Two primers are used—one for each of the complementary DNA strands (see Figure 10-11 on page 294). The primers will anneal, or bind, to the complementary strand. Primers mark the PCR starting and stopping points.

Polymerization

After the primers are in place, the scientist raises the temperature of the sample to 72°C. DNA synthesis begins at the primers. Taq DNA polymerase in the solution is an enzyme that is used to copy the part of the

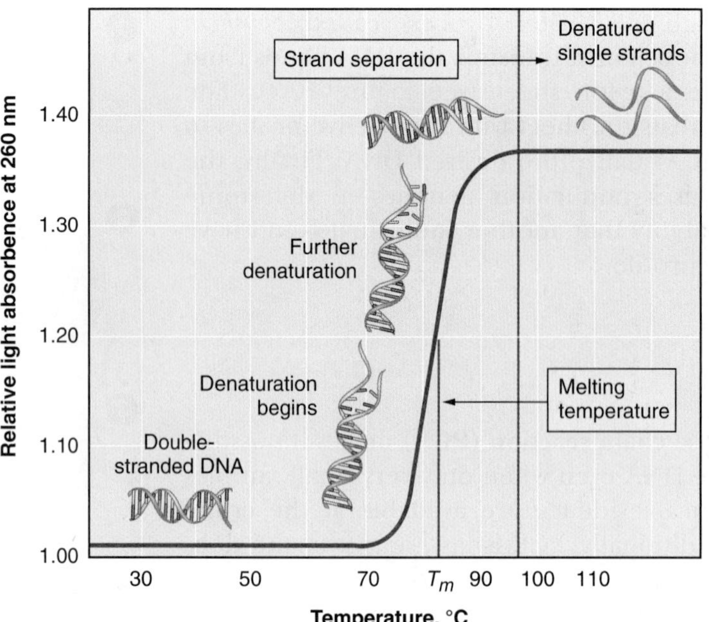

Figure 10-10. *In the first step of PCR, the strands of DNA are separated—the DNA is denatured.*

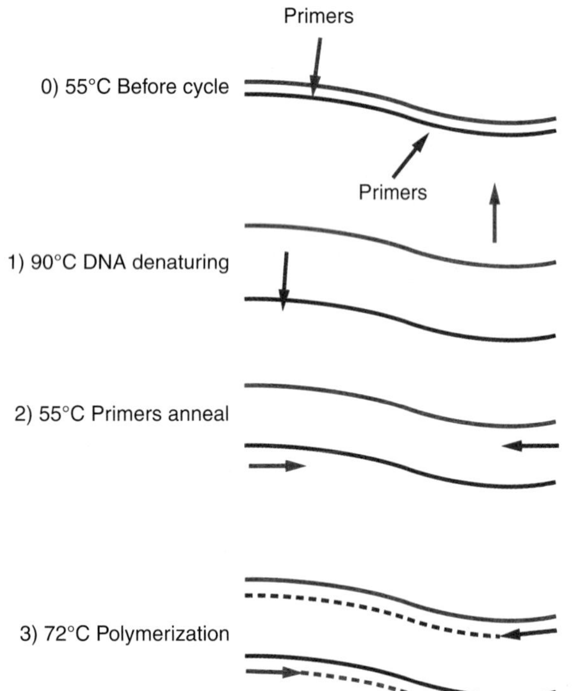

Figure 10-11. *The three steps of PCR. (1) Denaturing, (2) Primer Annealing, and (3) Polymerization.*

DNA between the two primers. A nucleotide solution is added to the sample. These nucleotides pair with the DNA strand to form a complementary strand. For example, a thymine will always pair with an adenine. The new strand is built one nucleotide at a time, 15 to 20 times per second. The entire segment of the targeted DNA is copied in each cycle of PCR (see Figure 10-12).

STR

Although RFLP analysis was the original method for developing a DNA fingerprint, currently, the most common PCR-based method of establishing a DNA fingerprint examines **short tandem repeats (STRs)**. An STR is a short segment of DNA in which the same sequence of two to six base pairs is repeated many times. Because they are shorter, it is less difficult and quicker to copy STRs through the PCR process than it is to copy VNTRs. The repeating pattern of STRs is highly variable from one person to another. Like all portions of chromosomes, every person has two inherited copies of STR—one from the mother and one from the father. Due to the natural variation that exists within STRs, the likelihood of two random individuals matching at multiple STRs has an incredibly low probability.

In 1996 the FBI, in partnership with the forensic community, implemented a nationwide effort to develop an electronic database of DNA profiles. This database is called the Combined DNA Index System (CODIS), and it stores the DNA fingerprint of individuals who have been convicted

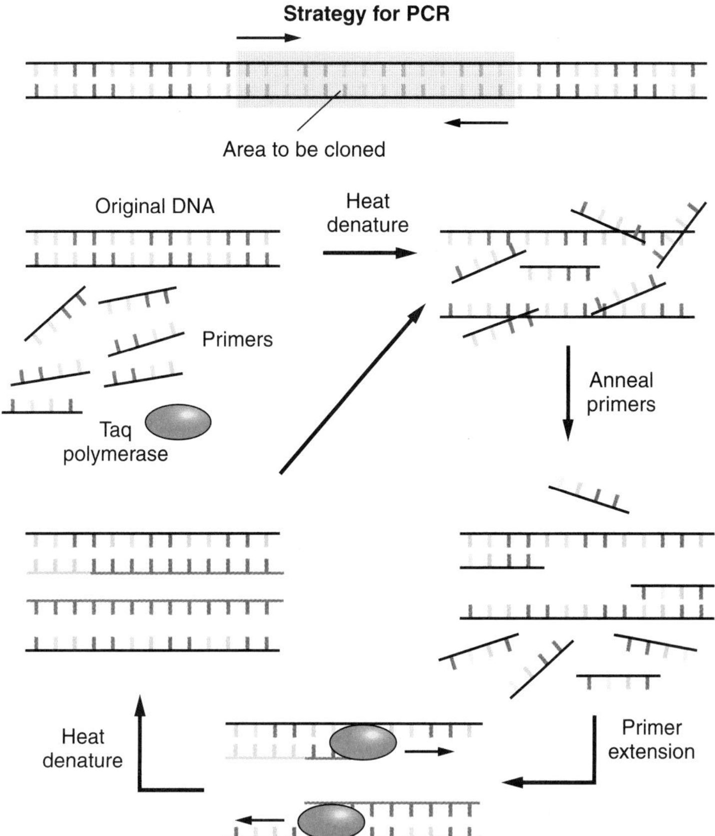

Figure 10-12. The entire PCR process.

of certain crimes, including rape and murder. The database allows forensic scientists to compare the DNA fingerprint of suspects based on 13 core STR locations (see Figure 10-13 on page 296).

Examining the 13 regions on an individual's DNA and determining a DNA profile is a three-step process. First, the DNA must be isolated and copied using PCR. After the DNA has been amplified, the scientists use gel electrophoresis to determine how many STRs exist on the DNA. The final stage is to evaluate the results and determine the probability that another person could have an identical DNA profile.

Digging Deeper
with Forensic Science e-Collection

With the use of technology and databases, DNA fingerprints are now being stored in computers. There are advantages and disadvantages to CODIS and the new technology being developed to store and retrieve DNA fingerprints. Using the Gale Forensic Science e-Collection at www.cengage.com/school/forensicscienceadv, research CODIS and the newest technology. After completing your research, write an informational pamphlet that discusses the forensic importance of CODIS and the new technology that will aid in identification and confirmation of criminals.

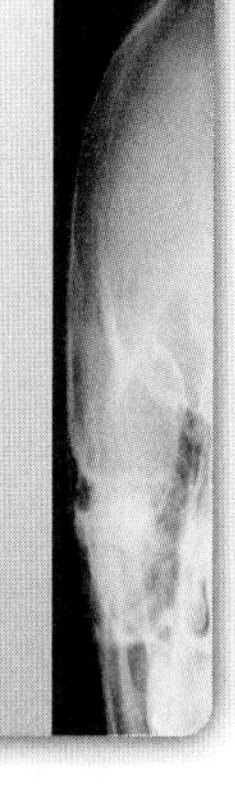

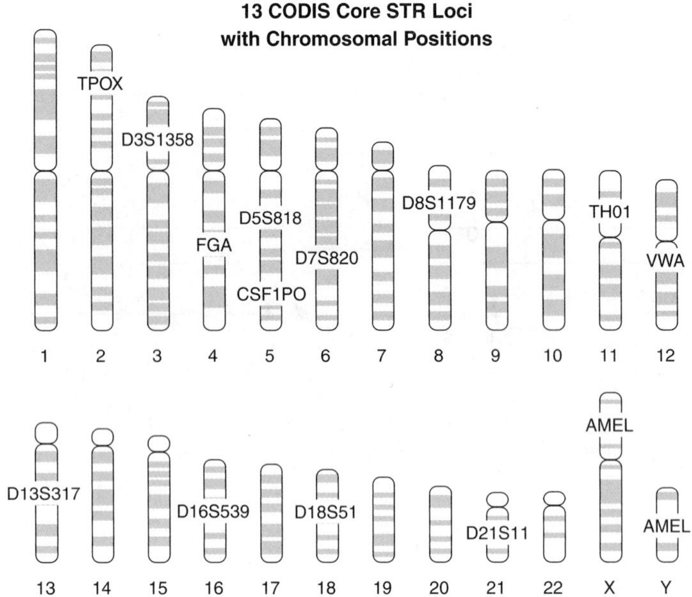

Figure 10-13. *Chromosomal positions of 13 CODIS core STR loci.*

Obj. 10.5

COLLECTING DNA EVIDENCE

When a crime-scene investigator begins to systematically and thoroughly process a crime scene, any biological evidence will be collected, bagged, and sent to the crime lab for analysis. Semen and blood are the most common biological evidence found at crime scenes. Crime-scene investigators are very careful when collecting biological evidence because once the evidence gets exposed to the environment, risk of DNA breakdown and contamination occurs. When investigators collect biological evidence, they keep the evidence on its original surface whenever possible. If, however, evidence is found on a stationary object, such as a wall or car, the CSI swabs the surface to collect the evidence. Reference samples are also collected for comparative analysis.

Once exposed to an environment outside the body, the cell opens and the DNA can begin to degrade. The degradation can affect the ability of the forensic biologist to collect and analyze useful data. RFLP analysis is especially sensitive to DNA degradation; this analysis technique requires relatively large segments of DNA. However, even degraded material can provide a useful genetic profile. This profile is very unlikely to be consistent with the genetic profile of more than one person. Therefore, although a degraded sample can hinder the results, it is not likely to completely invalidate the results.

DNA is especially likely to degrade when exposed to certain enzymes and microorganisms. Long-term exposure to warm, moist environments outside of the body increases the chances of exposure to these enzymes and microorganisms.

It is very important to stop the degradation of biological evidence and prevent any additional deterioration. The best way to stop further breakdown of the DNA is to remove moisture by drying out the sample and freezing it as soon as possible.

Whenever possible, crime-scene investigators collect biological evidence by taking the item, such as clothing or jewelry, that bears the evidence directly to the lab. If it is impossible to deliver the entire object to the lab, the CSI transfers the evidence to a medium, such as a test tube or cotton swab, for transportation to the lab.

COLLECTING BIOLOGICAL EVIDENCE AT A CRIME SCENE

The ultimate goal of any crime-scene investigation is to connect or exclude a person or people from involvement in the crime. Because it is very unlikely that two people exhibit the same genetic profile, DNA evidence is very compelling in court. A perpetrator's DNA can be left at a crime scene by direct contact. For example, the perpetrator's skin cells, blood, or other biological evidence may be left on an object at the scene or on the victim's clothing or body. The victim's DNA can also be important to the investigation. Skin cells, blood, or other biological evidence found on a suspect might connect the suspect to the victim. DNA evidence can also be used to identify the victim.

When collecting DNA evidence, extra precautions must be taken to ensure that it is kept in the best possible condition. The following are considerations to be taken when collecting biological evidence at a crime scene:

- Never use bare hands when collecting evidence. Avoid sneezing and coughing near the evidence.
- Always use clean protective gloves. Gloves should be discarded after each piece of evidence is collected.
- Always make sure each piece of evidence is packaged separately.
- When evidence such as blood and semen is collected, the object that bears it, such as clothing or bedsheets, must be air dried.
- After drying, the evidence should be placed into an envelope or paper bag. Plastic bags should be avoided because they allow moisture to build up, making it more likely that the DNA will degrade. A chain of custody must be established for each item.
- When transferring evidence from an object that cannot be removed, use a cotton swab and water, if needed (see Figure 10-14). The swabbed sample will need to air dry before it can be placed into an envelope. Each swab must be stored separately.

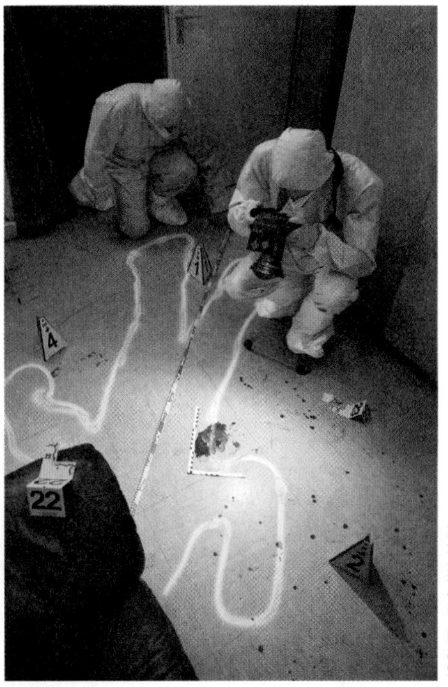

JOCHEN TACK/ALAMY

Figure 10-14. *CSIs use cotton swabs to collect blood from floors, walls, and other large objects.*

Collecting Reference Samples

When biological evidence is collected from a crime scene, the origin is unknown—the investigators do not know whose blood or hair it is. In order to find a match, a link to the crime scene, *reference samples,* or known samples, must be taken from suspects and victims. Forensic scientists compare the known DNA to the questioned sample. This evidence can potentially link a suspect to a crime scene or exonerate a suspect. Reference samples are also taken from family members in cases where someone is missing and during mass disaster investigations.

Testing for the Presence of Blood, Semen, and Saliva at the Crime Scene

Crime-scene investigators use presumptive tests to screen for blood, semen, and saliva at the crime scene.

Presumptive tests for blood detect hemoglobin. Hemoglobin is attached to red blood cells and is responsible for transporting oxygen

Figure 10-15. *Blood stain visible with luminol.*

to cells and carrying carbon dioxide back to the lungs. *Luminol* is the most common presumptive test for blood (see Figure 10-15). Luminol is a liquid that is sprayed on a large area that is being investigated to find bloodstains. The sprayed area needs to be darkened to allow the luminescence, or glow, to be more easily seen. Luminol is so effective because it can pick up the presence of blood that has been diluted 10 million times. Also, using luminol on the blood does not destroy the evidence and, therefore, does not affect DNA testing of STRs.

More than half of all DNA cases in crime laboratories involve evidence collected from a sexual assault. Various presumptive tests screen for semen present at a crime scene. Acid phosphatase is an enzyme secreted by the prostate gland and found in very high concentrations in seminal fluid. Chemicals and fluorescent dyes are used to visualize the fluid. Large areas can be tested at one time. The prostate-specific antigen (PSA) is found in high quantities in seminal fluid. Kits will test for the presence of this protein. If the test is positive, it tells the forensic scientist the sample might be semen. Investigators complete a microscopic examination of a sample to detect the presence of sperm. This is not always effective because the suspect may have no sperm in the seminal fluid or a very low sperm count, making identification difficult. This is why presumptive tests that rely on enzymes and proteins found in high quantities in seminal fluid are very useful to investigators.

Saliva at a crime scene can be found in many places, such as a cigarette butt, bite marks, or a soda can. Amylase is a digestive enzyme that breaks down starches in the mouth. Forensic biologists use the Phadebas test or a starch iodine test to detect amylase. The Phadebas reagent is a starch that is connected to a dye. When amylase comes in contact with the reagent, the dye is released, indicating the presence of amylase. In the starch iodine test, iodine turns blue in the presence of starch.

MITOCHONDRIAL ANALYSIS

With the exception of red blood cells, nuclear DNA is found in the nucleus of all eukaryotic cells. Additional circular DNA is found in mitochondria. Mitochondria are the cellular organelles that generate ATP molecules. The bonds of ATP molecules store energy for the cell. Nuclear DNA is inherited from both parents, but mitochondrial DNA is inherited only from the mother. Mitochondrial DNA (mtDNA) has genes that code for functions related to cellular respiration. Also, each cell has hundreds of copies of mtDNA. Therefore, it is easier to extract mtDNA than it is to extract nuclear DNA. Also, mtDNA is less likely to degrade because it is protected by the mitochondrion.

At fertilization, only the nucleus of the sperm will penetrate the egg, forming a zygote. As the zygote multiplies and a blastocyst forms, the cytoplasm and other cellular components are the same as the mother's egg cell. This process is why mitochondrial DNA is passed down through maternal lineage. Unlike nuclear DNA, each person's mtDNA is not unique. For example, all of a mother's children will share the same mtDNA. By understanding the way in which mtDNA is inherited, forensic investigators can use the information in missing-persons cases and mass disaster investigations.

mtDNA, however, is less useful in strictly forensic cases because it does not provide individual evidence.

The most common types of biological evidence from which mtDNA is extracted are teeth, hair, and bones. The most effective bones for mtDNA sequencing are the femurs. The extraction process is done to separate the material from *PCR inhibitors,* anything that diminishes the effectiveness of the PCR process such as dyes, soil, and hemoglobin. Because a forensic anthropologist will need to evaluate bone evidence at the same time mtDNA analysis is being completed, the forensic biologist collects samples from the bone without altering the overall physical features of the bone. These considerations are also taken when examining teeth.

When hair shafts are analyzed, microscopic examination is an effective screening tool. The physical comparison of the unknown hair to a known sample might provide enough information to rule a suspect out. In that case, the mtDNA analysis will not be completed. Additionally, the entire physical examination of the hair sample must be completed before mtDNA testing because the DNA extraction process destroys the evidence.

Did You Know?

In the late 1980s, the FBI began using mtDNA for the purpose of missing persons identification. Since the mid-1990s, the FBI has been using mtDNA in forensic cases—mostly working with hair samples. In 2005, four satellite offices were opened, more than doubling the number of mtDNA cases the FBI could analyze.

Digging Deeper
with Forensic Science e-Collection

There are vast differences between nuclear DNA and mtDNA. However, collecting and processing evidence, using databases, and presenting findings in court are all steps in working with both types of DNA. Using the Gale Forensic Science e-Collection at www.cengagecom/school/forensicscienceadv, research the similarities and differences in nuclear and mtDNA. Also, be sure to discuss the use of both types of DNA analysis in the courtroom. Discuss the advantages and disadvantages of both analyses.

Y-CHROMOSOME ANALYSIS

The Y chromosome is passed down through the paternal lineage. In other words, a son receives his Y chromosome from his father. This can make differentiating between two individuals more difficult. If a match is found using Y-chromosome analysis, it can be a match for a father, brother, or any paternal relative. The following are areas in which Y-chromosome analysis is beneficial:

- Sexual assault
- Missing persons
- Genealogy

Even though Y-chromosome analysis does not provide individual evidence, the Y chromosome possesses significant forensic value. Most sexual assaults involve a male perpetrator. Sometimes, there is very little DNA left by the perpetrator or it is too difficult to separate it from the victim's DNA. This may happen when the perpetrator has a very low sperm count or has had a vasectomy or when the victim has been assaulted by several men. In such cases, DNA tests that specifically analyze the genetic content on the Y chromosome can be forensically valuable.

CHAPTER SUMMARY

- The discovery of DNA's existence, structure, and function has a rich history dating back to the 1800s.
- DNA, shaped like a double helix, is made up of a sugar and phosphate backbone with nucleotides on the inside. The four nitrogenous bases are adenine, guanine, cytosine, and thymine.
- The genetic code is nearly universal. The code is always read three bases at a time and codes for amino acids, which combine to form proteins.
- DNA isolation and extraction from biological evidence such as blood, semen, and saliva must be done prior to completing a DNA profile.
- Polymerase chain reaction provides forensic scientists the ability to collect very minute and degraded samples. The three steps of this process are denaturing, annealing, and extending.
- Short tandem repeats are very short segments of DNA in which the same sequence of two to six base pairs is repeated several times. The repeating pattern that occurs is highly variable from one person to another.
- Once DNA is exposed to an environment outside the body, the DNA can begin to degrade. This can affect the ability of a forensic biologist to retrieve a useful result.
- When collecting DNA evidence, extra precautions must be taken to ensure it is kept in the best possible condition.
- Presumptive tests exist for the three most common types of body fluids found at a crime scene: blood, semen, and saliva.
- Mitochondrial DNA is inherited from the mother.
- A Y chromosome is inherited by males from their father. Y-chromosome analysis is beneficial in cases involving sexual assault, missing persons, and genealogy. Most sexual assaults involve a male as the perpetrator. DNA tests that specifically test the genetic content on the Y chromosome can be forensically valuable.

CASE STUDIES

AP PHOTO/VICTORIA AROCHO

Figure 10-16. *Kenneth Waters at a news conference with his sister, Betty Ann Waters (left) and Representative Patrick Kennedy (right).*

Kenneth Waters (1980)

On the morning of May 21, 1980, Katherina Reitz Brow was stabbed to death in her home. When the police arrived at the scene, they discovered that many of her valuables had been taken and there were bloodstains and fingerprints throughout the house. Investigators suspected that several of these bloodstains and fingerprints had come from the perpetrator. Kenneth Waters (see Figure 10-16) was Brow's next-door neighbor. Two years after the murder took place, Waters' ex-girlfriend told police that he had confessed to her that he had killed Brow. Waters' blood type matched the blood found at the crime scene, but his fingerprints did not match. However, the jury that convicted Waters was never told about the existence of the fingerprints. After

Waters' conviction, his ex-girlfriend recanted her testimony. Waters' sister, Betty Ann Waters, was determined to prove her brother's innocence, so she got her GED and put herself through college and law school. Working with the Innocence Project, Betty Ann Waters convinced the district attorney to reopen the case and perform DNA testing on the perpetrator's blood found at the crime scene. The test proved that the blood did not come from Waters; he was released from jail in 2001 after spending 17.5 years in prison. Waters died six months after his release.

Think Critically **1. On what evidence was Waters convicted of the murder?**

2. Why is DNA stronger evidence than blood types?

3. If Waters had been tried today, do you think he would have been convicted? Explain.

Identification of Human Remains: Air France 447 Crash (2009)

At 7:03 P.M. on June 1, 2009, Air France Flight 447 (see Figure 10-17) departed Rio de Janeiro, Brazil. While flying over the Atlantic Ocean on its way to Paris, Flight 447 disappeared off the radar. It soon became evident that the airplane had crashed and that none of the 228 people onboard had survived. Wreckage from the plane was found littered over the surface of the mid-Atlantic Ocean. Because of the location of the crash, it was difficult to recover the bodies of the victims. Five days after the crash, only two bodies had been found. Twenty-six days after the crash, only an additional 49 bodies had been recovered, and search efforts were called off, leaving 179 victims unaccounted for. Many of the recovered bodies had been floating in the ocean for days and could not be identified through photos alone. DNA samples of relatives of all the crash victims were collected and used to identify the remains retrieved from the crash location. DNA testing allowed all of the recovered remains to be returned to the families of the deceased.

PHOTO BY FORCA AEREA BRASILEIRA VIA LATINCONTENT/GETTY IMAGES

Figure 10-17. *The crash of Air France 447 was the deadliest accident in Air France's history.*

Think Critically **1. Why were DNA samples collected from the relatives of the crash victims after the accident?**

2. In many airplane crashes, only body parts instead of full bodies are found. Why is DNA testing particularly useful in this situation?

3. In what situations could DNA testing *not* be used to identify a plane crash victim?

Bibliography

Books and Journals

Butler, John M. (2005). *Forensic DNA Typing: Biology, Technology, and Genetics of STR Markers* (2nd ed.). New York: Academic Press.

Clayton, Julie, and Carina Dennis, eds. (2003). *50 years of DNA*. New York: Palgrave Macmillan.

Gaensslen, R. E., Howard A. Harris, and Henry Lee. (2008). *Introduction to Forensic Science & Criminalistics*. New York: McGraw-Hill.

Lerner, K. Lee, and Brenda Wilmoth Lerner, eds. (2006). *The World of Forensic Science*. New York: Gale Group.

Nickell, Joe, and John F. Fischer. (1999). *Crime Scene: Methods of Forensic Detection*. Lexington, KY: The University Press of Kentucky.

Petraco, Nicholas, and Hal Sherman. (2006). *Illustrated Guide to Crime Scene Investigation*, Boca Raton, FL: CRC Press.

Transnational College of LEX. (2003). *What Is DNA? A Biology Adventure*. Detroit: Language Research Foundation.

Rudin, Norah, and Keith Inman. (1997). *An Introduction to Forensic DNA Analysis*. Boca Raton, FL: CRC Press.

Buckleton, John, Christopher M. Triggs, and Simon J. Walsh. (2005). *Forensic DNA Evidence Interpretation*. Boca Raton, FL: CRC Press.

Websites

http://academy.asm.org
www.actionbioscience.org/newfrontiers/salyersarticle.html
www.suite101.com/content/identifying-bacteria-with-microbial-forensics-a181435
www.dhs.gov
www.enotes.com/forensic-science/str-short-tandem-repeat-analysis
www.forensicdnacenter.com/dna-str.html
www.forensicmag.com
http://molecular.roche.com/roche_pcr/roche_pcr_process.html
www.ojp.usdoj.gov/nij/training/firearms-training/module12/fir_m12_t05_03_a.htm
www.tsa.gov
www.twgfex.org

CAREERS IN FORENSICS

Science, Technology, Engineering & Mathematics

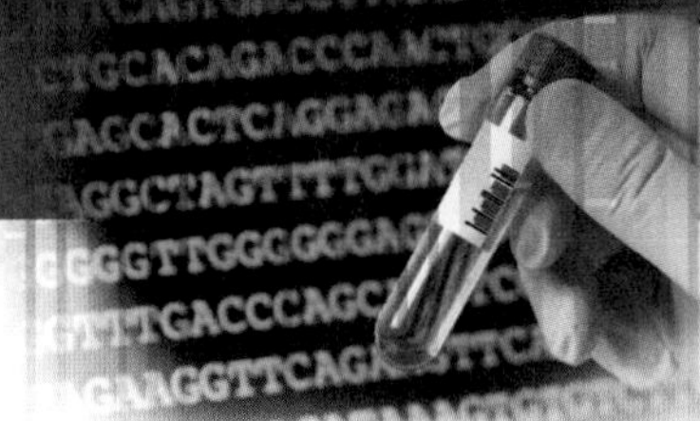

Abigail A. Salyers, Ph.D.: Forensic Microbiologist

Microbial forensics is a subspecialty of microbiology that focuses on the use of microorganisms as weapons of war or terrorist attacks, as well as other, more common criminal uses.

There have been reported instances of individuals who knowingly infected others with the HIV virus. Cases of neglect or lax hygiene can lead to hospital-acquired-infections. Lack of hygiene in food-processing plants can result in mass outbreaks of food poisoning. The U.S. Center for Disease Control has a system for tracking the DNA fingerprints of disease-causing microorganisms—those that might be found in hospitals or that are food-borne, such as *Salmonella*, but the technology needs to take another step forward.

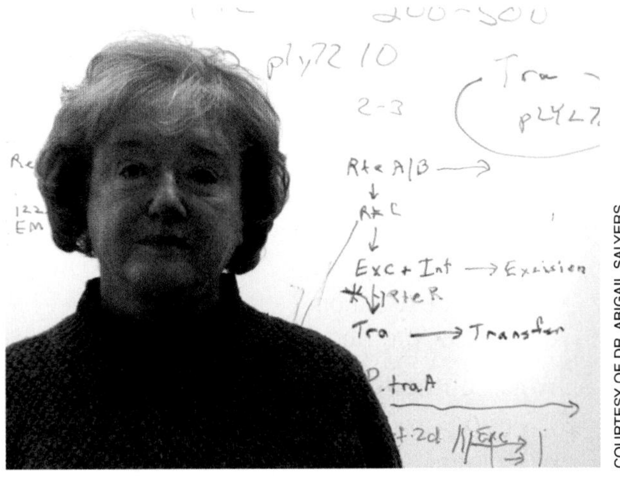

COURTESY OF DR. ABIGAIL SALYERS

Figure 10-18. *Abigail A. Salyers, Ph.D.*

The forensic microbiologist provides DNA evidence in court. The evidence can be used to identify the microorganism and to connect suspects to the crime in which a microorganism was used to cause harm.

Abigail A. Salyers, Ph.D., is Professor of Microbiology in the Department of Microbiology at the University of Illinois at Urbana-Champaign (see Figure 10-18). She is a leader in microbial forensics. As Dr. Salyers explains, a good model for the problems a forensic microbiologist solves is the presentation and resolution of the anthrax attack of 2001, a case involving the spores of the bacterium *Bacillus anthracis*. When inhaled, this bacteria causes the potentially fatal anthrax infection. The 2001 spore release resulted in the death of five people.

The case remains open to this day. If a suspect is arrested and tried for the anthrax attack, spores will be key physical evidence. However, there are difficulties with using spores for evidence: 1) It must be proven that the spores introduced as evidence are in fact the same spores mailed in the envelopes. 2) The spores of *B. anthracis* are commonly found in the soil of southern U.S. states. Therefore, prosecutors must prove that the spores introduced as evidence were not merely carried into the lab on clothing or shoes—that they are original and not contaminates. Because the burden of proof lies with the prosecution, there is no prosecution possible without further testing and cataloging.

Dr. Salyers was one of 35 scientists of the American Academy of Microbiology who met in 2002 to discuss how to handle the testing and setting of standards for the new challenges in bioterrorism. Anthrax was a key topic, as was intentional HIV infection, hospital infection outbreaks, and widespread food contamination.

The scientists are in favor of an open database of information within the scientific community. Because there are so many microbes, and so many possible contaminates, Dr. Saylers and many other scientists think an open database is the only way to stay on top of the vast amount of data. However, there are those, especially politicians, who are not in favor of openly sharing scientific data of any kind. The question remains: Is the benefit worth the risk?

Learn More About It

To learn more about forensic psychiatry, go to www.cengage.com/school/forensicscienceadv.

CHAPTER 10 REVIEW

Multiple Choice

1. A medical examiner is completing an autopsy on the body of a 45-year-old murder victim. On the victim's clothing, there are very minute traces of blood. The blood is collected, but the sample is too small to isolate the DNA by using traditional methods. Which method of DNA extraction is best for isolating the DNA from this sample? *Obj. 10.4*
   a. differential extraction
   b. organic extraction
   c. Chelex extraction
   d. none of the above
2. Which of the following is *not* a step in DNA fingerprinting? *Obj. 10.3*
   a. cutting the DNA using restriction enzymes
   b. isolating the DNA
   c. dissolving the DNA in an organic solvent
   d. sorting DNA bands according to size
3. With the PCR technique, what signifies the starting point of replication on the DNA sample? *Obj. 10.3*
   a. primers
   b. anneals
   c. restriction enzymes
   d. extensions
4. While collecting evidence at a homicide scene, crime-scene investigators notice blood spatter on a computer desk. Once the scene has been photographed, how should the investigators proceed in collecting the biological evidence? *Obj. 10.5*
   a. The entire desk should be collected and taken into the lab for processing.
   b. A sterile cotton swab should be used to collect the blood evidence which should then be placed into a test tube.
   c. The section of the desk that bears the blood evidence should be cut out and sent to the crime lab for processing.
   d. The evidence should be processed by the blood analyst at the scene.
5. A known drug dealer was found dead in his home. Blood evidence not belonging to the victim has been discovered on the victim. Four suspects have been identified. Each suspect is ordered to give a DNA sample. What is this sample called? *Obj. 10.5*
   a. a presumptive sample
   b. a paternity sample
   c. an unknown sample
   d. a reference sample

Short Answer

6. Diagram two strands of a DNA molecule. Label the nitrogenous bases, deoxyribose, phosphate groups, and the hydrogen bonds. *Obj. 10.1*

7. Explain how the base-pair sequence is related to the DNA fingerprint. *Obj. 10.1*

____

____

____

____

8. Is nuclear DNA evidence considered to be class or individual evidence? Why? *Obj. 10.2*

____

____

____

____

9. Explain PCR. Discuss the steps of PCR. Discuss the advantages of the technique. *Obj. 10.3*

10. Compare and contrast VNTR and STR. *Obj. 10.3*

11. Explain the function of CODIS, and describe how this database has benefited the forensic community. *Obj. 10.3*

12. Discuss the significance of contamination and degradation of DNA evidence. *Obj. 10.3*

13. What is the value of presumptive tests? Give two examples of presumptive tests and discuss how they are used. *Obj. 10.5*

14. Compare and contrast mtDNA and Y-chromosome analysis. Be sure to include advantages and disadvantages of each kind of analysis. *Obj. 10.4*

Think Critically

15. A 35-year-old woman is raped. A rape kit is completed, and two DNA profiles are discovered. One profile matches the husband's genetic profile, while the other is consistent with a suspect in the case. Is there enough DNA evidence to convict the suspect? Explain. If the woman were raped by two men, could DNA evidence be used to convict or exonerate potential suspects?. ***Obj. 10.3***

16. Men convicted of sex crimes must give a blood sample. A profile of the convicted sex offender is stored in CODIS. Investigators compare his profile to profiles of perpetrators in unsolved cases. When a sex crime occurs, states are able to use the database to search for a match. Discuss the merits and disadvantages of this protocol. ***Obj. 10.3***

Jealousy

By: **Jaynell Oyomire and Jessica Jiemenez**

East Ridge High School
Clermont, FL

On a chilly Saturday night, a young couple was walking their dog (see Figure 10-19). Suddenly, the dog broke free of the leash, and ran down an alley. The couple chased the dog and found a dead woman hanging over a wire fence. The couple immediately called 911. Police and CSI were on the scene just moments later. About a block away, CSI found the victim's purse. She was Robin Davis, 22 years old. She was attending the University of Nevada and working part-time as a cocktail waitress at the Rio Bella Casino. As CSI were completing an initial evaluation of the scene, death investigators were doing a preliminary assessment of the victim. She had scratches on her face and bruises on her body, indicating she likely tried to fight off her attacker. Later, the medical examiner did a thorough external and internal exam. Skin was found under her fingernails. She also had two fractured ribs and a hairline fracture on her left wrist.

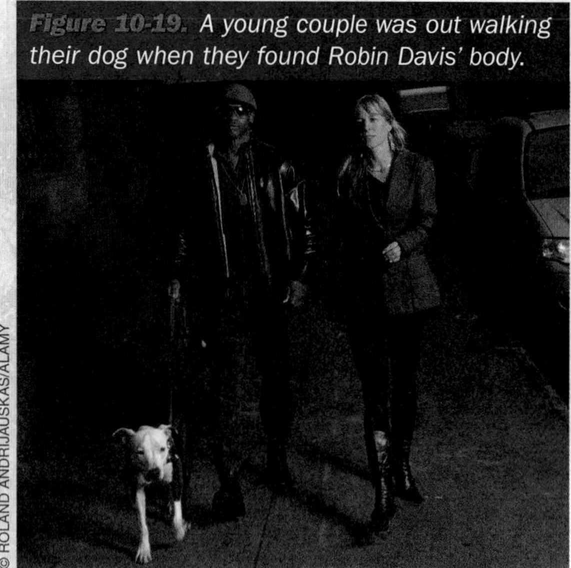

Figure 10-19. *A young couple was out walking their dog when they found Robin Davis' body.*

Later, the detective who was on the scene, Jose Rodriguez, went to the casino to ask her manager a few questions. Amanda Ruiz, the cocktail waitress manager, told Rodriguez that the victim had worked on Saturday night, but she had left early because she said her wrist was hurting. Ruiz recommended that the detective talk to friend, Valerie Torres. Ruiz found Torres and brought her to Detective Rodriguez. He asked Torres if she knew Robin Davis, and she stated she did. The detective told Torres that Robin had been found dead the night before. Immediately, Torres asked if Robin's boyfriend, Rocky Gates, had done it. The detective stated he was not at liberty to discuss the details of the case. Torres proceeded to say that Amanda had fired Robin. Amanda told everyone that Robin had been caught stealing, but it was rumored that Amanda's husband was involved with Robin.

Later, back at the station, Rodriguez discovered that Rocky Gates had two prior arrests—one for assaulting an officer and one for domestic abuse. Gates was brought in for questioning. Gates thought he was there regarding his priors. Rodriguez asked Gates where he had been on Saturday night, and Gates proceeded to say that he had been with his girlfriend. He told the detective that he could ask her himself, she was at work. Rodriguez told Gates that she was dead—murdered. Rodriguez encouraged Gates to come clean regarding his whereabouts that night. Gates was hesitant but finally admitted he had been drinking heavily and passed out. The detective told Gates that he had a warrant to collect DNA from him. A buccal (cheek) swab was taken and sent to the lab for testing (see Figure 10-20). Gates was released, but was told not to leave the area.

Next, Rodriguez called in Mr. and Mrs. Ruiz. Rodriguez asked about Robin's wrist

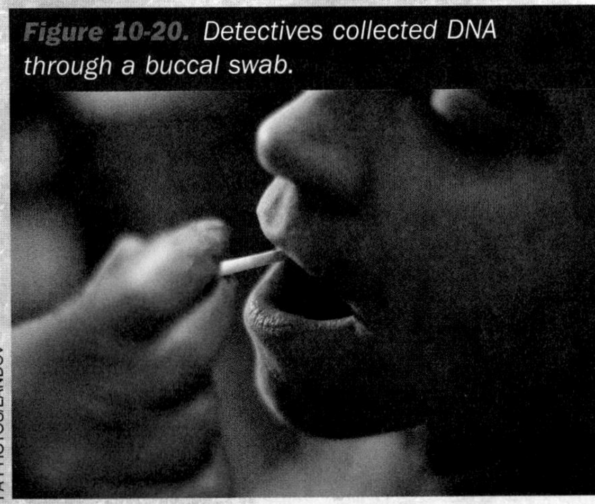

Figure 10-20. *Detectives collected DNA through a buccal swab.*

hurting. Amanda told the detective that Robin had been complaining about it all night, so she let her go home. He asked her about the conflicting statement that the waitress gave. He asked whether she had fired Robin for stealing and for the suspected affair with her husband. Mrs. Ruiz strongly denied the idea that her husband was involved with Robin. Mr. Ruiz looked confused and denied he was involved with Robin. The detective did not believe either of them. He had a warrant to collect DNA from both of them. After submitting a DNA sample, they both left—angry.

Rodriguez decided to go back to the casino and ask a few more questions. He spoke to several other waitresses. He walked up to a tall, blonde waitress, Michelle Bronson, and asked her if he could ask her a few questions. She agreed. He asked if she thought anyone could have had a grudge against Robin. Bronson said that Carla Lucksman and Robin had gotten into a heated argument a couple of days ago because Robin was stealing good "tippers." During the argument, Carla told Robin she was going to kill her.

Ms. Lucksman was called in for questioning. The detective asked if she knew Robin Davis. Lucksman immediately went on a rant about trying to get her fired and not fighting like a real woman. Detective Rodriguez told her that Robin had been murdered. She said, "She's dead? Who woulda thought that?" The detective asked her why she did not seem remotely upset. Lucksman told the detective that she would have liked to have killed Robin, but she didn't. The detective requested a DNA sample. Lucksman provided the sample. He asked where she had been that Saturday night, and she said she'd been working until 1:00 in the morning. He continued by asking for the names of the witnesses, and he wanted to know why Robin left work early that night. Lucksman stated that she had been complaining about her wrist hurting. Valerie had taken her home.

Several days later, Rodriguez got back the DNA tests and a toxicology report. A DNA match was found and a date rape drug was found in Robin's body. He conferred with the medical examiner. A rape kit was collected and sent off for analysis, but no genital trauma was noted during the autopsy.

Activity:

Answer the following questions based on information in the Crime Scene S.P.O.T.

1. Who killed Robin?
2. How would the DNA be analyzed in this case?
3. What type of extraction would be done on the DNA collected from the rape kit? Why?

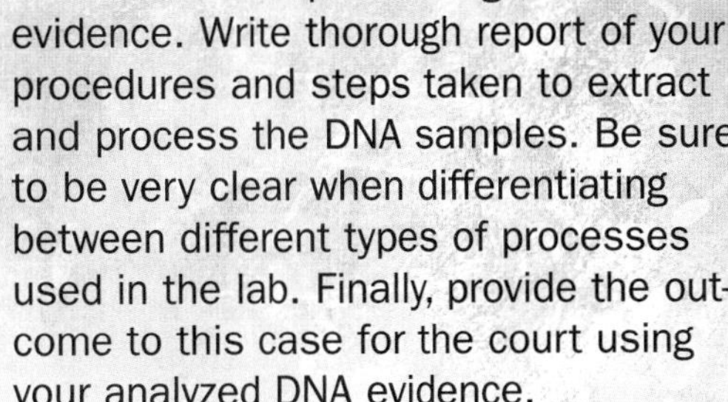

WRITING 4. You are the DNA expert on this case. This case has gone to trial and you must defend your procedures for processing the DNA evidence. Write thorough report of your procedures and steps taken to extract and process the DNA samples. Be sure to be very clear when differentiating between different types of processes used in the lab. Finally, provide the outcome to this case for the court using your analyzed DNA evidence.

Preventing Adolescent Crime Together™ (P.A.C.T)

Introduction:

In this activity, you will demonstrate the fingerprinting and DNA collection techniques you have mastered to build "identity kits" for children in your community. Hair follicles contain nuclear DNA. When DNA is collected from hair, the hair must be pulled directly from the root and stored properly. Additional identifiable information from the parent/child, including surgical scars and old injuries, will be recorded on the kit for identification purposes in case there is ever a need.

Procedure:

1. Secure a high-traffic area such as a retail establishment or restaurant. You may need to talk to the manager of the store or restaurant or write a letter regarding your request.
2. Role play with other students the steps you will take to approach a parent with an offer to complete an ID kit. You will need to practice filling in the written portion of the kit as well as collecting the DNA and rolling fingerprints.
3. Practice with another student how you will educate a parent on using the kit. Be sure to include how to secure the hair follicles collected and stored in the plastic bag. The follicular DNA must be stored in the freezer; the kit itself should be stored in a location that is easily accessible, if needed.
4. On the day of the event:
   a. Set up in location approved by the manager.
   b. Use what you have learned and practiced in class to build ID kits for children and their parents.
   c. When the kit is finished, educate the parent(s) on how to use the kit in case the need arises.
   d. Give the parent a PACT bracelet and thank everyone for participating.

This is intended to be a free service; however, many parents may wish to make a donation. It is recommended that a portion of the money collected be donated to a crime-related charity decided upon by your classmates. The remainder of the donation money should go to purchasing supplies for the forensics class.

Figure 10-21. *An adult is taking this child's fingerprints to be saved in an ID kit.*

ACTIVITY 10-1 *Ch. Obj. 10.1, 10.3, 10.4*
DNA EXTRACTION LAB

Objectives:

By the end of this activity, you will be able to:

1. Identify the steps in DNA extraction of a strawberry.
2. Discuss the purpose of each chemical involved in isolating the DNA of a strawberry.
3. Examine the usefulness of this lab in understanding the role of DNA analysis in identifying suspects.

Background:

DNA is found in the nucleus of most human cells. Extracting nuclear DNA is the first step in developing a DNA profile of a suspect. DNA must be separated from the rest of the cellular components in order to get a pure sample of DNA. Detergent dissolves the phospholipid bilayer of the cell membrane and the organelles. The salt keeps the DNA intact after it has been separated from the rest of the cellular components. The DNA will be dispersed in the solution. DNA is not soluble in alcohol, so the DNA will clump together and become visible. Once the DNA is isolated, examination of the quality of DNA is completed. The quality of the sample will determine the technique used to process the DNA.

Materials:

(per one or two students)
1/2 cup of strawberries
1/8 teaspoon table salt
1 cup of cold water
1 blender
1 strainer
1 1-cup measuring cup or beaker
2 tablespoons of liquid detergent
1 test tube
meat tenderizer (enzymes)
isopropyl alcohol (70 percent to 90 percent)

Procedure:

1. Place 1/2 cup of strawberries into the blender.
2. Add 1/8 teaspoon table salt.
3. Add 1 cup of cold water.
4. Blend on high for 15 seconds.
5. Pour the "strawberry puree" through a strainer into another container (such as a measuring cup or beaker).
6. Add 2 tablespoons of liquid detergent and swirl to mix.
7. Let mixture sit for 5 to 10 minutes.
8. Pour mixture into test tubes or other small containers, each about 1/3 full.
9. Add a pinch of enzymes to each test tube and stir gently. Be careful not to stir too hard because the DNA may break up, making it harder to see.
10. Tilt your test tube and slowly pour rubbing alcohol into the test tube down the side so that it forms a layer on top of the strawberry mixture. Pour until you have about the same amount of alcohol in the tube as strawberry mixture.
11. Alcohol is less dense than water, so it floats on top. Look for clumps of white strings where the water and alcohol layers meet.
12. Use a wooden stick or a straw to collect the DNA. If you want to save your DNA, transfer it to a small container filled with alcohol.

Questions:

1. Why is it necessary to isolate DNA prior to analyzing it?
2. What is the significance of placing the strawberries in the blender?
3. Describe how each step in the process simulates isolating DNA.
4. Evaluate the effectiveness of this lab in demonstrating the techniques used by DNA analysts to isolate DNA prior to processing.

Extension:

Complete the same process using peas. Compare the process of isolating DNA from fruits and vegetables. Is there a difference? Explain.

ACTIVITY 10-2 *Ch. Obj. 10.5*
TO CATCH A CRIMINAL

Objective:

By the end of this activity, you will be able to:
Examine the effectiveness of luminol in detecting the presence of blood at a crime scene.

Background:

Presumptive tests are screening tools that crime-scene investigators use to detect the presence of various body fluids such as blood and semen. Luminol is a presumptive test used to detect minute amounts of blood left behind at a crime scene. Luminol detects hemoglobin, a protein in red blood cells that transports oxygen. Luminol can detect blood that has been diluted 10 million times.

Introduction:

Your group of four or five students will be designing a micro-crime scene. You will be required to write a scenario to provide an overview of your scene. It can be any type of crime as long as blood is left behind as a result of the crime. You will set up your micro-crime scene and then move from station to station completing various tasks.

Materials:

(per group of four or five students)
1 spray bottle of "blood"
1 spray bottle of luminol
rags, paper towels, water, soap

Procedure:

Day One

1. On paper, design a micro-crime scene with your group.
2. Get approval of your micro-crime scene from your teacher. Your proposal must also contain a list of materials needed for the micro-crime scene. Decide who is bringing in which supplies and ask the teacher for any supplies that are not readily available.
3. Write the scenario of your micro-crime scene. Be sure to identify the victim. Explain the type of crime (for example, stabbing, strangulation, gun-shot wound).
4. Take your group's scenario to the teacher for approval.

Day Two

1. As a group, set up your micro-crime scene in the designated area. Make sure items in your scene are placed properly and are representative of your scenario.
2. At your micro-crime scene, leave a copy of the scenario for other groups to read while they are completing different tasks.
3. After every group has set up their crime scenes, rotate clockwise to the next group's crime scene, read the scenario, and look closely at the crime scene.
4. Based on the type of crime, using the "blood" provided, spray the crime scene with the appropriate amount of blood in the appropriate locations.
5. Move clockwise to the next crime scene. Read the scenario and view the crime scene.
6. You will "clean up" the blood evidence completely. You do not want any trace of blood evidence left at this crime scene. As a team, you will develop a strategy for eliminating the presence of any blood found at that crime scene.
7. After your group has a strategy set, clean up the blood evidence at the crime scene.
8. Move clockwise to the next station. Read the scenario, view the crime scene, and try to make any visual observations of blood.
9. Turn out the lights. In the dark, spray the luminol on the micro-crime scene. Record your observations.
10. Return to your original crime-scene station. Was blood detected?

Questions:

1. Explain the purpose of luminol.
2. After spraying the luminol, did you detect blood?
3. Was blood found at the crime scene you cleaned up? Did you think you cleaned it thoroughly enough not to find blood? Explain.
4. Discuss the value of presumptive tests at a crime scene.
5. Evaluate the effectiveness of this activity in demonstrating the benefits of luminol.

ACTIVITY 10-3 *Ch. Obj. 10.1, 10.3*
DNA PROFILING LAB

Objectives:

By the end of this activity, you will be able to:

1. Build a DNA model.
2. Simulate the processes of building a DNA fingerprint.
3. Based on the evidence, determine which DNA fingerprint is consistent with the sample found at the crime scene.

Background:

After DNA has been isolated from cellular material, the DNA is tested to establish a DNA fingerprint. When unknown body fluids are found at a crime scene, the fluids are collected and processed—a profile is established. This profile can also be used to prove that the fluid does not belong to the victim. When suspects are arrested, they are often required to submit a DNA sample. This may be in the form of blood, hair, or a buccal (cheek) swab. The DNA sample is processed. If the DNA found at the crime scene is consistent with a suspect's DNA sample, the suspect is often indicted and may go to trial for the crime. Because DNA is individual evidence and can narrow down suspects to near 100 percent certainty, DNA is the most indisputable evidence a prosecutor can have at trial.

You may recall from previous coursework in biology that the complementary strands of DNA run in opposite directions. Biochemists use the term *antiparallel* to describe this structure. The carbons of the deoxyribose are numbered 1' (1 prime) through 5' (5 prime). The 5' carbon attaches to the phosphate group. DNA replication moves in a 5' to 3' direction (see Figure 1).

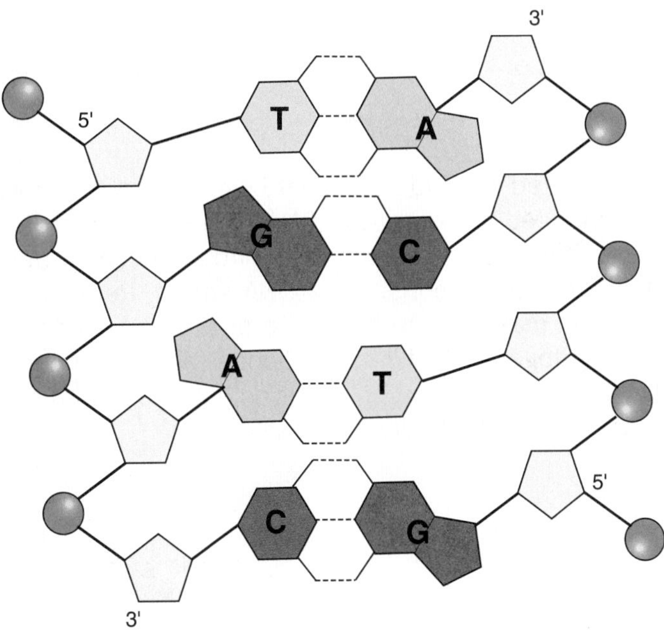

Figure 1. *DNA replication moves from the 5' end of the molecule to the 3' end.*

Introduction:

Last night, James Daniels, 43, was found dead in an alley. Upon initial exam, it was determined that Daniels was stabbed to death. CSI says it appears to be a mugging gone bad. It is likely that Daniels did not want to give the mugger his money, so he tried to fight off the perpetrator. During the physical altercation, the perpetrator stabbed Daniels multiple times and fled the scene with Daniel's wallet, watch, and wedding band. During the external exam, the medical examiner found blood on the victim's clothing. The DNA was collected for processing. According to statements taken from acquaintances, the victim had upset some clients because of how he managed their investments. Also, Daniels had a gambling problem and owed people money. After a thorough investigation, five suspects were ordered to submit DNA samples for testing. The suspects are as follows:

Tom Anderson: A client of Daniels; recently lost $250,000 in bad investments

Ethan Matthews: A client of Daniels; unable to account for $100,000 in his investment portfolio

Connor Stevens: A client of Daniels; unable to account for $50,000 in his investment portfolio

Liam Roark: A loan shark; loaned Daniels $100,000 a month ago; the loan was due three days ago

Paul Ellwood: A loan shark; loaned Daniels $50,000 six weeks ago; the loan was due two weeks ago

Materials:

(per group of three or four students)

6 colors of pop beads—(60 for the phosphate, 60 for the sugar, and 25 for each nucleotide); the colors for each component of the DNA will be determined by your teacher.

40 plastic connectors

2 enzyme cards (JSDI and RMBII)

1 electrophoresis lane

1 suspect, murderer, or victim DNA strip (chosen by your teacher)

Procedure:

1. DNA was collected from your suspect by court order. The white blood cells were ruptured to extract the DNA. The DNA strip you have is the sequence of base pairs found in the white blood cells. Using your suspect's DNA sample, construct a DNA strand. Remember, sugar and phosphate make up the backbone of the DNA, and the nucleotides are held together by hydrogen bonds (plastic connectors).
2. Once the DNA strand is built, place it on your desk with the 5' end on the top left side, as shown in Figure 2.

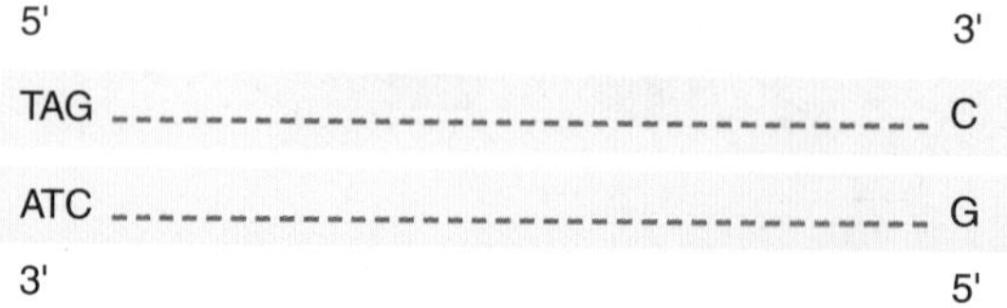

Figure 2. *Placement of DNA strand.*

3. Use the restriction enzyme cards to cut the DNA. Restriction enzymes are shaped specifically to identify specific base sequences. Use the restriction enzyme cards to move across your DNA strand from the 5' end to the 3' end and cutting the DNA the way it is shown on the card. Start with JSDI and then use RMBII. Keep the DNA sections oriented properly—with the 5' side up.
4. Count the number of nucleotides on the longest side of each fragment. Place each fragment on the number marked on the gel electrophoresis lane.
5. Separate the DNA at the hydrogen bonds. Unzipping the DNA will make it single stranded. (In a real electrophoresis, the single stranded DNA will be transferred to a nylon fiber. This is called *Southern blotting*.)
6. Draw lines across the lanes on the electrophoresis next to the number of nucleotides. For example, if you have pieces of DNA that are 5, 15, and 10 fragments long, your "lane" would look like Figure 3.

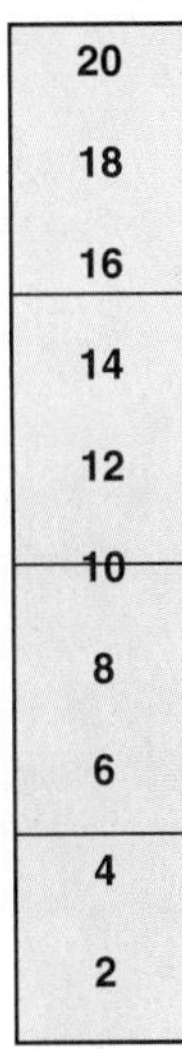

Figure 3. *Sample lane.*

7. When your DNA fingerprint is complete, ask your teacher to check your work. If your profile is correct, go to the board and report your results. If you are incorrect, review your work.
8. After every group has completed the process and has reported their results on the board, complete the questions.

Questions:

1. Which suspect committed the murder? Explain.
2. Why was it necessary to run DNA analysis on the victim?
3. What role did restriction enzymes play in analyzing the DNA?
4. Write a one-paragraph explanation about what you think happened during this crime now that the murderer is known.
5. If you were an expert witness on the stand in this case, how would you present your findings?

CHAPTER 11

Forensic Odontology

THE BITE MARKS THAT BROUGHT DOWN A SERIAL KILLER

Figure 11-1. Ted Bundy admitted to killing 30 women.

In the spring of 1974, a serial killer was on the loose in the Northwest United States. Young women were disappearing from college campuses at the rate of one per month. Soon after each disappearance, the woman was found bludgeoned and strangled to death. The killer left no fingerprints, body fluids, or hair, so the cases remained unsolved.

Soon, women began vanishing in Colorado and Utah. Again, police had few leads on the identity of the killer. Then, on August 16, 1975, a man named Theodore Bundy was arrested outside of Salt Lake City for failure to stop for a police officer. The arresting officer found a ski mask, another mask made from pantyhose, a crowbar, handcuffs, trash bags, a rope, and an ice pick in Bundy's car.

A search of Bundy's apartment revealed evidence connecting him to two murdered women. While in jail awaiting his trials, Bundy escaped to Florida. Early on the morning of January 15, 1978, Bundy broke into the dorm room of two Florida State University women and bludgeoned and strangled them to death.

Bundy left two bite marks on one of the women—Lisa Levy. These bite marks would be the evidence that finally brought Bundy's murder spree to an end. A month later, a police officer pulled Bundy over for a routine traffic stop. When the officer checked the license plate of the car Bundy was driving, he learned that the car had been stolen. Bundy was arrested. His fingerprints revealed that he was an escaped murder suspect. Florida police quickly realized that Bundy likely also committed the Florida State murders. Forensic odontologist Dr. Richard Souviron was able to match the bite marks on Levy to Bundy's teeth. Bundy was convicted of the murders. Overall, police think Bundy murdered between 30 and 100 women.

Figure 11-2. Lisa Levy (left) and Margaret Bowman (right) were Bundy's last victims. Bite marks left on Levy's body led to his conviction.

OBJECTIVES

By the end of this chapter, you will be able to:

11.1 Describe the structure of a typical tooth.

11.2 Compare and contrast permanent and deciduous human dentition.

11.3 Recognize the value of odontology in forensic investigations.

11.4 Explain how teeth and craniofacial features are helpful in estimating age, ancestry, and sex.

11.5 Differentiate between the dentition of humans and other animals.

VOCABULARY

cementum - bonelike covering of the portion of the tooth that extends into the bone (the root); attaches the tooth to the periodontal ligament, a connective tissue that anchors the tooth to the bone

crown - the portion of the tooth that is covered in enamel and is situated above the gum

dentin - hardened connective tissue that makes up the majority of a tooth; surrounds the pulp cavity and is covered by enamel in the crown and by cementum in the root

dentition pattern - the pattern made by a particular set of teeth

enamel - the outer covering of the crown of a tooth, made up of calcium carbonate and calcium phosphate

neck - area between the root and the crown of the tooth; also known as the *cementoenamel junction* (where the enamel and cementum meet)

odontology - the study of the anatomy and growth of teeth and diseases associated with the teeth and gums

pulp - softer connective tissue that composes the innermost portion of the tooth; contains nerves and blood vessels

root - the portion of the tooth that extends into the tooth socket and is covered with cementum

Did You Know?

In 1776, a young dentist by the name of Paul Revere identified the bodies of soldiers who were buried on the battlefields during the Revolutionary War by examining their teeth and dental work. Dr. Joseph Warren, the man who sent Revere on his famous ride, was identified by a bridge of silver wire and ivory that Revere had made two years earlier.

INTRODUCTION

Most people in the United States and other developed countries have visited a dentist and had dental X-rays (see Figure 11-3) taken. These X-rays, notes from the dentist, and any photographs or casts made of your teeth become your dental record. When a victim cannot be identified easily, dental records can help. If the gums and other soft tissues of the mouth are gone and teeth loosen or fall out of the jaw, forensic odontologists might still be able to match the individual teeth to dental records. Dental X-rays also provide information about the skeletal structure around the teeth. Forensic odontologists can use the structure of the jaws and sinuses to identify the victim even when no teeth remain. Even when no dental records exist, photographs of a person smiling can aid in making identifications.

Odontology is the study of the anatomy and growth of teeth and diseases associated with the teeth and gums. A forensic odontologist uses knowledge of the teeth to identify victims of mass disasters, help police in criminal investigations, and verify signs of abuse. A forensic odontologist begins his or her career as a dentist and then specializes in the field of forensic odontology.

The parts of the body on which bite marks are found are very often indicative of the type of attack. Bites from domestic violence or sexual assaults are most commonly found on the neck, chest, and shoulders of the victims. Bite marks on the arms or buttocks are sometimes seen in cases of child abuse. Bite marks found on arms and hands generally indicate defensive wounds.

STRUCTURE AND FUNCTION OF TEETH

Obj. 11.1, 11.2

BIOLOGY

Digestion begins in the mouth. Enzymes in the saliva chemically break down complex carbohydrates into simpler molecules. Teeth mechanically grind and crush large pieces of food into pieces that may be swallowed.

DAVID ACOSTA ALLELY/SHUTTERSTOCK

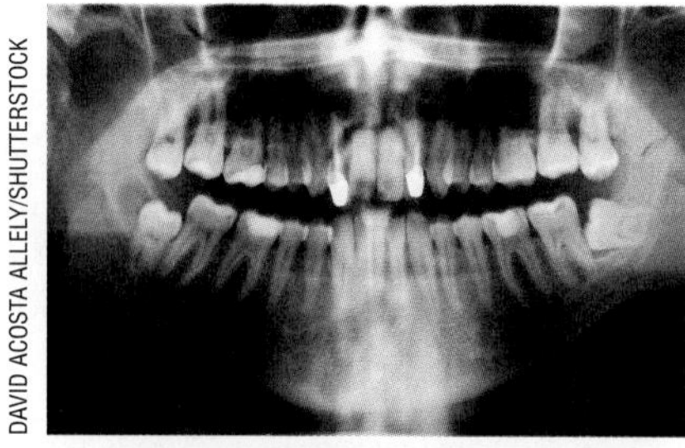

(A) Panoramic X-ray.

© JEREMY SUESS/ISTOCKPHOTO

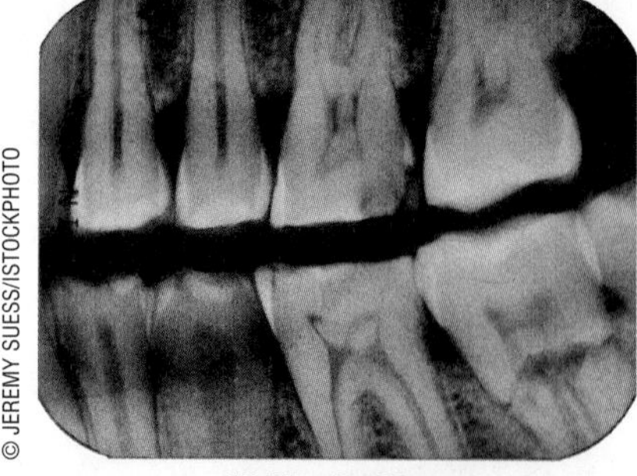

(B) Bitewing X-ray.

ISTVAN CSAK/SHUTTERSTOCK

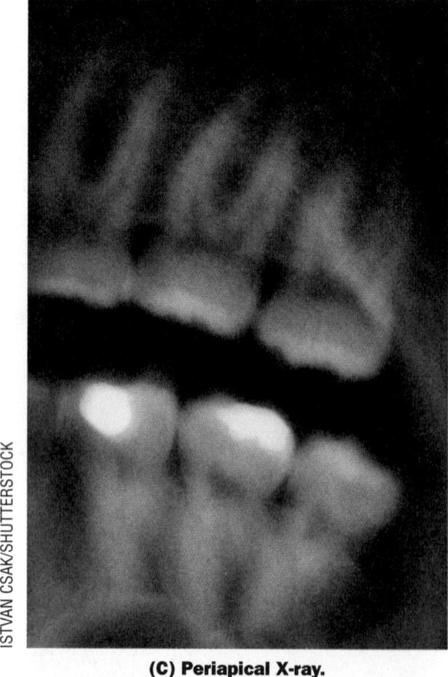

(C) Periapical X-ray.

Figure 11-3. *Typical dental X-rays.*

A typical tooth (see Figure 11-4) consists of three regions: the crown, the neck, and the root. The **crown** of the tooth is above the gum line. The **neck** is where the crown and the root meet at the gum line. The **root** is embedded in the bony socket of the jawbones.

The bones of the face that contain the dentition are the maxilla (upper dentition) and mandible (lower dentition). The maxilla and the mandible each have an alveolar process, which is the portion of each of those bones that contains the teeth.

Teeth are composed mostly of dentin. **Dentin** is a type of connective tissue that has *calcified* or hardened, and gives teeth their basic shape. The inside of the tooth is filled with a softer connective tissue called **pulp**. The pulp contains nerves and blood vessels. On the crown, the dentin is covered by a layer of **enamel** that is made up of calcium carbonate and calcium phosphate. Dentin is easily dissolved by acids, and the enamel acts as a barrier to protect the dentin. In the root of the tooth, the dentin is covered by **cementum**—a bonelike substance that attaches the tooth to the *periodontal ligament*, a connective tissue that anchors the tooth to the bone, helps keep teeth in the proper position, and acts as a shock absorber when you are chewing.

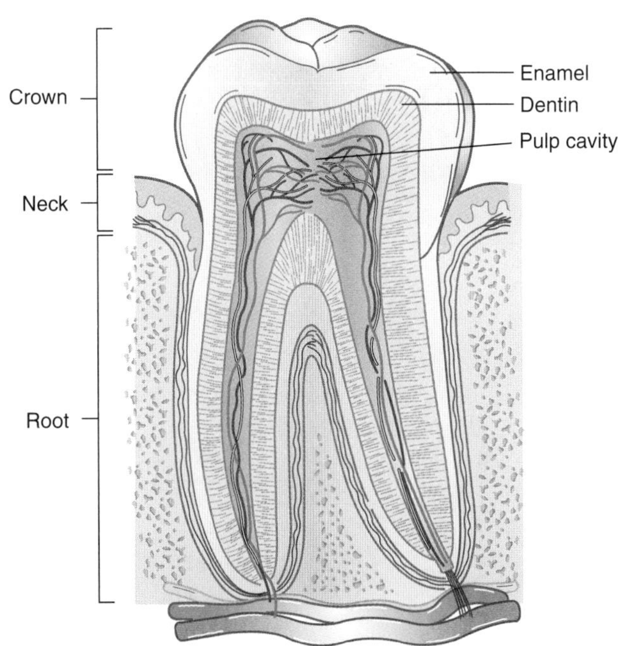

Figure 11-4. *Anatomy of a typical tooth.*

Humans have 20 deciduous (baby) teeth that begin to erupt from the gum when the child is about six months old. The deciduous teeth begin to fall out when the child is around six years old and are replaced by 32 permanent (adult) teeth. Figure 11-5 illustrates the differences between deciduous teeth and permanent teeth.

Contrary to popular belief, George Washington's dentures were not made of wood. The teeth were actually constructed from animal teeth as well as walrus and elephant tusks.

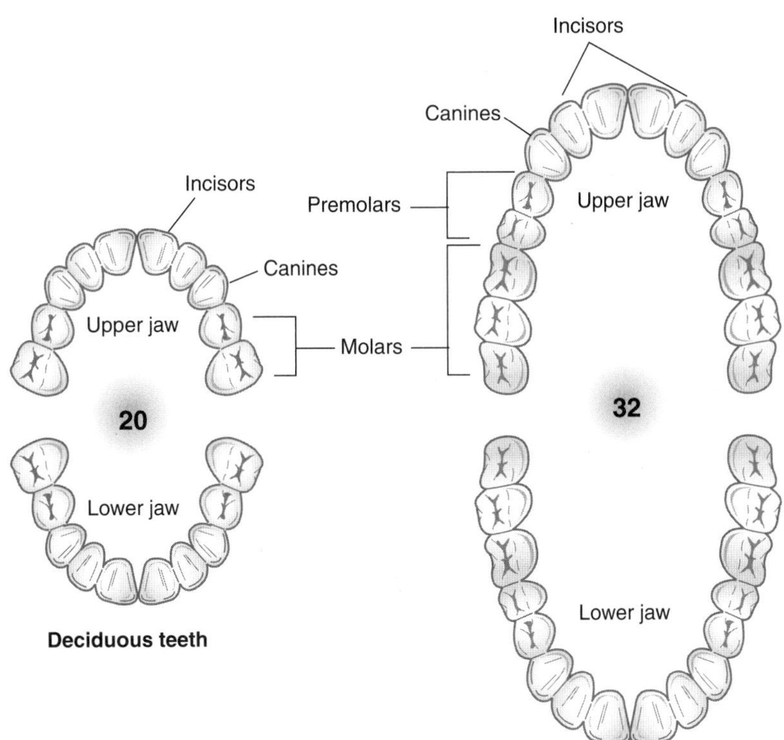

Figure 11-5. *Full sets of deciduous and permanent teeth.*

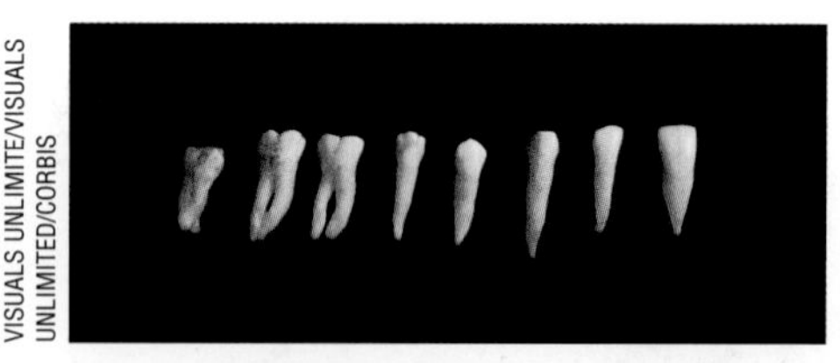
© RALPH HUTCHINGS/VISUALS UNLIMITE/VISUALS UNLIMITED/CORBIS

Figure 11-6. *The shape and size of each tooth reflects its function.*

The shape and size of each tooth reflects its function. *Incisors* are the four sharp cutting teeth in the front. Next to the incisors are the *canines*, also called the *cuspids*. Canines have one pointed surface and are used for tearing and shredding food. The *premolars* have two pointed surfaces with which to grind and crush. The broad flat teeth in the back of the mouth, the *molars*, are also used to crush and grind food. Figure 11-6 shows a full set of permanent teeth from the upper right jaw.

Obj. 11.3, 11.4

USE OF TEETH TO ESTIMATE PHYSICAL CHARACTERISTICS

Teeth are more resistant to burning and decomposition than other tissues in the body. Even after the flesh has been damaged or has decomposed, teeth and dental alterations, such as dental fillings, caps, bridgework, and dentures, often remain intact.

When a forensic odontologist is asked to help identify an unknown deceased person, he or she compares information recorded after the unknown person's death with records taken during a known person's life. Forensic odontologists compare the shape of the teeth as well as any gaps between the teeth. The placement of fillings or dental work, missing or broken teeth, and tooth alignment are important characteristics of comparison. The odontologist also observes supranumerary teeth (extra teeth), congenitally missing or anomalous teeth, and whether any of the teeth are rotated. The pattern made by a particular set of teeth is called the **dentition pattern**. Characteristics of the teeth that may indicate a person's personal habits include staining, enamel erosion, and occlusal surface wear (wear on the surface of the teeth where they rub against one another during the chewing process). For example, coffee drinkers and smokers are likely to have teeth that are stained. Erosion on the teeth can indicate habits such as alcohol abuse, bulimia, or substance abuse. Information about the unidentified decedent's personal habits helps investigators develop a biological profile of the decedent that may be matched to that of a missing person.

Identical twins have the same DNA, but just like fingerprints, they have different dentition patterns.

AGE ESTIMATION

Human teeth emerge before age 25 in a predictable pattern. Forensic odontologists are able to use these emergence patterns to develop relatively accurate estimates of the age of victims who are younger than 25 years old. The first teeth appear in babies around the age of six to nine months. Normally, the first to emerge are the lower incisors, followed by the upper incisors. Teething continues until the second set of molars comes in, somewhere around two years of age. Children usually begin losing the deciduous teeth by age six. The teeth tend to loosen and fall out in the same order they emerged.

The permanent teeth begin emerging at about the same time the deciduous teeth are coming out. Usually by age seven, a child will have begun developing the permanent set of teeth. Figure 11-7 charts the ages of the emergence of the permanent teeth.

If the victim is more than 25 years old, age estimation is more difficult because all of the teeth have usually fully emerged. Wear and tear on the

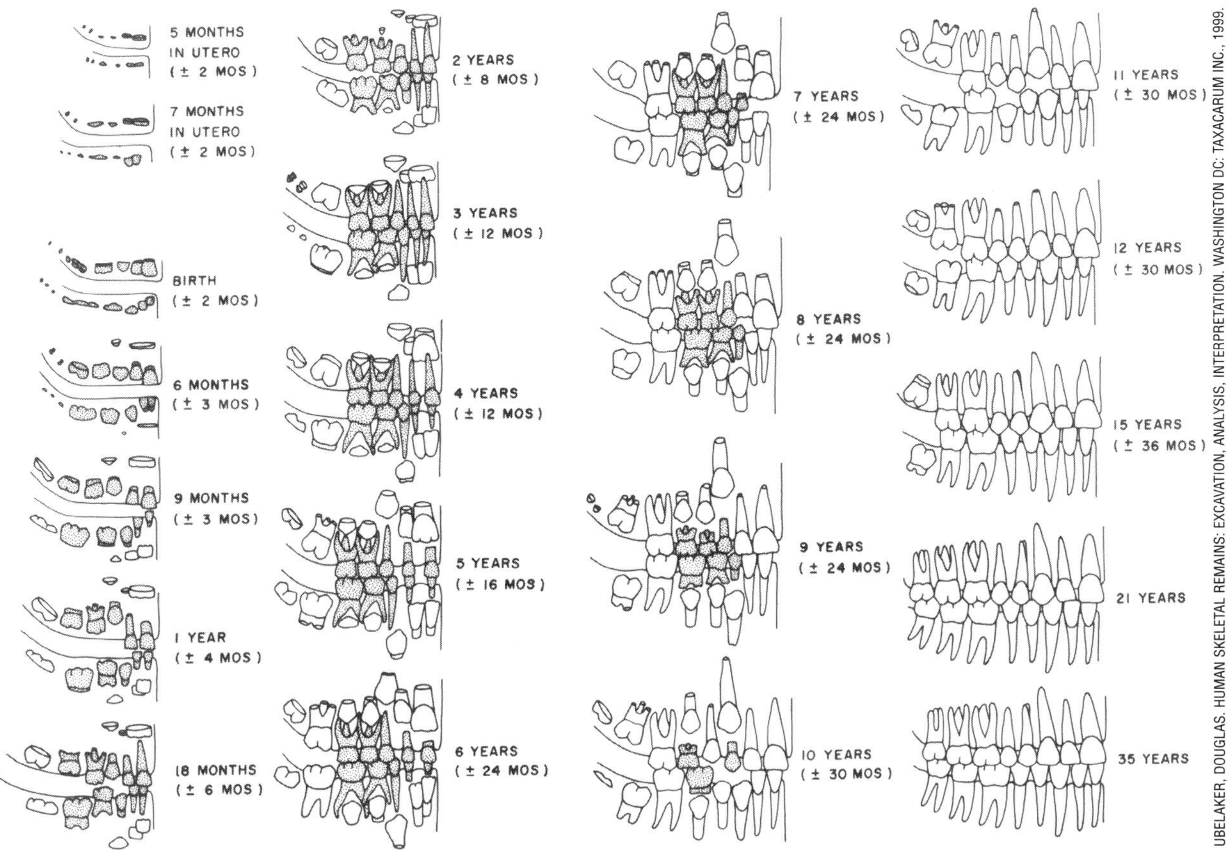

Figure 11-7. *Ubelaker's Chart of Dental Development shows the emergence pattern of human teeth.*

teeth from eating habits, *bruxism* (teeth grinding), dental work, and receding of the gums are useful in estimating age, but these characteristics provide a less-accurate estimate—usually only within four years.

ANCESTRY ESTIMATION

Forensic anthropologists generally rely on a combination of craniofacial characteristics to estimate ancestry. However, there is no way to use physical characteristics to determine absolutely an unidentified person's ancestry. Certain characteristics are more common within certain populations, and forensic anthropologists and odontologists use these characteristics to estimate a person's ancestry.

Prognathism is forward projection of the alveolar processes that contain dentition (upper and lower). People from Australia and those of African descent tend to exhibit prognathism. The absence of prognathism is called *orthognathism.* People of European descent are more likely to exhibit orthognathism. Anthropologists examine alveolar prognathism, proportions of the eye orbits, the nasal spine (a bony projection at the base of the nose), cheekbone contours, palate shape, and the shape of the central incisors when attempting to estimate ancestry. Figure 11-8 highlights the nasal spine.

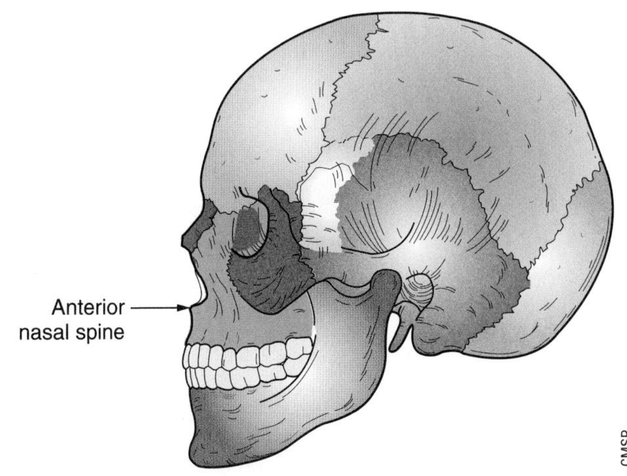

Figure 11-8. *The nasal spine is usually much more prominent in people of European descent than in people of African descent.*

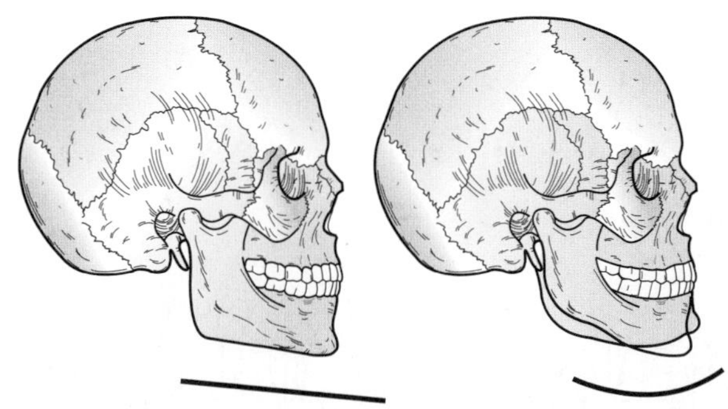

Figure 11-9. *Australian aborigines and some South Pacific islanders often exhibit rocker jaw.*

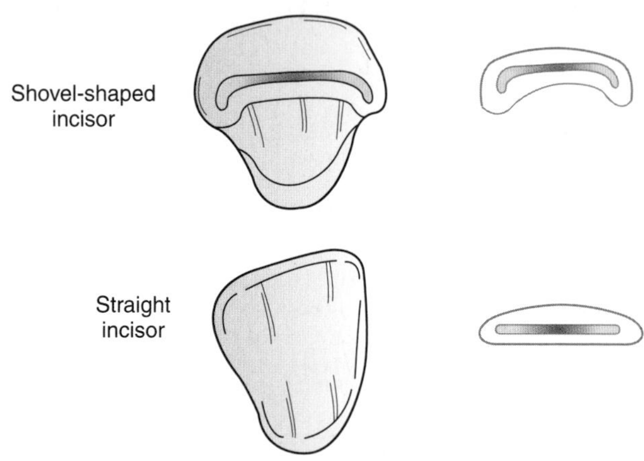

Figure 11-10. *Forensic odontologists are able to use the shape of the decedent's incisors to estimate ancestry.*

The chins of people of European and Asian descent tend to project, while Australian aborigines and some South Pacific islanders tend to have more rounded, nearly receding chins. When the jawbone is placed on a horizontal surface, it tends to rock in a motion similar to that of a cradle. For this reason, the characteristic is called a *rocker jaw* (see Figure 11-9). People of African ancestry often have chins that are more intermediate.

The size and shape of certain teeth are useful in making an estimation of ancestry. Even though teeth provide some indication of ancestry, it is, however, not conclusive identification. The shape of the decedent's incisors (see Figure 11-10) can be a useful feature when estimating his or her ancestry. Fewer than 10 percent of people of European or African descent have shovel-shaped incisors, but many people of Asian descent have them.

Cultural differences are often reflected in the teeth. For example, consumption of betel nuts in Indonesia tends to produce black stains on the teeth. Modification of the anterior teeth (teeth in the front of the mouth) is also indicative of geographical or ethnic regions as well as sex. The practice of decorating the teeth with gems was common among ancient Maya. These features found in skeletal remains aid investigators in determining the age and ancestry of the bones.

SEX ESTIMATION

Determining the sex solely by examination of the teeth is difficult. For the most part, sex determination is based on the size and shape of the teeth. In general, males have larger teeth than females, and the canines in females tend to be more pointed than those in males. If the teeth are still embedded in the bone, craniofacial differences make the determination more accurate.

Digging Deeper

with Forensic Science e-Collection

Bite mark analysis has existed for a very long time. In recent years, it has been used in court to convict criminals of violent attacks such as rape and child abuse. Even before bite mark analysis became a generally accepted and highly specialized field of forensics, it was used to solve cases in which no other evidence was available. Using the Gale Forensic Science e-Collection at www.cengage.com/school/forensicscienceadv, develop a timeline of the history of bite mark analysis. Also, review the completed study and develop a graph of the findings of the study. Finally, write a one-page analysis of the study and how the study has helped advance the field of forensic dentistry.

USE OF TEETH TO DETERMINE POSITIVE IDENTIFICATION

Obj. 11.3, 11.5

Personal effects, such as wallets, and family identifications can be used to make a presumptive identification. Investigators can also use the location of the body to make a presumptive identification—for example, if it is in a car or a house known to belong to a missing person. However, positive identifications are made using fingerprints, DNA analysis, and the comparison of medical and dental X-rays.

If the remains have become skeletonized or the body is badly decomposed or burned beyond recognition, other methods of identification may be necessary. Mass fatalities present other complications because the bodies may be fragmented, intermingled, or buried under rubble. Identifications using dental records and photographs showing teeth were made following the bombings of the World Trade Center, Hurricane Katrina, and the tsunami in Indonesia. Figure 11-11 shows a jawbone that was found in Pompeii, Italy. Pompeii was buried in a layer of ash and lava when Mt. Vesuvius erupted in 79 A.D.

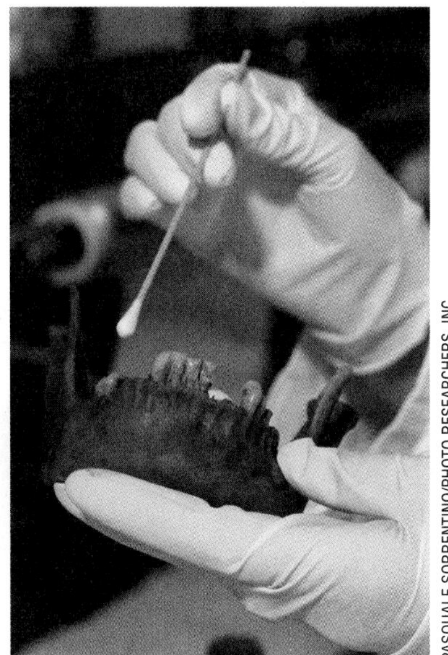

PASQUALE SORRENTINO/PHOTO RESEARCHERS, INC.

Figure 11-11. *Expert examining the jawbone from a skeleton found in the ruins of Pompeii, Italy.*

DENTAL RECORDS

When dental records are required for positive identification, the medical examiner sends the teeth, skull, or jawbones to the forensic odontologist. In most cases, the medical examiner already has a presumptive identity. The medical examiner orders the dental records of the person suspected to be the decedent. The odontologist compares the antemortem records (taken during life) with the remains and the postmortem records (recorded after death) in an attempt to make a match between them and identify the unknown decedent. If the remains are found at the site of a plane crash or catastrophic disaster, investigators gather the dental records of possible victims from individual dentists. Computer imaging of dental records has helped speed the process, but it is still a somewhat tedious task to make comparisons of hundreds and sometimes thousands of dental records.

Fillings, bridgework, and dental implants can be especially useful in making dental comparisons. Dental implants are artificial teeth that are anchored into the gums or jaw to replace missing teeth. Dental implants and other dental work often have unique serial numbers (see Figure 11-12) that can be traced to the patient. Fillings in specific teeth will have a unique shape that is visible on a dental X-ray. Even if only the filling is found without the tooth, its shape can sometimes allow dental matching to a specific dental record.

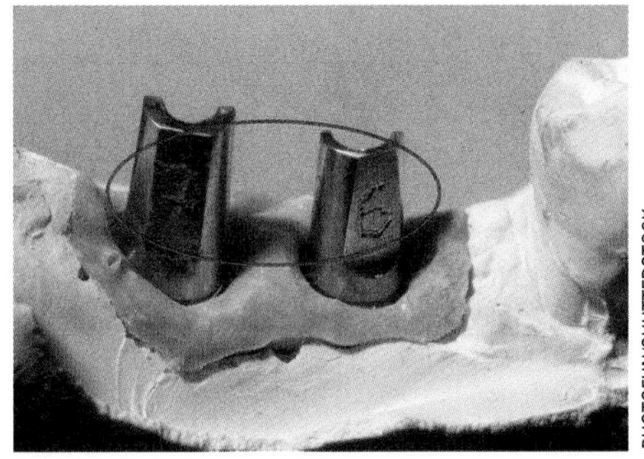

PHOTOFUN/SHUTTERSTOCK

Figure 11-12. *The serial numbers of the dental implants are circled in red.*

HUMAN BITE MARKS

The marks left behind by teeth are referred to as *bite marks*. Bite marks can be useful in narrowing down a list of suspects. Because flesh is soft and stretchy, a bite mark in flesh will look different from a bite mark in something like a pencil or a piece of cheese. If the person was alive when he or she was bitten, the area around the bite mark bruises and swells. If the bite occurs postmortem, the area will not bruise or swell. A pathologist swabs the bite mark to collect traces of saliva. The presence of saliva is important

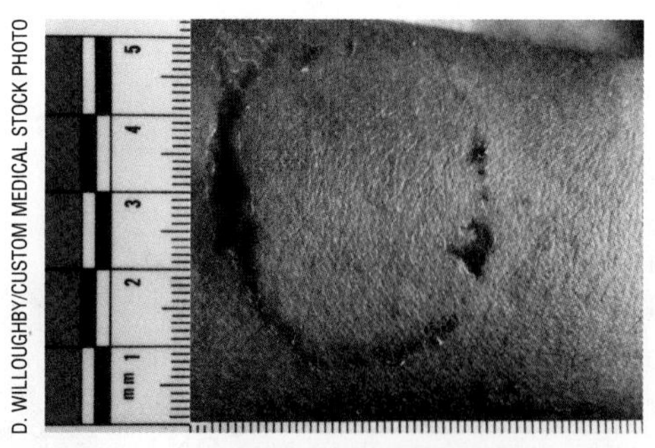
D. WILLOUGHBY/CUSTOM MEDICAL STOCK PHOTO

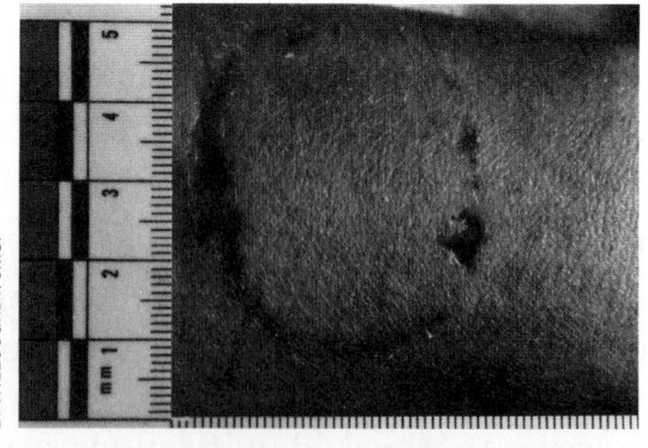
D. WILLOUGHBY/CMSP

Figure 11-13. *Odontologists review color and black-and-white photos of each bite mark. They also make careful dimensional measurements.*

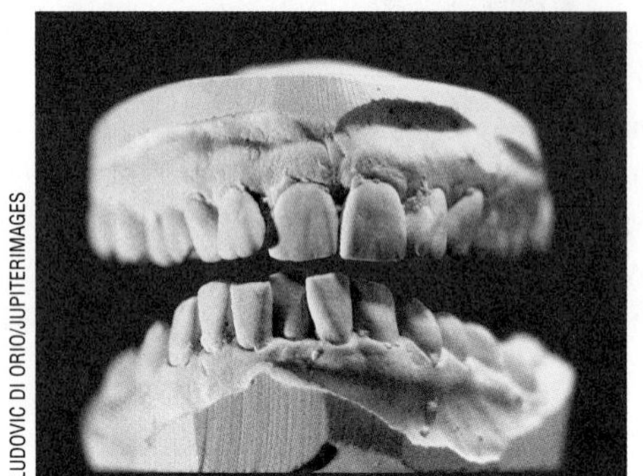
LUDOVIC DI ORIO/JUPITERIMAGES

Figure 11-14. *A casting of human teeth that can be compared to a bite mark.*

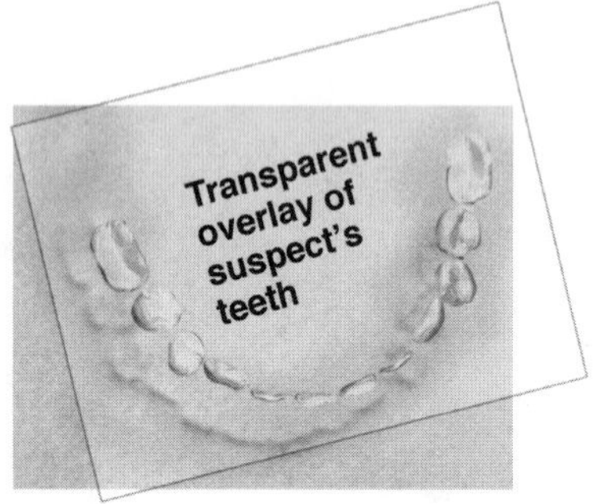

FORENSIC DENTISTRY ONLINE, DR. JOHN KENNEY. GRAPHIC: CHICAGO TRIBUNE. MCT/NEWSCOM

Figure 11-15. *Overlay of a bite mark.*

in confirming that the mark was caused by a bite. Additionally, the saliva contains DNA, which can be used to identify the biter.

To make a comparison of bite marks, odontologists look at the overall shape of the bite and measure the dimensions of the mark (see Figure 11-13). A typical human bite mark has a double horseshoe pattern. Marks left by the six most central teeth of both the upper and lower jaws are usually more well-defined than those left by other teeth. A bite mark is photographed from several angles in black and white as well as in color. Taking both black and white as well as color photographs is preferable to one or the other to ensure there is no distortion due to shading or discoloration. The various angles allow investigators to accurately capture the dimensions of the arch of the bite.

In a living victim, inflammation of the site of the bite mark can cause blurring of the pattern and make comparisons more difficult. The inflammation is caused by the tissue damage when the teeth break or place excessive pressure on the skin. Large amounts of blood will accumulate and interstitial fluids—fluids that fill in the spaces between cells—leak into the bite area. For this reason, it is very important to get the photographs before the swelling reaches its greatest extent—within eight hours after the occurrence.

If the bite is deep, a casting (see Figure 11-14) can be made. A *casting* is a three-dimensional replica of an impression made by pouring dental stone or plaster of Paris into the impression and allowing it to harden. An overlay (see Figure 11-15)

Digging Deeper

with Forensic Science e-Collection

Forensic odontologists must consider various factors when examining the teeth from a cadaver. Often when teeth are being examined, the body has been burned or decomposed beyond recognition, and teeth are the best method for developing a profile of the victim. Many times, dental records are not available and other means of making an identification are necessary. Using the Gale Forensic Science e-Collection at www.cengage.com/school/forensicscienceadv and working in groups of two or three students, develop a graphic organizer that details the multiple ways in which teeth can be evaluated for the purpose of identification. After completing your graphic organizer, write a one-paragraph summary of your findings. Present your research to the class.

is also made of the bite mark. To do this, a thin transparent piece of plastic is placed over the bite mark and the mark is manually traced onto a transparency or a photograph is scanned into the computer to produce a transparency. Forensic odontologists can then compare the overlay or the casting to other bite marks or impressions from a known suspect.

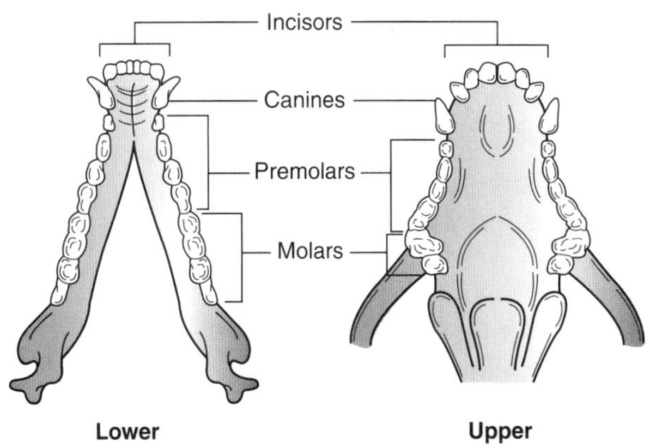

Figure 11-16. *A dog's jaws have a long, narrow arch.*

ANIMAL BITE MARKS

When a pet attacks a person, the owner is usually liable (financially responsible) for the medical bills. Sometimes, there are also criminal charges associated with the attack. In these cases, DNA may be extracted from the saliva and bite mark patterns may be compared to the pet's dentition pattern.

Nonhuman bite marks have a very different dentition pattern than human bite marks. An adult dog has 42 teeth, 10 more than a full set of teeth in adult humans. Figure 11-16 is a diagram of the typical teeth of a dog. Not only are there more teeth, but the shape of the mouth is also clearly distinguishable from the shape of a human mouth. A bite mark from a dog has a long, narrow arch compared to the C- or U-shaped arch of human bite marks. Injuries from rodents, usually rat bites, may be indicative of cases of child neglect. The teeth of a rat are shown in Figure 11-17. The prominent front teeth make rodent bites easy to identify. Children in homes that are infested with rats are often bitten while they sleep.

Figure 11-17. *The front teeth of a rodent, such as this rat, are quite prominent. Bite marks from rats are relatively easy to identify.*

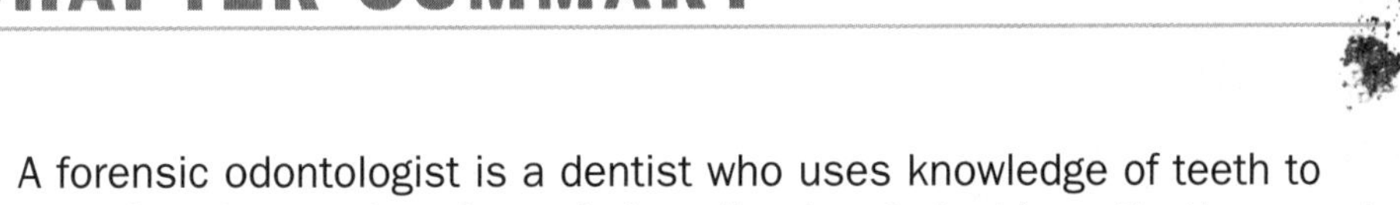

CHAPTER SUMMARY

- A forensic odontologist is a dentist who uses knowledge of teeth to identify unknown decedents, help police in criminal investigations, and verify signs of abuse.
- Teeth are used in the mechanical digestion of food. A human has two sets of teeth in a lifetime. The shape of teeth varies depending on function.
- A typical tooth consists of the crown, the neck, and the root. The crown is the portion of the tooth that is above the gum line.
- Family identification, personal effects, and location of the body may provide information for a presumptive identification of a body. Fingerprints, DNA analysis, or comparison of medical or dental X-rays are necessary for a positive identification.
- Teeth and craniofacial features provide clues to a person's age, ancestry, and sex.

Some small rodents, such as rats, mice, and hamsters, have incisors that grow continuously throughout their lives. Their constant gnawing keeps the teeth the proper length.

- Teeth and dental alterations are more resistant to the forces of decomposition than other tissues of the body. This makes teeth an important tool in identification of burned or badly decomposed remains.
- Bite marks alone are not conclusive evidence, but they are useful in narrowing down a list of suspects. Traces of DNA from saliva in a bite mark may help investigators make a positive identification of the perpetrator.
- A typical human bite mark has a double horseshoe pattern. The marks left by the six most central teeth of the upper and lower jaws are the most evident. Nonhuman bite marks have very different dentition patterns than those of humans.

CASE STUDIES

Helle Crafts (1986)

After suffering for years in an abusive marriage, Helle Crafts (see Figure 11-18) had finally had enough. She filed for divorce during the summer of 1986. A few months later, Crafts disappeared. Her friends filed a missing person report, but Crafts's husband, Richard Crafts, insisted variously that his wife was on vacation, visiting relatives, or spending time alone.

Police suspected foul play, but with no body, it was impossible to prove that a homicide had occurred. Police learned, however, that Richard Crafts had purchased several items, including new carpeting, bedding, and a large freezer, near the time of Helle's disappearance. Additionally, he had rented a woodchipper. A witness came forward, claiming that he had seen a man with a woodchipper standing on a bridge over a lake near the Crafts's home. Police searched the shoreline and found approximately five drops of blood, human nails, small bits of bone, and a tooth with a cap.

AP PHOTO

Figure 11-18. *Helle Crafts*

The tooth provided important evidence in the case. Forensic odontologists compared the tooth to Helle Crafts's dental records and identified the tooth as belonging to Helle. Richard Crafts was convicted of his wife's murder. His conviction was the first murder conviction in Connecticut in which a body was never found.

Think Critically

1. **Why was the discovery of Helle Crafts's tooth so important in convicting her husband of murder?**
2. **What do you think most likely happened to the rest of Crafts's body?**
3. **This case occurred in 1986. If it had occurred today, what other evidence from the recovered human remains do you think investigators could have used?**

War in the Socialist Federal Republic of Yugoslavia (1991–1995)

After the fall of the Soviet Union, violence quickly erupted in various regions of Yugoslavia, including Bosnia-Herzegovina, Croatia, Serbia, and Kosovo. Wars broke out between the different ethnic and religious groups in these regions as each group tried to claim independence. More than 100,000 civilians lost their lives.

Many of these civilians had been murdered by militias and buried in mass graves (see Figure 11-19). At the end of hostilities, forensic teams were sent to identify the bodies excavated from the mass graves. Working from long lists of missing people, the teams used DNA, skeletal, and dental evidence to return the remains to the victims' families. Dental records were invaluable in the identification of almost half of the individuals. Investigators set up a database containing all of the dental records available for the reported missing people. Once a possible match was found, DNA was used to confirm the identification so that family members of the victim could claim the body.

Figure 11-19. *Mass grave in Croatia.*

Think Critically

1. **Why did investigators rely so heavily on dental evidence to identify the remains of the victims?**
2. **What difficulties would a forensic investigator face when trying to identify thousands of remains?**
3. **Evidence from the remains was also collected to prove that war crimes had taken place. What type of evidence do you think would help prove this?**

Bibliography

Books and Journals

Conklin, Barbara, Robert Gardner, and Dennis Shortelle. (2002). *The Encyclopedia of Forensic Science*. Westport, CT: Onyx Press.

Dorion, Robert B. J., ed. (2005). *Bitemark Evidence*. Monticello, NY: Marcel Dekker.

Lerner, K. Lee, and Brenda Wilmoth Lerner, eds. (2006). *The World of Forensic Science*. New York: Gale Group.

Newton, Michael. (2008). *The Encyclopedia of Crime Scene Investigation*. New York: Info Base Publishing, pp. 125–126.

Ricciuti, Edward R. (2007). *Science 101: Forensics*. New York: HarperCollins.

Stimson, Paul G., and Curtis A. Metz, eds. (1997). *Forensic Dentistry*. Boca Raton, FL: CRC Press.

Websites

www.cengage.com/school/forensicscienceadv

www.chincare.com/HealthLifestyle/DHdocs/rodentdentistry.pdf

www.drtomoconnor.com/3210/3210lect02b.htm

www.forensicdentistryonline.org/forensichomepage.htm

www.msnbc.msn.com/id/6875436

www.nytimes.com/2009/10/11/opinion/11beevor.html

CAREERS IN FORENSICS

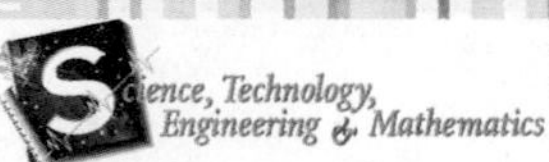

Forensic odontologists are highly trained dentists who use their medical and scientific training and experience to provide testimony and evidence in legal matters. They have pursued advanced studies in chemistry, biology, and medical subjects, as well as training in legal procedures and investigative methodology. Forensic odontologists must have a working knowledge of the law-enforcement and judicial practices in their area.

One of the most important roles of the forensic odontologist is the identification of human remains. Sometimes, a body has suffered so much trauma or gone through so much decomposition that it is difficult to identify the decedent. In these cases, a forensic odontologist can compare impressions of teeth and jaws with existing dental records to make an identification.

After the earthquake in Haiti in 2010, forensic odontologists and the dentists who were assisting them were under a great deal of pressure to make identifications as quickly as possible under difficult conditions. Many of the people of Haiti had no dental records, and many of the records that had existed were destroyed during the earthquake. To prevent the spread of disease, bodies had to be buried quickly. Still, rescue and recovery personnel had to identify victims and notify families whenever possible. Forensic odontologists identified as many victims as they could.

Forensic odontologists also examine and assess bite mark injuries in cases of assault or suspected abuse. Bite marks left in flesh begin to change shape soon after they are made. Odontologists rely on photography, molds of living skin, removal of skin from a corpse, and examination under ultraviolet light to reveal bruising under the skin. Bite marks on certain inanimate objects, however, hold their shape and can demonstrate a much clearer match to a dental mold or X-rays.

Occasionally, a forensic odontologist assists a medical examiner or forensic anthropologist in identifying a body. In the absence of dental records, DNA samples extracted from just one tooth can help identify the individual. Forensic odontologists gather information from dentures and dental implants. Even if trauma to the body has destroyed all of the teeth, the odontologist can examine X-rays of the skull and mouth to collect valuable information about the person's identity. The odontologist estimates the age of the individual by examining important markers on the teeth, including wear and the condition of the enamel.

Forensic odontology is a broad field. A dentist who has an interest in this area and takes the extra classes and required training can choose among many subspecialties.

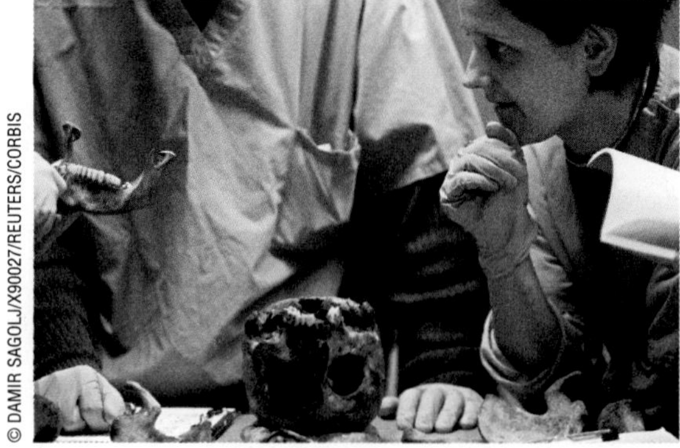

Figure 11-20. *Forensic experts try to identify the remains of a victim of a massacre that took place in Srebrenica, Bosnia in 1995.*

Learn More About It

To learn more about forensic science careers, go to www.cengage.com/school/forensicscienceadv.

CHAPTER 11 REVIEW

Multiple Choice

1. Bite marks are frequently seen _____. *Obj. 11.3*
   a. in cases of child abuse
   b. as defensive wounds
   c. as a result of sexual abuse
   d. all of the above
2. A typical dog bite mark has a _____ pattern. *Obj. 11.5*
   a. long, narrow arch
   b. C-shaped arch
   c. double horseshoe
   d. U-shaped arch
3. Which of the following statements about teeth is *not* true? *Obj. 11.1, 11.4*
   a. Teeth are more resistant to decomposition than other parts of the body.
   b. Individual teeth cannot be used to make identification.
   c. Placement of fillings or dental work is an important characteristic for comparison.
   d. Characteristics of teeth can indicate a person's personal habits.
4. Dental records and teeth may be used for identification _________. *Obj. 11.3*
   a. when the body is badly burned or decomposed beyond recognition
   b. when fingerprints are unavailable
   c. in mass fatalities when there are many fragmented, intermingled bodies
   d. all of the above
5. A common characteristic of people of _______ descent is rocker jaw. *Obj. 11.4*
   a. Hawaiian
   b. African
   c. Australian
   d. none of the above

Short Answer

6. Describe the structure of a typical tooth. *Obj. 11.1, 11.4*

   ____

   ____

   ____

7. Explain the process of making dental comparisons in an attempt to identify human remains. *Obj. 11.3, 11.4*

   ____

   ____

   ____

8. Describe how the shapes of various types of teeth are suited for their function. *Obj. 11.1*

---

9. Explain why it is important to photograph bite marks on living victims within eight hours after the attack occurs. *Obj. 11.3*

---

10. How might teeth reflect cultural differences? *Obj. 11.4*

---

11. Compare and contrast the bite mark pattern of an animal with that of a human. *Obj. 11.5*

Think Critically

12. A woman was assaulted. After the perpetrator ran away, the woman went to the hospital. She was bitten on her left arm during the attack. How will a forensic nurse process the evidence of the bite mark? What forensic value could the evidence provide? *Obj. 11.3*

13. A four-year-old child is bitten by a dog. The dog's owner states that the attack was not caused by his dog but by another dog in the neighborhood that looks like his dog. How can investigators determine which dog attacked the child? *Obj. 11.3, 11.5*

14. Skeletal remains are uncovered in an abandoned lot. Several teeth are intact. A forensic odontologist examines the remains. The lower jaw exhibits slight rocker jaw characteristics and significant prognathism. The nasal spine is minimal, and the canines are pointed. What might this information indicate about the remains? *Obj. 11.4*

15. Look at Figure 11-7. Deciduous teeth do not emerge in front-to-back order. The first molars come in at about the same time the incisors come in. What is your hypothesis for why this might happen? *Obj. 11.2*

CRIME SCENE  Student-Prepared Original Titles

The Hitman

By: **Steven Martinez and Kevin Tufts**

East Ridge High School
Clermont, Florida

Three months ago, Miguel Suarez, CEO of Hinders Inc., was assassinated during a public speech about his new invention. Since then, there have been seven additional suspicious deaths of high-powered executives. Today, apparently, there was another.

My name is Richard Rider. I work for the FBI. The latest crime scene was in San Diego. The victim was Andrew Thomas, the owner of Steamboat's, a popular local restaurant. The crime scene was especially bad. There was blood all over, and the odor of decomposition was strong. My colleagues and I gathered what evidence we could—blood on a broken window, a picture of the victim and an unknown woman, and a single fingerprint.

Our prime suspect was Amanda Thomas, the soon-to-be ex-wife of the victim. It turned out that the picture of our mysterious woman was not of Mrs. Thomas, but of Melissa Long. After we verified her alibi, Amanda Thomas was released. Melissa Long admitted she was dating Andrew and they planned to get married when both of their divorces were finalized. Long also had an ironclad alibi.

Four months into the investigation, another murder was committed that fit the profile of all the others. We were convinced that all the murders had been committed by the same person (see Figure 11-21). But so far, the only link was the lack of evidence found at the crime scenes. We decided to go back and interview all of the previous witnesses. There had to be something we'd missed.

The next day we went to reinterview Ms. Long, only to find her dead. Two bullet wounds were in the back of her head, and she had what appeared to be bite marks on her right arm. "Hopefully, the medical

Figure 11-21. *Could someone be hiring an assassin to commit these murders?*

ISTOCKPHOTO.COM/NATHAN GLEAVE

examiner can find some DNA from the biter's saliva in those wounds," I told my partner. "Maybe we can find a link between this murder and the others. We need to find her estranged husband."

The husband, Grant Hoskins, is a retired Marine Corps platoon leader. When he was questioned regarding his wife's murder, he claimed he had not seen her in months. He said he had been at his business partner's home the night before. The partner, Curtis Nyhus, had been a soldier in Hoskins' platoon. While we checked his alibi, Hoskins asked for something to drink. After he finished his soda, I sent the can to trace.

Nyhus backed up Hoskins' alibi, so Hoskins was released. However, after checking them both out further, some disturbing information was uncovered. They had worked together in the field for 12 years until they were discharged under suspicion of attempted murder. The charges were eventually dropped due to lack of evidence. Afterwards, it was rumored that they joined a private operative hitman service called "T1419." This just might be the break we had been hoping for.

By the time the DNA from the soda can was compared to the DNA from the saliva taken from Ms. Long, Hoskins and his partner were gone. An all points bulletin was issued for both men, and undercover agents were placed at the airport and bus terminal. Later the same day, a package addressed to me was delivered. It was postmarked the same day as the Long murder. It contained a note that said, "I found all of this when I packed up my stuff after the divorce." There was a list of names—names of the murdered executives we had been investigating. There were also names of people who had paid to have the executives killed. The most important piece of the puzzle was an audiotape of Hoskins and Nyhus negotiating a contract on one of the victims. Now it all fit. We had the evidence we needed to arrest both men and hopefully get a conviction.

Activity:

Answer the following questions based on information in the Crime Scene S.P.O.T.

1. Ms. Long did not fit the profile of the other victims. Why do you think she was murdered?
2. How would the bite mark evidence be processed?

WRITING

3. If saliva had not been recovered, how might investigators still have been able to link the murder to Ms. Long's ex-husband? Write a one-page paper explaining the process.

Preventing Adolescent Crime Together™ (P.A.C.T.)

Introduction:

Animal attacks usually involve animal bites. In most cases, the owners of the attacking animals are held accountable for the attack. The animals may be quarantined or even destroyed as a result. The owners may be fined or, in some cases, incarcerated. In 2009, a two-year-old girl was killed by a 12½-foot albino python while asleep in her crib. The snake's owner, also a resident in the home, discovered the girl's body. She had a bite mark on her forehead and had been strangled. The owner was later charged with her death.

Procedure:

1. Discuss the importance of public awareness concerning dangerous pets.
2. Research state laws and local ordinances pertaining to owner responsibilities for pets, including exotic pets.
3. Find out the numbers and types of animal attacks that have been reported over the past year in your state or community. How were the incidents resolved? What happened to each animal? Was the owner ticketed, fined, or charged with a crime?
4. Design an informational brochure or poster that includes:
    a. Laws and ordinances governing what types of animals may or may not be kept
    b. Responsibilities of animal ownership
    c. The consequences for the animal if it attacks
    d. The legal implications an attack has for the owner
    e. Statistics on animal attacks

ISTOCKPHOTO.COM/MELISSA LAWRENCE

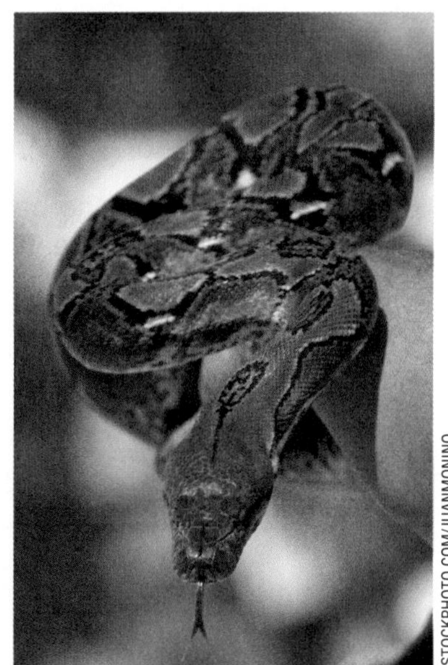

ISTOCKPHOTO.COM/JUANMONINO

Figure 11-22. *Owners of exotic or aggressive animals have a responsibility to keep the animals under control at all times.*

ACTIVITY 11-1 *Ch. Obj: 11.3, 11.4*
CASTINGS

Objectives:

By the end of this activity, you will be able to:

1. Make castings of teeth impressions.
2. Create transparency correlations of bite marks and teeth.

Materials:

(per student)
apple (2 quarters)
plastic wrap
dental stone or plaster
250-mL beaker
plastic bag (quart size)
water
plastic tray
transparency (10 cm square)
transparency marker

Safety Precautions:

Wear safety goggles and protective gloves.

Procedure:

Part 1

1. Bite firmly into an apple quarter, but don't bite all the way through. Make a second bite impression in the other apple quarter. Wrap one tightly in plastic wrap to prevent discoloring. Set aside.
2. Put 240 mL of plaster or dental stone into the plastic bag. Add 60 mL of water. Squeeze out most of the air and seal the bag. Squish the bag until the mixture is smooth. Be careful not to let the material set before pouring.
3. Snip one corner of the bag to use for pouring. Carefully pour the casting material into the teeth indentations. Pour casting material into the plastic tray to a depth of one to two centimeters. Push the teeth casting gently onto the surface of the casting material in the tray and let it sit until hardened.
4. Gently separate the casting from the apple.

Part 2

1. Place a piece of transparency film over the casting of the teeth.
2. Using a black transparency marker, carefully outline the teeth.
3. Unwrap the reserved apple quarter.
4. Compare the bite mark in the apple to the transparency you made of your teeth. Note similarities and differences.

Part 3

1. Sort the castings into two groups—male and female.
2. Look at the incisors. Make a note of any differences between the sexes.
3. Note any differences in size of the teeth.

Questions:

1. How closely did the casting and the transparency match the bite mark?
2. If bite marks were to be found in an item of food, what possible explanation could there be for differences in the bite mark and a casting of a suspect's teeth?
3. If there are differences, what other methods could be used to determine that the bite mark was indeed made by the suspect?
4. What, if any, noticeable differences did you see between the bite marks of males and females? Explain.

Extension:

Design an experiment to determine which of three or four materials (such as athletic mouth guards, wooden pencils, and hard cheese) produce bite marks that would have the highest forensic value. Be sure to include the scientific questions you hope to answer, a testable hypothesis, and a list of materials and procedures. Perform your experiment, recording qualitative and quantitative observations. Draw a conclusion and cite evidence from your experiment to support your conclusion.

ACTIVITY 11-2 *Ch. Obj: 11.3, 11.4*
BITE MARK COMPARISONS

Objectives:

By the end of this activity, you will be able to:

1. Properly examine and record bite mark information.
2. Compare bite marks and castings to dental records in order to identify the suspect.

Materials:

(per student)	*(per group)*
small foam plate	bite mark reports (from suspects)
scissors	casting from victim
pencil or pen	metric ruler
paper towels	
plastic bag	
foam cup	
permanent marker	

Safety Precautions:

Protective gloves must be worn when handling dental impressions and castings.

Wash hands thoroughly at the end of each class period.

Part 1: Day 1

Scenario: When leaving the movie theater, you and your friends discover the unconscious victim of a brutal attack. The man has been beaten and bitten. After the initial investigation, police discover the bite marks have left bruises on the victim, so the decision is made to consult with a forensic odontologist. Because the man is not able to identify his attacker(s), the odontologist decides to get bite impressions from everyone who was at the theater.

Procedure:

1. Cut the center from the foam plate. Discard the outer edges. Cut the center in half to make two half circles.
2. Stack the two halves and bite down firmly—but not all the way through the foam.
3. Mark the impressions as top and bottom.
4. Place on paper towels to dry. When dry, place the foam inside the plastic bag and label the bag with your name.
5. Put your hand inside the cup and bite firmly (but not all the way through the cup) along the side—not the rim. This will simulate a bite mark to an arm or leg. Write your name on the bottom of the cup and give the cup to your teacher.

Part 2: Day 1

Scenario: You now are working in the lab for the forensic odontologist, trying to get preliminary results of the bite plates. Wear gloves when handling the impressions.

Procedure:

1. From your own impressions, use the metric ruler to make the following measurements: (Refer to Figure 1. Measure from the center of each tooth.)
   a. second molar on the left side to the second molar on the right—top
   b. second molar on the left side to the first premolar on the right—top
   c. second molar on the left side to the central incisor on the right—top
   d. right canine to left canine—top
   e. second molar on the left side to the second molar on the right—bottom
   f. second molar on the left side to the first premolar on the right—bottom
   g. second molar on the left side to the central incisor on the right—bottom
   h. right canine to left canine—bottom
2. Record your results.
3. Make sure to note any interesting characteristics of the teeth. Some characteristics to look for are arch width (from canine to canine), shape of dental arch (C-shape, oval, U-shape), tooth alignment, rotational position (twisted or straight), spacing, width of tooth, thickness of tooth, curvatures of biting edges of the teeth, and wear patterns.
4. Make the same measurements using the impressions of three of your classmates.
5. Record the names and measurements.
6. Compare your measurements with the others in your group. If there is a considerable difference, recheck the measurements.

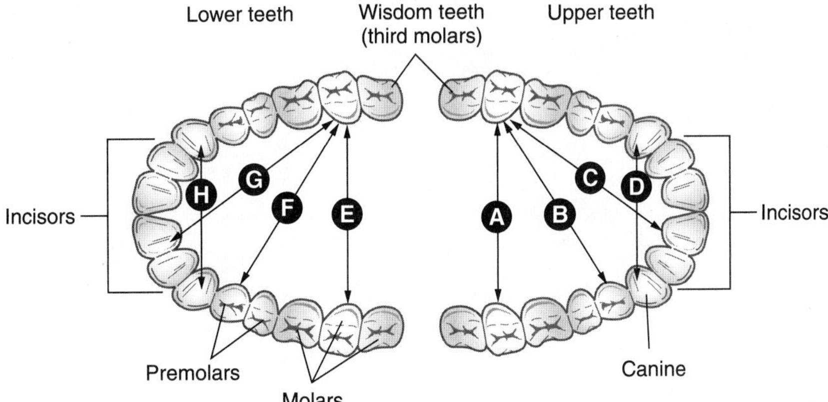

Figure 1. *Full set of adult human teeth.*

Part 3: Day 2

Scenario: Your team has been given a casting of a bite mark from the victim. Your job is to determine who is responsible for the attack by comparing the bite mark casting with the bite impressions taken from the suspects.

Procedure:

1. Measure the teeth (as you did in Part 2) on the casting. Record your results in a second data table.
2. After comparing the reports, draw a conclusion about the identity of the perpetrator.
3. Once you have made the determination, compare the casting with the actual foam impression.

Questions:

1. Who do you suspect the attacker to be? Explain your answer.
2. What unique characteristics helped you identify the perpetrator? Explain.
3. You are called in as an expert witness at the trial of the suspect. How would you defend your conclusion?

Extension:

WRITING  Research emerging techniques and technologies in forensic odontology. Be sure to gather information from primary and secondary sources, relying on the most valid and scientifically accurate sources. Write a one- to three-page paper describing one of the techniques or technologies in detail, including its potential effect on society.

ACTIVITY 11-3 *Ch. Obj: 11.3, 11.5*
COMPARISON OF ANIMAL TEETH

Objectives:

By the end of this activity, you will be able to:

1. Identify different types of teeth.
2. Determine an animal's diet based on its teeth.

Materials:

(per group)
various animal teeth
skulls from several animals
metric ruler
hand lens
photographs of dentition patterns of various animals
reference materials or computers

Safety Precautions:

Protective gloves must be worn when handling teeth, castings, and bones.

Procedure:

1. Examine each of the specimens carefully.
2. Construct a data table for the observations and measurements.
3. Record your observations on the data table. Describe the shape of the tooth (if a single tooth) or the shape of the tooth pattern (if teeth are embedded in the jaw bone).
4. Measure the relative size of the teeth. Record your data.

Questions:

1. Based on the shape of each animal's teeth, formulate a hypothesis about what types of food the animal eats.
2. If an animal jaw has a variety of teeth, what might you conclude about the animal's diet?
3. Try to identify the animal from the size and shape of the teeth and/or jaw. Use reference materials provided by your teacher to help you.
4. Based on the information gathered in the lab and from the reference materials, what other types of animals would you expect to have similar teeth and diets?
5. Why is it important for forensic odontologists to have knowledge of animal teeth?

Extension:

Research other types of animals, their teeth, the shape of the jaws, and the foods they eat. Present your findings using a multimedia format, poster, or brochure.

CHAPTER 12

Forensic Entomology

BUGS DON'T LIE

On July 9, 1997, Kevin Neal was at home in Champaign County, Ohio, with his stepchildren, Cody McGraw (4 years) and India Smith (11 years). Their mother (and his wife), Sue, was at work. According to Neal, the children went out to play in the yard around noon. At 1:30, he became concerned because he no longer heard them playing. After searching for the children for half an hour, he called 911.

Neal told the police dispatcher that the children were missing and asked the dispatcher to call his wife. He said she might be less upset if the news came from them rather than him because the couple was planning to separate and had argued that morning.

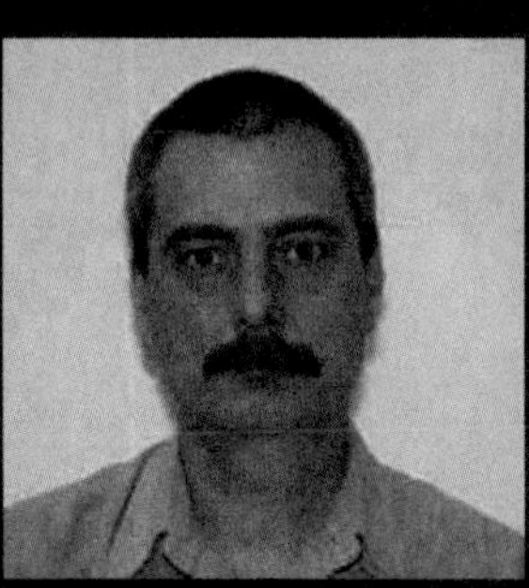

Figure 12-1. Kevin Neal

When Mrs. Neal received the news, she rushed home. She was hysterical and began to shout accusations at her husband and had to be restrained by deputies. An extensive search for the children using volunteers, dogs, and local law-enforcement officers lasted four days. On July 13, the unsuccessful search was called off.

Investigators collected evidence, much of it pointing to Neal's involvement. However, prosecutors agreed there was not enough evidence to arrest him. Questions included inconsistencies in the timeline of July 9, a lack of witnesses to corroborate Neal's version of events, and the lack of toys in the yard where the children had supposedly been playing. Neal also told the police that his car did not have brakes and had not been driven for more than a month. However, investigators found the brakes had been temporarily fixed with vise grips. Results of polygraph tests for both Kevin and Sue Neal were inconclusive.

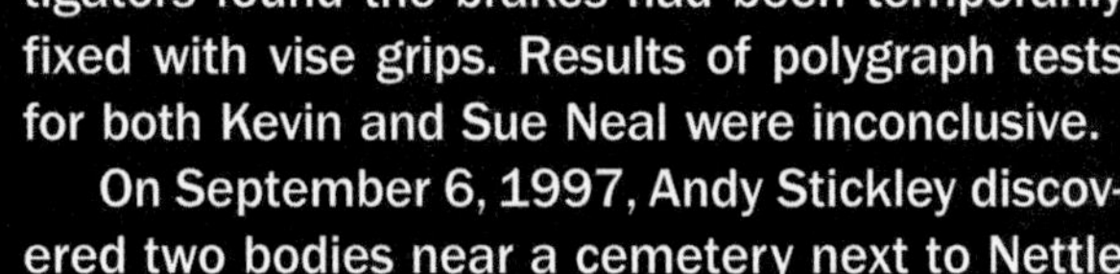

On September 6, 1997, Andy Stickley discovered two bodies near a cemetery next to Nettle Creek Farm. At first, he thought the remains were from a decomposing deer, but as he got closer he saw two small human skulls and called the police. After the bodies were identified as Cody and India, Kevin Neal was arrested and charged with two counts of murder.

Although the bodies were in the late stages of decomposition and had been exposed to the elements throughout the summer, investigators collected important evidence. The autopsies revealed maggot infestation in the abdomens. Blowflies and screwworms were absent. Screwworms repopulate Ohio every summer. The prosecution suggested that they begin to arrive in mid-July every year, suggesting that the children had died before the middle of July. Additionally, the cheese skipper fly was in its third instar stage. The cheese skipper fly arrives at a decomposing body several weeks or months after death. Using these clues, the prosecution argued that the time of death was between July 9 and July 14.

With extensive help from a forensic entomologist, Kevin Neal was convicted of the murders of Cody McGraw and India Smith. He was sentenced to life in prison.

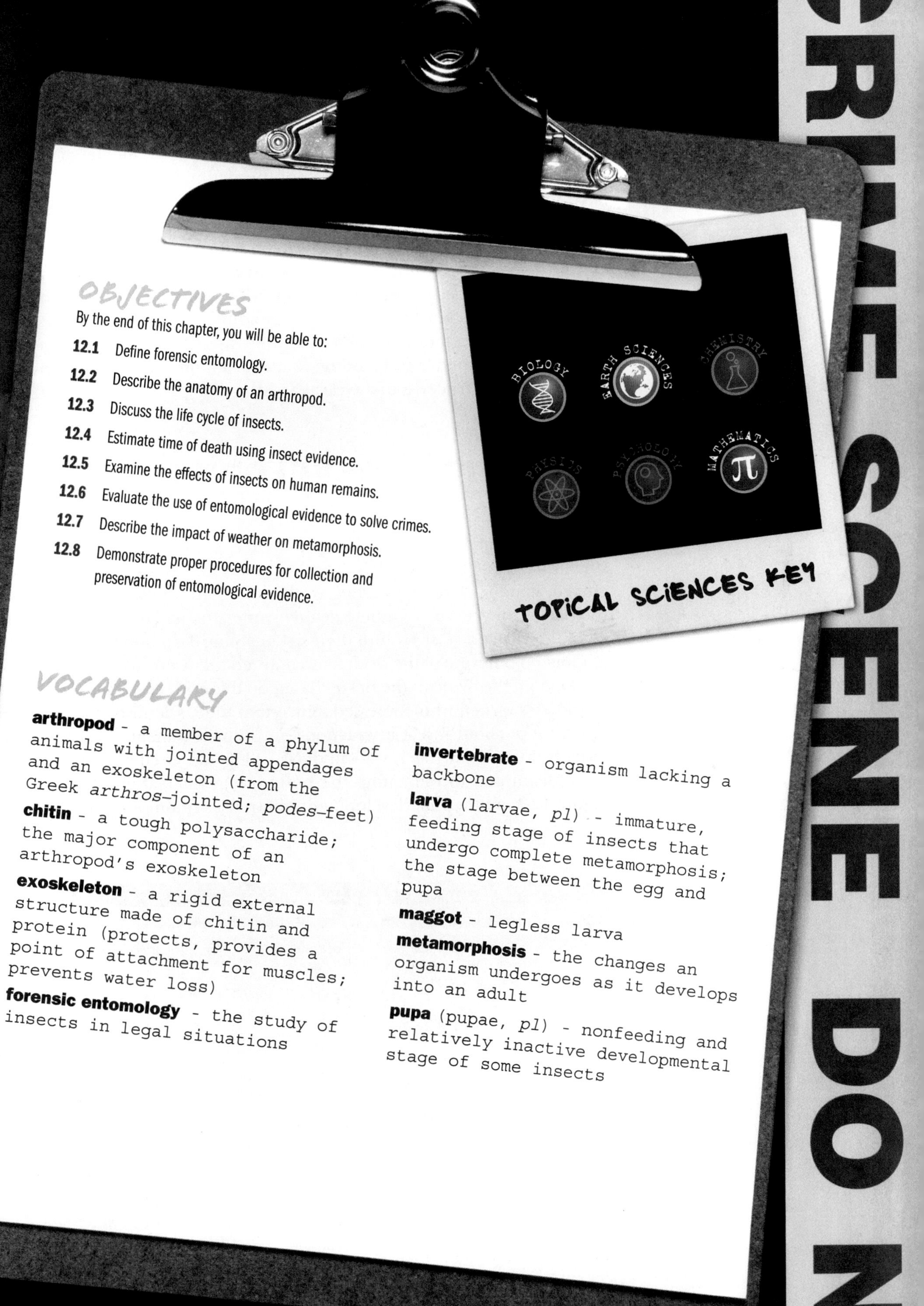

OBJECTIVES

By the end of this chapter, you will be able to:

12.1 Define forensic entomology.

12.2 Describe the anatomy of an arthropod.

12.3 Discuss the life cycle of insects.

12.4 Estimate time of death using insect evidence.

12.5 Examine the effects of insects on human remains.

12.6 Evaluate the use of entomological evidence to solve crimes.

12.7 Describe the impact of weather on metamorphosis.

12.8 Demonstrate proper procedures for collection and preservation of entomological evidence.

VOCABULARY

arthropod - a member of a phylum of animals with jointed appendages and an exoskeleton (from the Greek *arthros*–jointed; *podes*–feet)

chitin - a tough polysaccharide; the major component of an arthropod's exoskeleton

exoskeleton - a rigid external structure made of chitin and protein (protects, provides a point of attachment for muscles; prevents water loss)

forensic entomology - the study of insects in legal situations

invertebrate - organism lacking a backbone

larva (larvae, *pl*) - immature, feeding stage of insects that undergo complete metamorphosis; the stage between the egg and pupa

maggot - legless larva

metamorphosis - the changes an organism undergoes as it develops into an adult

pupa (pupae, *pl*) - nonfeeding and relatively inactive developmental stage of some insects

Obj. 12.1, 12.2

INTRODUCTION

When investigators enter a crime scene that involves a death, they discover various clues left behind by the perpetrator and the victim. These clues assist the investigators in establishing a timeline of the events that led to the death of the victim. One clue is the presence or absence of insects. The use of insect evidence in legal investigations is called **forensic entomology**.

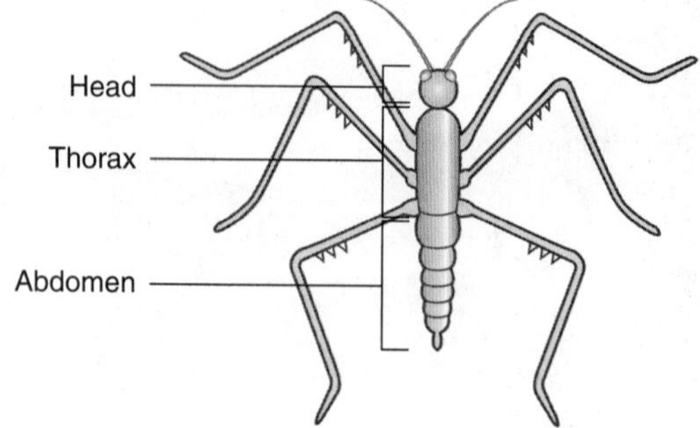

Figure 12-2. *Diagram of an arthropod.*

Insects belong to the phylum Arthropoda, which includes but is not limited to spiders, scorpions, crayfish, and millipedes. An **arthropod** is an **invertebrate**, an organism lacking a backbone. All adult arthropods have externally segmented bodies, jointed appendages (legs), and hardened **exoskeletons** made of **chitin**. Figure 12-2 shows the structure of a typical arthropod.

HISTORICAL DEVELOPMENT

The first documented use of insect evidence to solve a crime occurred in a small village in China. In 1235 a local man was slashed to death by a sickle (a curved, handheld tool for cutting vegetation). After questioning everyone in the village, Sung Tz'u, a Chinese death investigator, had no suspects. He asked all of the villagers to show him their sickles. Sung Tz'u noticed that flies were landing on only one of the sickles and realized they were attracted by the odor of blood on the blade of the sickle. Based on this evidence, Sung Tz'u accused the sickle's owner, who confessed. Since that time, scientists have discovered much more about insect interaction with human remains. Today, analyzing insect evidence can be key to understanding the circumstances surrounding a death and to estimating the postmortem interval and time of death. Figure 12-3 outlines the development of forensic entomology as a science.

Year	Name	Contribution
1235	Sung Tz'u	First to use insect evidence to solve a crime
1668	Francesco Redi	First to prove maggots arise from eggs laid by flies, discrediting the theory that maggots grew out of rotting meat (abiogenesis)
1855	Begeret d'Arbois	First to use entomology to estimate postmortem interval
1881	Hermann Reinhard	First to systematically study exhumed bodies and the insects associated with them
1894	Jean Pierre Megnin	First to identify eight stages of human decomposition
1960s	Jerry Payne	Reduced the number of stages of decomposition to six
1984		Accepted as a professional field of study
1996		Routinely used in legal investigations, especially those involving death

Figure 12-3. *Historical landmarks in forensic entomology.*

INSECTS AND FORENSICS

Obj. 12.2

The bodies of insects are divided into three main segments: head, thorax, and abdomen. Each main segment contains additional anatomical parts that allow the insect to function. The head consists of the brain, the antennae, eyes, and mouthparts. The thorax contains three pairs of legs. If the species also has wings, the thorax contains two pairs of wings in addition to the legs. The abdomen is the center of movement and contains reproductive organs and the digestive system. The self-moisturizing exoskeleton provides the rigid external structure. Figure 12-4 shows a typical insect with the body parts labeled.

There are 29 orders, or related groups, of insects. Two orders are very important to forensic investigators—flies and beetles. Flies belong to the order Diptera and live in almost every environment. Investigators often find

Did You Know?

A component of the exoskeleton of the lac beetle is a source of shellac, the hard coating used on furniture.

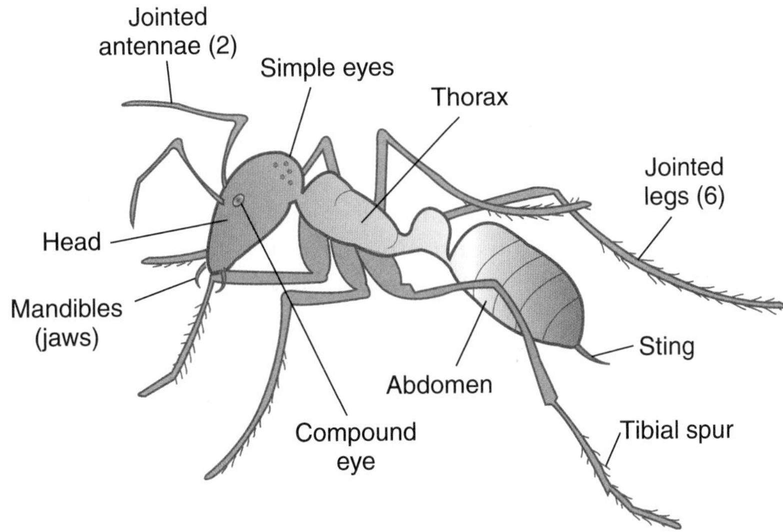

Figure 12-4. *A typical insect with body parts labeled.*

Digging Deeper

with Forensic Science e-Collection

One crucial function of forensic entomologists is estimating the time of death based on insect evidence. However, even with the availability of supporting evidence such as weather reports, crime-scene sketches, and additional insect evidence, entomologists are unable to determine an exact time of death. How do you think the uncertainty of this finding of entomological evidence might affect a jury's verdict? Using the Gale Forensic Science e-Collection at www.cengage.com/school/forensicscienceadv, research forensic entomology and the various pieces of entomological evidence used to determine time of death. Write a brief summary of your findings, making sure to cite your sources.

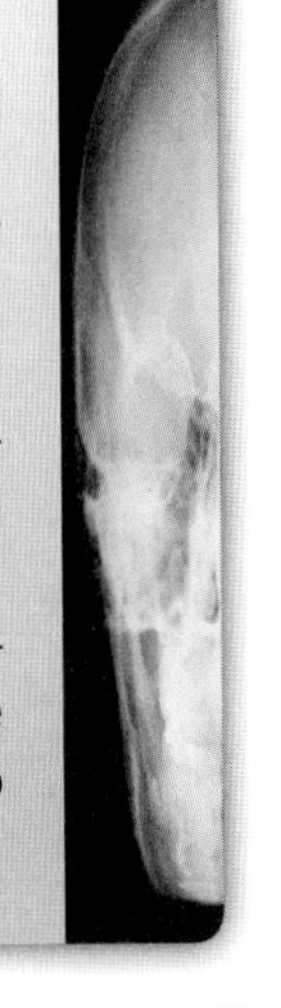

Order	Family	Common Name	Description
Diptera	*Calliphoridae*	Blowfly/Bottle Fly/Screwworm Fly	6–14 mm long; metallic green, blue, bronze, or black bodies
	Sarcophagidae	Flesh Fly	2–14 mm long; gray and black striping on thorax, checkerboard abdomen
	Muscidae	Housefly/Stable Fly/Latrine Fly	3–10 mm long; dull gray or dark, a few are metallic-looking
Coleoptera	*Silphidae*	Carrion Beetle	10–35 mm long; wings tend to be short and leave body segments exposed, black body marked with patches of color
	Histeridae	Clown Beetle	less than 10 mm long; oval shiny black or metallic green
	Staphylinidae	Rove Beetle	1–25 mm long; slender body, very short wings
	Cleridae	Checkered Beetle	3–12 mm long; head wider than area where wings start, bodies covered with bristly hairs
	Dermestidae	Skin Beetle/Hide Beetle/ Carpet Beetle	2–12 mm long; rounded to oval in shape, covered in scales forming colorful or unique patterns

Figure 12-5. *The most important insects in forensic science.*

blowflies at the first stages of decomposition, laying their eggs in body openings. By the bloated stage of decomposition, the eggs have hatched into maggots. After the maggots move away from the body to pupate, beetles move in and continue the decay process.

Beetles (order Coleoptera) are the most prominent insects in the late stages of decomposition. During the final stage of decomposition, the skeletal stage, beetles are joined by other soil-dwelling insects. For this reason, a complete investigation might require the collection of soil samples from under the body. Figure 12-5 shows the most important insects in forensic entomology.

INSECTS AT WORK

Obj. 12.3, 12.5

The life cycle of many insects is standard, allowing scientists to use it as evidence. The typical life cycle has four stages: egg, larva, pupa, and adult. The maturation process—from egg to adult—is called **metamorphosis**. During a typical life cycle, eggs hatch into **maggots**, or **larvae.** Upon completing this legless immature feeding stage, the insect enters the **pupa** stage, a relatively inactive and nonfeeding developmental stage. The insect emerges from this stage as an adult. Figure 12-6 provides information about the metamorphosis of a typical insect. Using knowledge of the developmental stages, forensic entomologists can estimate the time of death and assist in solving crimes.

Abiotic (nonliving) and biotic (living) factors interact to maintain balance in an ecosystem. Abiotic factors include the sun, atmosphere, and weather; biotic factors consist of the organisms found

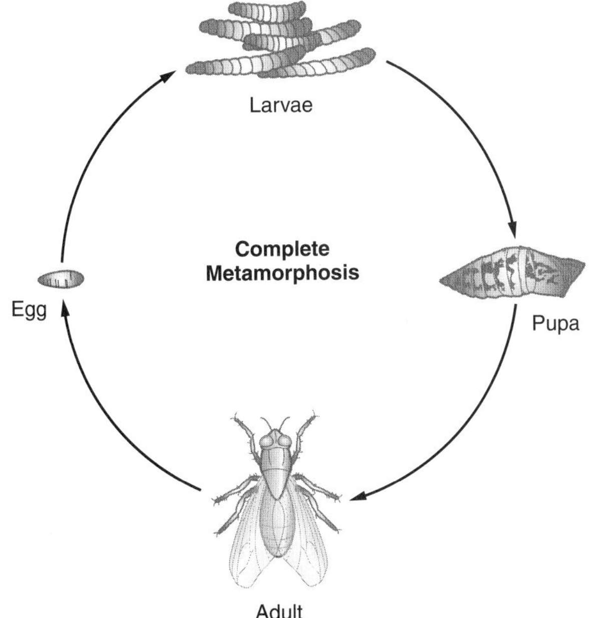

Figure 12-6. *The stages in the complete metamorphosis for a typical insect.*

Did You Know?

Some insects such as grasshoppers undergo incomplete metamorphosis.

in the ecosystem. There are more than a million species of insects on Earth. Most insects have a short lifespan and produce large numbers of eggs. Insects contribute to the ecosystem by loosening soil, which allows air to circulate, pollinating flowering plants, and consuming plant pests. Insects participate in the decomposition of dead bodies of all species. The interaction of insects on dead human bodies provides clues that assist the forensic entomologist in the death investigation. Figure 12-7 shows insects on a decomposing body.

Figure 12-7. *Maggots on the decomposing body of an animal in the wilderness.*

Digging Deeper
with Forensic Science e-Collection

Several highly trained professionals process a crime scene with a decomposing body. Forensic entomologists collect insect evidence, examining the stages of development to estimate the postmortem interval. In some instances, climatologists, meteorologists, odontologists, and other experts collaborate with the entomologists to interpret the evidence. Using multiple experts renders a more comprehensive evaluation of the crime scene. Using the Gale Forensic Science e-Collection at www.cengage.com/school/forensicscienceadv, research scenarios in which other experts were called in to assist an entomologist. What contributions might other experts provide in assisting the entomologist? Write a brief summary, citing your sources.

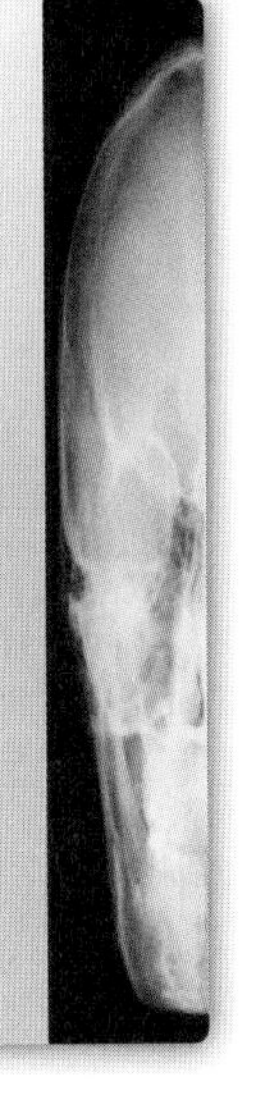

Obj. 12.4, 12.6

POSTMORTEM INTERVAL (PMI)

A key to determining what happened to a dead body is establishing a timeline of events. The postmortem interval is an important element of the timeline. For a body found more than 72 hours after death, forensic entomology can be essential to determining the postmortem interval. As you learned in Chapter 8, in the first 48 hours after death physiological changes such as algor mortis and rigor mortis occur. During that time, insect activity also takes place. Typically, the female blowfly (see Figure 12-8) arrives at the decomposing body first. She lays eggs in moist openings such as the nasal passages, mouth, eyes, ears, groin areas, and around wounds. Then the typical insect life cycle unfolds. Because the eggs hatch in the sequence they were laid, the uneven pattern allows investigators to estimate a timeline of decomposition and the postmortem interval (PMI).

STUDIOTOUCH/SHUTTERSTOCK

Figure 12-8. *A female blowfly.*

Obj. 12.6, 12.7, 12.8

COLLECTING AND PRESERVING ENTOMOLOGICAL EVIDENCE

Female blowflies do not lay eggs at night.

Investigators observe and record all pertinent factors at the death scene. This includes weather elements such as rainfall and temperature that can affect the life cycle of insects. Most crime scenes are processed without a trained forensic entomologist present. Instead, other team members collect the entomological evidence and the forensic entomologist works in the lab. The team uses care and caution when collecting evidence. They approach the body carefully to avoid disturbing insects near it. They also limit the access of other people to the area near the body until they finish the assessment; people moving about a crime scene threaten the integrity of the evidence. Without care, insect evidence could be destroyed or disturbed.

Investigators begin assessing insect evidence several feet from the body. As they move closer to the corpse, they determine the types of insects present and the stages of development. Next, an investigator measures the distances from the body to various insects, assesses current weather data for the scene, and collects samples of all insect stages on, near, and under the body. At the forensics laboratory, forensic entomologists raise samples of all insects collected in conditions similar to the conditions in which they were found at the scene.

Investigators also identify weather data for the scene and check those factors for the days before the discovery of the body. The temperature, humidity, type and amount of precipitation, type and amount of light, and amount and direction of wind impact how quickly insects progress through their life cycles. All of these data can be essential to the best possible estimate of the PMI.

The entomologist needs to know the status of windows or doors—open or closed—and other access points for insects. Both weather and access can affect insect activity and impact calculations of the postmortem interval. The forensic entomologist or another team member completes a forensic entomology data form (see Figure 12-9) as he or she conducts the investigation. The analysis of all factors allows investigators to recreate the

FORENSIC ENTOMOLOGY DATA FORM

DATE: ______________ **CASE NUMBER:** ______________________________

COUNTY/STATE: ____________________ **AGENCY:** ____________________

DECEDENT: ______________________________ **AGE:** ______ **SEX:** ______

Last Seen Alive: ____________________ **Date and Time Found:** ____________

Date Reported Missing: ________________ **Time Removed from Scene:** ________

Site Description:

____

____

____

Death Scene Area:
Rural: forest_______ field_______ pasture_______ brush_______ roadside_____
barren area_______ closed building_______ open building_______
other ____

Urban/suburban: closed building_______ open building_______
vacant lot_______ pavement_______ trash container_______
other ____

Aquatic habitat: pond_______ lake_______ creek_______ small river_______
large river_______ irrigation canal_______ ditch_______ gulf_______
swampy area_______ drainage ditch _______ salt water_______
fresh water_______ brackish water_______
other ____

Exposure: Open air_______ burial/depth_______
Clothing: entire _______ partial _______ nude _______
portion of body clothed ____
description of clothing ____
type of debris on body ____

Stage of decomposition: fresh_______ bloat_______ active decay_______
advanced decay_______ skeletonization _______ saponification__________
mummification_______ dismemberment_______
other: ____

Evidence of scavengers:

____

Possible traumatic injury sites: (Comment or draw below)

____

____

Scene temperatures: ambient:_______ ambient (1ft) _______ body surface_______
ground surface_______ under-body interface_______ maggot mass________
water temp, if aquatic_______ enclosed structure_______ AC/Heat- on/off_______
ceiling fan- on/off_______ soil temperature- 10 cm_______ 20 cm_______ Number of
preserved samples ________________ Number of live samples ______________
NOTE: Record all temperatures periodically each day at the site for 3-5 days after body recovery.

Figure 12-9. A sample forensic entomology data form.

Paint, acids, and other chemicals can delay colonization and are often used in an attempt to interfere with the natural process of decomposition.

crime-scene timeline and perhaps identify some of the activities related to the death. The data form provides space to document the following information:

1. Estimates on the types and number of insects
2. Key insect colonizations on and near the body
3. Exact positioning of the body using fixed points, if possible: placement of the arms and legs, the position of the head, and parts of the body in the shade and in the sunlight
4. Locations of eggs, larvae, and pupae on the body
5. List of opportunistic feeders and other insects (such rove beetles, ants, wasps, and insect parasites)
6. Insect movements (within 20 feet of the body)
7. Natural and artificial influences that might affect the speed of decomposition, such as burning, covering, or burying the body

The complete documentation of a crime scene also includes sketches, photographs, or videotapes. This supplementary evidence might further contribute to piecing together what happened.

FACTORS AFFECTING INSECT ACTIVITY

Many factors affect insect activity. Entomologists categorize death scene habitats based on four broad elements of the environment and light. Each description identifies factors that impact the rate of insect metamorphosis. The four habitats are:

Natural environment and natural light: The body is found in a natural environment (e.g., the woods) and in natural light (sunlight). There are no artificial influences such as air-conditioning associated with this environment.

Natural environment and artificial light: The body is found in a natural environment, but the light is from an artificial source, such as incandescent light bulbs.

Artificial environment and natural light: The body is found in an unnatural environment, such as inside with the air-conditioning on, and in natural light, such as sunlight (or moonlight).

Artificial environment and artificial light: The body is found in an unnatural environment (inside with air-conditioning) near an artificial light source (or no light). The body is not affected by natural light.

The types of insects that decompose a body are basically the same regardless of the habitat. The one exception is extreme cold; flies do not colonize bodies in frigid settings, delaying the decomposition process. In other settings, the rate of insect metamorphosis varies with the conditions of the habitat in which the body is decomposing. As a general rule the higher the temperature, the faster the rate of insect metamorphosis; the lower the temperature, the slower the rate. For example, insect metamorphosis in a corpse outside in Texas summer heat occurs faster than in a body inside with the air-conditioner on with no direct sunlight. Entomologists use their knowledge of local weather and climate data and collaborate with a meteorologist or climatologist as they complete their analysis (see Figure 12-10).

Figure 12-10. *Each species of insect progresses through metamorphosis at a certain time of year, in response to such variables as temperature and light.*

Because temperature has a major effect on metamorphosis, processing a crime scene requires technicians to take a number of temperature readings, including temperature of the ambient air (1 foot and 4 feet above the body) and temperature of the soil (on the surface, 10 centimeters, and 20 centimeters below the surface). When a large infestation of larvae is present, the investigator also takes the temperature at the center of the infestation. All of the readings are carefully recorded on the standard data form.

Did You Know?

Maggot masses can produce enough heat to raise the temperature of the body by several degrees.

MORE ON COLLECTING AND PRESERVING EVIDENCE

Because insects are living and moving, collecting insect evidence presents specific challenges. Investigators must use special techniques and materials. For example, entomologists collect mature flies with insect nets using several back and forth strokes in a figure-eight pattern. Once collected, the flies are placed in a jar at the end of the net. The investigator then uses either plaster of Paris or cotton balls soaked with ethyl acetate to kill the flies, which takes about five minutes. The insects can then be pinned or transferred to vials to be preserved with 80 percent alcohol, 20 percent water solution. (Isopropyl and ethyl alcohol both work well as preservatives.) Eggs are preserved in a similar solution. When available, 30 to 60 of the largest larvae also are preserved in the solution. The vials are immediately labeled, in pencil (ink washes off more easily), with the case number, collection time, date, geographic location, the location of the insects on the victim, and the initials of the investigator. As a precaution, it is customary to make two labels: one for inside the vial and one for the outside. Investigators follow this and similar redundant procedures to reduce the risk of compromising the integrity of the evidence. Of course, the procedures investigators use to collect and process the evidence might be modified to accommodate the requirements of the local laboratory in which the entomologist works.

CHAPTER SUMMARY

- Forensic entomology is the analysis of insect evidence in legal cases.
- Insects belong to the phylum Arthropoda, a group of organisms with jointed appendages, segmented bodies, and an exoskeleton.
- Since the first documented use of insects to solve a homicide case in China in AD 1235, scientists have added to the knowledge of insect behavior and life cycle.
- Insect evidence is valuable in estimating the postmortem interval.
- There are more than 700,000 known species of insects, but only a few have forensic value.
- Female blowflies are usually the first insects to colonize a dead body.
- The extent of decomposition—especially if it is unevenly distributed on the body—provides information for estimating the time (and perhaps manner) of death.
- Temperature, humidity, and other weather and climate elements affect the insects that colonize remains as well as the progression of the insect life cycle. For this reason, entomologists record and analyze climate and weather information related to the insect samples.
- Processing and preserving insect evidence requires special care, techniques, and materials.

CASE STUDIES

Lori Ann Auker (1989)

On June 12, 1989, a young woman walking near her grandparents' house noticed a foul odor and found a decomposing body wearing a jacket, jeans, and sneakers. She hurried home and called the police. Authorities used dental records to identify the body as Lori Ann Auker.

The pathologist, Dr. Isidore Mihalakis, ruled the death as a homicide caused by multiple stab wounds to the back and chest. A forensic entomologist was consulted to estimate the approximate time of death. The entomologist testified that he was able to estimate that the death occurred 19 to 25 days before the body was discovered. He told the court he used the climate report for the local area, the autopsy report, and the description of the scene to evaluate the stages of development of the insects on and in the body and the stage of decomposition based on the average temperature and weather conditions during the period of time the victim had been missing.

The only suspect in the case was the victim's ex-husband, Robert Donald Auker. Using the substantial entomological evidence, cat and human hair found on the body and in Robert Auker's car, and video footage from an automated teller machine (ATM) from May 24, the last day Lori Auker was seen alive, Robert Auker was arrested, eventually convicted, and sentenced to death.

In this case, entomological evidence was integral in establishing the time of death. The estimated time of death helped investigators link Robert Donald Auker to the crime, leading to his conviction.

Think Critically **1. Lori Ann Auker was killed in May in Pennsylvania. Her body was found about two weeks later. How would the insect evidence have been different had she been killed in January and her body found about two weeks after her death?**

2. How would the rate of decomposition and the type of insect evidence have changed had Robert Auker sealed Lori Ann's body in the trunk of the car?

3. Hairs that were consistent with Lori Ann's and with her pet cat's were found in Robert Auker's car. How might the defense explain this?

Sylvia Hunt (1986)

On September 21, 1986, a woman's decomposing body was discovered near Interstate Highway 95 (I-95) near Greenwich, Connecticut. It was wrapped in a foul-smelling carpet. Authorities estimated the woman, who was identified as Sylvia Hunt, had been dead five to seven days. The autopsy revealed Hunt had been stabbed 15 times. When her body was discovered, larvae from blowflies were feeding on the open areas of the body. Entomologists scoured the victim and the area around the body and collected more than 4,000 larvae to study and process in the laboratory.

Forensic entomologist William Krinsky analyzed the insect evidence collected on September 21 and 22; climatological data from Bridgeport, Connecticut, from September 1 to 23; and the maggot infestation on the victim's body. Eighty specimens of blowflies in the pupal stage were collected from the carpet on September 22.

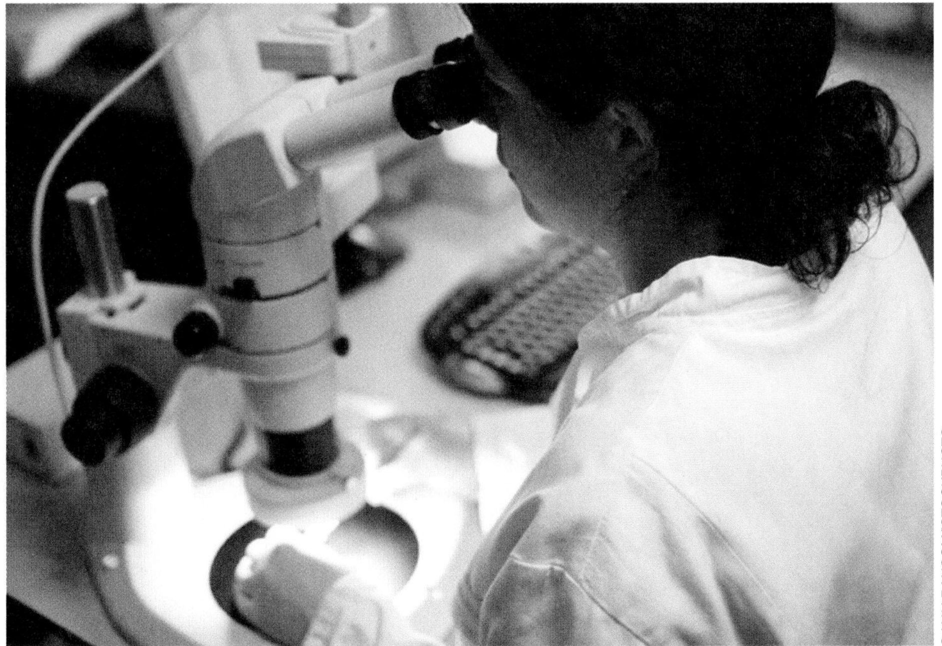

Figure 12-11. *To help determine who killed Sylvia Hunt, forensic scientists studied insect evidence and carpet fibers.*

CHAPTER 1 2 3 4 5 6 7 8 9 10 11 12 13 14 15

The samples were placed in a temperature-controlled chamber. Krinsky identified the adult flies that emerged on September 25 as black blowflies, *Phormia regina*. Krinsky determined that the eggs had probably been deposited on the body on the morning of September 15. Based on this estimate, he reasoned the body had probably been placed on the side of I-95 on the night of September 14. It should be noted that Krinsky estimated when the body had been moved to the interstate but not the time of death.

One suspect emerged. Yuri Hernandez, 29, told a friend a month before the murder that he planned to kill women because he was disillusioned with them. Investigators collected information and evidence, eventually matching the pattern of the carpet the body was wrapped in to the carpet in a room in Hernandez's house. Based on the combination of the carpet and entomological evidence, Hernandez was convicted of first-degree murder. Prosecutors learned he had killed Hunt in his apartment and kept her body between a mattress and box spring for several days before dumping her alongside the interstate.

Think Critically

1. **Why did entomological evidence help investigators estimate when Sylvia Hunt's body was deposited by the side of the highway but not the time of death?**
2. **Other than the carpet, what evidence might investigators have found to suggest that Hernandez had killed Hunt in his apartment?**

Bibliography

Books and Journals

Byrd, Jason H., and James L. Castner, eds. (2010). *Forensic Entomology: The Utility of Arthropods in Legal Investigations*. Boca Raton, FL: CRC Press.

Greenburg, Bernard, and John C. Kunich. (2002). *Entomology and the Law: Flies as Forensic Indicators*. Cambridge, NY: University Press.

Websites

www.bijlmakers.com/entomology/classification/Insect_classification.htm
www.cengage.com/school/forensicscienceadv
www.earthlife.net/insects/anatomy.html
www.forensicentomologist.org
www.insectidentification.org/anatomy.asp
www.riverdeep.net/current/2002/03/030402t_insects.jhtml

CAREERS IN FORENSICS

Science, Technology, Engineering & Mathematics

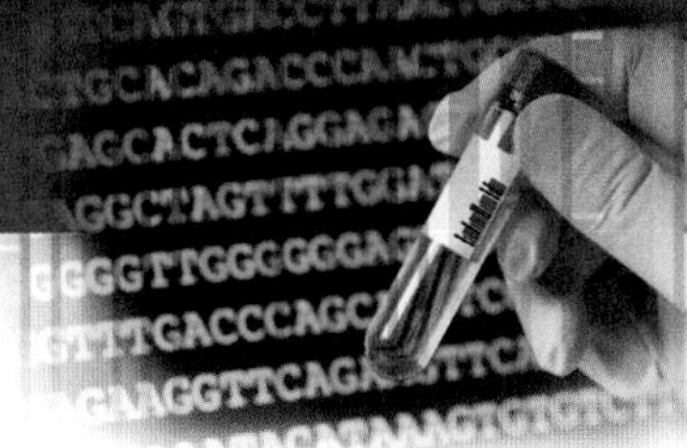

Carlton-Jane Beck, M.S.: Forensic Entomologist

Carlton-Jane Beck (see Figure 12-12) is a forensic entomologist for the Lake County Sheriff's Office Criminal Investigations Unit in Lake County, Florida. She has worked for the department for more than four years. Beck graduated from the University of Florida with a B.S. degree in entomology and an M.S. degree in forensic toxicology.

Figure 12-12. *Carlton-Jane Beck*

According to Beck, most entomologists work for either a government agency or an educational institution. Beck is involved with both in her career. She investigates crime scenes, consults on cases that have insect evidence, and serves as a guest speaker for schools in her community.

Beck always knew she wanted a profession that combined the skills of a doctor with the skills of a police officer, and she found forensics to be a perfect fit. She has worked in the Lake County Medical Examiner's office processing the "strictly buggy cases," and now she is a full-time crime-scene investigator. Although entomology is not her main responsibility, Florida weather means year-round insect activity that keeps her busy.

Beck is often called to a death scene to estimate the postmortem interval using insect clues. In one recent case, she found that flies had colonized a body shortly after sunrise, so she estimated the victim had probably died the previous evening. After a lengthy investigation, the suspects confessed that the crime had been committed between 11:00 P.M. and 1:00 A.M. the night before the body was found. "Not perfect, but the bugs told me what I needed to know," said Beck. "One of the most rewarding components of my job is providing closure to the family and friends of the victims."

Forensic entomologists may serve as expert witnesses and thus testify regarding insect evidence crucial to a case. Beck has been deposed (testified on the record during the evidence-collection stage of a case before trial) many times, but has not testified in court yet. She says of those cases, "I guess the bugs put the nail in the coffin and court was no longer necessary. They (the defendants) always seem to plead out (admit their guilt without going to trial)."

Learn More About It

To learn more about forensic science careers, go to www.cengage.com/school/forensicscienceadv.

CHAPTER 12 REVIEW

Fill-In-the-Blank

1. The study of insects in legal cases is called ______________. *Obj. 12.1*
2. Insects belong to the phylum ______________ and the class ______________. *Obj. 12.2*
3. ______________ are generally the first insects to colonize a body. *Obj. 12.3*
4. Maggots are the ______________ stage of flies. *Obj. 12.3*
5. Beetles belong to the order ______________. *Obj. 12.2*
6. The ______________ is self-moisturizing and keeps the insect from drying out. *Obj. 12.2*
7. Entomologists often collaborate with ______________ to analyze weather data. *Obj. 12.1*
8. The reproductive organs of an insect are found in the ______________. *Obj. 12.2*
9. Insects search for ______________ areas on the body to lay eggs. *Obj. 12.3*
10. ______________ and ______________ are additional types of cases that rely on entomological evidence. *Obj. 12.6*

Matching

11. polysaccharide and protein component of an exoskeleton *Obj. 12.2*
12. feeding stage of insects *Obj. 12.3*
13. time between death and discovery of the body *Obj. 12.4*
14. inactive developmental stage of some insects *Obj. 12.3*
15. animals with jointed appendages, segmented bodies, and skeletons *Obj. 12.2*

a. larva
b. arthropod
c. pupa
d. chitin
e. postmortem interval

Multiple Choice

16. Which of the following is *not* a basic body part of most insects? *Obj. 12.2*

    a. head
    b. thorax
    c. cephalothorax
    d. abdomen

17. Insects help forensic investigators determine _____. *Obj. 12.4*

    a. time of death
    b. whether the body sustained trauma
    c. whether the body has been moved
    d. all of these

18. The technique of _____ is most likely to be used when collecting adult insect samples. *Obj. 12.8*

    a. euthanizing
    b. swatting
    c. netting
    d. tabbing

19. Which of the following organisms is *not* a member of the phylum Arthropoda? *Obj. 12.2*

    a. tarantula
    b. butterfly
    c. sea star
    d. blowfly

20. Weather (temperature and humidity) affects _____. *Obj. 12.7*

    a. the types of insects that colonize remains
    b. the amount of time an insect takes to develop from egg to adult
    c. the size of the adult insect
    d. all of the above

21. _____ factors include sun, atmosphere, and weather. *Obj. 12.7*

    a. Abiotic
    b. Anaerobic
    c. Biotic
    d. all of the above

22. Currently there are more than _____ known species of insects. *Obj. 12.2*

    a. 300,000
    b. 500,000
    c. 700,000
    d. 1,000,000

23. The field of forensic entomology has been recognized as a profession since _____. *Obj. 12.1*

    a. 1974
    b. 1984
    c. 1994
    d. 2004

24. Which insect is the most prominent insect found in the post-decay stage of decomposition? *Obj. 12.6*

    a. beetles
    b. blowflies
    c. bottle flies
    d. all of these

25. Various species of beetles belong to the order _____. *Obj. 12.2*

    a. Chilopoda
    b. Diptera
    c. Coleoptera
    d. Diplopoda

Short Answer

26. Briefly describe the process of collecting and preserving entomological evidence. *Obj. 12.8*

27. Describe complete metamorphosis. *Obj. 12.3*

28. What factors determine how quickly an insect will complete its life cycle? *Obj. 12.3*

Think Critically

29. Explain why two labels are used to identify insect evidence. Why is it important to write the labels in pencil? *Obj. 12.8*

30. Why are only soil-dwelling insects found on a body in the skeletal stage of decomposition? *Obj. 12.5*

CRIME SCENE  Student-Prepared Original Titles

Model Down
By: Ashley Longo

Tavares High School,
Tavares, Florida

Sirens were blaring. A new fashion designer, Jon Le'Muzy, was standing at the entrance to one of the tents of New York's Fashion Week (see Figure 12-13). Le'Muzy had just signed top model Lexi Dillon for the show, only to find her dead on his runway the day before the show was to open.

Detective Jenna Meadows arrived with her partner, Victor Shells. They met with Le'Muzy, a small, overwhelmed French man. Detectives Meadows and Shells took Le'Muzy backstage to the privacy of the makeup area and asked the designer what he knew. Le'Muzy took a deep breath and began.

Figure 12-13. *Fashion week in New York City.*

LEV RADIN/SHUTTERSTOCK.COM

He had wanted to make some last-minute changes to his show and had gone to his runway. When he got there, he saw Lexi Dillon lying on the runway under the balcony. He rushed over and checked her pulse, but she was already dead. He called the police, and the homicide department was notified. Meadows and Shells then asked Le'Muzy why Dillon would be on the runway. Le'Muzy said he did not know, but Dillon could have been doing anything. She was Lexi Dillon. No one would question her.

Detective Meadows went to check the body, while Detective Shells talked with another model and Dillon's close friend, Riley Morgans. Security reported seeing Morgans near the locked balcony earlier in the day. As the detectives walked away, Le'Muzy suddenly shouted the name "Francis Gultersi." Gultersi, another designer and Le'Muzy's main competitor, had made Dillon a star, and she had been his top model until Le'Muzy lured her away. Le'Muzy had seen him earlier, but the two shows were being presented back-to-back in the tent, so he had a reason to be around the area.

Detective Meadows found the body of 23-year-old Lexi Dillon on the runway under a balcony. She lay on her stomach with blood around her head. It appeared she had fallen to her death. However, Meadows did not think this was a suicide. The Crime-Scene Investigation team was finishing their photographs when they noticed bruises on Dillon's forearm and told Meadows they could be from drug use or a struggle.

Meanwhile, Shells found Morgans sitting in the audience chairs a few rows back from Dillon's body. The detective hoped Dillon's best friend knew something, but Morgans seemed only slightly troubled. She was not even crying. Shells asked if Morgans knew whether Dillon was dating anyone, and Morgans said yes, Scott Hillsmen. Shells noticed Morgans had fresh scratches on her arms, but when he asked where they came from, Morgans said they were from working in her rose garden. Shells then asked about the bruises on Dillon's arm, and Morgans claimed they were from using heroin. Shells had not noted needle marks, but then he knew models could cover up anything. Morgans said Le'Muzy was planning to fire Dillon because of her drug use. Le'Muzy could not be responsible for a model on drugs. That could ruin his career. Morgans told the detectives that she was going to take Dillon's place in the show.

The detectives then investigated the balcony. The spotlights used for the show were placed on the balcony. The door was locked, but the investigators found Scott Hillsmen, the man in charge of the lights, on the platform next to the balcony. He told the investigators that he was finishing up some last-minute details. He insisted the balcony was only unlocked for rehearsals and shows. Meadows and Shells interviewed Hillsmen and found out he had briefly dated Morgans before Dillon joined Le'Muzy's team. Hillsmen quickly found out that Dillon was not interested in him and stopped seeing her.

The detectives realized Riley Morgans had not told them about Hillsmen and Dillon. Was she hiding something about the situation? Meadows wanted to talk with Morgans again. She asked Morgans why she said Hillsmen was Dillon's boyfriend. Morgans said nothing. Detective Meadows then demanded to know what Morgans was hiding. Morgans claimed she had no idea that Dillon and Hillsmen had stopped dating.

Meadows and Shells were determined to solve the case. Shells thought maybe Dillon could not handle the pressure of being the top model and had jumped because of it. Meadows did not think that was right: If the balcony was off limits and locked, how did Lexi Dillon jump from it? After thinking about the evidence for a few minutes, Meadows ran over to Shells and told him she knew who the killer was!

Activity:

Answer the following questions based on information in the Crime Scene S.P.O.T.

1. Did Lexi Dillon jump, or was she killed?
2. Who are all the potential suspects?
3. What were the motives of each suspect?

WRITING

4. Who do you think killed Lexi Dillon? In two or three paragraphs, justify your answer with evidence from the story.

WRITING

5. Rewrite the story to incorporate entomological evidence. Imagine that Dillon's body was not found until weeks after the show—perhaps her body was hidden under the stage or carried to another location. Be sure to explain how the temperature, light source, and other environmental factors affected the evidence.

Preventing Adolescent Crime Together™ (P.A.C.T.)

Introduction:

Understanding the life cycle of various insects helps crime-scene investigators and forensic entomologists determine the postmortem interval. Entomology is a specialized field that studies insects. Living people are susceptible to many insect-borne illnesses. Insects are more abundant in tropical climates. Insect-borne illness deaths are not very common in industrialized nations where health care and medicine are readily available. However, underdeveloped nations are vulnerable to insect-borne illness in almost epidemic proportions. One to two million people die each year of malaria alone.

Procedure:

1. Contact a nonprofit organization that works internationally. Discuss your interest in working on an awareness campaign at your school and in raising funds to help combat deaths caused by insect-borne illness.
2. Working in groups of three or four students, develop a school-wide campaign. If possible, contact other high schools in the area that offer this course and see if they are interested in a collaborative effort to raise awareness and funds for this cause.
3. The campaign should include the following:
    - A compilation of some of the most deadly insect-borne diseases, including the geographic region where the disease is most prevalent, the signs and symptoms of the disease, treatment and cures, and the reason people die from the disease
    - A method of distributing your information, such as designing posters to hang around the school campus, flyers, informational pamphlets, or a video documentary that can be shown school-wide through a video feed
4. Set a goal for the amount of money you wish to raise for your cause. Be realistic. Set a timeline for how long the fundraiser will last. Develop a creative means for raising money.
5. Make a progress chart to monitor your fundraising efforts.
6. Upon completion of your awareness and fundraising campaign, contact the organization for which you raised money. Invite a representative to come to your school to collect a check. You may even wish to invite the local news or newspaper to come to the school to showcase a great effort for an international cause.
7. Write thank-you notes to anyone who assisted in helping you campaign and raise funds for your project.

ACTIVITY 12-1 *Ch. Obj: 12.6, 12.7, 12.8*
BODY BUGS

Objective:

By the end of this activity, you will be able to:
Demonstrate proper techniques for collecting, preserving, and identifying insects.

Materials:

(per group of four students)

2 chickens (whole or pieces)
2 plastic plates (preferably white)
2 pieces of wire mesh, with 1-inch mesh
tent spikes
hammer
collection bottles
isopropyl (rubbing) alcohol
hand lens
forceps
stereomicroscope
2 insect nets
Mason jars, with nylon stocking material in the lid
metric ruler

Safety Precautions:

Wear safety goggles.
Use disposable gloves as an extra precaution.
Wash hands after the lab exercise.

Procedure:

Day 1

1. Place a chicken or chicken parts onto two plastic plates.
2. In an appropriate place outside, put one plate in a sunny location and the other plate in a location shielded from direct sunlight.
3. Cover the plates with the wire mesh and secure the wire with the fence spikes to prevent animals from eating the "evidence."

Day 2

1. Label two collection bottles with descriptions of your "evidence" locations.
2. Carefully pour about 5 mL of alcohol into each bottle.
3. Outside, use a hand lens to carefully search the "evidence" for insect eggs. Using forceps, collect a few of the eggs and place them in the proper collection bottle. (Note: There could be several different types of eggs; make sure to collect representative samples of all types.)
4. After returning to the laboratory, use a stereomicroscope to observe the eggs. Record a description and measurements for each specimen on Data Table 1. Make a sketch of each specimen.

Days 3–25

1. Return to the "evidence" every day to collect larvae, pupae, and adult insects.
2. Using nets, capture adult insects. Using jars, collect the pupae. Allow the adults to emerge and die, and then preserve them.
3. Repeat steps 3 and 4 from Day 2 with any new samples you collect. Adult flies stuck in the nets should be carefully removed and placed in alcohol.

Data Table 1. Insect evidence.

Date of Collection	Measurements and Description of Specimens	Sketch of Specimen

Questions:

1. Based on the data, are insects more active on a body in the sun or not in the sun?
2. Were the types of insects the same in the two locations? Describe any differences you observed.
3. Using the data, draw a complete life cycle for one of the insects you discovered in each location.
4. Describe changes that occurred in the chicken itself during the experiment.

Extension:

The body of a female is found behind an abandoned gas station. There is significant colonization of insects on the remains, especially in the openings on the face. The larvae vary in length—as small as 6 millimeters and as large as 15 millimeters. The autopsy reveals the woman was elderly and might have died of natural causes.

1. How long do you think the woman has been dead?
2. If the woman died of natural causes, what is one explanation for the location of the body behind the abandoned gas station?
3. Why were notable infestations located in the eyes, nose, and mouth?
4. Where would you expect to find insect colonies if the woman had died of a gunshot wound to the chest? Why?

ACTIVITY 12-2 *Ch. Obj: 12.2, 12.3, 12.6*
A WORLD OF INSECTS

Objectives:

By the end of this activity, you will be able to:

1. Research insects native to various parts of the world.
2. Assess the importance of insects to forensics.
3. Produce a multimedia project illustrating a research topic.

Materials:

Computers with Publisher®, PowerPoint®, and/or Flash® Movie

Procedure:

1. After choosing a geographical area, research insects of forensic importance in that part of the world. Information should include the anatomy, life cycle, habitat, where they are found, food source(s), position in the food chain (predators, prey, etc.), forensic importance, and any other "fun facts" you may find.
2. Choose the format to showcase your research:
   a. PowerPoint presentation
   b. Flash Movie
   c. Brochure

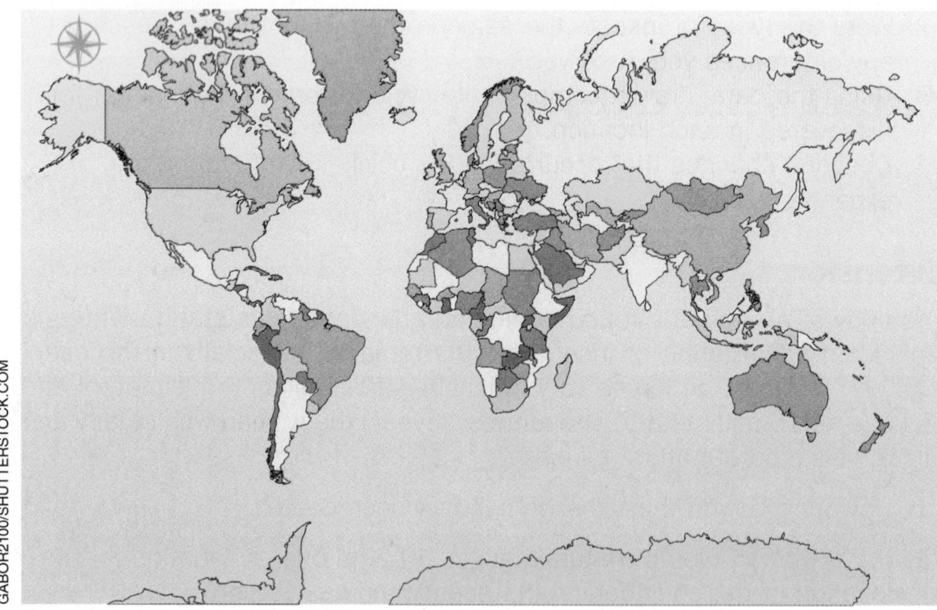

Figure 1. *World map.*

ACTIVITY 12-3 *Ch. Obj: 12.8*
INSECT COLLECTION

Objective

By the end of this activity, you will be able to:
Demonstrate proper techniques for collecting, preserving, and displaying insects.

Materials:

cotton balls
acetone or ammonia
collection bottles
soft-touch tweezers/forceps
insect pins (labeled 1, 2, and 3)
paper tags
plastic foam squares
insect field guide
moth balls
shadow box (optional)
shoe box

Safety Precautions:

Consult MSDS Sheets for recommended handling precautions for the chemicals used in this activity.
Wear goggles and protective gloves when handing chemicals.
Use caution when pinning insects to the plastic foam.
Do not inhale chemical fumes.
Wash hands after the lab exercise.

Procedure:

Day 1

1. Locate nine insect specimens. (These can be insects that are already dead.) Moisten a cotton ball with acetone and drop it in the collection bottle.
2. Using forceps, carefully pick up the insect and place it on top of the cotton in the collection bottle. (Note: This relaxes the insect without promoting mold growth.) Leave in the bottle for three days.

Day 4

1. Remove insects from bottles and carefully pin them to the plastic foam (see Figure 1).

Figure 1. Specimen placement in the Styrofoam.

Specimen 1	Specimen 2	Specimen 3
Specimen 4	Specimen 5	Specimen 6
Specimen 7	Specimen 8	Specimen 9

a. Place pins through the thorax, slightly to the right of the midline (see Figure 2).

Figure 2. Pin insects through the thorax, just to the right of the midline.

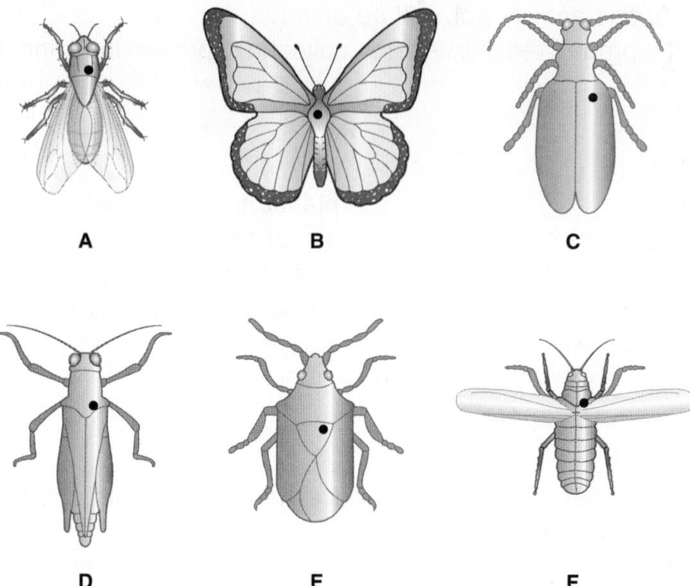

b. For insects with large wings (such as a butterfly), additional pins may be necessary.

c. Look in a field guide for other suggestions for mounting the insects.

2. Using a field guide, identify the insect species. On a label, write the name (common name and scientific name) of the insect, date of collection, and location of collection. Place the label below the specimen on the grid.
3. Consider placing the plastic foam in a shadow box for display.

ACTIVITY 12-4 *Ch. Obj: 12.6, 12.7*
WHAT "BUGS" BUGS

Objective:

By the end of this activity, you will be able to:
Describe the insect evidence found at certain crime scenes.

Procedure:

1. Using Data Table 1, design a crime-scene scenario for each quadrant. Be specific in the development of your crime scene.

Data Table 1. Crime-scene characteristics.

	Natural Temperature	**Controlled Temperature**
Natural Light		
Artificial Light		

2. Based on each scenario, discuss the insect development. Be sure to justify your answers.

Questions:

1. In what type of geographic area would insects develop most rapidly?
2. How does rate of insect development relate to rate of decay?
3. How is light related to insect development?

CHAPTER 13

Crime and Accident Reconstruction

HOW MANY MOTIVES FOR A MURDER?

Early on July 4, 1954, Spencer and Esther Houk were awakened by a frantic phone call from their neighbor, Dr. Samuel Sheppard. Sheppard claimed that intruders had broken into his home and killed his wife, Marilyn. Upon arriving at Sheppard's home, the Houks discovered Marilyn's body in her upstairs bedroom and called the police.

When the police arrived at the Sheppard residence, they quickly suspected that the crime scene was staged. First, the position of Marilyn's body and clothing indicated that the murderer had sexually assaulted her. However, a medical examination proved that no sexual assault had taken place. Additionally, Dr. Sheppard reported that a few vials of morphine had been taken. Dr. Sheppard's medical bag had been emptied onto the bedroom floor, and several drugs were plainly visible. Yet, investigators thought it was odd that a thief would take the time to read the labels, taking only morphine. Instead, investigators would expect a thief intent on stealing drugs to take all of the drugs from the bag.

Figure 13-1. Marilyn and Samuel Sheppard.

The unusual crime scene and the discovery that Dr. Sheppard had been having an affair convinced police that the motive for the crime was not robbery or sexual assault but murder. Dr. Sheppard was tried and convicted of his wife's murder. However, the defense hired Dr. Paul Kirk in 1966 to help with the appeal. Dr. Kirk performed a detailed reconstruction of the crime scene. Based on several factors, including blood-spatter evidence, Dr. Kirk argued that there had been a third person in the room and that Dr. Sheppard could not have killed his wife. Dr. Sheppard's conviction was overturned. A popular TV show, *The Fugitive*, and later a movie with the same title were inspired by Dr. Sheppard's stories of the intruder he claimed was the actual murderer.

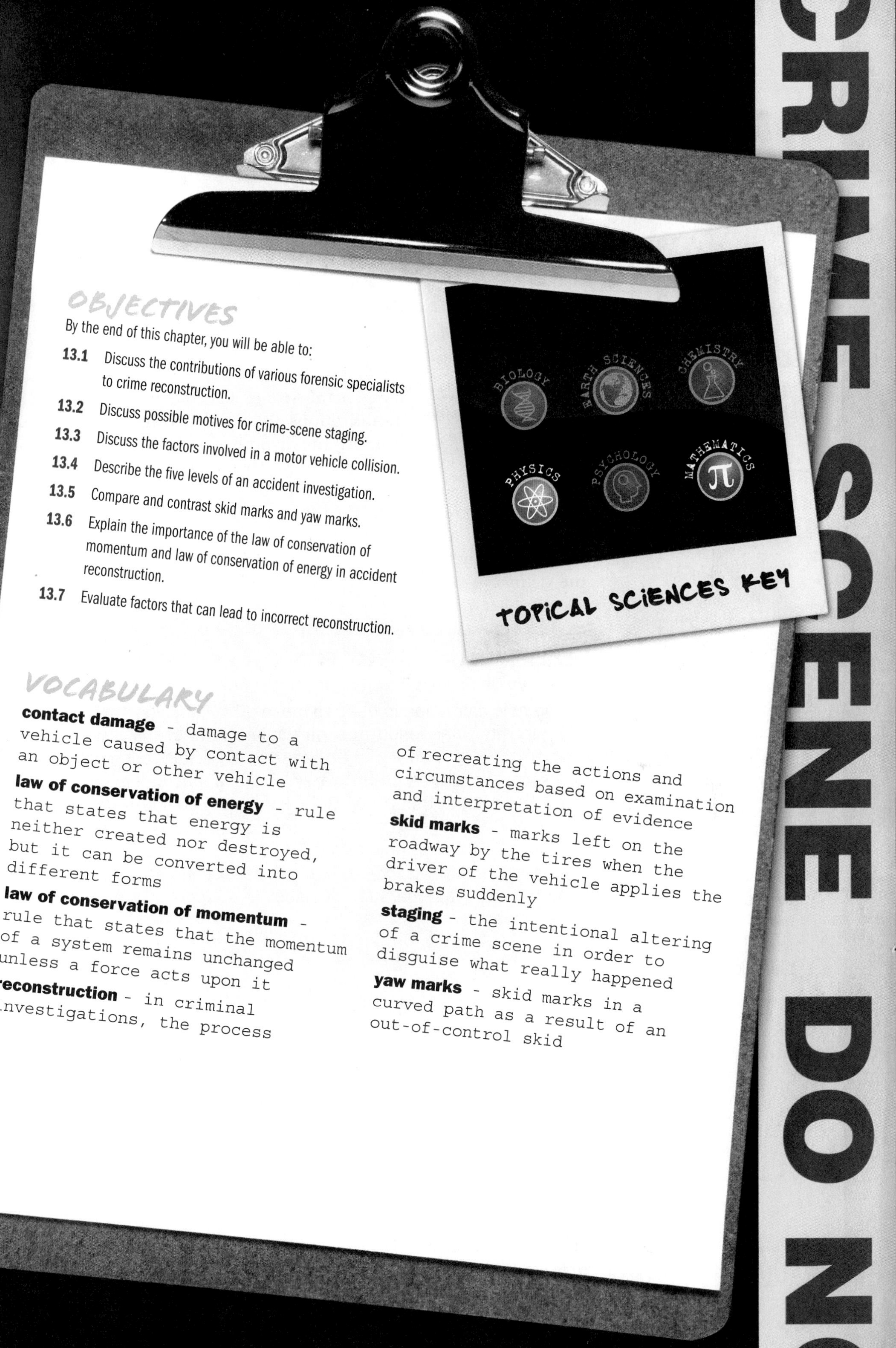

OBJECTIVES

By the end of this chapter, you will be able to:

13.1 Discuss the contributions of various forensic specialists to crime reconstruction.

13.2 Discuss possible motives for crime-scene staging.

13.3 Discuss the factors involved in a motor vehicle collision.

13.4 Describe the five levels of an accident investigation.

13.5 Compare and contrast skid marks and yaw marks.

13.6 Explain the importance of the law of conservation of momentum and law of conservation of energy in accident reconstruction.

13.7 Evaluate factors that can lead to incorrect reconstruction.

VOCABULARY

contact damage - damage to a vehicle caused by contact with an object or other vehicle

law of conservation of energy - rule that states that energy is neither created nor destroyed, but it can be converted into different forms

law of conservation of momentum - rule that states that the momentum of a system remains unchanged unless a force acts upon it

reconstruction - in criminal investigations, the process of recreating the actions and circumstances based on examination and interpretation of evidence

skid marks - marks left on the roadway by the tires when the driver of the vehicle applies the brakes suddenly

staging - the intentional altering of a crime scene in order to disguise what really happened

yaw marks - skid marks in a curved path as a result of an out-of-control skid

INTRODUCTION

EDDIES IMAGES/SHUTTERSTOCK.COM

Figure 13-2. *A shoe print at a crime scene must be photographed and measured. The shoe print may be used to estimate the person's size, gait, and pace.*

When law-enforcement personnel arrive at an accident or crime scene, one of the first tasks is to determine what happened. Evidence at the scene can help law enforcement develop a timeline of events leading up to the accident or crime. They use this information to recreate the accident or crime.

Reconstruction is the process of reproducing the actions and circumstances of an accident or crime based on examination and interpretation of evidence. Forensic investigators and scientists must formulate a hypothesis based on careful observations at the scene. The reconstruction may require modifications if the evidence and information do not support the hypothesis. Some reconstructions are relatively simple, but others require extensive analysis of evidence and interviews with many witnesses. For example, a shoe print (see Figure 13-2) might require only a comparison with a suspect's shoe, whereas a double homicide reconstruction could require several types of lab analysis and interviews of friends, relatives, and neighbors of the victims.

Did You Know?

The first forensic science degree covering all the major subjects in criminalistics was offered in 1909 at the University of Lausanne in Switzerland. The degree, offered through the Institut de Police Scientifique et de Criminologie, required courses in forensic photography, crime-scene investigation, and identification.

Digging Deeper
with Forensic Science e-Collection

Forensic animation is used to make a "mini-movie" that shows a jury what happened at a particular crime scene. Animation is used in a wide variety of cases, especially crimes involving car accidents. Controversy exists about using this type of technology. Using the Gale Forensic Science e-Collection at www.cengage.com/school/forensicscienceadv, read the uses of forensic animation along with the benefits, risks, and expenses associated with the technology. Afterward, write a one-page persuasive paper to support one side of the controversy.

Obj. 13.1

HISTORY OF CRIME RECONSTRUCTION

Crime reconstruction requires methodical and scien-tific examination of the relationships among the various evidence. Throughout the years, several individuals contributed research, publications, and techniques valuable to the understanding of crime reconstruction. Though they all entered the forensic science profession for different reasons, all worked to establish crime reconstruction as a scientifically based process through which evidence is logically and objectively examined and interpreted.

DR. HANS GROSS (1847–1915)

During his tenure as Examining Magistrate of the Criminal Court at Czernovitz, Austria, Dr. Hans Gross became aware of the need for a systematic method of analyzing the facts of a criminal case. In 1906, he wrote the *Hanbuch fur Untersuchunrichter, als System der Kriminalistik* (see Figure 13-3), a textbook for magistrates, police officers, and lawyers. Gross wrote of the importance of objectivity and reliance on science as opposed to eyewitnesses and intuition. His book quickly became the standard for scientific investigation and forensic analysis as well as crime reconstruction.

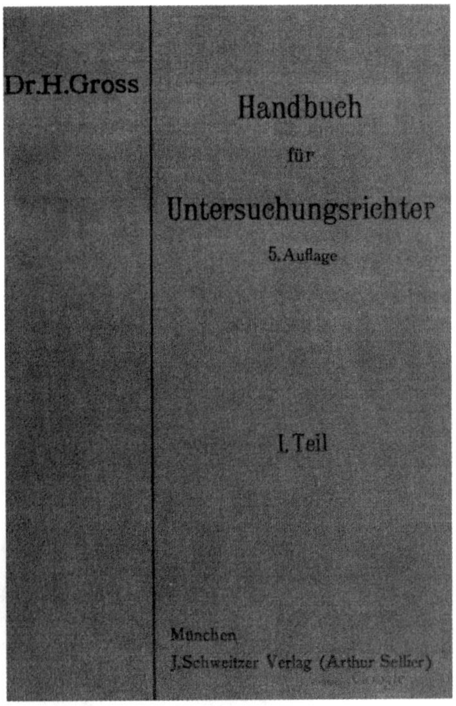

Figure 13-3. *This text by Dr. Hans Gross emphasized the importance of an objective, scientific approach to the evidence.*

DR. ALEXANDRE LACASSAGNE (1843–1924)

In 1880, Dr. Alexandre Lacassagne became the director of the Lyons Institute of Forensic Medicine in Lyon, France. He advocated for combining science with criminology. As a professor of forensic medicine at the University of Lyon, Lacassagne instructed his students on the importance of what is now referred to as trace and transfer evidence. Lacassagne proposed that such evidence could be used to link suspects to crimes, provide clues to a person's identity, or reconstruct a crime.

DR. EDMOND LOCARD (1872–1966)

One of Lacassagne's students, Dr. Edmond Locard (see Figure 13-4) is best known for Locard's exchange principle. Recall from Chapter 1 that every contact leaves a trace. Locard analyzed and interpreted bloodstains, fingerprints, dust, and handwriting. His ultimate goal, however, was to use a systematic method of analyzing physical evidence to reconstruct crime scenes.

Figure 13-4. *Edmond Locard is best known for his exchange principle.*

EDWARD OSCAR HEINRICH (1881–1953)

Edward Heinrich began his career working as a chemist and assisting law-enforcement personnel and the coroner's office with their investigations. Heinrich soon realized that specializing in only chemistry limited his ability to fully reconstruct crimes, so he began studying other areas such as geology, physics, hair analysis, and photography. He thought that a forensic scientist should have knowledge in as many fields as possible. Heinrich also thought that evidence was the only reliable witness to a crime and that it was the forensic investigator's responsibility to find and interpret that evidence. According to Heinrich, crime reconstruction was a systematic and scientific evaluation of the clues to provide links between the crimes and the suspects.

DR. PAUL LELAND KIRK (1902–1970)

In 1953, Dr. Paul Kirk (see Figure 13-5) published his dissertation titled *Crime Investigation.* The paper described principles of crime investigation

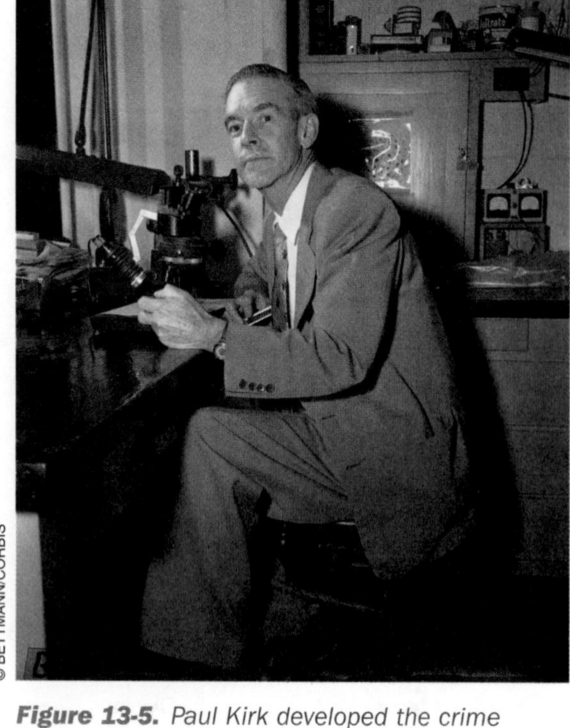
© BETTMANN/CORBIS

Figure 13-5. Paul Kirk developed the crime reconstruction report that was pivotal in the Sam Sheppard case.

and forensic examination as well as crime reconstruction. Kirk gained fame when he testified as an expert witness for the defense in the case against Sam Sheppard, described in the opening scenario.

Kirk visited the Sheppard home to examine the crime scene and collect evidence in order to prepare an appeal. In the crime reconstruction report, he stated that the blood-spatter analysis indicated that a third person was present at the scene. The information eventually led to Sheppard's release after the conviction was overturned.

THE ROLE OF EVIDENCE IN CRIME RECONSTRUCTION

Reconstructionists rely on crime-scene evidence to help them establish a timeline of events and relationships between the evidence and the victims and suspects involved in a crime. For example, a shoe print at a crime scene can yield several different types of information. The size and shape of the impression made by a shoe indicates the sex of the person who wore the shoe. Females tend to have smaller, narrower feet than males. The tread pattern on the shoe (see Figure 13-6) might even identify the type of shoe—a running shoe, work shoe, or basketball shoe. The depth of the tread helps investigators calculate the approximate weight of the individual who left the impression. The distance between an individual's footprints indicates the person's height. If there are several impressions, all will be compared to determine whether multiple people were involved. Overlapping shoe prints or shoe prints that overlap some other type of impression, such as tire tracks, establish the sequence of the impressions.

Did You Know?

Paul Kirk was not allowed to join the American Academy of Forensic Sciences (AAFS) because of his involvement in the Sam Sheppard case. A powerful member of the AAFS believed Sheppard was guilty, and he used his influence to keep Kirk out. Kirk continued to break ground in forensic science and write publications for investigators. Today, the highest honor bestowed on criminalists by the AAFS is called the Paul L. Kirk award.

TROPININA OLGA/SHUTTERSTOCK.COM

Figure 13-6. This shoe has a very distinctive tread pattern.

Other types of evidence also yield valuable information about the events of a crime. A fingerprint inside a room where a crime was committed proves the person was in the room at some time. Even though the print does not prove the individual committed the crime, it puts him or her at the scene. Trace evidence, such as hair or fibers, that is transferred from one person to another indicates that the two people have come into contact with one another. The presence of two cups of coffee on a table suggests that at least two people were present. If blood spatter at a homicide scene is found under a piece of broken window glass, investigators conclude that the victim was injured or killed prior to the window break. Anything found underneath the broken glass was there before the glass was broken. Anything found on top of the glass was left behind after it was broken.

As the evidence is analyzed and a sequence of events is constructed, reconstructionists must be open to the possibility of alternative explanations. Ultimately, witness statements may corroborate a hypothesis, but the evidence must be evaluated scientifically and be sufficient to adequately support the reconstruction.

Did You Know?

Automobile accidents are also staged. Insurance fraud, mental illness, depression, and marital problems are often reasons for an individual to stage a motor vehicle accident.

CRIME-SCENE STAGING

Obj. 13.2

Whenever a crime scene is being evaluated and the crime is being reconstructed, investigators must be aware of the possibility that the scene was staged. **Staging** is the intentional altering of a crime scene in order to disguise what really happened.

Typically, staged crime scenes involve an attempt to make a murder appear to be a suicide or an accident. For example, a gun is placed in a victim's hand to indicate the wound was self-inflicted. However, investigators will look at the location of the wound to determine whether it is consistent with suicide. The presence or absence of gunshot residue or primer residue also helps investigators properly reconstruct the crime. You may wish to refer to Chapter 3 for methods of detecting gunshot residue. A body discovered at the bottom of a flight of stairs does not necessarily prove a victim fell to his death. Investigators must look at all of the evidence to determine whether the death was accidental or a homicide. Trace evidence overlooked by an assailant might be discovered through forensic analysis.

Another common staging technique is to make a murder scene appear to be a burglary. A death that is particularly violent, such as by multiple stab wounds, is usually an indication of a crime of passion, not a crime of opportunity. In cases in which a murder and apparent burglary have occurred, investigators must consider the possibility of a staged crime scene.

A burglary scene where windows are broken, furniture is turned over, and drawers are emptied tends to be excessive and raises a red flag for investigators (see Figure 13-7). If something of value is reported missing, investigators may suspect insurance fraud. Presence or absence of trace evidence from a robber would be of significant value in providing investigators clues that the scene was staged.

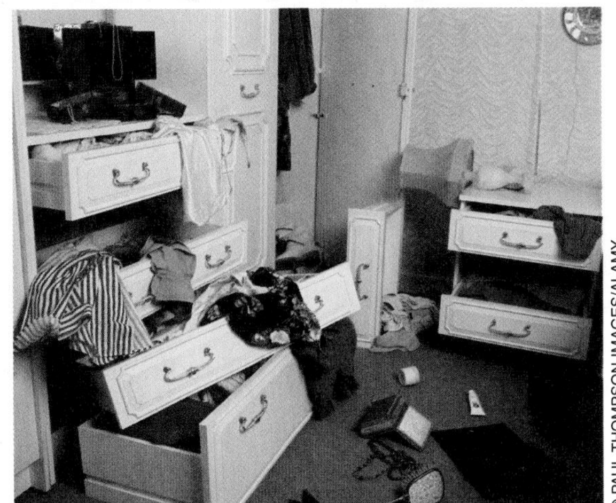

Figure 13-7. *Excessive damage at a crime scene often indicates that the scene was staged.*

Digging Deeper
with Forensic Science e-Collection

A strong background in physics and math assists reconstructionists in gathering very detailed information regarding the events that occurred prior to and during the commission of a crime. Math and physics are especially useful in cases involving the interpretation of bullet trajectories and blood spatter. Using the Gale Forensic Science e-Collection at www.cengage.com/school/forensicscienceadv, discuss the math and physics principles used while interpreting a crime scene. Also, discuss how additional factors make the interpretation of bullet trajectory and blood spatter more difficult. What is the Pythagorean theorem, and how does it specifically assist investigators in interpreting evidence?

Obj. 13.3 ACCIDENT INVESTIGATION

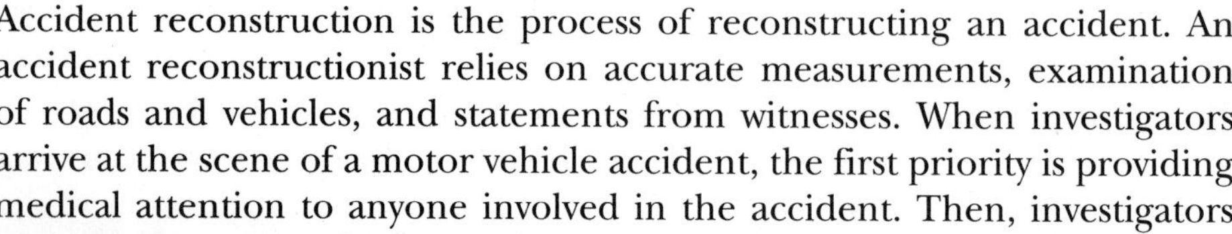

Accident reconstruction is the process of reconstructing an accident. An accident reconstructionist relies on accurate measurements, examination of roads and vehicles, and statements from witnesses. When investigators arrive at the scene of a motor vehicle accident, the first priority is providing medical attention to anyone involved in the accident. Then, investigators focus on determining the cause of the accident. If there is pending legal action, it is the responsibility of the investigators to evaluate the scene and ultimately to reconstruct the events that led up to the accident.

Figure 13-8. *Depending on the circumstances, an automobile accident might require only a report or a complete investigation.*

Most traffic accidents need only a basic report, usually for insurance purposes. However, in cases of fatalities or a criminal investigation such as a hit and run, a more thorough investigation must be conducted (see Figure 13-8). An accident investigation consists of as many as five levels:

- Level 1: Reporting
- Level 2: At-the-scene investigation
- Level 3: Technical follow-up
- Level 4: Accident reconstruction
- Level 5: Cause analysis

Obj. 13.4 REPORTING

The first level of an investigation into a traffic accident is simply reporting the accident. Level one consists of collecting basic information about the persons and vehicles involved. The report includes only factual information and no opinions. All accidents, no matter how minor, should be reported at this level. Figure 13-9 is an example of a typical accident report. If the accident results in a serious injury or fatality, the investigators will go to level two.

REPORT OF INCIDENT/ACCIDENT
(PROPERY DAMAGE)

Date Prepared ______________

See Management Directive 630.2
Send completed report immediately. If additional space is needed for any item, attach 8 1/2 x 11 sheet referring to item number.

1. TIME AND LOCATION

INCIDENT/ACCIDENT DATE	TIME: AM ☐ PM ☐	LOCATION (STREET & NUMBER, BUILDING/INSTITUTION, CITY, COUNTY, STATE)

2. PERSONS INJURED

NAME	ADDRESS & TELEPHONE NO.	AGE	EXTENT OF INJURY

3. PROPERTY DAMAGE

OWNER	ADDRESS & TELEPHONE NO.	ESTIMATED DAMAGE $

PROPERTY DESCRIPTION	DAMAGE DESCRIPTION

4. DESCRIPTION OF INCIDENT/ACCIDENT

5. WITNESS

NAME, ADDRESS & TELEPHONE NO.

a.

b.

c.

6. CLAIM INFORMATION (If no State employee is involved, disregard thie section)

NAME OF EMPLOYEE INVOLDED	WORKING TITLE	ADDRESS & TELEPHONE NO.
AGENCY	IMMEDIATE SUPERVISOR	SUPERVISOR'S BUSINESS ADDRESS & TELEPHONE NO.

7. NOTIFICATION OF POSSIBLE CLAIM

HOW NOTIFIED?	IS CLAIM BEING MADE?
☐ LETTER ☐ PHONE ☐ IN PERSON	☐ YES ☐ NO ☐ UNKNOWN

8. REPORTED BY

AGENCY	BUREAU/INSTITUTION/FIELD OFFICE		
PREPARING REPORT	NAME (PRINT)	WORKING TITLE	BUSINESS TELEPHONE NO.

1. DEPARTMENT OF GENERAL SERVICES, BUREAU OF RISK AND INSURANCE MANAGEMENT

Figure 13-9. *A typical accident form states only facts surrounding the accident and no opinions.*

Obj. 13.4, 13.5 AT-THE-SCENE INVESTIGATION

Level two requires collection of extra data at the scene. This level of investigation is generally done for accidents involving injuries or fatalities. At this level, trained investigators collect facts, such as measurements of tire and skid marks. Police officers might also perform such tests as a field sobriety test. Investigators gather as much information as possible, because some evidence will not be available later. For example, the driver's blood alcohol level cannot be determined more than a few hours after the accident. After the data are assessed, investigators determine whether a criminal act occurred. If so, or if any litigation is likely, investigators will move to level three.

© TED SOQUI/CORBIS

Figure 13-10. Tire marks left on the road by a Porsche that was involved in a hit-and-run accident.

PHOTOGRAPHS

Investigators take photographs immediately. They photograph any debris left on the roadway from the collision and marks—especially tire marks (see Figure 13-10)—on the road. They also photograph any external damage to the vehicles involved in the accident. If any other objects were damaged, investigators will photograph the damage and the surrounding area.

© DAVID R. FRAZIER PHOTOLIBRARY, INC./ALAMY

Figure 13-11. Accident investigators take measurements at the scene of an accident.

MEASUREMENTS

After all of the photographs are taken, investigators must make careful measurements (see Figure 13-11), recording the locations and distances between tire marks on the road and the debris from the vehicles. They also record the final positions of vehicles and bodies of any victims. Investigators try to match tire marks on the roadway to the vehicles. This allows proper placement of vehicles for the reconstruction of the accident.

COEFFICIENT OF FRICTION

The *coefficient of friction* is a quantitative value, determined through experimentation that measures the roughness of a surface. Surfaces such as concrete will have a higher coefficient of friction than a nonstick surface. Hypothetically, if no friction existed, the coefficient of friction would be zero.

Marks left on the roadway by tires when the driver of a vehicle applies the brakes suddenly are called **skid marks**. The length of the skid marks helps investigators determine the speed at which the vehicle was traveling before the driver applied the brakes. Speed can be calculated using the following equation:

$$s^2 = 255df$$

In this equation, s represents the speed of travel (in km/h), d is the distance or length of the skid mark (in meters), and f is the coefficient of friction based on the surface of the road. The number 255 is a constant used to adjust the units. Figure 13-12 lists some common coefficients of friction for a variety of surfaces or road conditions. Scientists have determined a range of coefficients of friction for each surface. For this reason, investigators can only estimate the speed at which the vehicle was traveling. This estimate falls into a certain range based on the coefficient of friction.

For example, imagine that investigators discovered skid marks on an asphalt road that were 45 m long. The vehicle's range of speed can be calculated using the information in Figure 13-12 as follows:

Road Material or Surface Condition	Range of Coefficient of Friction
Cement	0.55–1.2
Asphalt	0.50–0.90
Gravel	0.40–0.80
Ice	0.10–0.25
Snow	0.10–0.55

Figure 13-12. *Coefficients of friction used in speed determinations.*

$$s^2 = 255df$$

$$s^2 = 255\ (45)(.5) \qquad s^2 = 255\ (45)(.9)$$

$$s^2 = 5737.5 \qquad s^2 = 10328$$

$$s = 75.7\ \text{km/h} \qquad s = 101.6\ \text{km/h}$$

The vehicle was traveling between 75.7 km/h and 101.6 km/h.

Notice that the outside temperature and the mass of the vehicle are not factors in the speed equation. Temperature has a minimal and generally negligible effect on the friction coefficient. Once in a skid, all vehicles, regardless of the size of the vehicle, take the same distance to stop. It is also important to note that this equation is applicable only wnen the skid marks are in straight lines.

Figure 13-13. *Yaw marks are made when a vehicle skids sideways.*

Sometimes, a vehicle will skid sideways along a curved path. The resulting skid marks are called **yaw marks** (see Figure 13-13). Because the primary distinction between skid marks and yaw marks is that yaw marks are curved, an accident reconstructionist can calculate the radius (r) of the yaw mark by drawing an imaginary line to complete the circle. The equation for calculating speed using yaw marks includes the radius (r) as well as the elevation (e) of the road:

MATHEMATICS π

$$s^2 = \frac{255[r(f +/- e)]}{2}$$

If the car was moving uphill during the skid, the elevation (e) is added to the coefficient of friction (f). If the car was moving downhill, the elevation (e) is subtracted from the coefficient of friction (f). If the surface is flat, there is no elevation and e equals zero.

Imagine that investigators found that the radius of the yaw marks made on a flat asphalt surface during a certain accident is 22 meters long. You can use the equation to calculate the range of velocity of the vehicle:

$$s^2 = \frac{255[r(f +/- e)]}{2}$$

$$s = \frac{255(22)(.9)}{2} \qquad s = \frac{255(22)(.5)}{2}$$

$$s = 2524.5 \qquad s^2 = 5610$$

$$s = 50.2\ \text{km/h} \qquad s = 74.9\ \text{km/h}$$

The vehicle was traveling between 50.2 km/h and 74.9 km/h when the driver applied the brakes.

Other measurements include locations of debris, victims, vehicles, bystanders, and witnesses. The measurements may be used later in level three to create a map for legal purposes. Each item of evidence must be measured from two reference points, a process similar to measurements made for crime-scene sketches. Each reference point must be a permanent or fixed structure. This assures that if investigators—or reconstructionists—need to return to the scene at a later time, they will be able to determine accurately the placement of all relevant items.

EXAMINATIONS

During the early stages of the investigation, preliminary tests are done to determine whether alcohol was a factor in the accident. Field sobriety tests (refer to Chapter 9) may be conducted, and witnesses may be interviewed.

Vehicles involved in the collision will be examined for damage. Investigators will attempt to match the location of the damage on one vehicle to damage on the other. This information will aid in determining the sequence of events for the reconstruction process. Damage on a vehicle made by another object or vehicle is referred to as **contact damage**. This type of damage is often indicated by imprints from objects such as vehicle bumpers or door handles, crumpled metal, scratches, and transferred paint. Investigators conduct a preliminary check of the condition of the each vehicle's equipment to determine whether equipment failure or malfunction was a factor. They examine occupant restraints, such as seat belts and child safety seats, to determine whether they were installed and working properly at the time of the collision. Investigators also examine the condition of headlights and taillights, the inflation pressure of the tires, the amount of tread on the tires, and the condition of the steering and brake mechanisms.

Obj. 13.4 TECHNICAL FOLLOW-UP

Technical follow-up is the third level of an accident investigation. Investigators generally proceed to level three only when a crime has been committed. In such cases, investigators complete levels three, four, and five. In addition to the information collected at level two, investigators prepare a map of the accident site, interview witnesses, and assess the visibility at the site.

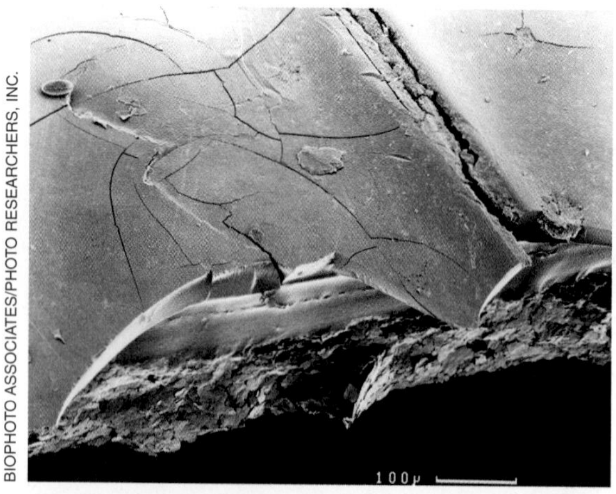

BIOPHOTO ASSOCIATES/PHOTO RESEARCHERS, INC.

Figure 13-14. A paint chip often consists of multiple layers of paints.

Many specialists are involved at this level. For example, samples of paint from the vehicles must be processed in a lab using special equipment such as spectrometers. Because paint is a mixture of pigment, binder, and solvents as well as chemicals that add glossiness, flexibility, or durability, it can provide clues about the make and model of a specific vehicle. Figure 13-14 shows a scanning electron micrograph of a vehicle paint chip. To analyze the paint, the layers must be separated and individually inspected. Infrared (IR) spectrometry provides information about the molecular structure of a single layer of paint, allowing investigators to compare control samples and questioned paint samples.

Lab technicians conduct IR spectrometry on the paint samples to classify the pigments and binders in paints. Investigators compare the results to spectrographs of known paints in vehicle databases. The comparisons provide information such as the make, model, and year of a vehicle.

If there is a fatality, the forensic pathologist performs a postmortem examination and files an autopsy report. The report will include results from toxicology to help investigators determine whether alcohol or drugs were contributing factors in the accident. If the driver was not killed, law-enforcement personnel will request toxicology reports to determine whether alcohol or drugs were involved. Any of the specialists involved in processing the evidence may be called to testify if the case goes to court. After the investigator has completed the legal examination of facts and data, he or she will move to level four in order to reconstruct the events of the accident. If level three is warranted, then investigators also perform levels four and five.

ACCIDENT RECONSTRUCTION

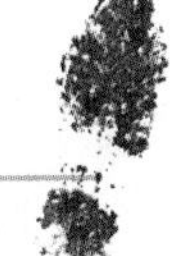

Obj. 13.4, 13.6

Accident reconstruction is level four of an accident investigation. Most accidents involve a vehicle colliding with another vehicle or a stationary object, such as a tree or building. In order to complete an analysis of the collision, investigators rely on the laws of physics. Specifically, investigators rely on the law of conservation of energy, Newton's Laws of Motion, and the law of conservation of momentum.

THE LAW OF CONSERVATION OF ENERGY

The **law of conservation of energy** states that energy cannot be created or destroyed, but can be converted from one form to another. When an object, including a vehicle, is moving, it has kinetic energy. At rest the same object

Digging Deeper
with Forensic Science e-Collection

Once crime-scene investigators get to a crime scene, many steps are taken to ensure that the area is secured, photographed, and processed. Eventually, the crime scene must be cleaned up, and the area where the crime occurred must be restored to living/working conditions. When the crime scene no longer exists, it is the responsibility of the experts to put the pieces together and let the evidence tell the story. Also, often when experts discuss a case at trial, they rely on demonstrating their hypothesis by using traditional boards and cutting-edge software programs to reconstruct the events of the crime. Reconstruction is based on the information provided from the crime scene, including photos, blood spatter, and trace evidence. Using the Gale Forensic Science e-Collection at www.cengage.com/school/forensicscienceadv, discuss how crime-scene investigators use a scientific approach to reconstruct the crime scene. How is each piece of evidence integral in explaining to the jury what happened? Why is it important for an investigator to remain free from bias in reconstructing a crime scene for a jury?

Figure 13-15. *Sir Isaac Newton is credited for discovering the laws governing motion.*

has no kinetic energy, only potential energy. What happens to the energy as the vehicle comes to a stop? When the brakes are applied, some of the energy is converted into heat due to friction between the brakes and the wheels and between the tires and the road.

When two cars collide, one of the cars will move the other one. This involves "work." The scientific definition of *work* (w) is the amount of energy transferred by a force (F) acting through a distance (d). It is represented by the equation $w = F \times d$. During a collision, a force (caused by one moving vehicle) acts on the metal (on the other car) and moves the metal inward a certain distance. The first car has done work on the second. Work is also done on the cars themselves. Both cars have kinetic energy before the collision, and they exert force on each other upon impact. The distance one car is moved by the other is a result of work.

NEWTON'S LAWS OF MOTION

When investigators reconstruct an accident, they rely on a thorough understanding of Newton's three laws of motion. Sir Isaac Newton (see Figure 13-15) first described his laws of motion in his series of books called *Principia,* which was first published in 1687. Accident reconstructionists examine the speed at which a vehicle was traveling prior to a collision. In the case of multiple vehicles, the speed at which each vehicle was going at the time of the collision is integral to the investigation. Newton's three laws are as follows:

- ***First Law*** An object at rest will remain at rest, and an object in motion will remain in motion with a constant velocity unless some external force acts on the object. This law can also be called the *Law of Inertia. Inertia* is the tendency of objects to resist a change in motion. In other words, objects tend to stay in motion if they are moving or to stay at rest if they are at rest.
- ***Second Law*** Newton's second law of motion describes the relationships among force, mass, and acceleration. Recall from Chapter 7 that force is equal to mass times acceleration. The acceleration is directly proportional to the force acting on the object and is inversely proportional to the mass of the object. Newton first described this law using the following equation:

$$\text{Acceleration } (a) = \frac{\text{Net force } (F)}{\text{Mass } (m)}$$

- ***Third Law*** For every action, there is an equal and opposite reaction. For example, when a moving vehicle hits a stationary object, such as a tree, the vehicle exerts an action force on the tree. The tree exerts an equal and opposite reaction force on the car. In this case, the action force might cause damage to the tree. The reaction force is very likely to damage the vehicle.

THE LAW OF CONSERVATION OF MOMENTUM

Moving objects have momentum. Momentum is equal to mass (in kilograms) times velocity (in meters per second). Objects with greater momentum are more difficult to stop than objects with less momentum. The **law of conservation of momentum** states that the momentum of a system remains unchanged unless a force acts upon it. The momentum of each vehicle involved in a collision is dependent upon the mass (m) of the vehicle and the velocity (v) at which it was traveling. Velocity is speed and direction. The equation for momentum (p) is: $p = m \times v$.

Imagine that a small car with a mass of 1,000 kilograms (about 2,204 lb) is driving southbound at 20 meters per second (about 45 mph) when it collides with a 2,000-kilogram (about 4,408 lb) truck traveling at the same speed but in the opposite direction. The truck will have more momentum because it is more massive than the car. When the collision occurs, the momentum of each will change, but the total momentum of the system will remain the same (see Figure 13-16).

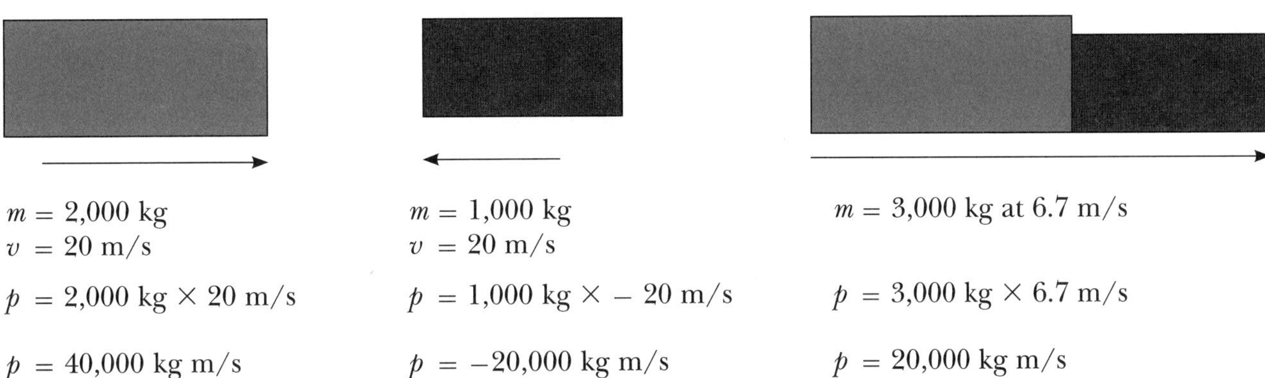

$m = 2{,}000$ kg $v = 20$ m/s	$m = 1{,}000$ kg $v = 20$ m/s	$m = 3{,}000$ kg at 6.7 m/s
$p = 2{,}000 \text{ kg} \times 20 \text{ m/s}$	$p = 1{,}000 \text{ kg} \times -20 \text{ m/s}$	$p = 3{,}000 \text{ kg} \times 6.7 \text{ m/s}$
$p = 40{,}000$ kg m/s	$p = -20{,}000$ kg m/s	$p = 20{,}000$ kg m/s

Figure 13-16. *The total momentum of the vehicles before the collision is equal to the total momentum of the vehicles after the collision. Note that, because the vehicles are traveling in opposite directions, the speed of the smaller vehicle is indicated by a negative number.*

The total momentum of the car and the truck before the crash is also 20,000 kg m/s: 40,000 kg m/s – 20,000 kg m/s.

CAUSE ANALYSIS

Obj. 13.4

Cause analysis, the fifth level of an investigation, is done only for the purpose of research. Investigators attempt to determine the indirect cause or conditions that resulted in the accident. In this level, experts share in court, based on facts and research, how and why an accident occurred. Although cause analysis is fact based, the interpretation by the expert witness is an opinion. If multiple experts review the same facts, it is possible to get multiple perspectives. The jury must then consider the testimony of all experts to determine whether fault is to be assessed in the case.

Imagine the following case: A 19-year-old driver got into a head-on collision in which everyone in the second car was killed. Police administered a breathalyzer test, which showed that the 19-year-old driver's BAC was 0.11. Police charge the driver with driving under the influence and vehicular manslaughter. Accident reconstruction experts examine evidence at the scene to determine whether additional factors, such as speeding, were

Cause	Description
Environmental conditions	• Weather such as rain, snow, heavy winds, sun glare, and fog can increase the likelihood of an accident. • Accidents in rain commonly happen at night. • In snow, as visibility decreases, risk for collision increases. When only a small amount of snow falls, and freezes, the risk of collision and injury increases. • Heavy winds affect large trucks and buses more than other vehicles due to the risk of the vehicle being overturned or spinning out of control. • Fog and sun glare decrease visibility, causing the motorist to slow down. This increases the risk of the motorist being rear-ended.
Human error	• Many motor vehicle accidents are a result of negligence, recklessness, illness, and/or poor decision making. • The following actions result in the most motor vehicle accidents: Driving over the posted speed limit Driving under the influence of drugs and/or alcohol Driving while severely fatigued Driving while distracted Driving aggressively Driving with little experience due to age
Vehicle equipment malfunction	• Accidents may happen because brakes do not work or accelerators stick. • When tires are too old or the air pressure is incorrect, tires may not respond as expected to road conditions, causing the driver to lose control of the vehicle.

Figure 13-17. *The three most common causes of collisions.*

involved. If this case goes to trial, accident reconstruction experts will recreate the accident scene for the jury, based on evidence such as skid and yaw marks left at the scene, how and why this collision occurred, and who was responsible.

Most accidents are caused by three general categories of factors: environmental conditions, human error, and vehicle equipment malfunction. An accident also may be caused by some combination of two or more of the factors. Figure 13-17 describes these three common causes of collisions.

Obj. 13.7

POTENTIAL PITFALLS OF RECONSTRUCTION

Reconstructing a crime or accident involves proper recognition of evidence, logic, reasoning, and a moral obligation to get to the truth. The reconstruction must also be free of bias. In order to ensure the validity of the reconstruction, the investigator must avoid potential pitfalls that could lead to weak or incorrect conclusions. Figure 13-18 summarizes the possible flaws in logic that might lead to poor reconstruction.

Flaw	Description
Fallacies of logic	False logic based on weak inductive reasoning, ambiguity, or generalizations
Deliberate deception	Investigators claim a suspect is implicated by the reconstruction of the scene in order to elicit a confession
Fraud	Someone who is caught falsifying the evidence, misrepresenting the truth, providing false testimony, or engaging in unethical behavior
Haste	Mistakes made due to carelessness or a lack of thoroughness when reconstruction is done hastily
Inexperience	Someone who lacks the expertise in a particular area or has little experience or training is asked to reconstruct a crime or play a significant role in the reconstruction
Pressure	Pressure by various agencies or individuals, in an attempt to advocate for their client or cause, may influence the outcome of crime reconstruction
Unevaluated surmise	Crime reconstruction is not based on fact but on the experience of the person reconstructing without verifying, testing, or evaluating thoroughly

Figure 13-18. *Potential flaws in crime reconstruction.*

CHAPTER SUMMARY

- Several forensic investigators contributed to the establishment and acceptance of crime reconstruction as a scientific, logical, and objective process.
- Evidence helps crime reconstructionists establish a sequence of events leading up to the crime as well as relationships between the scene, the victim, and the suspect.
- Forensic investigators and scientists rely on analysis of evidence as well as witness accounts to recreate or reconstruct a chain of events.
- Staged crime scenes are usually an attempt to cover up a murder or insurance fraud.
- All accidents must be reported at level one, but accidents that require research and potential litigation require all five levels of investigation.
- Skid marks and yaw marks help investigators determine the speed at which a vehicle was traveling prior to a collision or accident.
- Investigators can use the law of conservation of momentum and the law of conservation of energy to determine initial velocity of the collided vehicles.
- Environmental factors, human factors, and vehicular malfunction are the three main causes of accidents.
- To ensure the validity of reconstruction, investigators must avoid flaws in logic.

CASE STUDIES

Earl and Maryella Morris (2009)

On June 25, 2009, Alan Pearsall was heading home after visiting with his brother. Pearsall saw a bicyclist ahead of him on the shoulder and swerved to avoid the bicyclist. At this point, Pearsall's Toyota collided head on with a motorcycle being ridden by Earl and Maryella Morris. Even though both the Morrises were wearing helmets, Earl Morris died instantly of head injuries. Maryella Morris was taken by emergency helicopter to the nearest hospital, but she soon died from internal injuries.

Witnesses at the scene reported that directly after the accident Pearsall insisted that he had been in his lane when the collision occurred and the accident was not his fault. Accident reconstruction experts noted two pieces of evidence, though, that contradicted Pearsall's statement. First, the front wheels of Pearsall's vehicle were turned toward the right. This indicated that Pearsall was attempting to steer back into his lane at the time the collision occurred. Second, skid marks from the motorcycle indicated that Morris had not crossed over into oncoming traffic and had attempted to avoid the oncoming Toyota when it had veered into his lane. Both of these pieces of evidence led to the charging of Pearsall with vehicular manslaughter.

Think Critically

1. **How did the evidence from the accident reconstruction investigation contradict Pearsall's original statements?**
2. **What type of evidence would you expect to see if Pearsall had remained in his lane?**
3. **What other evidence do you think the investigators looked at while reconstructing this accident?**

Natan Luehrmann-Cowen (2010)

On March 17, 2010, police were called to the scene of a fatal accident involving a 13-year-old boy, Natan Luehrmann-Cowen, and a Ford Ranger pickup truck. The boy's body was found on the corner of an intersection near where the truck had come to a stop. Initially, police thought the death was just an accident because the driver of the truck, Andrew Richard Meyers, reported that the boy had ridden his skateboard into the road. Meyers said that he had tried to swerve out of the way, but had struck the boy by accident.

Blood tests indicated, though, that Meyers had a prescription anti-anxiety medication in his system along with two different illegal drugs. Meyers was also a known heroin addict. As investigators begin reconstructing the accident, they realized that the drugs Meyers had taken had altered his perception of reality and his reaction time. First, skid marks indicated that Meyers had been driving 63 miles per hour in a 25-mile-per-hour zone. Furthermore, a time–distance analysis proved that Luehrmann-Cowen had already safely crossed to the opposite side of the road from Meyers's vehicle when

Meyers struck him. However, because he was under the influence and driving at such high speeds, Meyers overreacted when he saw the boy in the street, skidded out of control, and hit him. Meyers (see Figure 13-19) has been charged with vehicular homicide for his role in the boy's death.

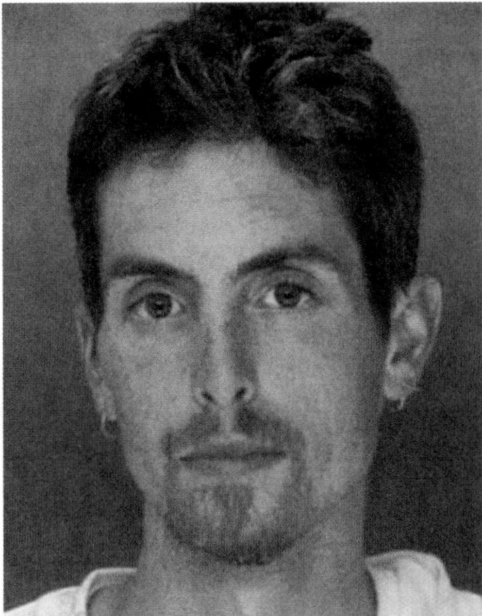

Figure 13-19. *Andrew Meyers was charged with vehicular homicide for striking and killing a 13-year-old boy while under the influence of drugs.*

Think Critically

1. What do you think a time–distance analysis involves?

2. How did drugs play a role in this accident?

3. Explain how police can use skid marks to determine the speed of a vehicle.

Bibliography

Books and Journals

Baker, J. Stannard, and Lynne B. Fricke. (1986). *The Traffic-Accident Investigation Manual: At-Scene Investigation and Technical Follow-Up*. Evanston, IL: Northwestern University Traffic Institute.

Chisum, W. Jerry, and Brent E. Turvey. (2007). *Crime Reconstruction*. Burlington, MA: Elsevier Academic Press.

Conklin, Barbara, Robert Gardner, and Dennis Shortelle. (2002). *The Encyclopedia of Forensic Science*. Westport, CT: Onyx Press.

Lerner, K. Lee, and Brenda Wilmoth Lerner, eds. (2006). *The World of Forensic Science*. New York: Gale Group.

Newton, Michael. (2008). *The Encyclopedia of Crime Scene Investigation*. New York: Info Base Publishing, pp. 125–126.

Rivers, R.W. (2006). *Evidence in Traffic Crash Investigation and Reconstruction: Identification, Interpretation and Analysis of Evidence, and the Traffic Crash Investigation and Reconstruction Process*. Springfield, IL: Thomas Books.

Websites

www.cengage.com/school/forensicscienceadv

www.iihs.org/research/fatality_facts_2009/teenagers.html

www.swgmat.org/IR_Guideline_061408_taj_combo_rev_final%20voted%20on%20comments%20 added%20040909c.pdf

www.waltersforensic.com/articles/accident_reconstruction/vol1-no8.htm

CAREERS IN FORENSICS

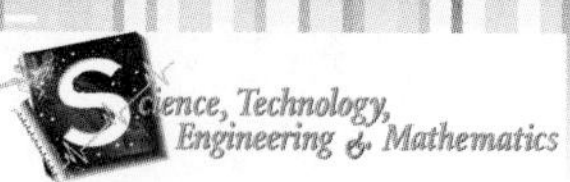

Ralph Cunningham: Accident Reconstruction Specialist

No two days are the same for Accident Reconstruction Specialist Ralph Cunningham (see Figure 13-20). On some days, his work involves civil litigations. He investigates the facts relating to an accident and then presents an unbiased conclusion and opinion concerning what happened in an accident. Much of this work is done for insurance companies or private parties in hopes that a fair settlement can be reached without the expense and emotional drain of a trial.

On other days, Cunningham testifies in criminal defense cases. He gives expert testimony about the facts in a case in which someone has been charged with homicide because they had been driving at excessive speeds, causing a fatal accident.

On still other days, Cunningham spends time analyzing unexpected details about a case. For example, Ralph once evaluated the wear on the tires of a tractor-trailer. Inadequate depth of tread on the rear tires combined with wet road and a steep decline resulted in the driver's inability to stop when a pickup truck driving in the opposite direction spun out of control and directly into the path of the tractor-trailer. Because the rear tires of the tractor-trailer had not been maintained, the vehicle was unable to stop to avoid the pickup truck. The driver of the pickup truck did not survive the crash.

COURTESY OF RALPH CUNNINGHAM

Figure 13-20. *Ralph Cunningham*

When asked about the requirements of his profession, Cunningham said, "Common sense is fundamental. An understanding of mathematics and physics of vehicle motions (primarily kinematics and dynamics) and collision interactions (momentum and energy principles) is essential. There are many tangential aspects which build on the basics: filament examination to determine if lights were on or off at impact; crash data retrieval to extract and interpret data contained in on-board event recorders; an understanding of the mechanical components of motor vehicles—how they work together, and what to look for when they fail to work together properly; human factors in the driving environment—what normal or typical reaction times are, what driver expectations are, the differences between the driving environment and a test environment or a pedestrian environment, and other aspects of human performance. It is a challenging field which requires a well-rounded skill set and provides new challenges with each assignment—no two collisions are alike."

Accident reconstructions can work in the private sector or in law enforcement. Either way, most accident reconstructionists begin their training as police officers. They then go on to specialize in accident reconstruction through special assignment and by taking extra classes and certification courses. Cunningham emphasizes that this field "requires dedication, study, and the willingness to keep learning, because the science and art of accident reconstruction continue to evolve and mature."

Learn More About It

To learn more about a career in accident reconstruction, go to www.cengage.com/school/forensicscienceadv.

CHAPTER 13 REVIEW

Multiple Choice

1. A teenager is driving along a road on a rainy night. Although he is driving at the speed limit, he loses control when the car hydroplanes and he crashes into a tree. Would the investigation progress beyond the first level of accident investigation? *Obj. 13.4*
   a. yes, because all accidents require multiple levels of investigation
   b. yes, because all accidents involving teen drivers require technical follow-up
   c. no, because there were no fatalities
   d. no, because there was no criminal activity

2. After a minor automobile accident, a report must be written. An investigator's report will include _____. *Obj. 13.4*
   a. basic information about the persons and vehicles involved
   b. measurements of any skid or yaw marks
   c. results from field sobriety tests
   d. all of these

3. Which of the following is *not* part of the level two protocol of an investigation? *Obj. 13.4*
   a. taking photographs of the scene
   b. measuring tire marks
   c. rolling the driver's fingerprints
   d. conducting a field sobriety test

4. The coefficient of friction for asphalt is 0.50–0.90. If the asphalt is wet after a rainstorm, what would likely happen to the coefficient of friction? *Obj. 13.5*
   a. The coefficient of friction would remain the same.
   b. The coefficient of friction would increase.
   c. The coefficient of friction would decrease.
   d. The change in coefficient of friction would be negligible.

5. Paint samples collected at a crash site will be subjected to spectrometry testing. Spectrometry is useful for which of the following? *Obj. 13.3*
   a. classifying pigments in the paint
   b. determining molecular structure of the paint
   c. classifying solvents in the paint
   d. all of these

Short Answer

6. Discuss how accident investigators rely on the conservation of energy when developing a reconstruction of a motor vehicle accident. *Obj. 13.6*

7. Provide an example of how Newton's laws of motion are used for accident reconstruction. *Obj. 13.6*

8. Explain the rationale behind staging a crime scene. Include things investigators look for that indicate staging. *Obj. 13.2*

9. Reconstructionists are working on an accident reconstruction to prepare for an upcoming court case. Time is short and their superiors want to get the case closed. Explain some potential pitfalls the reconstructionists must avoid in order to ensure lack of bias in the process. *Obj. 13.7*

10. Discuss how environmental factors can contribute to motor vehicle accidents. *Obj. 13.3*

11. Create a detailed scenario of a motor vehicle accident. Include at least two of the three common causes of accidents. Identify all of the contributing factors involved. *Obj. 13.3*

12. Identify and explain the significant key contributors to the process of crime reconstruction. *Obj. 13.1*

13. Compare and contrast tire marks, skid marks, and yaw marks. *Obj. 13.5*

14. A two-car accident occurred on a flat stretch of gravel road. Two distinct sets of tire marks are discovered. Car A made skid marks measuring 35 m. Car B made yaw marks with a measured radius of 29.5 m. The speed limit on the road is 55 km/h. Calculate the speeds of each vehicle. Were either of the cars speeding? Explain. *Obj. 13.5*

MATHEMATICS

PHYSICS

15. A car lost control and left the flat cement road as the driver was trying to drive around a curve. The radius of the yaw mark is 74 m. The speed limit on this road was 65 km/h. What was the car's speed? Was the car speeding when it left the road? *Obj. 13.5*

16. What is the momentum of a 1,500 kg vehicle driving at 30 m/s? *Obj. 13.6*

17. Imagine that Car A, with a mass of 1,000 kg, is driving southbound at 25 m/s when it collides with Car B, with a mass of 2,000 kg. Car B is traveling at the same speed but in the opposite direction. *Obj. 13.6*

a. What is the momentum of each vehicle just before the collision?

____

____

____

b. What is the total momentum after the collision?

____

____

____

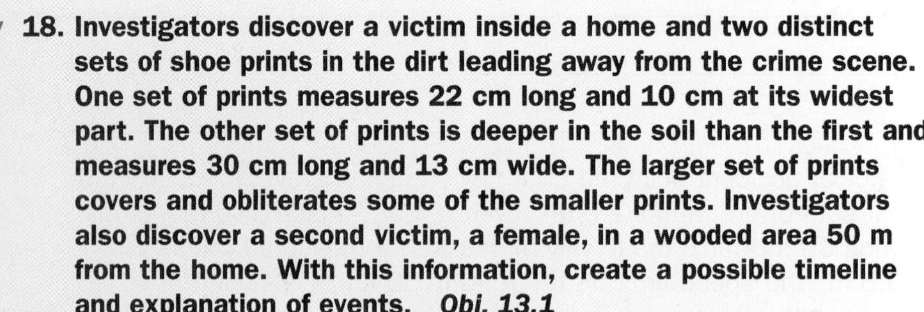

Think Critically

18. Investigators discover a victim inside a home and two distinct sets of shoe prints in the dirt leading away from the crime scene. One set of prints measures 22 cm long and 10 cm at its widest part. The other set of prints is deeper in the soil than the first and measures 30 cm long and 13 cm wide. The larger set of prints covers and obliterates some of the smaller prints. Investigators also discover a second victim, a female, in a wooded area 50 m from the home. With this information, create a possible timeline and explanation of events. ***Obj. 13.1***

19. A local theme park has a public transport system. As a bus was pulling away from the depot, a collision occurred between that bus and a taxi. A nearby pedestrian was injured. Explain the steps in the process of investigating this accident. Include any and all levels that may be required to complete the investigation and the rationale behind each. ***Obj. 13.4***

CRIME SCENE

Student-Prepared Original Titles

Three Bears and the Curious Girl

By: **Skye Ratter**

East Ridge High School
Clermont, Florida

Once upon a time in a large forest, there stood a cottage where the Bear family lived. Papa Bear was the Forest Law-Enforcement Sheriff.

Papa Bear was very large, Mama Bear was medium sized, and Baby Bear was very small. Inside the cottage, each bear had his or her own bed—Papa Bear's was the largest, Mama Bear's was middle in size, and Baby Bear had a little bed. In the dining room, there were three china bowls laid out on the kitchen table—a large bowl for Papa Bear, a smaller one for Mama Bear, and the smallest for Baby Bear.

One day, the Beaver family asked the Bears to come over for an afternoon visit. They happily agreed. Mama Bear decided to get supper ready before visiting with the Beaver family. She cooked up her famous porridge and placed the porridge in the bowls on the table; it would take hours for the porridge to cool. The family then headed over to their friends' house.

A young girl from a nearby village was picking berries. She began to smell the most delicious aroma coming from the Bears' home. She dropped her berries, even accidentally squishing a few in her excitement, and ran up to the Bear Family house.

Once inside, she followed the delicious scent into the kitchen, where she saw the three bowls left lying out on the table. She sampled the largest bowl—it was extremely hot. She then sampled the medium bowl; it was too cold. Finally, she tried the smallest bowl. It was just right. She ate the entire bowl.

After she ate, she felt tired. She found the bedroom with the three beds. The first one was too hard and the middle one was too soft. In all of her frustration and rolling around to get comfortable, her hair got caught on a piece of wood and ripped out several strands on Mama Bear's bed. However, she tried the smallest one, and found that it was just right.

It started to get late, and the Bear family decided that it was time to go home and eat supper. When the three of them got home, they saw that the door had been opened—there was an intruder in their home! Papa Bear carefully searched the house, looking for the culprit. As he was searching, the girl fled the house through an open window and ran home (see Figure 13-21).

Figure 13-21. *The intruder fled the home before the Bear family could catch her.*

After Papa Bear discovered the intruder was gone, he scanned their home for any evidence that could be use to find who had broken in. He noticed small, muddy footprints leading to the kitchen. He examined the prints as well as the impressions that had been left outside in the soil. The pattern was flowers and stripes, and the impression did not go deep into the ground. He made a casting of the prints. Based on the small size, the flowery pattern, and the lack of weight of the shoe, Papa Bear determined that a female had entered their home.

He searched the kitchen looking for anything but had no luck. Just as he was about to leave, he noticed that the bowls of porridge had been touched, and Baby Bear's bowl was empty! Papa Bear realized there must be fingerprints on the bowl! He picked up the bowls and found a set of red fingerprints on all three of the bowls. Papa Bear got his magnifying glass to examine the print. He dusted the print, photographed it, and lifted the print. He found ridge characteristics, including bifurcations, bridges, deltas, and ridge endings. Papa Bear noted all of his findings. He bagged the evidence to take to the station after he was finished processing the rest of the house.

Papa Bear had found some hair samples in all three beds. As he collected and bagged the hair samples, he noticed the hair was from someone other than the Bear Family. It was a long blonde strand, something belonging only to a human. Papa Bear knew that there was a chance that DNA could be on one of the strands. He bagged the evidence, took it downstairs, and added it to the fingerprint samples and the footprint evidence.

At the police station later that night, Papa Bear scanned the prints into AFIS to search for a suspect. Because the village was very small, it did not take long before he found the individual whose prints matched.

Papa Bear sent the hair samples to the DNA lab. The first couple of samples tested had no DNA. The third sample tested had a follicular tag. The DNA from the follicle was tested and a match was found. The DNA match was the same individual whose prints were found on the bowls. They had their culprit.

The next day, Papa Bear interrogated and arrested the individual who had broken into their home and used their personal belongings. The young girl was charged with breaking and entering. She accepted a plea agreement and got 6 months' probation. She was never allowed to enter the nonhuman part of the village again.

Activity:

Answer the following questions based on information in the Crime Scene S.P.O.T.

1. You are an expert crime reconstructionist and were asked by the prosecutor to develop a reconstruction of the events in this crime. Based on the details in the story and your deductive reasoning skills, develop a courtroom-ready sketch to explain the order in which the crime occurred.

WRITING

2. As an expert you will have to communicate your data to the prosecutor to make a final determination as to whether to proceed with a trial or a plea deal. Write a one-page brief explaining the details of the events of the crime drawn in the crime-scene sketch. Use proper terminology. Also, incorporate the evidence recovered and processed at the crime scene in your evaluation of the scene.

Preventing Adolescent Crime Together™ (P.A.C.T.)

Introduction:

According to the Insurance Institute for Highway Safety, approximately 3,500 teens were killed in car accidents in the United States in 2009. Motor vehicle crashes are the leading cause of death among 13- to 19-year-olds in the United States. For every mile driven, teens are four times more likely to get in an accident than any other population group. Approximately 26 percent of 16- to 19-year-olds who die in car accidents have a blood alcohol concentration above .08. Also, teens are more likely to engage in other reckless behavior while driving, such as not wearing a seat belt, texting while driving, and speeding (see Figure 13-22.)

Procedure:

1. Select three other people to work with—you should be in a group of four.
2. Research teen-related car accidents. Focus on a specific behavior, such as speeding or texting while driving, that can lead to a car accident. Each group must focus on a different behavior.
3. Your group will be responsible for writing a scenario describing events leading up to an accident. The accident must be caused directly by the teen driver participating in the behavior you are highlighting.
4. Develop scene boards of the story. What will the characters look like? What will the car look like? Will it be daylight or nighttime? Is the driver male or female? Are there fatalities? If so, how many?
5. In collaboration with the computer graphics department, share your scenario and work with the computer graphics students to develop a "mini-movie" or documentary.
6. A short movie can be done on any multimedia you wish to use as long as the features of the scenario are in included.
7. As a class, combine all of the movies onto a single CD or flash drive.
8. Working with administration, try to get the movie footage broadcast across the school. Perhaps you can showcase one movie per day. You may also set up the movies in one location such as the media center and invite students to view the movies. Have the student body vote on the one they felt was most effective.
9. Offer to share your movies with other high schools in the county. The more teens are educated on the dangers of reckless behavior while driving, the greater the chance they will not engage in the negligent acts.

WAVEBREAKMEDIA LTD./SHUTTERSTOCK.COM

Figure 13-22. *Texting while driving causes many accidents each year.*

ACTIVITY 13-1 *Ch. Obj: 13.6*
CRASH TEST GUMMIES

Objectives:

By the end of this activity, you will be able to:

1. Explain the scientific principles behind seat belt usage.
2. Identify variables that affect the speed of the vehicle.

Materials:

(per group of two students)
toy car or cart
gummy-style bear-shaped candies (2 sizes)
ramp
meter stick
wooden blocks
string
scissors

Safety Precautions

Safety goggles should be worn at all times because the "crash victims" have the potential to become airborne.

Procedure

Test 1

1. Place one end of the ramp on a single block or book on the desktop or floor. Record the elevation of the ramp.
2 At the other end of the ramp, stack some blocks or books to serve as the point of impact for the "crash."
3. Place a driver on the cart or car.
4. Set the vehicle at the top of the ramp and release.
5. Observe what happens to the "driver" when the vehicle hits the barricade at the bottom of the ramp.
6. Measure the distance the "driver" was thrown from the vehicle.
7. Construct a chart to record your observations and measurements.
8. Repeat for a total of at least three trials. Calculate the average distance traveled by the driver.

Test 2

1. Place a "passenger" in the backseat of the vehicle. (Use the smaller candy if two sizes are available.)
2. Repeat Test 1 with two occupants in the vehicle.
3. Record the results.

Test 3

1. Add another block to increase the elevation of the ramp. Record the new elevation.
2. Repeat Tests 1 and 2.
3. Record your observations and measurements.

Test 4

1. Strap the occupants into the vehicle using pieces of string to simulate a seat belt.
2. Repeat Tests 1, 2, and 3.
2. Record the results.

Questions:

1. How did the elevation of the ramp affect the distance the "driver" of the vehicle was thrown?
2. How did the addition of "seat belts" change the outcome of the crash?
3. Why do you think infants are restrained in rear-facing child safety seats?
4. How could you modify this activity (using the same car or cart) to test the effect of a larger vehicle on the same "road"?
5. Explain how a reconstructionist could use this concept to determine the events of a motor vehicle accident.

Extension:

Hypothesize ways to change the speed of the vehicle without changing the elevation of the ramp. Complete the experiment to test your hypothesis. Be sure to record data and write a lab report. The report should include the hypothesis, materials, procedure, data, analysis, and conclusion.

ACTIVITY 13-2 *Ch. Obj: 13.4, 13.5, 13.6*
SCIENTIFIC INQUIRY AND RECONSTRUCTION

Objectives:

By the end of this activity, you will be able to:

1. Evaluate the evidence found at the scene of an accident.
2. Formulate a reconstruction of an accident based on the evidence recovered from the scene.

Materials:

(per group of two or three students)
paper
pencil
envelope containing evidence

Background:

Law-enforcement officers are called to the scene of an automobile accident. The driver, a male approximately 30 years of age, was found unconscious and transported by ambulance to the hospital. The airbag had deployed and various items of evidence were collected to be sent to the lab for evaluation and analysis.

Procedure:

1. Obtain an envelope from your teacher. Do *not* dump the contents.
2. At random, select three cards from the envelope.
3. Write a story based on the information on your cards.
4. After you have finished, select three more cards from the envelope at random.
5. Reevaluate your story based on the new information. Make any necessary modifications or changes to your story.
6. Repeat the procedure until you have evaluated all of the cards from the envelope.
7. Rewrite the story—the reconstruction—based on all of the information you now have.

Questions:

1. How did your reconstruction change as you gathered more information and/or evidence?
2. What impact would additional information or evidence have for the reconstructionist? Explain.
3. If another group of students was given the same envelope your group had, how likely is it that they would have developed the same or similar reconstruction? Why is it possible to have more than one explanation of an event when the evidence is the same?

Extensions:

1. Develop a similar crime-scene activity. Trade scenarios and evidence envelopes with another team. Once you have completed the reconstruction, make a sketch or build a three-dimensional model of the crime you have reconstructed.

WRITING

2. Write a short crime scene scenario. Based on your scenario, draw a thorough sketch of your crime scene. If you have computer software to design crime scenes, you may use that. Otherwise, your crime-scene sketch should include all the components of a traditional crime-scene sketch. The crime can occur anywhere, but be very clear about the physical evidence and location of the body in the sketch. Next, your teacher will collect the sketches and scenarios. Your teacher will pass back the sketches randomly. Review the sketch you are given. Based on the information provided in the sketch, can you determine the events leading up to and during the crime? Afterward, write a scenario of what you think happened in your assigned sketch. After you have finished, ask your teacher to see the original writer's scenario and see how close your perspective was to your peer. If there was a difference in perspectives, discuss it with the original writer. What does it mean to you if your scenario matches that of the original writer?

CHAPTER 14

Cyber Crimes

IT TAKES A HACKER TO CATCH A HACKER

DAN CALLISTER/NEWSMAKERS/GETTY IMAGES

Figure 14-1. Kevin Mitnick

While at school, Kevin Mitnick never stood out as a promising student. His grades were below average and he had few friends. But there was one place where Mitnick excelled—he had an instinctive understanding of computer networks and how to manipulate them.

As a young man, Mitnick started his hacking career by taking over the digital central office switches of telephone companies. At first, he used this skill to make free telephone calls, but eventually Mitnick became more aggressive. He started eavesdropping on people's phone calls, breaking into government databases, stealing credit card numbers, and much more.

The FBI tried to pin Mitnick down, but he could easily use his computer skills to evade them. Mitnick's downfall came when in 1994 he hacked into the computer of Tsutomu Shimomura, a physicist and brilliant computer engineer. Shimomura took the theft of his files personally and decided to help the FBI catch Mitnick.

Shimomura started by setting up software "monitoring posts" to track hackers' activities on the Internet. Once he found an attack that had Mitnick's signature, he was able to trace Mitnick's computer to Raleigh, North Carolina, and determine that Mitnick was using a cellular modem. Shimomura flew to North Carolina and worked with the phone company there to use cellular-frequency scanners to hone in on Mitnick's location.

Shimomura and phone company technicians drove around Raleigh until they picked up signals consistent with Mitnick's modem in a neighborhood near the airport. The FBI quickly zeroed in on Mitnick's apartment and placed him under arrest. He became the first person to be convicted of gaining access to an interstate computer for criminal purposes.

© NAJLAH FEANNY/CORBIS SABA

Figure 14-2. Tsutomu Shimomura

OBJECTIVES

By the end of this chapter, you will be able to:

14.1 Discuss typical uses for the Internet.

14.2 Differentiate among the three general categories of cyber crime.

14.3 Discuss the process of investigating and processing various types of computer evidence.

14.4 Distinguish among the four types of computer evidence presented at court.

14.5 Identify various types of evidence that can be collected at a cyber crime scene and the forensic value of each.

14.6 Discuss the importance of the various tools available to cyber crime investigators and experts.

14.7 Explain the importance of the expert witness in cyber crimes.

14.8 Examine how cyber evidence is documented.

14.9 Discuss concerns associated with the future of cyber crimes.

VOCABULARY

clone - a copy made in the same type of medium

computer forensics - the specialized practice of identifying, preserving, extracting, documenting, and interpreting electronic data that can be used as evidence

content spyware - software that is used to allow a hacker to access all the activity on an individual's personal or business computer

cyber-terrorism - hacking into a company's internal networking system for the purpose of demonstrating or protesting a political agenda

hacking - intentionally entering an unauthorized network system

Internet forensics - uses the same analysis techniques as computer forensics except the emphasis is placed on the Internet as a whole

malware - software designed to provide unauthorized access to a computer system

phishing - illegally gathering personal information

Trojan horse - software designed with the intention to harm a computer or the information on the computer

worm - self-replicating malware program that spreads through a computer system by sending copies of itself to networked computers

Obj. 14.1 INTRODUCTION

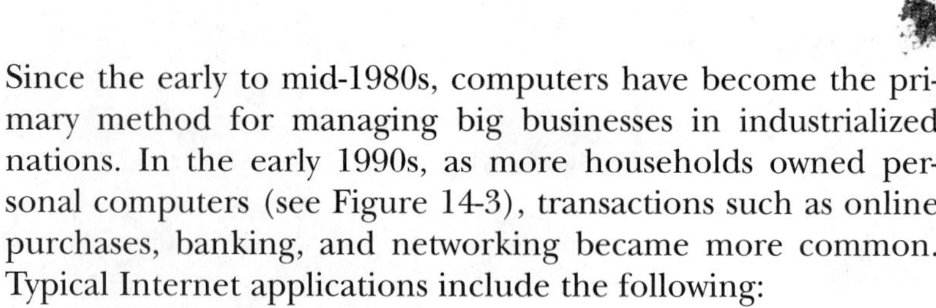

Figure 14-3. *This woman is using her home computer to make a purchase online.*

Since the early to mid-1980s, computers have become the primary method for managing big businesses in industrialized nations. In the early 1990s, as more households owned personal computers (see Figure 14-3), transactions such as online purchases, banking, and networking became more common. Typical Internet applications include the following:

- Information services
- Leisure and recreation
- Mobile access
- Education and training
- Distribution systems
- Social networking and communication

As industry and individuals continue to rely more and more on readily accessible information, the risk of someone or a group of people using the massive amounts of information illegally increases. Information may be invaded when someone destroys hardware, software, or communication accounts, such as e-mail or social networking accounts. Criminals might also attempt to retrieve financial information, such as bank statements and credit card numbers. The digital age has led to a relatively new, cutting-edge area of forensics called *computer forensics.* **Computer forensics** is the systematic identification, preservation, extraction, documentation, and analysis of electronic data that could potentially be used as evidence in court. In **Internet forensics**, additional emphasis is placed on the Internet as a whole.

Working in the field of computer forensics requires extensive knowledge of computer hardware and software. A careless error or lack of knowledge can destroy evidence or accidentally alter the data, making it inadmissible in court. Also, an individual working in computer forensics must be very familiar with legislation governing local, regional, state, national, and international laws.

IDENTITY THEFT

Figure 14-4. *Criminals can gather personal information from utility and credit card statements. They can use this information to steal an individual's identity.*

Because the Internet stores massive amounts of information, criminals are able to steal personal information and obtain credit cards and loans. The theft can take place in several ways. When individuals and businesses throw away unwanted documents, such as bank and credit card statements, a criminal can take them from the dumpster and gain full access to personal information, making it very easy to take the information and apply for credit cards. These types of documents (see Figure 14-4) should always be shredded before being placed in a trash can or a dumpster. Another, more common, method of stealing identity information is by **phishing**—defrauding the victim by sending

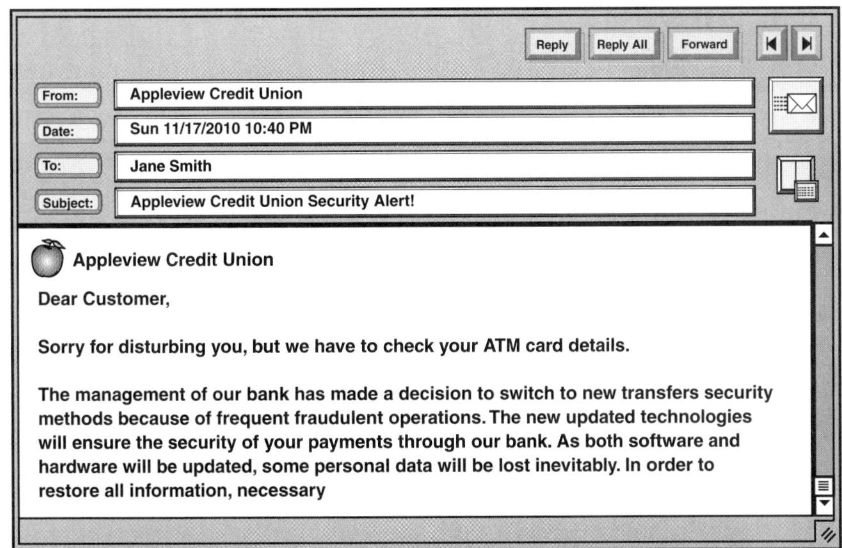

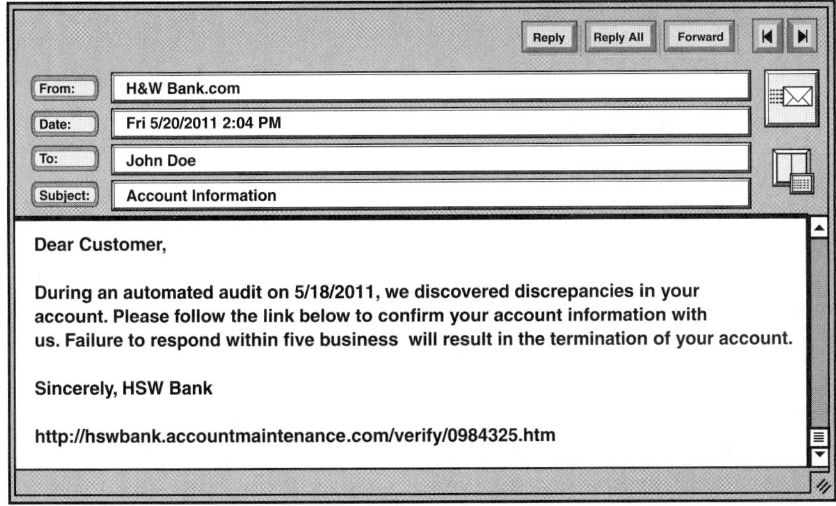

Figure 14-5. *Examples of phishing e-mails. Notice the generic greeting and request for personal information. Your financial institution should never request personal information through an e-mail.*

a fraudulent e-mail that looks remarkably real and asks the recipient to update his or her personal information. The e-mail (see Figure 14-5) usually looks like it came from the victim's banking institution or an online retailer. The e-mail states that if personal information is not updated, the victim's service or access will be temporarily denied until the information is updated. Unfortunately, criminals are generally able to develop new ways of accessing the information as quickly as technology experts are able to stop each scam.

Spyware is another tool criminals use to obtain personal information. Installing spyware on a victim's computer gives the criminal access to all the computer activity of the individual—including password information. The spyware gathers the information and sends it electronically to the criminal. Spyware is dangerous in many ways. It covertly sends nearly all personal information stored on the computer, including account numbers, passwords, and user IDs, to the criminal, and the victim has no idea the theft is occurring (see Figure 14-6). Spyware is described in further detail later in this chapter.

It is estimated that one in six people around the world has access to a computer.

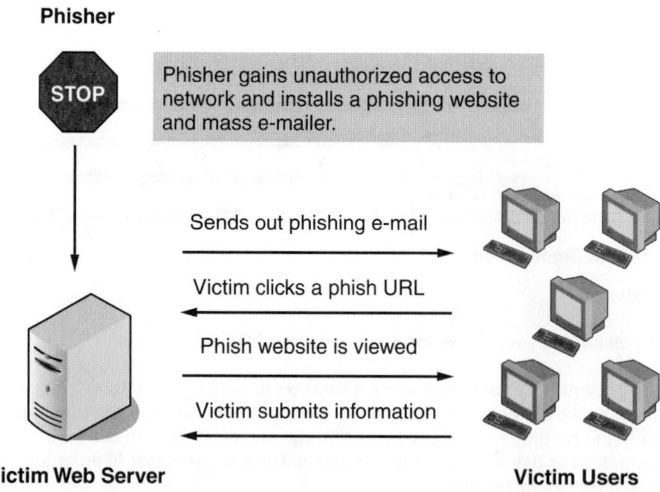
Phisher
STOP
Phisher gains unauthorized access to network and installs a phishing website and mass e-mailer.
Sends out phishing e-mail
Victim clicks a phish URL
Phish website is viewed
Victim submits information
Victim Web Server
Victim Users
Ways to Stop a Phishing Scam
• Secure the network before phisher gets access.
• Block the mass e-mail so it never reaches victims.
• Educate network users not to click suspicious web links.
• Prevent the suspicious website from opening on network computers.
• Educate network users not to submit personal account information.

Figure 14-6. *The flow of how phishing can reveal personal and business information.*

When individuals make online purchases, it is common to use a credit card as the form of payment. In the past, if networks were not secure enough, criminals could access the account information and use the credit card number to make fraudulent purchases. Now that credit cards have three- or four-digit validation codes, it is far more difficult for a credit card number to be used fraudulently online. However, with the expansion of the online gambling industry and deceptive Internet advertising banners, the risk of identity theft continues to exist. Ongoing rigorous efforts are necessary among technology experts to provide secure websites for consumers.

Obj. 14.2

TYPES OF CYBER CRIME

Cyber crimes—computer and Internet crimes—typically fall into at least one of three categories. Experts in analyzing evidence in each category must have knowledge and expertise in various types of law. The three general categories of cyber crime are:

- Computer integrity crimes
- Computer-assisted crimes
- Computer content crimes

Today, most computer and Internet crimes fall into more than one of these categories. For example, phishing falls into all three categories. The victim responds to a spam e-mail (computer integrity crime) by logging onto a fraudulent website that appears to be the victim's bank (computer content crime) and provides key personal information to the perpetrator (computer-assisted crime). Ultimately, even though all three categories are involved and multiple crimes are committed, the primary motive is to steal money from an unsuspecting victim.

COMPUTER INTEGRITY CRIMES

Computer integrity crimes are crimes that involve illegal access to data on a computer or network system. **Hacking** is intentionally entering an unauthorized network system. People who commit this crime are called *hackers*. By destroying the security of a network, the hacker compromises the integrity of the computer system and gains access to the previously protected information. Hackers might steal proprietary, commercial information, or personal identity data (see Figure 14-7). In the worst cases, significant destruction to the internal structure of a network can lead to a massive loss of corporate and personal information. This can cause extensive loss in business income and physical equipment.

```
Last login: Mar 12 07:03:29 on console
Welcome to os4!
> telnet -a -b ABSOLUT 188.142.200.2:8080
> enter login: #####
> enter passw: #####
> invalid passw ERROR (retype)
> retype passw #######
> OK you are SUCCESFULLY logged in
> cd/usr/ABSOLUT/SECRETS
> 1s -1 -a BACKDOORVIRUSES
-rwxr-xr--  TROJANHORSE#BF1 -    306 Mar 20:55
-r-xr-xr--  TROJANHORSE#CA0 -   1026 Mar 00:13
-r-xr-xr--  TROJANHORSE#CB9 -    716 Mar 14:15
-rwxrw-r--  TROJANHORSE#CFF -   4865 Feb 22:06
-r-xr--r--  TROJANHORSE#D2C -     48 Jan 17:24
-r-xr-xr--  TROJANHORSE#D8A -    512 Mar 02:22
-r-xr-xr-x  TROJANHORSE#DA6 -    512 Mar 04:46
-r-xr--r--  TROJANHORSE#DD7 -    642 Feb 01:58
-r-xr--r--  TROJANHORSE#DF2 -   1784 Dec 11:33
-rwxr-r--   TROJANHORSE#EA3 -   1256 Mar 14:56
-rwxrw--r-- TROJANHORSE#EB4 -   2873 Mar 08:17
-r-xr--r--  TROJANHORSE#ED8 -    255 Feb 10:45
-rxr--r--   TROJANHORSE#FA3 -    207 Feb 10:57
> sudo -sP  TROJANHORSE#D2C
System is about to reboot
Killing all proceses ........
```

Figure 14-7. *Hackers crack codes inside a personal or business computer system to steal information. They can then install software that gives them unauthorized access to private information.*

The term *hacker* is also used to describe a person who has been hired to test the strength and integrity of an organization's network security. These computer experts test the vulnerability of the system and help the organization improve security. People in the industry sometimes refer to this group of hackers as *ethical hackers*. As the Internet was beginning to develop and evolve, ethical hackers helped make security network systems safer and more protected. They ensured that the codes encrypted to protect the networks were of the highest standards possible.

Unfortunately, some people with the knowledge and skills to test computer security systems use these skills to commit crimes. They steal data for personal gain or they break into the system or website of an organization with which they disagree. In recent years, fear of cyber-terrorism has generated a lot of attention. **Cyber-terrorism** is hacking into a company's internal networking system for the purpose of demonstrating or protesting a political agenda. Often, cyber-terrorists cause a great deal of fear and insecurity among the network's users. The users generally fear the loss, destruction, or theft of stored data. Mass-transit companies, such as airlines and railroads, are prime targets for this sort of attack. Also, military and government agencies, such as the Pentagon, are also at great risk.

Digging Deeper
with Forensic Science e-Collection

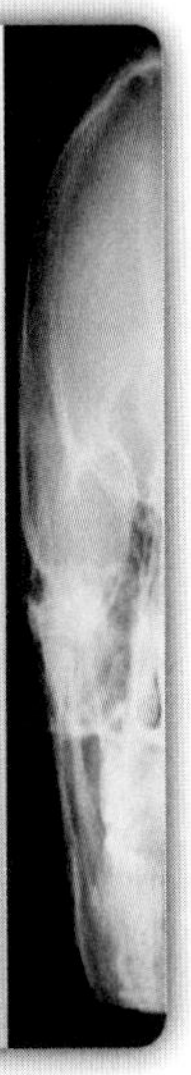

Nearly all businesses and agencies now use computerized databases to store massive amounts of information. Social security numbers, account balances, personal or medical histories, and contact information allow the businesses to operate efficiently. However, the availability of the data and the threat of a security breach may put clients and customers at risk of identity theft, loss of finances, damage to reputation, stalking, and even blackmail. Using the Gale Forensic Science e-Collection at www.cengage.com/school/forensicscienceadv, research ways in which organizations are safeguarding the privacy of electronic records. Write a one-page summary of the methods being used to ensure the integrity of personal information.

Cyber crimes can have significant financial impact on businesses. The ILOVEYOU virus cost users worldwide an estimated billion dollars. When the cost of lost data, losses in business, and decreased productivity were included, the estimates were as high as $6.9 billion!

Hackers rely on three specific methods to infiltrate a network system—social engineering, malware, and spyware. Additionally, hackers are often able to damage software and codes to infiltrate the network.

Social Engineering

Sometimes a hacker builds interpersonal relationships with specific employees at the target company. By establishing trust with key people, hackers can sometimes determine possible passwords the victim uses to enter the system. If developing a relationship with the target does not yield the desired result through conversation, the hacker may also search through discarded materials. This information might either provide the hacker with access information or further assist the hacker in building credibility with the victim.

Malware

Malware is software designed to provide unauthorized access to a computer system. Illegal spyware is an example of malware. Other examples include Trojan horses and worms. A **Trojan horse** is software that is designed with the intention to harm a computer or the information stored on the computer. Generally, a Trojan horse appears to be legitimate, useful software to the user, but when it is run or installed, it provides the hacker unauthorized access to data stored on the system. A **worm** is self-replicating malware that sends copies of itself to other computers on the network. Worms almost always damage the infected computer networks. A computer *virus* is similar to a worm, but it requires a specific command or file be executed or opened before it can attach itself and infect the computer.

Digging Deeper
with Forensic Science e-Collection

The first computer viruses were mostly the result of pranksters with a little too much time on their hands and some computer savvy. Today, it is a totally different story. Organized-crime networks are using viruses to infect computers worldwide in order to offer fake services or products to make a lot of money very quickly and then disappear. Viruses also immobilize many businesses and disable personal computers. Using the Gale Forensic Science e-Collection at www.cengage.com/school/forensicscienceadv, research how viruses are created, infect, and destroy. Write a one-page paper comparing and contrasting computer viruses with biological viruses.

Spyware

As described earlier in the chapter, *spyware*, a type of malware, is software that collects information about the computer's user without the user's knowledge. A relatively harmless method of spyware tracks an individual's search preferences and online purchases. Advertising analysts use this information to analyze online consumer searching and spending. This information also helps advertisers determine where to advertise a product or service. Conversely, a more malicious kind of spyware uses password-sniffing technology to determine passwords and record communication between users on the network system. **Content spyware** is software that allows the hacker to access all of the activity on an individual's personal or business computer.

The information gathered includes activity received or sent from the target computer.

COMPUTER-ASSISTED CRIMES

Figure 14-8 illustrates how far-reaching and connected the Internet has become. Sometimes, criminals use the computer to deceive an individual or business for the purpose of gaining information on a global scale. This category of crime is called *computer-assisted crime.* There are three subsections of computer-assisted crimes: the virtual bank robbery, the virtual sting, and virtual theft.

The Virtual Bank Robbery

Many individuals and business manage finances and perform financial transactions online. This activity makes the financial institutions particularly vulnerable to criminal activity. For example, a criminal might open bank and credit card accounts under false identities. Then, once a credit rating has been established under the false identity, the criminal might be given a loan for a large sum of money and never repay the loan.

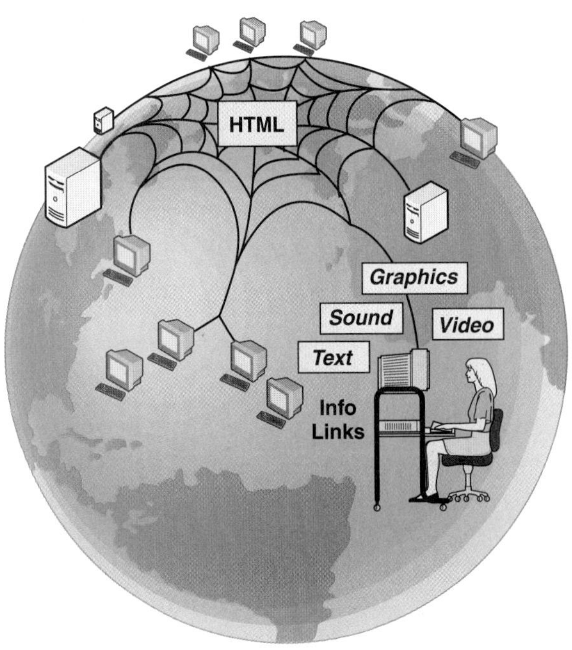

Figure 14-8. *The Internet connects individuals, organizations, and governments around the world. When hackers are able to breach the security, they can have a global reach.*

The Virtual Sting

In the virtual sting, a criminal exploits online financial systems to gain access to goods and services under false pretenses. There are many versions of this kind of Internet crime. A criminal might make purchases with stolen or falsified credit card information. In another version of the virtual sting, called *arbitrage,* goods or services that are illegal or restricted in one jurisdiction are purchased from another jurisdiction. For example, gambling is illegal in many places. However, people are able to circumvent the system and gamble online. In doing so, these people are committing a computer-assisted crime.

The FBI has formed teams of highly trained agents, analysts, and forensic computer and malicious code experts to travel the world to respond quickly to cyber threats. The teams are called Cyber Action Teams or CATs.

The Virtual Scam

In the virtual scam, victims are tricked into giving their money to the criminal. Many of these scams involve "get-rich-quick" schemes. Generally, the victim gives money under the pretenses of a direct investment or a below-market-value product. The victim ultimately receives no goods or services in return.

COMPUTER CONTENT CRIMES

Unfortunately, some of the content on the Internet is offensive and created maliciously. *Computer content crimes* are crimes that involve posting illegal content, such as child pornography or hate speech. The following examples of computer content crime continue to be difficult to control:

- Sexually explicit materials
- Child pornography
- Hateful or aggressive speech or text related to race and extreme politics
- Violent content

CHAPTER 1 2 3 4 5 6 7 8 9 10 11 12 13 14 15

- Distribution of information about making and using drugs and weapons, including bombs
- Sites for organizations to do harm
- Distasteful e-mails, chat rooms, and blogs

The victimization caused by virtual crime is not necessarily any less serious than that caused by physical crimes. Additionally, computer content crimes sometimes inspire other people to perform the acts portrayed or suggested by the text and images on screen. This causes further victimization of innocent people.

Obj. 14.3, 14.4, 14.5, 14.6, 14.7, 14.8

INVESTIGATING AND PROSECUTING CYBER CRIMES

The most common type of cyber crime is the unauthorized access and use of information. Less experienced cyber criminals will leave a trail (often tracing back to their own e-mail address), which makes tracking them down easier. More technologically advanced criminals can make it harder to find them. These criminals often use aliases or someone else's identity. They will also "loop" and "weave" their information across several jurisdictions and continuously encrypt the data. This makes it harder for law enforcement to identify suspects.

ENTERING THE CRIME SCENE

As with any crime, investigators working a computer crime scene collect as much evidence as possible (see Figure 14-9). The computer hardware and components may also be a source of valuable trace evidence. Investigators must establish a chain of custody for each piece of evidence collected. Figure 14-10 describes the forensic value of typical types of evidence recovered from a computer-related crime scene.

Figure 14-9. *Investigator searching for evidence within a central processing unit (CPU).*

PRESERVING THE EVIDENCE

The data collected during the investigation of a computer-related crime is carefully processed and analyzed to ensure that no changes to the original data have been made. If changes have been made, it will be more difficult or impossible to prove the origin of the data. A chain of custody (see Figure 14-11) must be established to ensure the integrity of the evidence, as is the case for all evidence.

Often, the first thing an investigator must decide when preserving computer evidence is how to turn off a system that is currently running.

Evidence	Examples	Forensic Value
Hardware	Keyboard, mouse, scanner lids, laptop case, touchpad.	Often the hardware is the origin of information that will be used in court—the data.
Removable storage	Removable media for data transport and installation programs. Zip disks, CDs, and USB drives are also common places where data is stored.	All removable media are collected and analyzed. Finding information may be a slow process but can produce evidence not available anywhere else.
Documents	Printed memos, reports, handwritten notes, drawings, data on a whiteboard, e-mail addresses, and passwords.	Helps accessing the data needed for court. Provides direction for investigators by narrowing down the scope of the search.

Figure 14-10. *Evidence typically collected from a computer-related crime scene.*

Determining whether to turn off the computer or simply pull the plug is not as easy as it seems. If the investigator wants to stop all processing immediately, he or she will pull the plug out of the wall; however, this technique can demolish anything in the memory. This can also damage all the files on the computer. Alternatively, the suspect could have installed cleansing software that will delete valuable evidence when the computer is shut down properly.

The next concern is how to turn on the suspect's computer. During startup, certain files are accessed immediately, changing the last access date and time stamp. Unfortunately, changing the date and time stamp destroys evidence about when the computer was last turned on. Newer computer forensic software allows investigators to take a live computer and access memory and disk data while the computer is still on. Several stations are necessary to complete this process, and permission or a warrant to install the software is needed. The software is usually carried on a USB drive and can be run straight from the USB drive. The software gives the investigator access to data and files significant to the case in question.

Figure 14-11. *A chain of custody must be established for each piece of computer evidence collected, such as hardware or files stored on a mass-storage device.*

ANALYZING THE EVIDENCE

Before analyzing any of the evidence collected, the investigator must first make an exact image of the drive. This is not merely a copy of the data, but a bit-by-bit image of the original. This kind of copy contains all data on the drive, including deleted files, e-mails, temporary files, and file fragments. A more traditional copy of the data, such as a backup copy, will contain only files that are visible and of known file types. All analysis is done on this exact copy, not on the original. Investigators must strategically look for evidence that may be subtle, hidden, or damaged. The process will depend on the type of investigation it is. For example, if files were collected in a movie-pirating case, the investigators are going to be focused on sound files. The most difficult part for investigators is sorting all the data and collecting and preserving only data pertinent to the case.

Hash codes are large numbers, specific to each file and each drive, that are computed mathematically. If a file or drive is changed, even in the smallest way, the hash code will also change. These hash codes are recomputed on the original and images at various points during the investigation in order to ensure that the examination process itself does not modify the image being examined.

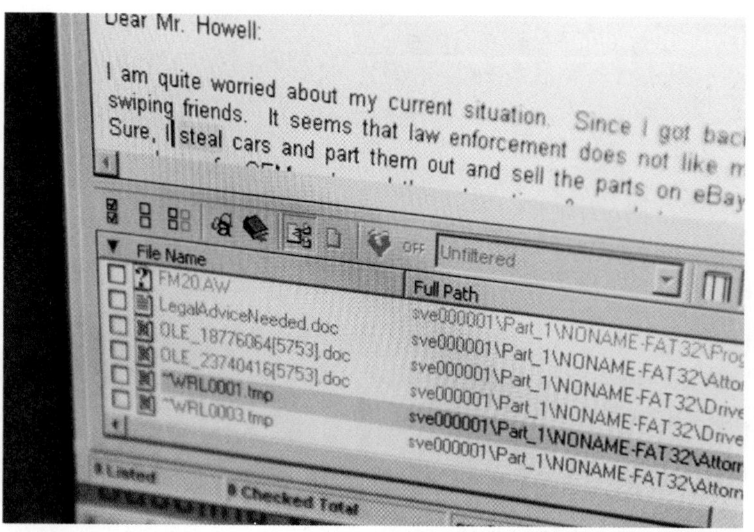

KIM KULISH

Figure 14-12. *Software programs sort and index computer evidence.*

Copying Files

Several types of computer copying software exist and allow for files to be **cloned**—copied in the same type of media, such as sound files or databases. Cloning software also allows for deleted files to be restored. These software programs are utilized because they are quick, and they allow easy access to search for and look at the copied data. File viewer software (see Figure 14-12) allows material to be viewed and examined in multiple formats. Software allows for portions of text and graphics to be selected, cut, and then pasted into the investigator's own computer application.

Computers hold a large amount of trace evidence. In computer forensics, *trace evidence* is essentially "hidden" evidence in deleted files. A computer's hard drive is made up of *sectors,* chunks of memory to store files and data. The sectors make up *clusters,* smaller segments of memory. As a file is saved, the file is given a designated amount of allocated space. For example, if the allocated space is 20 KB and the file being stored is 15 KB, 5 KB of space is unallocated. If the 15 KB file is deleted and replaced with a 10 KB file, 5 KB will still be unallocated, unused, but the other 5 KB will be *slack space,* a section of the cluster that retains information from the original file. The slack space contains fragments of data that are often very valuable to investigators when analyzing evidence retrieved from computers. Figures 14-13 and 14-14 illustrate how parts of deleted files may remain on a computer's hard drive or on a data disc.

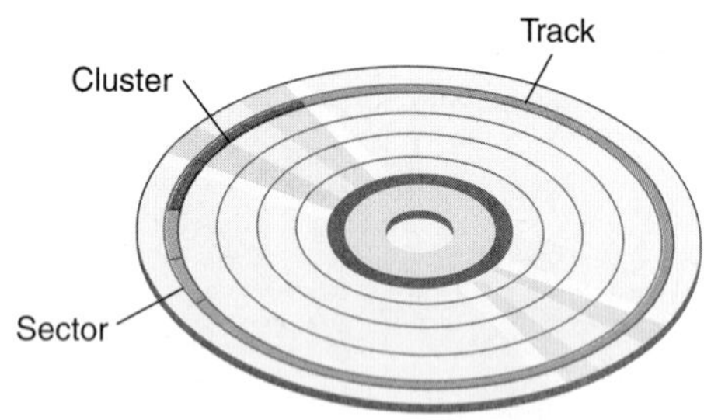

Figure 14-13. *Available space for data storage on a disc.*

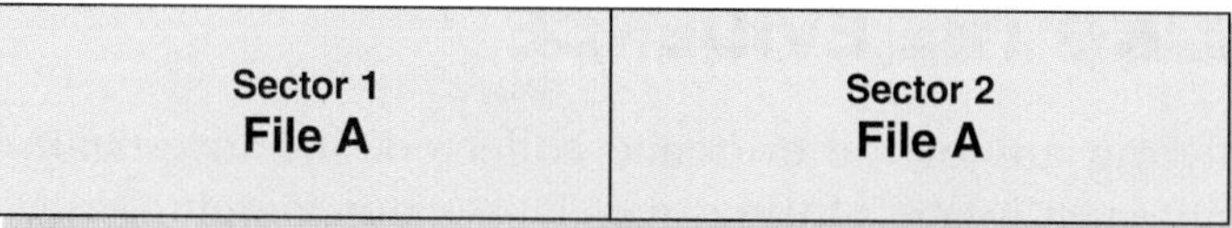

Sectors 1 and 2 are allocated to File A

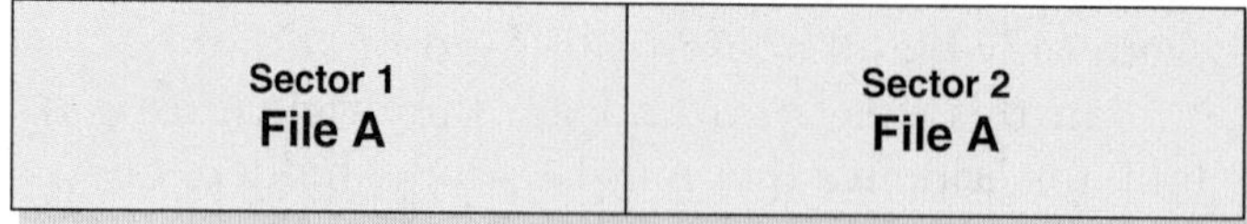

File A was deleted – Sectors marked as unallocated

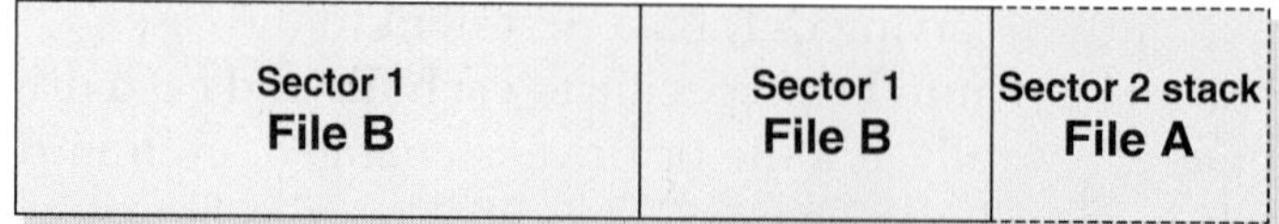

File B written to unallocated space

Figure 14-14. *How slack space is created from files that are stored and deleted.*

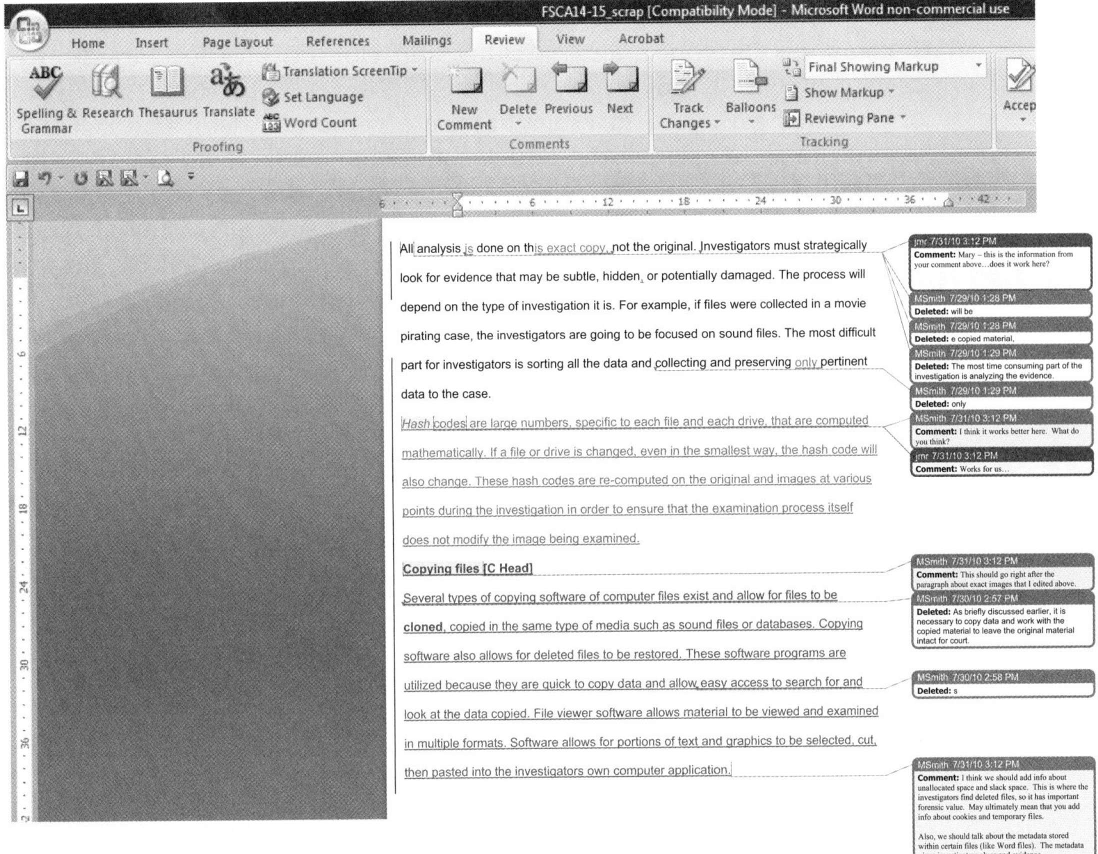

Figure 14-15. *Metadata recovered from a document.*

Information on computers can be hidden using enhancements on current software. For example, Microsoft Word has a tracking feature that allows documents to be edited and comments added for improvements or changes to be made to the original document. Essentially, a conversation exists among multiple parties while passing the document back and forth. Once the changes are made and accepted by both parties, the comments and changes are no longer visible. However, the hidden text can be retrieved. This text is called *metadata*. Metadata is all of the data that describes the content, including when the file was last saved and by whom, the file type, and the document title. Investigators can turn the tracked changes back on to see the conversations and edits made to the original document (see Figure 4-15). These clues can be very helpful in cyber crime cases.

Forensic Tools Used in Cyber Crimes

Because various types of computer crimes exist, investigators use several tools to extract evidence (see Figure 14-16). The following need to be taken into consideration when deciding what equipment will be needed:

- The type of investigation
- The type of evidence that will be presented at court
- The operating system being used to find evidence
- Extensive training and a strong background in the equipment being used
- The financial resources of the cyber crime department

Figure 14-16. *Cyber crime lab to process and analyze evidence collected from a crime scene.*

Hardware

Technology developed for searching and cloning files for forensic purposes can typically run on any computer. However, to preserve the integrity of evidence gathered, investigators use a computer dedicated solely for forensic purposes so that the evidence does not get contaminated by extraneous files or data. Companies have begun to develop full-functioning forensic computers in which the hardware and software are used specifically for cyber crimes. These computers are typically expensive, and the specific needs of the department must be considered before making the investment. For example,

- Will evidence more often be examined at the scene or at the lab?
- How much will the full-functioning forensic system be used?
- What kinds of operating systems and hardware are the investigators typically evaluating?
- Is the evidence generally used in court?

Every computer crime is different, with different needs. Evaluating those needs helps a computer crime division in a local jurisdiction determine the type of equipment required to fulfill those needs.

DOCUMENTING CYBER CRIME EVIDENCE

As in any forensic investigation, documentation of evidence and findings is pivotal in court. A forensic analysis report of the evidence is created and saved in a secure location (see Figure 14-17). Various types of software exist to help generate reports as well as index and cross-reference data. Well-developed reports share the following characteristics:

- The report must be relevant and fact based.
- The design, format, and delivery of the material must be easily understood.
- The individual must write clearly and concisely.
- The methods used to collect the evidence must be specifically stated.
- The results must be clearly worded.

When forensic computer experts document evidence, they document the who, what, when, where, and why of the case. Along with typical computer evidence, other evidence can include interviews, diagrams, videotapes, and photographs. The expert must also precisely document the procedures used to extract and analyze that evidence. These cases often take a year or longer to make it to trial. Therefore, the more detailed the report is during and right after the investigation, the better the

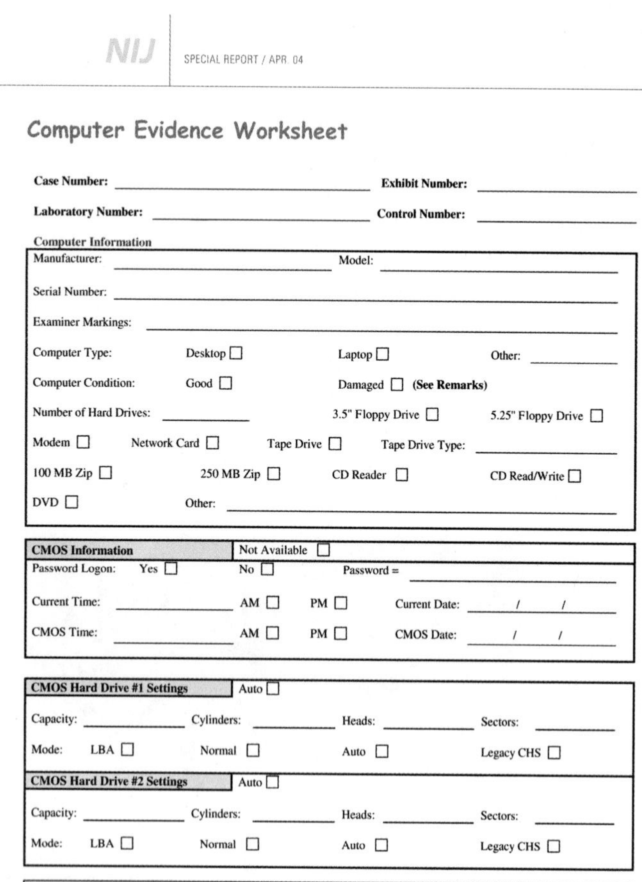
NIJ | SPECIAL REPORT / APR. 04

Computer Evidence Worksheet

Case Number: ________ Exhibit Number: ________

Laboratory Number: ________ Control Number: ________

Computer Information

Manufacturer: ________ Model: ________

Serial Number: ________

Examiner Markings: ________

Computer Type: Desktop ☐ Laptop ☐ Other: ________

Computer Condition: Good ☐ Damaged ☐ (See Remarks)

Number of Hard Drives: ________ 3.5" Floppy Drive ☐ 5.25" Floppy Drive ☐

Modem ☐ Network Card ☐ Tape Drive ☐ Tape Drive Type: ________

100 MB Zip ☐ 250 MB Zip ☐ CD Reader ☐ CD Read/Write ☐

DVD ☐ Other: ________

CMOS Information | Not Available ☐

Password Logon: Yes ☐ No ☐ Password = ________

Current Time: ________ AM ☐ PM ☐ Current Date: __/__/__

CMOS Time: ________ AM ☐ PM ☐ CMOS Date: __/__/__

CMOS Hard Drive #1 Settings | Auto ☐

Capacity: ________ Cylinders: ________ Heads: ________ Sectors: ________

Mode: LBA ☐ Normal ☐ Auto ☐ Legacy CHS ☐

CMOS Hard Drive #2 Settings | Auto ☐

Capacity: ________ Cylinders: ________ Heads: ________ Sectors: ________

Mode: LBA ☐ Normal ☐ Auto ☐ Legacy CHS ☐

Computer Evidence Worksheet — Page 1 of 2

US DEPARTMENT OF JUSTICE

Figure 14-17. *Forensic analysis report used by cyber crime experts to record findings and present in court.*

expert will be able to recall facts quickly and clearly during testimony in court. In court, as with all expert witnesses, the cyber crime expert needs to state education and degrees, professional training, certifications, length of experience in the field, and the number of times he or she has testified in court. This establishes the credibility of the expert. The expert states facts of the findings. Additionally, as an expert witness, the computer expert states his or her opinion about the evidence.

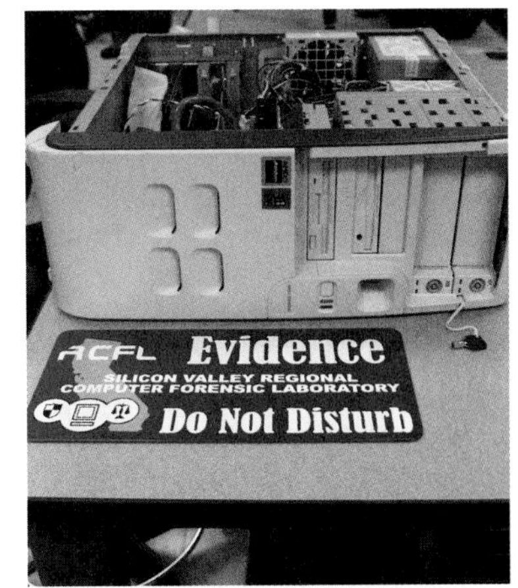

Figure 14-18. *Computer at Silicon Valley Regional Computer Forensics Lab in Menlo Park, CA, open for collection of the evidence.*

PRESENTING COMPUTER EVIDENCE IN COURT

In any court of law, evidence (see Figure 14-18) is necessary to prove that a crime has been committed. Various types of evidence exist, including blood, fibers, hairs, and glass; however, when a computer crime occurs, the computer itself and its contents are also evidence. Identifying, preserving, analyzing, and presenting evidence from a computer are different from working with traditional forensic evidence; however, the approach is still the same. Investigators rely on logic, reasoning, and the scientific method to prepare the evidence for trial. Exemplary scientific thinking and processing skills are invaluable assets to have when working in a cyber crime division, as it is in all other areas of forensics. Figure 14-19 shows the basic steps of collecting and analyzing evidence before presenting it in court. Notice that the evidence must be protected from contamination at every step.

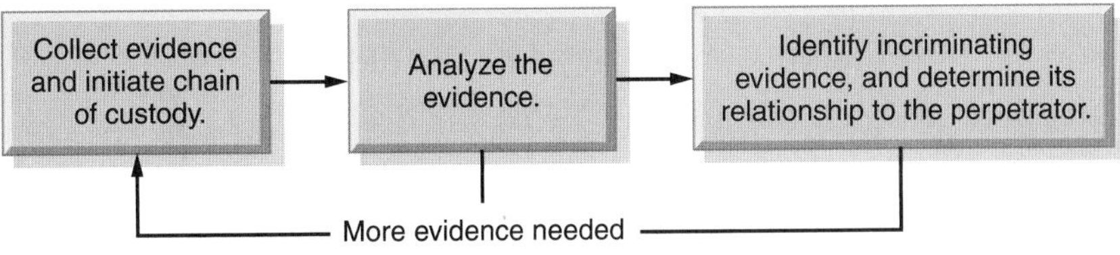

Figure 14-19. *Steps in collecting and analyzing computer forensic evidence.*

Four types of computer evidence may be presented in court. They are as follows:

- Real evidence
- Documentary evidence
- Testimonial evidence
- Demonstrative evidence

Real evidence is actual tangible evidence that can be brought into court. The evidence can be seen and touched. Examples of real evidence include cell phones, fingerprints on the computer linking the suspect to the computer in question, or the computer itself (see Figure 14-20). *Documentary evidence* is any type of written evidence, such as reports or data saved in files. Investigators must first confirm the authenticity of the data before it can be used in court. Investigators must ensure the

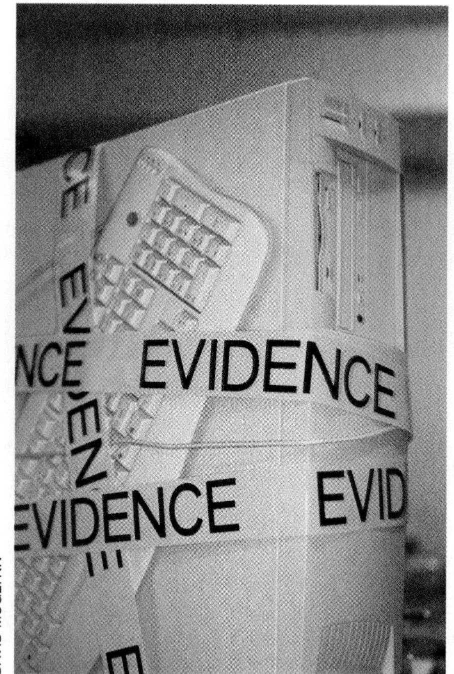

DAVID MCGLYNN

Figure 14-20. *This CPU and keyboard are examples of real evidence.*

data collected proves a crime was committed. When presenting documentary evidence, it is protocol to use the original document rather than a copy. *Testimonial evidence* is written or spoken testimony from a witness to the events surrounding the alleged crime. Most often testimonial evidence is presented in court. However, a witness might also provide testimonial evidence during a deposition. Often depositions are more effective because they occur long before a trial starts, and the witness has a better recollection of events. Whether the testimony is presented in court or during a deposition, the witness has been sworn in prior to making his or her statement. The witness is not allowed to offer any opinion on the case—only to describe facts about what he or she saw or heard. *Demonstrative evidence* is evidence that represents facts or objects. For example, X-rays, graphs, charts, and photographs are demonstrative evidence. These visual aids help clarify and put into context the technical explanations provided by witnesses. For example, in a case involving spam that made computers vulnerable to hacking, the lawyer might use a flowchart or diagram to show the jury how the process works. Demonstrative evidence cannot stand alone—it must be supported by real evidence based on fact.

EXPERT TESTIMONY

Once all the data is analyzed, the findings are presented in court. When a cyber crime expert testifies in court, the goal is to convince the jury that the evidence regarding the case is scientifically based and relevant. The processes used to collect and preserve the evidence will also be questioned. Four simple steps ensure factual based testimony:

1. The expert must tell the jury what he or she did.
2. The expert must tell the jury why he or she did it.
3. The expert must tell the jury how he or she did it.
4. The expert must tell the jury what the findings were.

Computer experts must make the delivery of their results fact based, yet easy to understand and clear for a jury. Often how well the testimony of the expert is presented can sway a jury. Computer jargon is hard for many jurors to follow. It is the responsibility of the expert to provide visuals and a simplified explanation of the evidence.

Obj. 14.9 FUTURE OF CYBER CRIME

As technology advances, society as a whole will continue to benefit from its conveniences. Conversely, as technology improves and the number of people who use and access computers grows, so does the potential pool of cyber criminals. Current and future law-enforcement efforts are centered on encouraging cyber ethics and empowering individuals with more knowledge to protect themselves, such as by adding anti-viral software and other enhanced software to protect businesses and banks from security breaches that allow criminals access to clients' personal information. Also, law-enforcement personnel make strenuous efforts to educate the public on the consequences associated with cyber crimes. Cyber crime divisions and other law-enforcement personnel work hard to keep up with cyber criminals. Often, as new cyber crimes surface, new technology must be developed. It is very difficult to be proactive with these types of crimes. As cyber crime divisions become more experienced and skilled in new technology, they will be able to protect the conveniences that the general public appreciates.

CHAPTER SUMMARY

- Individuals and businesses use the Internet to provide mobile access to data, to share information, for education, and for communication. The Internet is also important for many financial transactions, such as banking and online commerce.
- As industry and individuals continue to rely more and more on the advantages of readily accessible information through a massive worldwide network, the greater the risk of someone or a group of people gaining unauthorized access to the massive amounts of information available.
- The three categories of computer and Internet crimes are computer integrity crimes, computer-assisted crimes, and computer content crimes. Many crimes fall into more than one category.
- Hackers have strong skills in computers and expert knowledge regarding the computer system they are trying to expose.
- Phishing is defrauding the victim by sending a fraudulent e-mail that looks remarkably real, asking the recipient of the e-mail to update their personal information.
- All of the evidence collected during the investigation of a cyber crime must first be cloned.
- Documented evidence in a report must be concise and fact based.
- Four types of computer evidence are used in court—real evidence, documentary evidence, testimonial evidence, and demonstrative evidence.
- The collected data is typically the most compelling evidence provided in cyber crime trials; however, investigators must first prove that the integrity of the hardware was maintained when collecting the evidence.
- The expert witness is often key in the decision made by a jury. The witness must present fact-based evidence in a way that is clear and convincing to a jury.
- As technology improves and the number of people who use computers grows, so does the potential pool of cyber criminals.
- As technology continues to advance, so do the number of individuals who have access to the Internet, which forces law-enforcement agencies to establish procedures and methods for managing online activity.

According to the FBI's National Computer Crimes Squad, less than 15 percent of the attacks on computers are detected; of those, only about 19 percent are reported.

CASE STUDIES

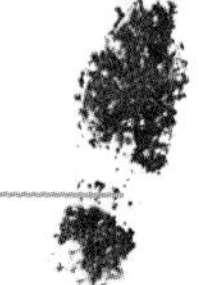

Sasser Worm (2004)

On May 1, 2004, computers around the world were hit by a worm—later named the Sasser Worm—that was able to invade networks directly through the Internet. The Sasser Worm (see Figure 14-21) caused an infected computer to crash and reboot continuously. The worm quickly spread to hundreds of thousands of computers, causing businesses and government agencies to shut down. In the United States alone, the worm caused about $15 billion in damages to computer networks.

The Sasser Worm affected computers running on Microsoft Windows XP and Windows 2000 operating systems. At first, the maker of the worm seemed untraceable, but Microsoft Corporation offered an award of $250,000 for information leading to the arrest and conviction of the worm's creator. Six days later, an informant came forward and identified Sven Jaschan, an 18-year-old German high school student, as the author of the Sasser Worm. Jaschan confessed to the crime, but because he had been a minor when he wrote the virus, he received only three years of probation and had to complete 30 hours community service.

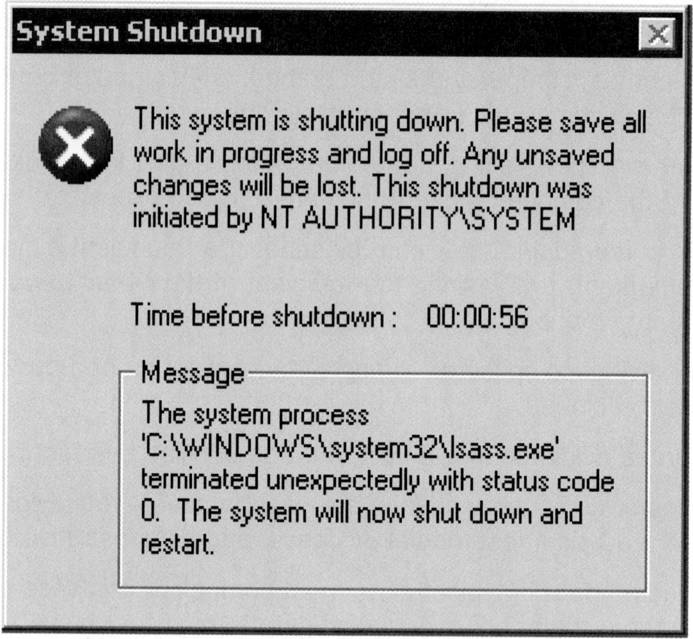

Figure 14-21. *The Sasser Worm caused computers to repeatedly crash and reboot.*

Think Critically

1. **Why can a computer virus disrupt an entire business?**
2. **Why is it difficult to identify the author of a virus?**
3. **What do you think Jaschan's motives were?**

Philip Cummings (2008)

Figure 14-22. *Philip Cummings used his position at a credit report bureau to help steal the identities of thousands of individuals.*

Like many companies that offer their customers credit, Ford Motor Credit (FMC) orders a credit report for any individual applying for a line of credit through them. In 2002, FMC realized that they had been billed for thousands of credit reports they had never ordered. FMC used Teledata Communications Inc. (TCI) for their credit reports. FMC and TCI contacted the FBI about the thousands of unexplained reports.

Soon, FBI investigators realized that many of TCI's other clients had also been billed for credit reports they hadn't ordered. Many of the fraudulent reports were linked to the computer of Philip Cummings—a help-desk employee at TCI (see Figure 14-22). Upon questioning, Cummings admitted that he had been paid $30 per stolen credit report, which he passed on to an identity theft ring. The ring was run by several Nigerian nationals, and the FBI estimated that they had stolen the identities of approximately 30,000 people. They used the stolen identities to open lines of credit, stealing more than $100 million from their victims.

Cummings identified the people who had purchased the credit reports, leading to their arrests. Because of his cooperation, Cummings was given only a 14-year prison sentence.

Think Critically

1. **How could a credit report be used to steal a person's identity?**
2. **How do computers make international identity theft rings possible?**
3. **How do you think the FBI was able to link the fraudulent reports to Philip Cummings's computer?**

Bibliography

Books and Journals

Furnell, Steven. (2002). *Cybercrime: Vandalizing the Information Society*. Great Britain: Pearson Education Limited.

Grabowsky, Peter. (2007). *Electronic Crime*. Upper Saddle River, NJ: Pearson.

Icove, David, Karl Seger, and William VonStorch. (1995). *Computer Crime: A Crimefighter's Handbook*. Sebastopol, CA: O'Reilly & Associates, Inc.

Jones, Robert. (2006). *Internet Forensics*. Sebastopol, CA: O'Reilly Media, Inc.

Nelson, Bill, Amelia Phillips, and Christopher Steuart. (2010). *Guide to Computer Forensics and Investigations* (4th ed.). Boston, MA: Cengage Learning.

Solomon, Michael G., Diane Barrett, and Neil Broom. (2005). *Computer Forensics: Jump Start*. Alameda, CA: Sybex, Inc.

Wall, David S. (2007). *Cybercrime*. Cambridge, UK: Polity Press.

Wall, David S. (2010). "Micro-Frauds: Virtual Robberies, Stings and Scams in the Information Age," *Corporate Hacking and Technology-Driven Crime: Social Dynamics and Implications*, T. Holt, B. Schell, (eds.) Hershey, PA: IGI Global.

Websites

www.cengage.com/school/forensicscienceadv

www.cybercrime.gov

www.expertlaw.com/library/forensic_evidence/computer_forensics_101.html

www.ncpc.org/newsroom/current-campaigns/cyberbullying

CAREERS IN 

Science, Technology, Engineering & Mathematics

Computer Forensics Specialist

A computer forensics specialist, or digital specialist, identifies the current state of a digital artifact: a computer system; a storage system, such as a hard drive or CD-ROM; or an electronic document, such an e-mail or digital image.

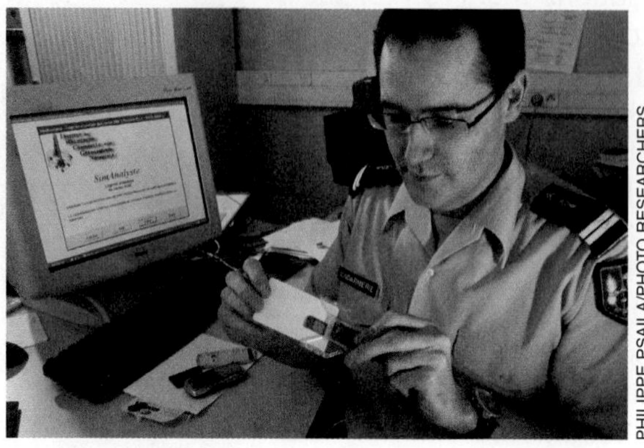

PHILIPPE PSAILA/PHOTO RESEARCHERS

Figure 14-23. *Computer forensics specialist collecting data from the SIM card of a suspect's mobile phone.*

In our increasingly paperless society, computer forensics specialists are in high demand in both criminal and civil cases where evidence may be stored on computers or their storage systems—in cases where such information may have formed a "paper trail" in the past. Examples of criminal cases include fraud, sexual harassment, theft, and intellectual property violations. Additionally, electronic communications may provide clues to other crimes such as murder, gang activity, or child pornography—the list is limitless. A computer forensics specialist can also determine how a hacker broke into a computer's system and then repair the weak entry spot. The specialist might be hired by a company to track an employee's computer activity during work hours for the purpose of determining whether the employee is compromising security or otherwise breaching company policy. Finally, a computer specialist is able to use technically advanced programs to retrieve lost data in the event of a system-wide crash.

The first step a computer forensics specialist takes is to collect data. Careful and precise execution of protocol is critical—if an error is made during data collection, it cannot be corrected and the evidence is invalidated. To protect the evidence, the specialist first "locks" the computer with software to make sure that no data is changed or lost during the inspection, as it must arrive in court in the exact same state as when it was seized for inspection. Then, the computer forensics specialist initiates the chain of custody. Documentation of each and every step is critical. The specialist will use only those tools and software that have been proven to be safe and accurate in the examination of data.

If the data is static (i.e., has been written to the hard drive), an imaging program is used to make an exact duplicate of all evidentiary media and the original hard drive is then stored to protect it from possible criminal tampering. If the data is live, it is volatile—the computer cannot be turned off or all data will be lost. Therefore, a program that can capture the live data must be used to capture data immediately, before changes are made or before there is a power-down.

Once the data has been collected, the analysis begins. Again, the computer forensics specialist will use only programs that have been proven to be reliable. In many cases, the computer forensics specialist will need to use more than one tool to confirm findings or to supplement information. An analysis will usually include a manual review of material on the media examined, registry review, password decoding, searches for key words related to the crime subject, and e-mail and picture extraction.

Finally, the computer forensics specialist reports the data. Depending on the client, the report may be a brief oral presentation, a complex written report, or testimony in court. Court testimony usually requires the support of detailed written documentation.

Because technology is in a constant state of flux, a computer forensics specialist must keep current on the latest programs and advances in the field. Many states require specialists to hold either a private investigator's license or other professional certification, such as the Global Information Assurance Certification (GIAC) Certified Forensic Analyst.

Learn More About It

To learn more about forensic science careers, go to www.cengage.com/school/forensicscienceadv.

CHAPTER 14 REVIEW

Multiple Choice

1. Mike is a college student who enjoys chatting with friends and playing video games. Which of the typical Internet applications is *not* applicable in this example? *Obj. 14.1*

   a. social networking b. distribution systems
   c. education/training d. leisure/recreation

2. A cyber crime investigator goes to a crime scene, collects computer data, and returns to the lab for analysis. Which type of forensics is being utilized? *Obj. 14.3*

   a. Internet forensics
   b. World Wide Web forensics
   c. computer forensics
   d. none of these

3. Julie received an e-mail notification stating that she must update her personal banking information. The website looked identical to her banking institution's website. Upon opening the e-mail, her personal information was stolen. She was the victim of ____. *Obj. 14.2*

   a. a worm b. a Trojan horse
   c. spamming d. phishing

4. Jennifer is a lab technician who specializes in finding hidden computer files. Before she begins any analysis, she must first make copies of all of the files on the computer's hard drive. This process is called _____. *Obj. 14.3*

   a. phishing b. cloning
   c. hacking d. downloading

5. A law-enforcement agency is expanding its small cyber crime unit. What criteria must be considered when deciding what types of hardware and software are needed to fulfill its needs? *Obj. 14.6*

   a. types of investigations the agency anticipates
   b. training and background required to operate new systems
   c. the financial resources of the agency
   d. all of these

Short Answer

6. Discuss the importance of documenting cyber crime evidence. What key elements are necessary for a thorough report? *Obj. 14.8*

   ____

   ____

   ____

CHAPTER 1 2 3 4 5 6 7 8 9 10 11 12 13 14 15

7. Investigators use cloned data files for forensic analysis. Explain why they do not use the original files. *Obj. 14.3*

8. What factors must be considered when entering a cyber crime scene? How is this similar to—or different from—any other crime scene? *Obj. 14.2*

9. Compare and contrast the three types of cyber crimes. Provide examples of how they might overlap. *Obj. 14.1*

10. Explain the three methods hackers use to infiltrate a network system. *Obj. 14.2*

11. What types of information can cyber criminals retrieve using spyware? *Obj. 14.2*

12. Discuss how phishing can have international consequences. *Obj. 14.2*

____

____

____

13. Differentiate between hardware and software. Explain the forensic value of each. *Obj. 14.5*

____

____

____

14. Examine the risks and benefits of the expansion and evolution of technology. *Obj. 14.9*

____

____

____

15. Describe the role of the expert witness in the court. What evidence does the witness present? *Obj. 14.7*

____

____

____

Think Critically

16. You have been assigned a research paper for a language arts class. As you begin the research, you recall writing a contest essay a couple of years ago that contained some of the same information. Unfortunately, you deleted the files from your hard drive when the contest was over. Explain in detail how you would go about retrieving that old file. ***Obj. 14.3***

17. Recall from previous chapters that Locard's exchange principle refers to the exchange of materials between persons or things that come into contact with each other. Explain how using a computer also leaves traces of the user behind. Discuss both electronic traces as well as physical evidence. ***Obj. 14.3, 14.4***

18. Create a scenario in which all four types of computer evidence are used in court. Include the importance of an expert's testimony in your scenario. ***Obj. 14.4, 14.7***

CRIME SCENE 
Student-Prepared Original Titles

Phishing for Trouble

By: **Chase Mills and John Wiffen**

East Ridge High School
Clermont, Florida

Loud ticks of the keyboard filled the room. Some people might find that annoying, but looking back, I thought nothing of it. In fact, it was music to my ears. Little did I know how that would come back to haunt me . . .

Technology has always fascinated me. Ever since I can remember, I would tinker with appliances—dismantling and reconstructing the living-room clock was a favorite. I even created an automatic dog feeder so I never had to feed him manually again. When I was 15, my parents bought me my own computer. I wasted no time. I taught myself various programming languages and I learned everything I could about computers. Then it began.

At 17, I learned how to keylog. You know—when you record the keys struck on a keyboard. I began hacking and soon found a way to record passwords to get into my grades. I changed some of my grades in a few classes, but until I graduated, I didn't do much with my newfound talents.

One afternoon, while sitting at an Internet café, I received a mysterious e-mail (see Figure 14-24). It stated simply, "We know you are one of us," and instructed me to meet "them" in the park. I knew immediately what the e-mail meant because it was written in a type of code hackers use. That was the beginning of my new life of crime.

We were a team, a family. Within the first month, I found myself changing. Hacking was no longer a hobby—something I did just for fun. It became a lifestyle and an obsession. I couldn't stop. At first, I spent my time phishing people's social networking accounts looking for personal information. I posted advertisements on their profiles for fictitious companies to make money. The more I hacked, the more

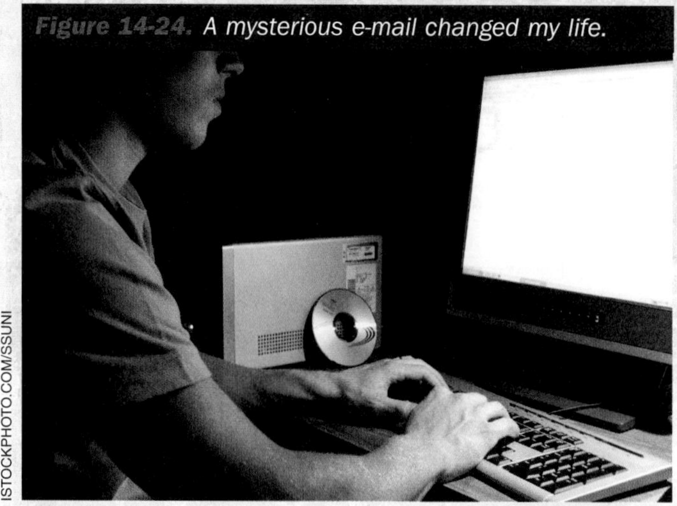
Figure 14-24. A mysterious e-mail changed my life.

ISTOCKPHOTO.COM/SSUNI

powerful I felt. So I moved on. I began working with the search engines and messing with people's e-mail accounts. I was having so much fun, I didn't care who I hurt.

Soon we were dreaming of bigger and better schemes. We devised a plan that would net us all enough money that we would be set for life. We carefully plotted how we would hack into a large company's database and "borrow" money overnight. During the night, it would sit in a hidden account and draw interest. In the morning, the money would be returned—minus the interest, of course. We tried it out a few times and no one came knocking on our doors to arrest us, so we decided to do it.

We pulled off the biggest heist in history. I hesitantly typed in the last key and pressed enter. All I had to do was wait—as soon as I put the phished page up, money began pouring into the account. I sat there all night watching the screen as thousands of dollars poured in. We made over $13 million each that night. We had done it!

We were hacker heroes. International criminals. Headlines in newspapers around the world reported the crime but also admitted having no leads as to the identity of the thief or thieves. We had covered our tracks well. There was no way for them to catch us.

I spent the next few days laying low—just in case. Eventually, I began to spend more time on the computer. Nothing big, just searching the Web and making online purchases. That was when it happened. I got a message on my computer. *We have traced your IP address.* The next day I was arrested. Me, Tyler Srodes, the hacker behind one of the biggest Internet crimes in history, was caught by downloading an illegal copy of a song!

Activity

Answer the following questions based on information in the Crime Scene S.P.O.T.

1. How would cyber crime investigators collect, preserve, and analyze the trace evidence in this case?
2. What types of cyber crimes are committed in this story? Explain.

WRITING

3. After reading the story, assume the role of an expert witness. Write a 400–600 word report detailing how you would present the four types of cyber crime evidence to a jury. Address your explanation to a potential juror who may not have an extensive knowledge of computers and the Internet.

Introduction:

Today's teenagers and young adults spend much of their time on cell phones, texting with friends, instant messaging acquaintances they have met online, downloading information from websites, and uploading pictures and videos to share with family and friends. But a nemesis from the past has now become an enemy online. The old-fashioned bully has surfaced on the Internet.

According to the National Crime Prevention Council, *cyberbullying*, the term given to bullying online, affects about half of all teenagers. Like its schoolyard counterpart, the cyber bully seeks out an intended victim and harasses, embarrasses, or harms the person in some way.

COURTESY OF DR. VIVIAN WRIGHT, UNIVERSITY OF ALABAMA

© RONNIE MCMILLAN/ALAMY

© DOUG STELEY C/ALAMY

Figure 14-25. *Cyberbullying comes in many forms.*

Procedure:

Part I: Research cyberbullying to find the answers to the following questions:

1. What constitutes cyberbullying?
2. How are teens cyberbullied?
3. What are some common reactions by the victim to cyberbullying?
4. How do onlookers perceive cyberbullying?
5. What precautions can be taken to protect yourself from cyberbullying?

Part II: After completing your research, choose one or more of the following:

1. Design an informational brochure or poster to make students in your school aware of the potential legal and personal dangers of cyberbullying.
2. Contact the feeder middle school and offer to supply posters concerning cyberbullying that can be placed in common areas around the school—the hallways, cafeteria, etc.
3. Create a skit to perform at a faculty meeting to inform teachers about concerns of cyberbullying and its effect on students.
4. Write and film a public service announcement to be streamed throughout the school on the morning news.

ACTIVITY 14-1 *Ch. Obj: 14.2, 14.5, 14.6*
PASSWORD PROTECTED?

Objectives:

By the end of this activity, you will be able to:

1. Determine the category of cyber crime committed in the scenario.
2. Use deductive reasoning skills to link passwords to an identity.

Materials:

(per group of four or five students)
list of passwords
completed surveys

Safety Precautions:

In order to safeguard your privacy, do not divulge a password you are currently using for e-mail, social networking, or any other accounts.

Background:

You are a part of group of cyber crime experts. A large company's computer has been hacked. Data and files have been stolen. It is believed to be an "inside" job. Experts have gone into each office and collected personal information that was easily accessible. Your task is to review the survey of each employee and match the survey with the password. This will save the cyber crime experts an enormous amount of time. If they have the passwords of each employee's computer, the experts will be able to quickly start searching for clues to who stole company data. An effort is being made not to "tip off" the hacker, so this is a covert effort.

Procedure:

1. Read through the set of surveys your consulting group was given.
2. Discuss the information provided on the survey and how it might be used to generate a password.
3. Compare the list of passwords to the surveys.
4. Once your group has reached a consensus on each of the passwords and the identity of the person who generated it, record the information in Data Table 1.

Data Table 1. Notes.

Password	Identity	Rationale

5. Once your group is sure the identities and passwords are matched correctly, check with your teacher.
6. If any of the matches is incorrect, your consulting group must reconvene and begin the process again.

Questions:

1. What information proved most valuable in matching the passwords with the identities? Explain.
2. Were some of the passwords easier to match than others? Why or why not?
3. What could be added to a password to make it less vulnerable to hacking or identity theft?
4. In addition to the procedure you performed in this activity, how might a cyber crime expert determine who hacked into the mainframe and stole the files and data?
5. Which category—or categories—of cyber crime does this activity illustrate? Explain your rationale.

Extension:

WRITING

People sometimes forget passwords and cannot access their own files. Research how companies are able to retrieve employee passwords for this purpose. Write a report describing the process and discussing other reasons a company might need access to its employees' passwords.

ACTIVITY 14-2 *Ch. Obj: 14.5, 14.6*
HIDDEN SECRETS

Objective:

By the end of this activity, you will be able to:
Decipher codes embedded in an e-mail address.

Materials:

(per groups of two or three students)
computer access
copy of compact e-mail header
copy of full e-mail header

Procedure:

1. With a partner, look at the details included in the compact header of the e-mail in Figure 1.

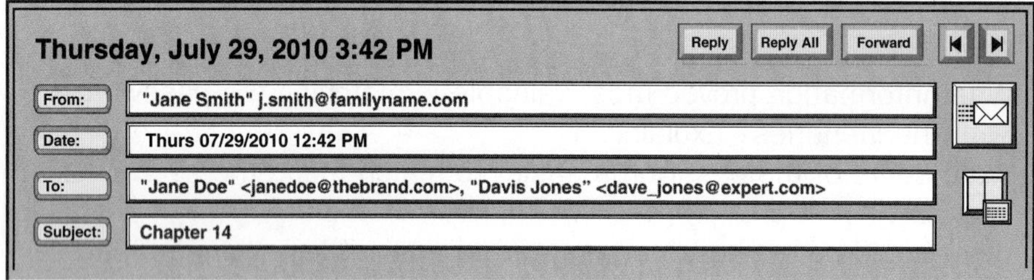

Thursday, July 29, 2010 3:42 PM
Reply
Reply All
Forward
From:
"Jane Smith" j.smith@familyname.com
Date:
Thurs 07/29/2010 12:42 PM
To:
"Jane Doe" <janedoe@thebrand.com>, "Davis Jones" <dave_jones@expert.com>
Subject:
Chapter 14

Figure 1. *Sample compact e-mail header.*

2. Now look at the details provided in the full header of the e-mail in Figure 2.
3. Compare the difference between the compact and full headers.
4. Research the information that is provided and what the terms mean.

Questions:

1. What information is provided in the compact header?
2. What information is provided in the full header?
3. How is the information provided in a full header of an e-mail useful to cyber crime investigators?

From Jane Smith Thu Jul 29 19:42:36 2010

X-Apparently-To:	janedoe@thebrand.com via 35.148.211.103; Thu, 29 Jul 2010 12:42:39 -0700
Return-Path:	<j.smith@familyname.com>
Received-SPF:	pass (mta136.mail.ac4.thebrand.com: domain of j.smith@familyname.com designates 283.91.164.158 as permitted sender)
X-YMailISG:	BDaE3EccZAoRYTQMfjFlBkMWVer2c1swxuMHBAYtBR69iOHN IjP68mJTbAaXsOrd1EGQ86xM_SEZz0iXex0c5EDkil8wIjmO9J27DFew9AVm i1GphQp1X6YXWMkLoevAswINAGCSfsosQV4zMIX.lzRPrkTBFMxy7gI3xWZi XArnf8hAB4M2Ue9Gwo0yLbi_yd9jRG2IS3cGBGysa2gFJZl_96iM.TRNniRI OeaI7ajzyOyOad_6hOomQW5iWOhWxiTsTJgP_HZ7fXoJpoprJdcmTNaxGyKl iEP3leGFuhvdAwY729xYzZTUZcUZDRJ_u4inwVbIc1DQ6PEl.QBpLHkN8JjN EG_SZnNijMILlVJU4MbJPwOmgN3.8imM0Ih9Kgw.oQwI9mllc63cai6bCsAc t_UAGrMyrXGtoPZC7X5s1kjjP30C6vzjoxOkyDZHGvzsqFvN91Hfyn.yJF8S zX3wvUNGOKRZM2Fs.EfZeVQ3yW2qXteWsVB0YUuEXCv47PTqFS.z.JoJ_goE c9PwpBqPOxCPpRms3bxNGq1VuYNLAphEZi7iZ1PxB47_nszE1PSiJujkxIbT zvsv7QNTuBYT3bwLiqKQNoY0ZNwv4B46TXR093tY6mMrdRYM96uDSwlCncB. 03RpCJZlWERhHO.hp5dSLJ_v7DyzEO7dVwxXNHA-
X-Originating-IP:	[283.91.164.158]
Authentication-Results:	mta136.mail.ac4.thebrand.com from=familyname.com; domainkeys=pass (ok); from=familyname.com; dkim=pass (ok)
Received:	from 145.0.1.1 (EHLO mail-gx0-f145.familyname.com) (283.91.164.158) by mta136.mail.ac4.thebrand.com with SMTP; Thu, 29 Jul 2010 12:42:39 -0700
Received:	by mta136.mail.ac4.thebrand.com with SMTP id 10so374836gyk.33 for <janedoe@thebrand.com>; Thu, 29 Jul 2010 12:42:38 -0700 (PDT)
DKIM-Signature:	v=1; a=rsa-sha256; c=relaxed/relaxed; d=familyname.com; s=gamma; h=domainkey-signature:mime-version:received:received:date:message-id :subject:from:to:content-type; bh=lRl56fIC57iW0USKxcbDI0Bo4ILapnYX7p4JcO8lD+o=; b=Eog49x/834Er8jhOnndgLONJCga2mcSHvV3Z8JwepOuwUm7Pzq8MQ0c8q6fiLoh+5U AlosYqC3TNK8ilQ10AzKMxFPtnnIZYnGMqVWdvVwzBv3yB9DEiDVAEjQSiluX0uqFCTM th/2VOiP0vRxdrPQgTqdNVs4YvZsz7kJ4dfd8=
DomainKey-Signature:	a=rsp-shr1; c=noghs; d=familyname.com; s=gamma; h=mime-version:date:message-id:subject:from:to:content-type; b=LMP9x+DV0/mNqB866N8TaB2VLM4V8RR75PgYcMDqr1PJvUz41MoyVOpxcDUj+ERhUS 9Q/nX3loXS6xj9pS8j/a7iYcEqtqxNVcK3bYp79FOtzM4s8XxqHYB4UpLg/biTN6Fj7h AAlCgb8VMqBFFjZXU0SnT/HsxVFpZrrvnrckE=
MIME-Version:	1.0
Received:	by 23.167.182.3 with SMTP id d2mr1889141zpf.398.1458432226765; Thu, 29 Jul 2010 12:42:36 -0700 (PDT)
Received:	by 23.012.505.197 with HTTP; Thu, 29 Jul 2010 12:42:36 -0700 (PDT)
Date:	Thu, 29 Jul 2010 14:42:36 -0500
Message-ID:	<AVPLkTamaR7vCcBvFqWLQMoYJY2iLWtpiwmneEKA+YswP@mail.familyname.com>
Subject:	Chapter 14
From:	This sender is DomainKeys verified Jane Smith <j.smith@familyname.com> View contact details
To:	Jane Doe < janedoe@thebrand.com >, Davis Jones <dave_jones@expert.com>
Content-Type:	multipart/alternative; boundary=000e0cd6gyx0b986f2048c8bef5y
Content-Length:	4183

Figure 2. *Sample full e-mail header.*

CHAPTER 15

Criminal Profiling

RETURNING TO THE SCENE OF THE CRIME

In 1989, police in Rochester, New York, realized they had a serial killer on their hands. Over the course of that year, bodies of several murdered women were found near the Genesee River. All of the women had been strangled and beaten. Several of them had been mutilated as well.

The river had washed away most of the evidence, and police had few leads as to the identity of the killer. Meanwhile, the body count continued to rise. The Rochester police contacted the FBI for assistance, and FBI profiler Gregg McCrary was assigned to the case. After analyzing the killer's modus operandi, McCrary predicted that the killer had some sort of sexual dysfunction and that murdering the women excited him. McCrary thought that the killer would be likely to revisit the bodies he dumped in order to prolong the pleasure he took from his crime.

AP PHOTO

Figure 15-1. Arthur Shawcross

Because of McCrary's profile, when the next murder victim was discovered floating in the Genesee River on January 3, 1990, the police decided not to remove it immediately. Instead, they set up surveillance around the body to see if the killer would actually return. The police did not have to wait long. A man named Arthur Shawcross turned up on the bridge near where the body had been dumped.

The police arrested Shawcross and soon discovered that he had already been arrested and convicted for molesting and murdering two children almost 20 years before. He had been given only a 25-year sentence for that double homicide. He had served 15 years of the sentence and had been released on parole in 1987 (against the advice of his parole officer, who thought he was a danger to society).

Shawcross was convicted of murdering a total of 11 women. He was given a sentence of 250 years for the crimes.

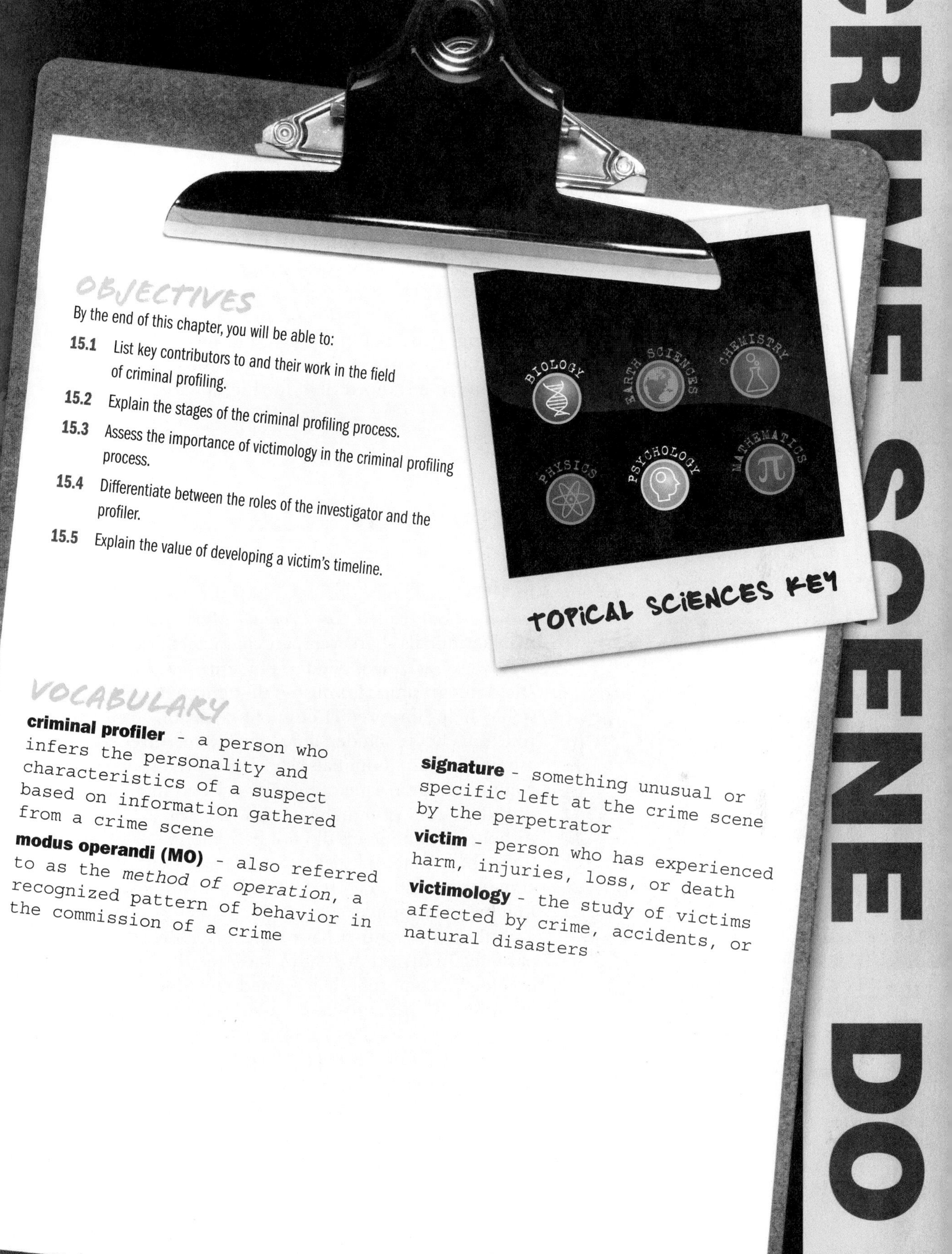

OBJECTIVES

By the end of this chapter, you will be able to:

15.1 List key contributors to and their work in the field of criminal profiling.

15.2 Explain the stages of the criminal profiling process.

15.3 Assess the importance of victimology in the criminal profiling process.

15.4 Differentiate between the roles of the investigator and the profiler.

15.5 Explain the value of developing a victim's timeline.

VOCABULARY

criminal profiler - a person who infers the personality and characteristics of a suspect based on information gathered from a crime scene

modus operandi (MO) - also referred to as the *method of operation*, a recognized pattern of behavior in the commission of a crime

signature - something unusual or specific left at the crime scene by the perpetrator

victim - person who has experienced harm, injuries, loss, or death

victimology - the study of victims affected by crime, accidents, or natural disasters

INTRODUCTION

Various experts collaborate to resolve criminal cases. Multiple experts work within their fields of expertise and share data gathered to build a case against an offender. Criminal profilers are part of that process. In order to help narrow down the list of suspects, **criminal profilers** study evidence collected and analyzed by crime-scene investigators to formulate a hypothesis about a perpetrator's age, personality, lifestyle, and social environment. Information about both the perpetrator and victim might help CSIs solve the crime. Most often, a person with experience as a profiler also has expertise in other forensic science fields. As the field continues to grow, however, it is likely that criminal profilers will eventually work solely on profiling. Criminal profiling dates back to the 1800s. Through advancements in behavioral science and forensic psychiatry, criminal profiling has become an advanced field for highly skilled scientists.

Obj. 15.1, 15.2

HISTORICAL DEVELOPMENT

Figure 15-2. *Cesare Lombroso*

In 1876, an Italian doctor, Cesare Lombroso (see Figure 15-2) wrote a book called *The Criminal Man*. In it, he suggested that many criminals shared certain characteristics. Factors he considered were race, age, gender, physique, educational background, and demographics. Lombroso thought that evaluating these traits would help him predict criminal behavior. After completing his research, he concluded that there were three general types of criminals: born criminals who were predisposed from birth to commit crime; insane criminals who had mental and or physical disorders; and criminaloids who did not exhibit specific traits. Although criminaloids did not have mental or physical disorders, their emotional and mental state made them likely offenders in certain situations. Figure 15-3 shows a sampling of characteristics Lombroso examined.

Although scientists have since disproved the hypothesis that particular physical attributes indicate that a person is a "born criminal," Lombroso has earned the title "father of scientific criminology" because he focused on the scientific method and because he was the first to suggest that certain factors increase the chances that an individual will commit crimes.

Hans Gross is considered to be one of the founding fathers of modern-day criminal profiling. With his aggressive research and data, he was able to suggest ways to profile behaviors of criminals, including murderers, arsonists, and thieves. A portion of his research was dedicated to studying women with children. In 1906, Dr. Gross published a book called *Criminal Investigation: A Practical Textbook for Magistrates, Police Officers, and Lawyers* that presented a systematic approach to crime-scene reconstruction and criminal profiling. He encouraged investigators to focus on science rather than on intuition.

In 1914, Gerald Fosbroke, an American character analyst, wrote a book called *Character Reading through Analysis of the Features*. He spent 30 years

Abnormal head size and shape in relation to the ethnicity of the criminal
Extremely large jaw and cheekbones
Large, pouty, pronounced lips
Extremely long arms
A diminished chin or one that is extremely long, short, or flat
Abnormal size, shape of ears

Figure 15-3. *Lombroso thought that certain characteristics, such as those listed above, indicated that a person is more likely to commit crimes.*

studying the physical features of an individual's face. He hypothesized that by examining the facial features, he could predict character. Again, the connection between physical characteristics and the probability that a person will become a criminal has been disproved. However, further research has made connections between experiences and behavioral attributes that are more common among people who commit certain crimes.

Although great emphasis was placed on studying male criminal behavior, Dr. Erich Wulffen, a German criminologist, wrote a book called *Woman as a Sexual Criminal* (1935). The work was dedicated solely to female criminal behavior. He studied only women and examined various potential causes of criminal behavior in women, such as psychological, social, moral, and biological factors. Throughout his 30 years of research, he determined that most female criminal behavior stems from various sexual dysfunctions. Since Wulffen's research, additional data has supported the idea that female offenders also suffer from poor self-worth, depression, and the inability to engage in normal social relationships.

In 1486, during the Spanish Inquisition, a manual was written that was used to profile "witches." The profiles typically consisted of women who would cause some type of harm that risked the welfare of communities and children. The profiles were used to catch and prosecute the alleged witches by burning them at the stake.

In the 1930s John O'Connell and Harry Soderman published multiple editions of the book *Modern Criminal Investigation.* They both had a background in law enforcement with an emphasis on establishing profiles of various types of criminals. Due to their law-enforcement background, a primary focus of their publication was the systematic approach given to crime-scene investigation. They thought that good physical evidence coupled with the knowledge of behavior of offenders would allow investigators and profilers to apprehend suspects more quickly.

In 1953, Dr. Paul L. Kirk (see Figure 15-4) took a very aggressive approach to criminal profiling in his text *Criminal Investigation,* noting that profiling is a natural step in the process of criminalistics. He appreciated the value of profiling criminal behavior in conjunction with evaluating evidence. His approach was considered progressive for his time. He later wrote a book about profiling arsonists called *Fire Investigations.*

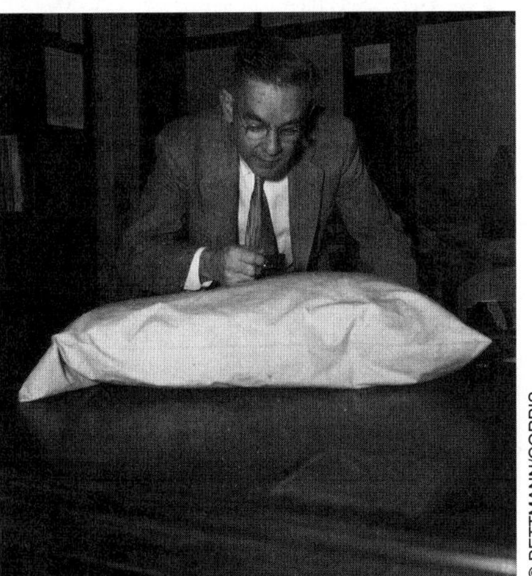
© BETTMANN/CORBIS

Figure 15-4. *Paul L. Kirk*

MODERN-DAY PROFILING

As the study of criminal behavior continued to evolve, forensic psychiatry became an important field in the criminal investigation process. A forensic psychiatrist has advanced training in the study of criminal behavior and mental illness. His or her training is focused on the application of behavioral science to legal cases (see Figure 15-5). A forensic psychiatrist

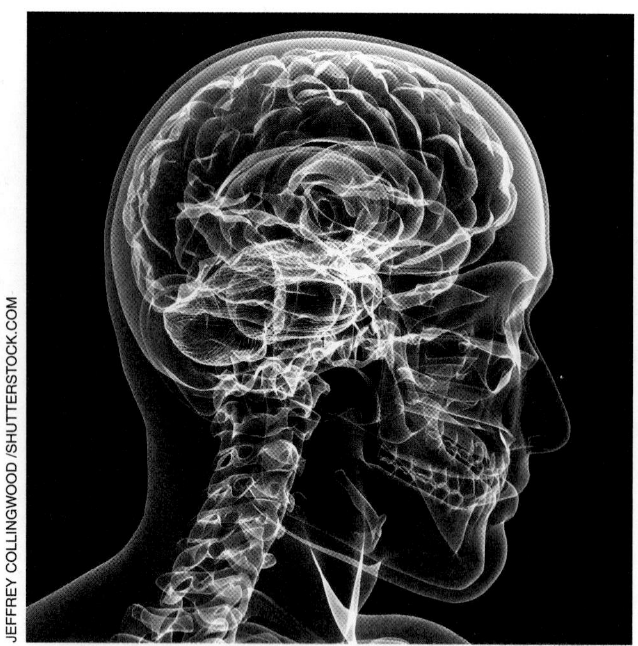

JEFFREY COLLINGWOOD /SHUTTERSTOCK.COM

Figure 15-5. *Forensic psychiatrists study the mind and the body of suspects and victims for the purpose of the law.*

interviews criminals and suspects, evaluates personal history, and administers personality tests. The data accumulated from these measurements assist the forensic psychiatrist in establishing an opinion of the mental well-being of the individual. As an expert witness, the forensic psychiatrist is allowed to present his or her opinion of the suspect from a psychiatric perspective as testimony in court.

As time progressed, the scientific community began to recognize the value of criminal profiling. In the 1950s, a psychiatrist named Dr. James Brussel took an investigative approach to criminal profiling (see Figure 15-6). He viewed profiling as a diagnostic tool and bridged the gap between crime-scene investigators and forensic psychiatrists. He hypothesized that by studying crime scenes he could predict potential behaviors in the unknown suspect. He drew from his experiences working with his own patients who had similar mental disorders.

In the 1960s, a police officer in California, Howard Teten, used profiling skills inspired by Dr. Paul L. Kirk. By taking a multidisciplinary approach to forensic science, death investigation, and criminal behavior, Teten became an exemplary investigator with impressive profiling skills. He eventually went on to work for the FBI and began a criminal profiling division in 1970. He taught agents how to use their investigative skills to improve their criminal profiling skills.

In 1972, Jack Kirsh, an FBI agent, opened the FBI's Behavioral Science Unit (BSU). Using skills taught by Teten, Kirsh helped solve several cases by developing criminal profiles of unknown suspects. The BSU was so effective in developing profiles that they were soon getting requests daily for profiles from law-enforcement agencies across the country. In the 1990s the BSU underwent a massive overhaul. The Behavioral Science Unit is now directed by the National Center for the Analysis of Violent Crime (NCAVC) located at the FBI Academy in Quantico, Virginia. Figure 15-7 shows the seal of the BSU.

Even into the 20th century, there was a common belief that physical characteristics played a role in a person's potential to become a criminal. It was thought that tall, thin people had the potential to become petty thieves and those who were short and fat were likely to become involved in deception and fraud crimes. Crimes of violence were associated with persons of athletic builds.

Digging Deeper

with Forensic Science e-Collection

Very often on television and in movies, the forensic evaluation of a criminal consists of a brief jailhouse interview and possibly a follow-up visit with family members. The truth, however, is very different. Using the Gale Forensic Science e-Collection at www.cengage.com/school/forensicscienceadv, research the methods of evaluating and assessing various types of offenders. Write a one- to two-page summary of the interview techniques as well as the types of testing procedures that are used to ensure a thorough and accurate assessment. Be sure to address the problems associated with a hastily completed assessment.

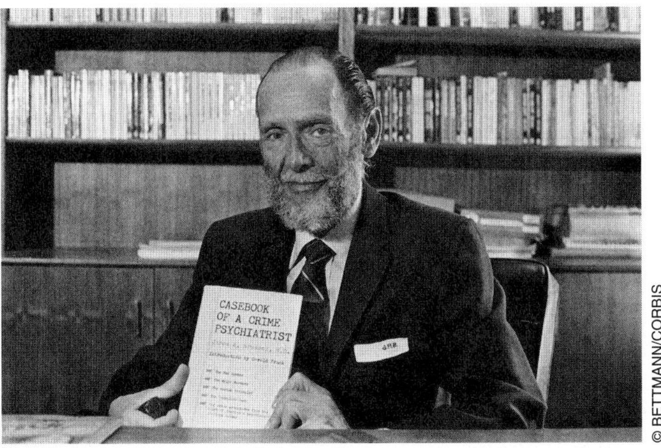

© BETTMANN/CORBIS

Figure 15-6. Dr. James Brussel

Figure 15-7. The seal of the Behavior Science Unit emphasizes the unit's role in training, consultation, and behavioral science research.

STAGES OF THE PROFILING PROCESS

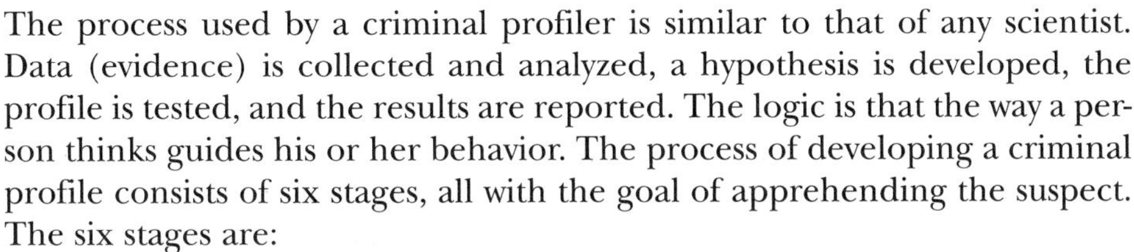

The process used by a criminal profiler is similar to that of any scientist. Data (evidence) is collected and analyzed, a hypothesis is developed, the profile is tested, and the results are reported. The logic is that the way a person thinks guides his or her behavior. The process of developing a criminal profile consists of six stages, all with the goal of apprehending the suspect. The six stages are:

- Input
- Decision process models
- Crime assessment
- Criminal profile
- Investigation
- Apprehension

INPUT

The input stage begins with collecting as much information about the crime as possible. The crime-scene photographs (see Figure 15-8) yield information such as the placement of the victim, weapons, and other physical evidence. Body position and patterns of evidence, such as blood spatter or broken glass, give investigators and profilers insight into circumstances surrounding the crime.

A complete history of the victim is extremely important in a homicide investigation and profile. The profiler will request information, such as the victim's employment history, social habits, hobbies and interests, and reputation among family and peers as well as fears and behaviors

AP PHOTO/THE PATRIOT-NEWS, JOE HERMITT

Figure 15-8. Forensic psychiatrists analyze evidence found at the crime scene to gain insight into circumstances surrounding the crime.

CHAPTER 1 2 3 4 5 6 7 8 9 10 11 12 13 14 15

Alphonse Bertillon developed a camera that was raised on a tripod high above a dead victim, allowing the entire body to be photographed. Photographs of crime scenes—important in all investigations—are studied by psychological profilers.

in social settings. Other information from autopsy photos, police and laboratory reports, and a medical examiner's autopsy report is useful in determining risk factors of the victim, the control exhibited by the offender, the emotional state of the offender, and even the offender's criminal savvy. When available, interviews with witnesses and the victim may reveal more information about the perpetrator's demeanor and behavior.

Though a profiler needs as much information as possible to help in creating an accurate and detailed profile, there are several pieces of information that should not be included. For example, if the profiler has access to the files of a police suspect, the profiler could subconsciously generate a profile matching the suspect. Therefore, profilers look only at evidence directly related to the crime. They do not look at information about suspects before developing their profile.

DECISION PROCESS MODELS

Once the profiler has gathered all pertinent information, the data must be organized and classified. The decision process model utilizes data from the input stage and looks for recognizable patterns. To do this, profilers use decision points, or models, to compare and differentiate the information and to form a basis for the profile.

Classifying the Crime

The first step in the decision process model is classifying the type of crime that has been committed. A profiler may be involved in a wide range of crimes, including arson, burglary, terrorist attacks, and homicide. If the crime is murder, it must be further classified by the number of victims and locations as well as by the number of events and the period of time between them. For example, if there are three victims at a single location at the same time, the murder is classified as a triple homicide. If, however, there are three victims at three locations at three different times and if there is evidence to suggest the same person committed all three crimes, then the homicides would be classified as serial or spree murders. The difference between spree and serial homicides is the amount of time between events. A killing spree takes place at more than one location, but there is not a lot of time between the events. A serial killer, on the other hand, may wait months between events. Figure 15-9 summarizes homicide classification. Ted Bundy was a serial killer. Figure 15-10 provides examples of spree killers.

	Victims	Locations	Events	Cooling-Off Period
Single homicide	1	1	1	does not apply
Double homicide	2	1	1	does not apply
Triple homicide	3	1	1	does not apply
Mass murder	4 or more	1	1	does not apply
Killing spree	2 or more	2 or more	1	None
Serial murders	3 or more	3 or more	3 or more	may be days, weeks, months, or more

Figure 15-9. *Classification of homicides based on the number of victims and the time between each event.*

Figure 15-10. *The shootings at Columbine High School and Virginia Tech were examples of spree killings—multiple shootings over a very short period of time.*

Determining Motives

After the crime has been classified, profilers look at the criminal's primary motive. For example, the perpetrator's intent is considered to be financial if crime is his or her livelihood or main source of income. This could include burglary, contract killing, or insurance scams. Sexually motivated crimes include not only rape and sexual assault, but also dismemberment and mutilation—especially of genitalia. Crimes committed during family disputes, for personal religious reasons, and as part of cults or fanatic organizations usually involve emotional motives.

Serial offenders are criminals who commit the same or similar crimes repeatedly. Regardless of the type of crime, these criminals often are motivated by unhealthy sexual fantasies, voyeurism, or sadism. Voyeurism is a psychological disorder in which a person enjoys spying on people who believe they are in a private setting, such as the bathroom or a bedroom. Sadism is a psychological disorder in which a person derives pleasure from inflicting pain on someone else. Sadistic criminals often torture their victims.

The Risk Level

Profilers must also take into account the risk level of the victim prior to the crime. Various factors, such as age, lifestyle, occupation, ability to resist, size, and location, are used in determining whether the victim was at high, medium, or low risk. Criminals often seek out victims who are the most vulnerable. A lone woman getting into a car in a dark, secluded parking lot has a much higher risk level than two men walking in a well-lit, public area (see Figure 15-11). The risk taken by the person who committed the crime is also assessed.

Criminal profilers also consider clues about the sequence of events leading up to the crime, whether a series of events exhibits a pattern of escalation, how long it took to complete the criminal acts, and all of the information about the location or locations of the crimes. Once all of the data is analyzed, the profiler will move into the next stage.

Figure 15-11. *Withdrawing money from an isolated cash machine at night increases a person's risk of becoming a victim of a crime.*

CRIME ASSESSMENT

During the crime assessment stage, the profiler completes a crime reconstruction based on the behaviors of the criminal

JIM VARNEY

Figure 15-12. *The location of the crime, the position of the victim, and the cause of death can all provide clues about the criminal.*

and the victim. The crime scene provides clues about the way a crime was committed, which, in turn, provides clues about the person who committed the crime. For example, a well-organized crime indicates that the criminal is clever and takes control over the victim. The crime is usually well planned and the victim is specifically targeted. A disorganized crime indicates a suspect who is impulsive and tends to select victims at random. The profiler also considers whether the crime scene was staged. For example, a murderer may try to mislead investigators by carefully placing evidence or by creating the appearance of a kidnapping.

Modus Operandi

The motivation behind the crime also helps in the crime assessment. A crime that is well planned and premeditated indicates a suspect who is capable of logical thinking and of creating and completing a plan of action. By contrast, the disorganized crime may be motivated by panic, stress, drugs or alcohol, or even mental illness.

Crime-scene dynamics are also important when reconstructing the crime. The profiler will consider the location of the crime as well as the methods used by the suspect. If the crime scene involves a murder, the profiler will also consider such information as the cause of death, the position of the body, trauma to the victim, the type of weapon used, and the location of the wounds. The profiler's interpretation is based on previous experience with similar cases and the method of operation in the case. Often, especially in the case of a serial criminal, there is a recognized pattern of behavior including the use of tools, weapon of choice, or preference of victims. This is referred to as the **modus operandi (MO)** or method of operation. Characteristic factors of the crime may be very valuable in leading to the identification of a suspect (see Figure 15-12).

Signature

Criminals who commit the same type of crime repeatedly also tend to repeat the same habits while committing the criminal act. If they were successful previously, the same pattern likely will lead to success again. However, the repetitive behavior is often what allows investigators to make the connection between a string of similar crimes and then anticipate the perpetrator's next move. When the MO includes something peculiar, the evidence is even stronger that the same perpetrator committed all of the crimes. For example, if a man enters a bank dressed in jeans, a baseball cap, and a sweatshirt and gives the teller a note stating he has a gun hidden in his pocket, his MO is fairly ordinary. He may or may not be the same person who has robbed several banks in a nearby town over the past year. If however, the robber also has a small plastic bandage on his cheek, the bandage is something odd that does not usually fit a common MO. This peculiarity would be very strong evidence if the perpetrator in the previous bank robberies also wore the bandage. It may also indicate to investigators that the perpetrator has a tattoo, scar, or mole in the area covered by the bandage. This will aid further in the physical description of the robber.

In some serial cases, the criminal also leaves behind a **signature**, something unusual or specific, at the crime scene. For example, a serial killer might position the bodies of his victims in a distinct fashion or leave behind the same piece of evidence at every scene. The Boston Strangler murdered 13 women over a period of a year and a half in the early 1960s. Each victim was strangled with an article of her own clothing, sexually assaulted, and posed—the MO. The killer also tied a bow under each victim's chin—his signature. The Zodiac killer, who killed seven people in northern California in 1968 and 1969, used the symbol in Figure 15-13 in his messages to local news agencies.

Figure 15-13. *Often serial killers will leave behind a signature after committing the crime.*

CRIMINAL PROFILE

The fourth stage of the profiling process involves developing a description of the suspect. A typical complete profile includes estimates of the criminal's race, sex, physical characteristics, habits, and values and beliefs. Statistically, specific types of crimes are committed by criminals who fit a particular pattern. For example, a typical arsonist is a white male between 18 and 27 years of age. He is usually a loner with a history of drug or alcohol abuse. He is likely to have a criminal record and a poor employment history. Serial murderers tend to be white males between the ages of 25 and 34. They are usually of at least average intelligence and possess charming personalities. If the scene appears to have been carefully planned or orchestrated, the killer is likely to be older and very intelligent. Many have been victims of child abuse. Profilers may also be able to hypothesize whether the criminal has committed crimes in the past.

Once the profile has been generated, it must be compared with the decision process models and the reconstruction of the crime to ensure the profile matches the data gathered about the crime.

Figure 15-14. *A suspect is placed under arrest.*

INVESTIGATION

The investigation stage begins when the profiler's written report is added to the information investigators have already gathered. Suspects matching the profiler's description are evaluated. Once the field has been narrowed and if there has been no new evidence, such as another crime, the case will move into the final phase of the investigation. If, however, another crime of the same nature with the same MO and signature is committed or if there has been no suspect identified, then the information will be reexamined and the profile will be reevaluated and validated.

APPREHENSION

The ultimate goal of the profiling process is apprehension of the person responsible for the crime or crimes. Investigators interview and interrogate the suspect. If the interrogation results in a confession or the physical evidence links the suspect to the crime, the suspect is arrested (see Figure 15-14).

VICTIMOLOGY

Obj. 15.4, 15.5, 15.6

In order to understand fully the nature of a crime and to determine a pool of suspects, as much emphasis should be placed on understanding the victim of the crime involved as on understanding the perpetrator. A **victim** is

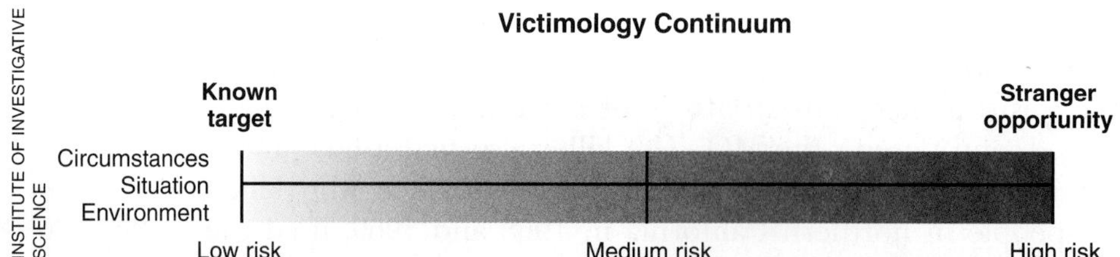

Figure 15-15. *Generally, a victim who exhibits a low level of risk is likely to know the perpetrator of the crime. On the other hand, if the victim exhibits a high level of risk, he or she is more likely to be a victim of opportunity.*

a person who has experienced harm, injuries, loss, or death. **Victimology** is the study of victims affected by crime, accidents, or natural disasters. During an investigation, a thorough analysis of the victim is often the most emotionally difficult part of the profiling process because it puts the investigator at risk of becoming consumed by the heinous nature of the crime. Although it is difficult for investigators to approach and get to know victims, it is an essential step in the investigative process.

By assessing a victim's lifestyle, preferences, family, relationships, and routines, the investigators can get a glimpse at the potential list of suspects who may have access to the victim. For example, if the victim's lifestyle rarely puts him or her at risk, then the perpetrator is more likely to be someone the victim knows (see Figure 15-15). If possible, profilers attempt to understand why a criminal may have selected the particular victim. This establishes a link between the victim and a potential suspect. Additionally, understanding why a victim was chosen for a particular crime may protect other potential victims from the same type of crime.

VICTIM RISK

When investigators evaluate a crime scene and interview the victim, they consider the degree of risk the victim was taking. This is a very subjective assessment made by the investigators and profilers, which, in many cases, is different from the perspective of the victim. Essentially, investigators try to determine if the location or events involved in the crime may have occurred due to the high risk the victim was taking. For example, if a female college student attends a party and leaves her drink on an end table to use the restroom, her risk increases. It is possible for someone to place a drug into her drink, which increases the chances she can be sexually assaulted. This scenario, in no way, intends to blame the victim. Instead, this assessment helps profilers and investigators determine the overall risk of the crime and the perpetrator's MO. Often, in these types of cases, the victims will blame themselves and the crimes usually go unreported. According to the National Institute of Justice, victim risk is categorized into three categories: low-risk victim, medium-risk victim, and high-risk victim. See Figure 15-16 for further explanation of victim risk.

When interviewing victims, it is necessary to establish their typical lifestyle and habits. By evaluating the victim's lifestyle, investigators can estimate the degree of risk that was present when the crime occurred. The victim's overall demeanor and personality can also contribute to risk. The following list shows a few examples of personality traits used to identify victim risk. The stronger the personality characteristic, the

Risk Level	Explanation
Low-risk victim	The lifestyle and social environment of the individual does not predispose him or her to being a victim of a crime.
Medium-risk victim	The lifestyle and social environment of the individual can increase his or her risk of being a victim of a crime.
High-risk victim	The lifestyle and social environment of the individual often puts him or her at risk of becoming the victim of a crime.

Figure 15-16. *Victim risk assessment.*

higher the risk. Risk is also higher in individuals who exhibit more than one characteristic. According to the National Institute of Justice, all risks are considered from the perspective of the victim's age, job, and criminal background.

- Aggressive nature
- Impulsive
- Anxious
- Passive
- Thrives on attention
- Self-inflicted injuries
- Poor self-image
- Negative
- Exhibits addictive behaviors

Digging Deeper
with Forensic Science e-Collection

Very often, a criminal chooses a target based on the vulnerability of the victim. Many of us subconsciously make ourselves easy marks by some of the behaviors we exhibit. Using the Gale Forensic Science e-Collection at www.cengage.com/school/forensicscienceadv, research how criminals study people to select the most vulnerable targets. Create a poster or informational brochure to educate others on how to protect themselves from becoming a crime victim. Be sure to include the behaviors that criminals look for in their selection of targets as well as some precautions that may be taken to lessen one's probability of becoming a victim.

VICTIMOLOGY ASSESSMENT

General guidelines have been established to serve investigators and profilers in gathering as much information as possible to narrow the pool of potential suspects. Although the data discussed in this section is a starting point, it in no way can be considered thorough because each case and victim will vary.

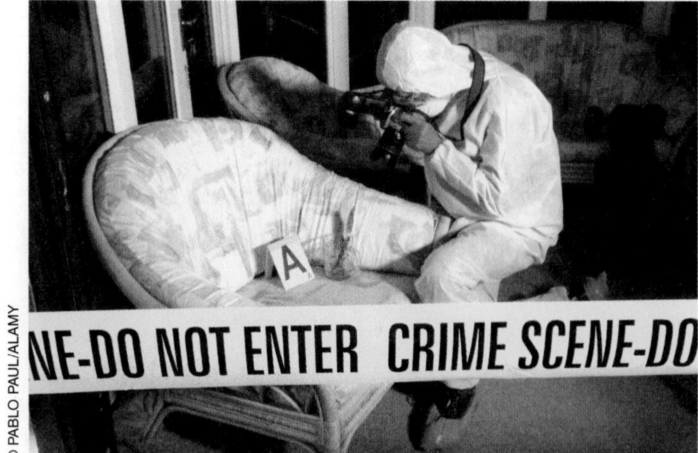

Figure 15-17. *Crime-scene investigators are responsible for collecting and analyzing physical evidence from the scene.*

At the Crime Scene

At the crime scene, investigators collect physical evidence (see Figure 15-17). A surviving victim needs to be interviewed as soon as possible. The investigators may begin with the following questions:

- Did the victim know the perpetrator?
- Does the victim suspect anyone in particular and why?
- Has the victim reported similar or other crimes in which he or she was the victim?
- Is the victim carrying a weapon? Does the victim own a weapon?
- Has the victim been in any other police reports in the past?

Forensics and Profiling Working Together

When investigators and profilers collaborate on a case, at the crime scene, physical evidence is collected and taken back to the lab for analysis by the investigator and the profiler collects data on the victim. As stated previously, no set list exists for information a profiler will collect regarding a victim; the determination is made on a case-by-case basis. After the initial interview process is complete, investigators and profilers may follow up with questions similar to the following:

- What are some of the victim's general physical characteristics, including eye color, hair color, weight, and ethnicity?
- Where does the victim work and what is his or her work schedule?
- Does the victim have a criminal history? If so, investigators will ask the victim to explain.
- What is the victim's daily routine?
- What is the contact information of the victim's family members, friends, and coworkers?
- What is the victim's medical history, including mental health records?
- What medications is the victim taking?
- What is the victim's education level?
- Where does the victim live and with whom?

Investigators and profilers may also ask the victim to provide bank and credit card statements and for access to the victim's computer (including e-mail, saved documents, and software). Investigators work with the victim to establish a timeline of events for the 24 hours leading up to the crime, including the route the victim took to arrive at the crime scene. Investigators then review any security cameras along the route.

A timeline of events leading up to the crime helps investigators determine who had access to the victim. Typically, the 24 hours just prior to the crime are most informative, but investigators might need to develop a timeline to include several days, depending on the nature of the crime and the lifestyle of the victim. By developing a timeline, a possible link can be established between a potential suspect and the victim. The criminal profiler needs this information to get a picture of the victim's daily routine and to make a determination of the location and time the perpetrator could have direct contact with the victim. It is often very difficult for a victim to provide these details. However, the timeline can provide important clues about who might have committed the crime.

CHAPTER SUMMARY

- Criminal profilers are responsible for estimating the characteristics and traits of a perpetrator to narrow the field of suspects.
- Criminal profiling has a deeply rooted history dating back to the 1800s beginning with Cesare Lombroso, who concluded that there were three types of criminals.
- In 1972, Jack Kirsh opened the FBI's Behavioral Science Unit and helped law enforcement solve cases by developing criminal profiles of unknown suspects.
- The process of criminal profiling consists of six stages with a goal of apprehension of a suspect. The stages are input, decision process models, crime assessment, criminal profile, investigation, and apprehension.
- A victim is a person who has experienced harm, injuries, loss, or death. Victimology is the study victims who have been affected by crime, accidents, or natural disasters.
- By assessing a victim's lifestyle, preferences, family, relationships, and routines, investigators may gather clues about potential suspects who had access to the victim.
- The investigator will collect and analyze physical evidence, and the profiler makes inferences about the personality and characteristics of a suspect based on information gathered from the scene.
- A timeline of the victim's events helps investigators determine who had access to the victim. The 24 hours leading up to the crime generally provide the most helpful information.

Did You Know?

Public interest in criminal profiling was piqued by the release of the 1992 movie *The Silence of the Lambs*. The movie was based on the Thomas Harris book by the same title. Harris enlisted the help of the FBI Behavioral Science Unit for technical advice when developing the characters. Jodie Foster played the role of an FBI trainee sent by the BSU to question a serial killer.

CASE STUDIES

AP PHOTO

Figure 15-18. *George Metesky, also known as New York's Mad Bomber, was found to be legally insane and was committed to an asylum.*

New York's Mad Bomber (1957)

For 16 years during the 1940s and 1950s, the residents of New York City were terrorized by a series of 33 bombings. The bombs were small and placed in various locations, such as under theater seats and in storage lockers, public restrooms, and telephone booths. Fortunately, none of the bombs killed anyone, but 15 people were injured and property was destroyed.

The bombings appeared to be completely random in both location and timing, so the police had very little to go on to identify a suspect. Then, in 1956, a police inspector decided to visit the office of Dr. James Brussel, a criminal psychologist, to ask for his ideas about the identity of the bomber. Brussel didn't think that the profile he drew up of the bomber could be of any help to the police, but he agreed to try nonetheless. In his profile, Brussel described the bomber as a middle-aged man who had likely been a former employee of Consolidated Edison (Con Ed)—where the first bomb was found on November 16, 1940. Brussel went on to describe the bomber as a meticulous man and predicted that he would be wearing a double-breasted suit when he was apprehended.

Police had Con Ed search their records for a former employee that had caused problems. The company turned up the name George Metesky (see Figure 15-18). Metesky had been injured in 1931 while working at one of Con Ed's generating plants. He sought workers' compensation and claimed disability. His claims were denied, and he became very angry and resentful of the company and its employees.

Upon questioning Metesky, police discovered that he was indeed their suspect. He was wearing a bathrobe when he first greeted police at the door. When they arrested him, they asked him to change clothes so they could take him down to the station. Metesky put on a double-breasted suit.

Metesky was found legally insane and incompetent to stand trial. On April 18, 1957, he was committed to Matteawan Hospital for the Criminally Insane in Beacon, New York. He was released in 1973 and died in 1994.

Think Critically

1. **What type of evidence would tell a profiler that a criminal was meticulous?**
2. **How did criminal profiling help solve this case?**

Atlanta Child Murders (1981)

Between 1979 and 1981, 25 black males from the same neighborhood in Atlanta, Georgia, were strangled, beaten, or smothered. The youngest of the victims was only nine years old. Several of the bodies had been found in the Chattahoochee River. Several of the bodies were also found with dog hair and similar fibers on them. Other than these leads, police had little to go on to identify the killer. However, because all of the victims were black, detectives thought the murders might have been racially motivated. They investigated a member of the Ku Klux Klan, but did not find evidence connecting him to the crime.

The Atlanta police contacted the FBI for assistance in solving the murders. John Douglas and Roy Hazelwood of the Behavioral Science Unit were assigned to the case. Douglas and Hazelwood examined the evidence and became familiar with the victims' neighborhood.

AP PHOTO/JOHN BAZEMORE, FILE

Figure 15-19. *Wayne Williams is now serving two life sentences for murder.*

Douglas and Hazelwood predicted that the murders were not racially motivated, as the police suspected. Instead, they thought the killer was black. In addition, Douglas and Hazelwood thought the killer was likely to strike again soon and would dump the body in the Chattahoochee River. Using this prediction, the police set up a stakeout around the river's bridges. On May 22, 1981, the stakeout patrol heard a splash in the river. They stopped a local man named Wayne Williams, who claimed he had just dumped some trash. However, the "trash" turned out to be a body. Later, investigators determined that hairs and fibers found on one of the victim's bodies were consistent with hairs and fibers found in Williams' home and car. Williams was arrested and convicted of two of the murders (see Figure 15-19).

Williams has continued to proclaim his innocence. However, in 2010, mitochondrial DNA testing found that hair found on one of the victims was consistent with Williams' hair.

Think Critically

1. **What are some of the things the victims of this case had in common?**
2. **Why do you think the police thought the 25 victims were murdered by the same person?**
3. **How did criminal profiling help solve this case?**

Bibliography

Books and Journals

Campbell, John H., and Don DeNevi, eds. (2004). *Profilers*. Amherst, NY: Prometheus Books.

Hicks, Scotia J., and Bruce D. Sales. (2006). *Criminal Profiling: Developing an Effective Science and Practice*. Washington, DC: American Psychological Association.

Innes, Brian. (2003). *Profile of a Criminal Mind: How Psychological Profiling Helps Solve True Crimes*. Pleasantville, NY: The Readers Digest Association, Inc.

Raine, Adrian. (1993). *The Psychopathology of Crime: Criminal Behavior as a Clinical Disorder*. San Diego, CA: Academic Press.

Turvey, Brent. (2002). *Criminal Profiling: An Introduction to Behavioral Evidence Analysis* (2nd ed.). San Diego, CA: Elsevier Academic Press.

Yochelson, Samuel, and Stanton E. Samenow. (1976). *The Criminal Personality*. Northvale, NJ: Jason Aronson, Inc.

Websites

www.apa.org/about/gr/issues/lgbt/anti-gay.aspx
www.cengage.com/school/forensicscienceadv
www.jblearning.com/samples/0763741159/Ch3_Female_Offenders_2e.pdf
www.pbs.org/wgbh/pages/frontline/shows/assault/roots/franklin.html

CAREERS IN FORENSICS

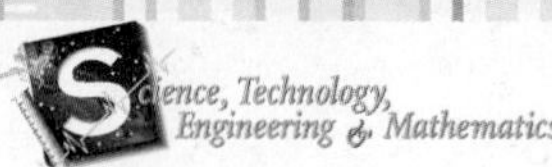

Forensic Psychologist

A psychologist is a scientist who studies the workings of the human mind and human behavior. Forensic psychologists have advanced training and education that qualifies them to apply psychological science to help resolve legal matters and court cases.

Meet Karen Franklin, Ph.D., licensed clinical psychologist, licensed private investigator, and forensic psychology teacher at the California School of Professional Psychology (see Figure 15-20). "My forensic practice focuses on adult and adolescent criminal defendants. In a typical week, I might conduct evaluations at several jails, meet with attorneys, review scientific research, and write reports. The forensic issues are varied. I evaluate mentally ill defendants to see if they are competent to stand trial, that is, do they understand what is going on in court? I may assess someone's mental state at the time of a crime, to see if he met the legal criteria for insanity. I evaluate people's risk for committing future violence or sex offenses."

COURTESY OF DR. KAREN FRANKLIN

Figure 15-20. Dr. Karen Franklin

The work is challenging and always in a state of flux—there are few "predictable days" for Dr. Franklin. She meets people from varied backgrounds and stages in life and tries to understand what motivates them, what drives them to do what they do. "I have gotten to know people who have committed all manner of horrendous crimes—serial killers, rapists, child molesters, mothers who have killed their children. The challenge is to apply the science of psychology to understanding people's behavior without moral judgment. My philosophy is that no one is all bad or all good. Even people who have done terrible things have some good in them.

"My most fulfilling cases have been when I have been able to help right a wrong. For example, I was able to help a high school boy who succumbed to police pressure and confessed to a crime he had not committed get acquitted and go home. Another time, I helped win the release of a 30-year-old man who had been locked in a mental hospital since he was 18 years old even though he was not mentally ill. The hardest cases for me have been 'cold cases' in which someone was charged with a crime based on DNA evidence. It is very hard to collect the data needed to determine someone's psychological state decades in the past."

Dr. Franklin stresses that a good forensic psychologist needs excellent writing skills, sharp investigative skills, and a "thick skin" to be able to cope with the stress of being publicly cross-examined in court.

"Forensic psychology is an ideal career for someone with a low threshold for boredom. I became interested in forensic psychology while working as a criminal investigator on death penalty cases. I realized that forensic psychologists got paid to do what I liked best—study what makes people tick."

Learn More About It

To learn more about forensic science careers, go to www.cengage.com/school/forensicscienceadv.

CHAPTER 15 REVIEW

Multiple Choice

1. Duties of a criminal profiler include all of the following *except:* *Obj. 15.4*
   a. inferring the characteristics of a suspect
   b. hypothesizing about a suspect's age, lifestyle, and social environment
   c. collecting physical evidence
   d. evaluating data prepared by investigators
2. What is the correct process in which a criminal profile is developed? *Obj. 15.2*
   a. decision process models, crime assessment, input, investigation, criminal profile, apprehension
   b. input, crime assessment, decision process model, investigation, criminal profile, apprehension
   c. input, investigation, crime assessment, decision process models, criminal profile, apprehension
   d. input, decision process models, crime assessment, criminal profile, investigation, apprehension
3. Mark is a criminal profiler. He has been thoroughly analyzing autopsy photos and police and lab reports related to a recent homicide. In which stage of the profiling process is Mark working? *Obj. 15.2*
   a. decision process model
   b. input
   c. criminal profile
   d. reconstruction
4. Adrianna is a college student driving back to her apartment from class at 9:00 PM. She stops at the gas station to fill up her tank. She walks into the store to pay for the gas. She gets in the car and drives away. About five miles down the road, a man leans forward from her backseat and holds her at knifepoint, makes her pull over, sexually assaults her, and then flees the scene. Who is the victim and what was his or her level of risk in this scenario? *Obj. 15.3*
   a. the sales clerk at the gas station; low risk
   b. Adrianna, the driver; moderate risk
   c. Adrianna, the driver; low risk
   d. the man with the knife; high risk
5. Four burglaries have occurred in a local community over the course of five days. Law enforcement knows the burglaries were committed by the same person because store employees have stated that the individual has a dragon tattoo on his or her left hand. Which of the following motives is most representative of someone who would commit this crime? *Obj. 15.2*
   a. addiction
   b. sexual satisfaction
   c. religion
   d. financial gain

CHAPTER 1 2 3 4 5 6 7 8 9 10 11 12 13 14 15

Short Answer

6. Develop a chart that represents key contributors to and their work in the field of criminal profiling. *Obj. 15.1*

7. Using the chart developed in the previous item, compare and contrast modern-day criminal profiling with original historic perspectives. *Obj. 15.1*

____

____

____

8. Discuss the importance of returning to the reconstruction of the crime and the decision process model once a profile has been generated. *Obj. 15.2*

____

____

____

9. Explain the difference between a serial crime and a crime spree. Give a brief example of each. *Obj. 15.2*

____

____

____

10. Differentiate between the profile of an arsonist and the profile of a murderer. Why is it possible to have different profiles for similar crimes? *Obj. 15.2*

____

____

____

11. How do profilers use MOs and signatures to identify suspects? Provide an example of each. *Obj. 15.2*

____

____

____

12. Discuss how a victim's level of risk affects criminal behavior. Write a short scenario in which each level of risk is represented. *Obj. 15.3*

____

____

____

13. Explain why the victim's daily routine is important to investigators and profilers in developing a timeline. *Obj. 15.3, 15.5*

____

____

____

Think Critically

14. Evaluate the importance of understanding the victim's profile as well as the perpetrator's profile. Also, why is it difficult for profilers to work with victims? *Obj. 15.3*

15. One evening, at around 2:00 AM, Marcus, 24, leaves a bar after having several drinks. He is severely intoxicated and is stumbling. He heads down a dark alley toward his apartment. On his way, he is held at gunpoint and mugged. In the process of the crime, he struggles with the perpetrator in an effort to retain his belongings. He turns the gun on the perpetrator and kills him. Evaluate this scenario and determine the level of risk associated with the criminal act (Marcus being mugged). Should Marcus be charged with murder? Justify your position. *Obj. 15.3, 15.5*

CRIME SCENE S.P.O.T. Student-Prepared Original Titles

Death at the Docks

By: **Dakota Swanson**

Tavares High School
Tavares, Florida

Figure 15-21. *When police arrived, the dock was still burning.*

LOULOUPHOTOS/SHUTTERSTOCK.COM

Detective Henry Stokes ducked under the yellow crime-scene tape, walking toward the commotion. The police had already taped off the surrounding area, and the death investigators were waiting to take the body back to the lab to be processed by the medical examiner.

Approaching the scene of the crime, Stokes could tell that this was not an ordinary crime scene—it was what was left of a wooden fishing dock with a boat launch slanting into the water on the left side. The dock had been badly burned and despite the fact that the fire marshal and his team had doused the remains with a hose, the remnants still smoldered. Much of the dock was submerged under water. The side of the dock that remained afloat was beyond repair.

On the shore lay the torso of a human body—badly charred and burned. It had been pulled from the water by the fire marshal, who explained that when he arrived at the scene, he saw what looked like a flaming body still floating in the water.

Detective Stokes could smell the accelerant in the air. "What happened here?" he asked a police officer.

"The victim was pulled from the water badly burned. Evidently, the victim and the dock were lit on fire using accelerant. Most probably, he or she was on the dock when it went up in flames. Then, when the dock gave way because of the fire, the victim fell into the water and stayed afloat until pulled to the shore by the marshal."

"Any suspects?" asked Stokes.

"None; not to mention we don't have any leads either. Well, except that the fire marshal said that when he arrived he saw some unusual markings in the sand right over there. But other than that, we don't have any leads."

"That's why the Captain called me in."

"Yes, it is," shouted the Captain, walking toward them. "Stokes is the best detective around; and he's good at profiling, which is exactly why I called him. We can't seem to find any leads, so we found the next best thing—a criminal profiler."

"The killer is smart," said Stokes, producing a clipboard from his trench coat with a "Profiling Report" clamped to it. As he wrote, he talked. "He burned the body and the entire scene of the murder—the dock. It would make more sense to kill a person and dispose of the evidence in the same place rather than to kill someone and transport the body—and risk being caught. This is why I think that the victim was killed on the dock itself.

"Also, if the killer is smart enough to know not to leave any clues and burn the evidence, then it was probably planned out. How would the killer have known to bring the accelerant in his car if it wasn't premeditated? That is, unless the murderer was a mechanic; but this crime does not seem to be an impulse crime—it appears almost professional."

"Yes," said the police officer, "Usually if it is a crime of sudden impulse the killer leaves trace evidence and clues. It wasn't on an impulse."

The detective went on. "Usually, arsonists have had some traumatic experience in their lifetime—a loved one who died tragically or physical abuse. The killer most likely lured the victim to this location with the motive to kill. He or she didn't overlook any factor. A person who takes the time to lead a person to a specific place with the intention to burn the entire crime scene wouldn't be careless."

Stokes walked over to the unusual marks in the soft, white sand on the shore, which the police officer had told him about. The marks led all the way from the dock to the end of the white sand and then stopped at the grass.

"Look at this," exclaimed Stokes, who stopped writing on his report for a moment. "The killer even covered his tracks by smoothing out the sand that he walked on so we wouldn't identify his shoeprints. I'm surprised the officers haven't covered it with their own tracks yet."

The Captain called over his shoulder to get a photo of the markings in the sand.

"That's interesting," Stokes went on. "He definitely took his time cleaning up all the loose ends. I would also say that the killer has no morals or ethics whatsoever. Probably a sociopath who distances himself from others. Whether he was born like this or developed this way is nearly impossible to say. I would guess that he developed it through a horrible accident of some kind—most arsonists light fires because of a traumatic event of some kind. Usually an arsonist commits a crime shortly after an accident. If we find the accident, we could find the killer."

The Captain asked, "Have any big accidents of any sort happened recently?"

The police officer started to go through a list of suspects. "It's not a big town and I remember three months ago, the brother of a college professor was shot by a gang member—it was all over the news. Then four months ago, the mayor's wife was in a fatal accident caused by a drunk driver. Last week, a college graduate's wife was killed in a burglary. . . . Oh, and that store clerk who was shot in a violent fight over a gun. His wife, who is also a clerk at the same store, was on the news talking about how she would like to find her husband's killer. That's about it though."

"So we have four possible suspects—the mayor, the college professor, the graduate, and the wife. How can we tell which one is the arsonist?"

"Only four big events have happened here in this little town and two of those suspects events can be ruled out immediately. To be a successful store owner, you have to be social and outgoing. The same goes for a mayor. That leaves us just two likely suspects and one of them just doesn't seem to fit. . . . The other is very interesting though . . ."

"What are you getting at?" asked the Captain.

"I think I have an idea who the killer is. Didn't that college graduate major in . . . forensics?"

Activity

Answer the following questions based on information in the Crime Scene S.P.O.T..

1. How did the profiler and the investigator work together to develop a list of suspects?
2. How did the investigator narrow down the pool of suspects to determine who committed the crime?
3. WRITING Rewrite the ending of this story to make one of the other suspects the criminal. Additional changes in the story may need to be made in order for a different criminal to work within the story. Be creative in your changes.

Preventing Adolescent Crime Together™ (P.A.C.T)

Introduction:

Our society is a "melting pot" of diverse individuals. Building relationships with people of different backgrounds and cultures is essential to building good character and a strong work ethic. Children, beginning typically in elementary school, start to recognize differences in other children. Partnerships already exist in many elementary schools between parents, teachers, and the community aimed at teaching tolerance, embracing differences, compassion, and service to the needs of others. Ultimately, tolerance is acknowledging, appreciating, and accepting the differences that exist among one another. The differences in one another allow for the uniqueness of each person to contribute to society.

Although criminal profiling is a highly skilled field of forensics, individuals often profile people based on various characteristics. For example, a teenage boy with a large baggy coat on might walk into a department store. Does that necessarily mean he is in there to steal something? Would he be more likely to steal from the store if he exhibited a specific ethnicity, a specific sexual orientation, or a gang-related bandana? Is it possible that the teenager went into the department store on a winter day to buy his sister a gift? Individuals not trained in profiling often profile an individual based on very superficial characteristics. This is not the foundation of criminal profiling. Your task in this activity will be to encourage elementary school students to understand that tolerance and acceptance are critical to existing and prospering in our society.

Procedure:

1. Working in groups of two or three students, research criminal profiling and "casual" profiling. Be sure to document the distinction in your research.
2. After your group has gathered enough research, organize the data collected to write a fictional book for an elementary student designed to discourage casual profiling. Your target audience should be fourth and fifth graders.
3. Pages in your book should be colorful with many pictures to demonstrate various ethnic and cultural differences. Also, make sure the illustrations match the text on the page to help visual learners see the message you are trying to convey.
4. The font should be large (somewhere between 14 to 20 points), and you should not overcrowd the page with text. Too much text might cause your audience to get lost or lose interest in the message.
5. Be creative in your story development. You may write your story from a first-person point of view where the individual being profiled is telling the story or you can tell the story in third person, through the eyes of the author.
6. Make small storyboard pages to show what content and images will be included on each page.
7. After you have developed your storyboard, show it to your teacher for approval.

8. Once you have your teacher's approval, you may begin to create your book. You may cut out pictures from magazines, create images on the computer, or work with a student trained in computer graphics.
9. After your book is completed, your teacher will laminate and bind the pages.
10. The books will either be delivered to your local feeder elementary school or you will travel to the school as a class to distribute the books personally.

MONKEY BUSINESS IMAGES/SHUTTERSTOCK.COM

Figure 15-22. *Racial profiling is not the same as criminal profiling. Strong communities develop respect and tolerance among all people.*

ACTIVITY 15-1
ONE MAN'S TRASH IS A PROFILER'S TREASURE!

Objectives:

By the end of this activity, you will be able to:
Create a profile based on items found in household trash.

Materials:

(per group of three or four students)
3 (or 4) forceps
3 (or 4) probes
newspaper or butcher paper
stereomicroscope
3 (or 4) hand lenses
1 bag of trash
camera (optional)

Safety Precautions:

Goggles and gloves must be worn to process the trash.
If you are allergic to latex, alert your teacher so that you can use alternative gloves.
Aprons should be used to protect your clothing.

Procedure:

1. Before opening the trash bag, spread newspaper or butcher paper on the table to protect the surface.
2. Carefully open the bag and remove the contents one item at a time.
3. Carefully examine each item.
4. Use a hand lens or stereomicroscope to see details or to examine small items.
5. Construct a data table to record a full description of each item as well as any other pertinent details.
6. After all items have been removed from the bag, decide as a group how the items can be classified. For example, if your bag contains only items belonging to an adult female, you might classify them based on how each item is used.
7. You may wish to photograph each item or category of items to include in your written report.

© ANGELA HAMPTON PICTURE LIBRARY/ALAMY

Figure 1. *Items found in trash may provide evidence about the people living in the household.*

Questions:

1. Based on what you found in the trash, describe the people who live in the household. Include relative ages, sex, hobbies or interests, and the career or profession of each person in the household.
2. Compare your "profile" to the photographs and descriptions of the families provided by your teacher. Once you think you have a match, verify with your teacher. What item(s) made the identification difficult? Which items were most helpful?
3. How might items found in a suspect's trash help a criminal profiler determine whether the suspect fits the criminal profile?

Extension:

Choose one of the cases highlighted in this book, and research how investigators identified the perpetrator. Describe how profilers may have played a role in identifying the perpetrator based on the MO, victimology, evidence, and type of crime.

PROJECT 1

BODY DUMP

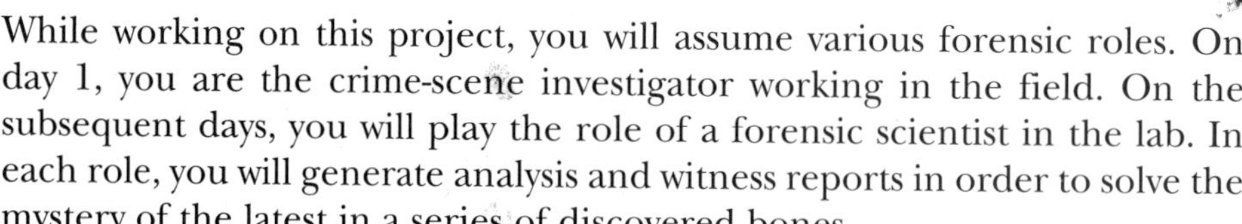

OBJECTIVES

By the end of this project, you will have:

- ✔ Used various methods to document a crime scene.
- ✔ Properly identified and collected physical evidence.
- ✔ Used metric measurements to calculate height from a long bone.
- ✔ Performed chemical analysis on evidence for the purpose of comparison and identification.

INTRODUCTION

While working on this project, you will assume various forensic roles. On day 1, you are the crime-scene investigator working in the field. On the subsequent days, you will play the role of a forensic scientist in the lab. In each role, you will generate analysis and witness reports in order to solve the mystery of the latest in a series of discovered bones.

You will use skills you developed throughout the course (Chapters 1–15) as you complete this project.

SCENARIO

Over the past 12 months, seven bodies have been found within a 50-mile radius of the city park. Most of them have been severely decomposed. The bones found most recently are completely skeletonized.

Investigators think that all of the victims were killed by the same perpetrator in part because a pink scarf was tied around the neck of each victim.

DAY 1: THE CRIME SCENE

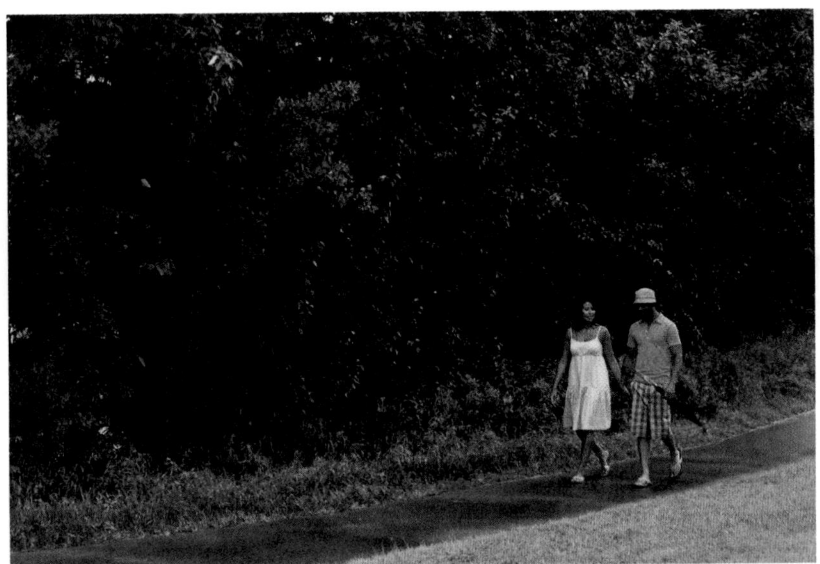

© FANCY/ALAMY

The call came in a little after noon on a typical Monday. The day, however, would not remain typical. The caller had stumbled upon a bone sticking out of a partially shredded black plastic garbage bag while hiking along a trail in the city park.

When investigators arrived at the scene, they immediately secured the area near the bone and began a search of the entire park and surrounding area. After several hours, more bones were recovered along with various types of physical evidence.

Though this is the first body to be discovered that is completely skeletonized, there is evidence to tie it to the others. The shoe at this scene is stained. Suspects from the previous crimes were wearing clothing smudged with a similar material. However, none of the other scenes contained evidence that could connect them. This could be the break investigators need to nail the murderer. You are the first crime-scene investigator to arrive at the scene.

MATERIALS

sketch pad
pencils
metric ruler
metric tape measure
crime-scene tape
camera
evidence collection materials (bags, boxes, plastic bags, vials, etc.)
forceps

SAFETY PRECAUTIONS

Protective gloves must be worn at all times.

If you are allergic to latex, alert your teacher so that you can use alternative gloves.

Aprons may be worn to protect clothing.

PROCEDURE

1. Outline the steps you would take from the time you arrive at the crime scene until you are ready to send evidence to the lab for processing. Develop a table for recording your data.
2. Refer to Chapters 1, 2, and 3 to ensure proper technique when handling and processing evidence.
3. Begin a rough draft of the report to submit to your superior investigator. Be sure to include your sketches and photographs. (The report will be a working document throughout the entire project. A final report will be submitted at the end of the project.)

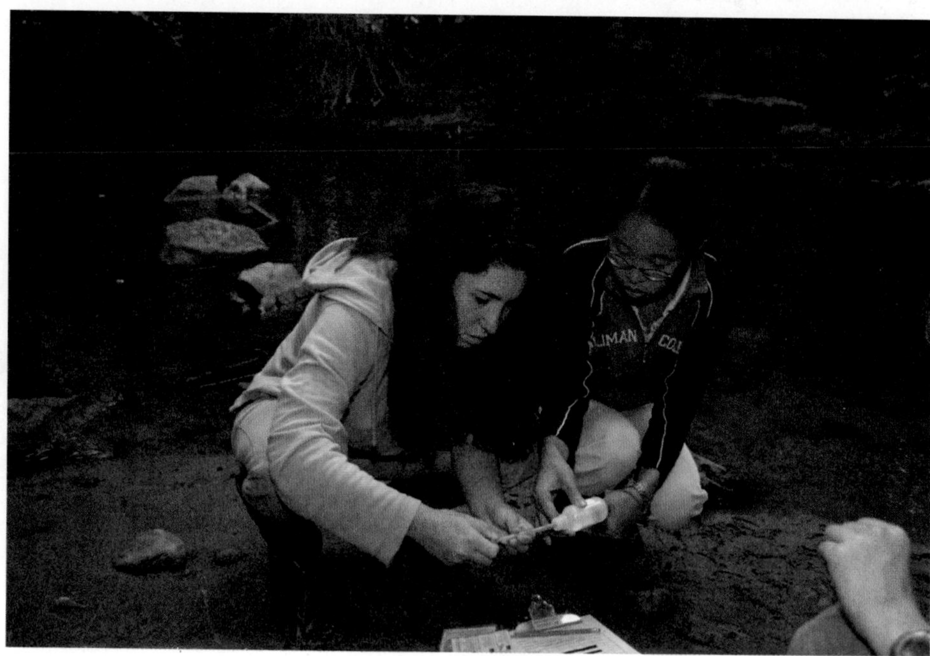

DAYS 2 AND 3: THE CRIME LAB

Today, you play the role of a crime-lab scientist. You will analyze the evidence you collected at the crime scene on Day 1.

MATERIALS

osteometric board (optional)
metric tape measure
chart of bone measurements
cotton swabs
solvents
scissors
ethanol
chromatography paper
stereomicroscope
forceps
evidence from the scene
missing person reports
cyanoacrylate glue
fuming chamber
hand lens

SAFETY PRECAUTIONS

Goggles and protective gloves must be worn at all times.
If you are allergic to latex, alert your teacher so that you can use alternative gloves.
Aprons may be worn to protect clothing.

PROCEDURE

The bones:

1. Inspect each of the bones for evidence of trauma.
2. Make measurements of the various bones. Construct a table to record your data.
3. Refer to the bone chart to determine the approximate size of the victim.
4. Compare the data to information on the missing person reports to make a preliminary identification.
5. Once you have a preliminary identification, ask your lab supervisor for the witness statements taken when that person disappeared.

The stains:

1. Perform a qualitative analysis to determine the R_f values of the components of the stain on the shoe.
2. Perform the same test on a sample of clothing for each of the suspects for comparisons. Is there a clear suspect?
3. Construct a data table to record your results.
4. After preliminary identification of the suspect, ask your lab supervisor for the envelope containing interrogation reports on each of the suspects.

The fabric:

1. Using a stereomicroscope, examine the piece of fabric found at the scene. Construct a table to record the information.
2. Compare the fabric to a piece of scarf found at one of the other scenes. What characteristics of the fabric and the scarf are the same? Different?

The garbage bag:

1. Examine the plastic bag for visible and latent fingerprints.
2. Record your observations.

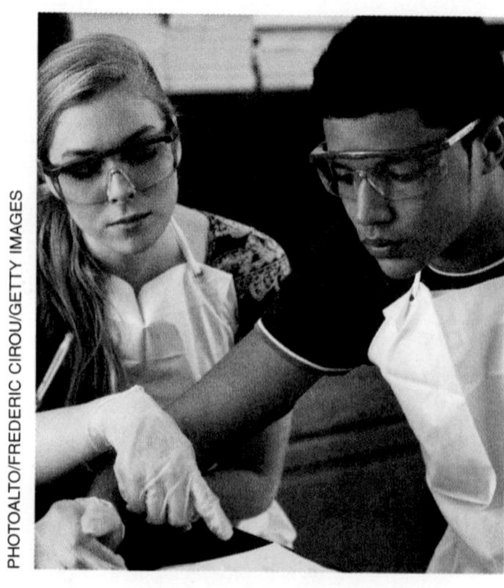

DAY 4: THE REPORT

© BOB DAEMMRICH/ALAMY

PROCEDURE

1. Prepare your final report.
2. Be sure to include data, sketches, photographs, and your analysis of the results.
3. The final section of the report should include a crime reconstruction report.

QUESTIONS

1. Who committed the crime? What evidence would you use to present your case to the prosecuting attorney?
2. Is this an isolated murder, a spree killing, or a serial murder? Justify your answer.
3. If this scene is part of a spree killing or a serial murder, what ties all of the murders together?
4. Based on the condition of all of the victims, is this victim the latest or the first? Explain.
5. What other evidence might an investigator look for to determine how long the body had been at the site?

PROJECT 2

ACTIVITY 2A: ACCIDENT OR ARSON?

OBJECTIVES

By the end of this project, you will have:

- ✔ Collected, processed, and analyzed evidence from a fire investigation.
- ✔ Developed a courtroom-ready report related to the fire investigation.

INTRODUCTION

In this activity, you will collect and analyze data as part of a fire investigation to determine the cause of a fire. You will use skills you developed in Chapters 1–5 as you complete this project.

PROCEDURE

1. Your teacher will provide you with an arson scenario.
2. In groups of two or three students, comb the scene for trace evidence and collect and process the evidence.
3. Determine how the fire started and the type of evidence found at the scene. Maintain proper evidence handling, collecting, and processing.

QUESTIONS

1. How did the fire start? What evidence did you find that supports your conclusion?
2. As a team, develop a courtroom-ready report that explains your analysis techniques and findings.

EXTENSION

WRITING

Imagine that law enforcement have decided to press charges against a suspect in this case. Write an essay describing your testimony as an expert witness. Be sure to explain the procedures and science you used to support your findings. Your response needs to be thorough and clear.

DARIN ECHELBERGER/SHUTTERSTOCK.COM

ACTIVITY 2B: CRIME TV ANALYSIS

OBJECTIVE

By the end of this project, you will have:

✔ Summarized the evidence collected during an autopsy and explained the conclusions drawn by the medical examiner.

INTRODUCTION

In this activity, you will observe the steps of an autopsy. You will need to listen carefully and take notes in order to provide a summary of the autopsy and of the evidence found during the autopsy. You will use skills you developed in Chapters 6–8 as you complete this project.

PROCEDURE

1. Watch an episode of *Dr. G.: Medical Examiner* (Discovery Heath).
2. While watching the episode, provide a summary of the victim's general appearance, including the victim's age, gender, and ethnicity.
3. Document clues discovered during autopsy and explain how evidence found on the victim assists Dr. G in determining the cause, manner, and mechanism of death.
4. Be sure to use proper technical terminology regarding trauma to the body and human anatomy throughout the activity.

ACTIVITY 2C: BEAT THE STREET

OBJECTIVES

By the end of this project, you will have:

- ✔ Researched the chemical composition and effects of illegal drugs.
- ✔ Summarized the relationship between drug abuse and crime.
- ✔ Outlined the role of the toxicological evidence in a high-profile case.

INTRODUCTION

In this activity, you will research the chemistry and effects of a particular illegal drug. You will then research the relationship between that drug and crime. This activity focuses on content and skills you learned in Chapter 9.

1. Your teacher will assign each group of two or three students a common illegal drug, such as heroin, cocaine, crack, crystal meth, or LSD, to research.
2. Develop a poster or a multimedia presentation that describes the relationship between the abuse of your assigned drug and crime and that answers the following questions.

   Is the drug a stimulant or depressant?

   How does the drug metabolize through the body?

   What are the risks of addiction and why?

   How is the drug detected in the body on a toxicology report?

3. Research a cutting-edge technique being used by toxicologists.
4. Discuss a high-profile case in which a toxicology screening was useful in determining whether foul play or an overdose was involved.

EXTENSION:

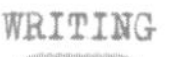

WRITING In your small group, debate the merits of doctors prescribing high doses of medications, that, when taken in conjunction with each other, can be lethal to the patient. Should doctors be held accountable for drug interactions if they did not prescribe all of the medications the patient was taking? Write a one- or two-page persuasive essay on the topic.

ACTIVITY 2D: THE ABCs OF PCR

OBJECTIVES

By the end of this project, you will have:

- ✔ Summarized and modeled the process of PCR.
- ✔ Outlined the role of PCR in a recent case.

INTRODUCTION

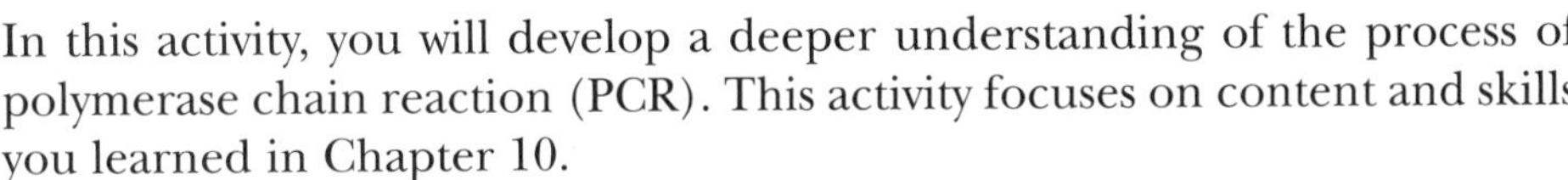

In this activity, you will develop a deeper understanding of the process of polymerase chain reaction (PCR). This activity focuses on content and skills you learned in Chapter 10.

1. In groups of three or four students, research additional information about the process of PCR.
2. Develop a simulation of the process.
3. Write a scenario in which PCR would be used. Explain each step of the process thoroughly. Describe the forensic importance of PCR.
4. If possible, collaborate with your computer gaming or design class on the development of this activity.
5. Research a current case (within the past three years) in which PCR was useful. Discuss how the evidence was presented in court.

ACTIVITY 2E: PARTNERSHIPS IN PRACTICE

OBJECTIVE

By the end of this project, you will have:

✔ Evaluated information from a crime scene to determine the role of various experts in the collection and analysis of relevant evidence.

INTRODUCTION

In this activity, you will review evidence from a crime scene and describe the roles various experts would play in collecting and analyzing information about the crime. This activity focuses on content and skills you learned in Chapters 11 and 12.

SCENARIO

The body of a male was discovered near a beach. The body was in the third stage of decomposition. The body was too badly decomposed to make an identification, and the victim had no identification on him. No immediate sign of trauma to the body was discovered.

QUESTIONS

1. How would crime-scene investigators collaborate with forensic odontologists and forensic entomologists in this case?
2. What role would each expert play in the process of identifying the body?
3. What are some specific types of key information each expert can provide?
4. Is it possible for these two experts to provide enough information to determine how this victim died? Justify your answer, using accurate scientific terminology.

PROJECT 3

PORTFOLIO ASSESSMENT

OBJECTIVES

By the end of this project, you will have:

✔ Assembled a personal portfolio that demonstrates mastery of techniques learned throughout this course.

This project represents the depth and breadth of the work you have accomplished throughout the course. You will demonstrate skills you developed in Chapters 1–15.

Develop a portfolio of at least six of the techniques demonstrated in these capstone projects and throughout the course. Although six is the minimum, your portfolio may demonstrate additional techniques. Your portfolio must be held in a binder and must meet the following criteria:

- Design six brief scenarios. Within the six scenarios, at least six different evidence collection, processing, and analysis techniques must be covered.
- Begin each new scenario and procedure on a new page.
- Include pictures and thorough documentation.
- Choose a case reviewed during the course that particularly resonated with you. On the next-to-last page of your portfolio, discuss the forensic concepts in the case and why the case was important to you.
- On the last page of your portfolio, write a reflection essay about why you chose which techniques to showcase, what you have learned in the course, and how you expect to apply these skills in the future.
- On the outside cover, write your name, date, class period, and "End of Year Portfolio."

HILL STREET STUDIOS/GETTY IMAGES

Glossary

A

abrasion: an injury in which the superficial, or top, layer of skin has been removed due to motion against a rough surface.

accelerant: in fire investigation, any material used to start or sustain a fire; the most common are combustible liquids.

algor mortis: postmortem (after death) cooling of the body.

arson: the intentional and illegal burning of property.

arthropod: a phylum of animals with jointed appendages and an exoskeleton (from the Greek *arthros*—jointed; *podes*—feet).

asphyxiation: a condition in which the amount of oxygen available to the lungs decreases sharply while the level of other gases, especially carbon monoxide, increases.

autopsy: a postmortem examination of the body, including dissection to determine cause of death.

C

cementum: bonelike covering of the portion of the tooth that extends into the bone (the root); attaches the tooth to the periodontal ligament, a connective tissue that anchors the tooth to the bone.

chain of custody: a list of all people who came into contact with an item of evidence.

chemical property: property of a substance that describes how it reacts in the presence of other substances.

chitin: a tough polysaccharide; the major component of an arthropod's exoskeleton.

chop wound: wounds that result in cuts (incised wounds) on the surface and deep internal injuries and/or fractures to bones.

chromatography: any of several processes used to separate a mixture into its individual components based on their attraction to a stationary liquid or solid.

class characteristics: properties of evidence that can be associated only with a group and never a single source.

clone: a copy made in the same type of medium.

combustion reaction: oxidation reaction that involves oxygen and that releases heat and light.

computer forensics: the specialized practice of identifying, preserving, extracting, documenting, and interpreting electronic data that can be used as evidence.

confirmatory test: test done to establish with certainty the characteristics of a substance.

contact damage: damage to a vehicle caused by contact with an object or other vehicle.

content spyware: software that is used to allow a hacker to access all the activity on an individual's personal/business computer.

contusion: a bruise caused by broken blood vessels below the skin.

criminal profiler: a person who infers the personality and characteristics of a suspect based on information gathered from a crime scene.

crown: the portion of the tooth that is covered in enamel and is situated above the gum.

cyber-terrorism: hacking into a company's internal networking system for the purpose of demonstrating or protesting a political agenda.

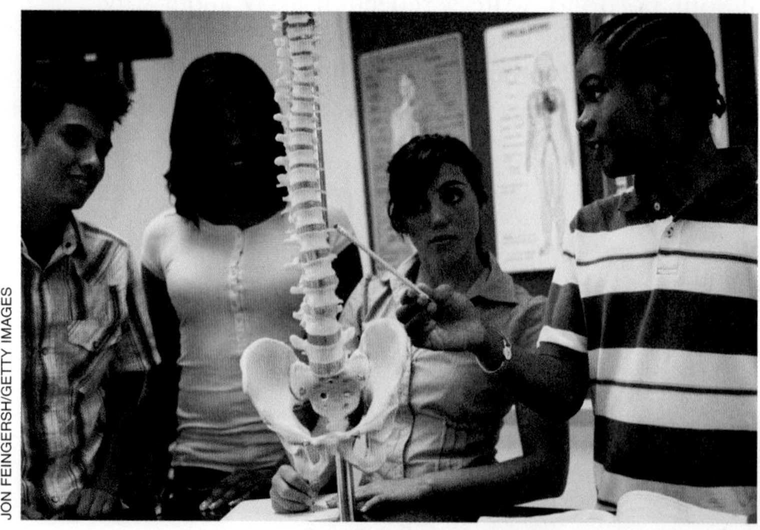

JON FEINGERSH/GETTY IMAGES

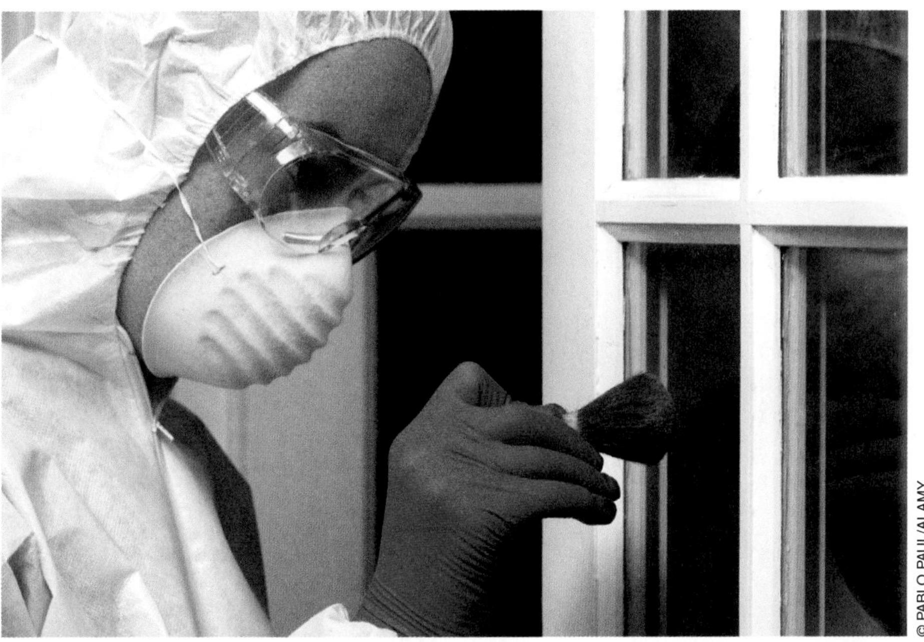

D

dentin: hardened connective tissue that makes up the majority of a tooth; surrounds the pulp cavity and is covered by enamel in the crown and by cementum in the root.

dentition pattern: the pattern made by a particular set of teeth.

depressant: a chemical that slows the heart rate and brain activity and causes drowsiness.

DNA extraction: process of removing DNA from a cell; procedures include Chelex extraction and organic extraction.

E

enamel: the outer covering of the crown of a tooth, made up of calcium carbonate and calcium phosphate.

erythrocyte: red blood cell.

ethics: a set of rules that define appropriate behavior in a situation.

exoskeleton: a rigid external structure made of chitin and protein (protects, provides a point of attachment for muscles; prevents water loss).

exothermic reaction: chemical reaction that releases heat.

explosion: the sudden release of chemical or mechanical energy caused by an oxidation or decomposition reaction that produces heat and a rapid expansion of gases.

F

force: a push or pull against an object; force equals mass times acceleration ($F = ma$).

forensic entomology: the study of insects in legal situations.

forensic science: the application of science to the law.

frequency: the number of waves that pass a specific point within a given time; usually expressed in cycles per second or hertz (Hz).

Frye Standard: rule of admissibility of evidence; evidence, procedures, and equipment presented at trial must be generally accepted by the relevant scientific community.

H

hacking: intentionally entering an unauthorized network system.

heat of combustion: excess heat that is given off in a combustion reaction.

hesitation marks: jagged and rough superficial wounds caused by someone attempting to take their own life, caused as the person responds to the pain.

high explosives: chemicals that oxidize extremely rapidly, producing heat, light, and a shock wave; will explode even when not confined.

hilt: protective piece where the blade meets the handle of a knife.

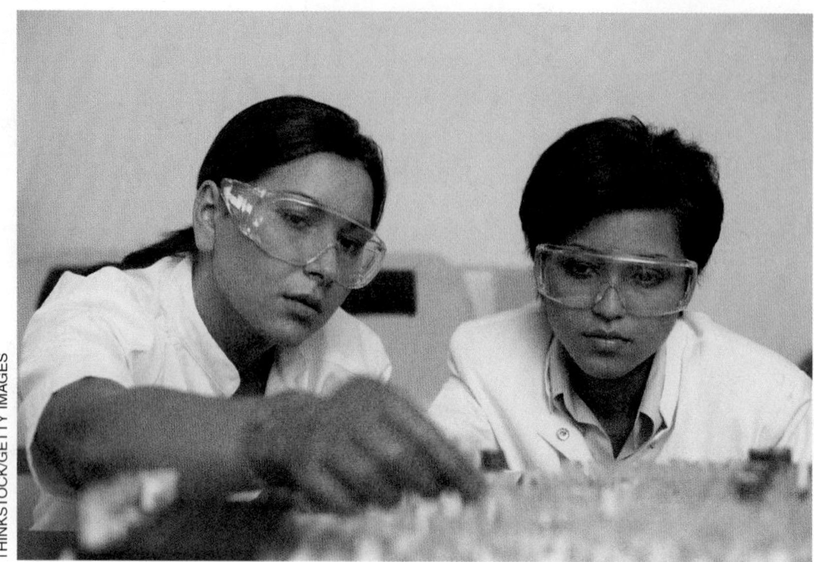

homeostasis: an organism's relatively stable internal conditions.

homologous chromosomes: pair of chromosomes in which one chromosome was inherited from the male parent and the other was inherited from the female parent.

hydrocarbon: any compound consisting only of hydrogen and carbon.

I

immunoassay: a test that relies on the antigen-antibody response.

incised wounds: cuts along the surface of the body produced by a sharp-edged object such as a knife, glass, metal, or even paper.

individual characteristics: properties of evidence that can be attributed to a common source with an extremely high degree of certainty.

Internet forensics: uses the same analysis techniques as computer forensics except the emphasis is placed on the Internet as a whole.

interrogation: official questioning of a suspect or witness by law enforcement.

interview: a question-and-answer session that does not accuse but is instead intended to gather information concerning a case and/or a suspect.

interviewer: a trained individual who questions witnesses or suspects and is able to interpret cues in verbal and physical behavior.

invertebrate: organism lacking a backbone.

K

kinetic molecular theory: a theory that states that the behavior of gases is predictable and explainable based on certain assumptions.

L

laceration: a tear in the tissue caused by sliding or crushing force.

larva (larvae, *pl*)**:** immature, feeding stage of insects that undergo complete metamorphosis; the stage between the egg and pupa.

law of conservation of energy: rule that states that energy is neither created nor destroyed, but it can be converted into different forms.

law of conservation of momentum: rule that states that the momentum of a system remains unchanged unless a force acts upon it.

leukocyte: white blood cell.

lividity: pooling of blood in the lowest portion of the body.

Locard's exchange principle: when two objects come into contact with one another, a cross-transfer of materials occurs.

low explosives: chemicals that oxidize rapidly, producing heat, light, and a pressure wave; will explode only when confined.

M

maggot: legless larva.

malware: software designed to provide unauthorized access to a computer system.

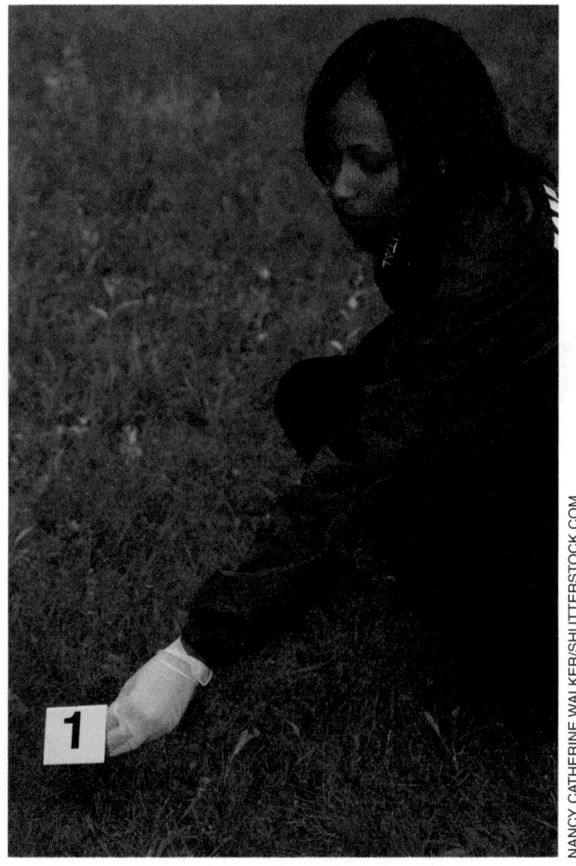

NANCY CATHERINE WALKER/SHUTTERSTOCK.COM

metamorphosis: the changes an organism undergoes as it develops into an adult.

modus operandi (MO): also referred to as the *method of operation*, a recognized pattern of behavior in the commission of a crime.

N

neck: area between the root and the crown of the tooth; also known as the *cementoenamel junction* (where the enamel and cementum meet).

nucleotide: subunit of DNA and other nucleic acids; made up of a 5-carbon sugar, a phosphate group, and a nitrogenous base.

nystagmus: involuntary jerking movement of the eyes.

O

objectivity: judgment that is not influenced by personal feelings or bias, focused on fact.

odontology: the study of the anatomy and growth of teeth and diseases associated with the teeth and gums.

oxidation reaction: the complete or partial loss of electrons or gain of oxygen.

P

phishing: illegally gathering personal information.

physical evidence: any object that can establish that a crime has been committed or can link a suspect to a victim or crime scene.

physical property: property of a substance that can be observed or measured without changing the chemical identity of the substance.

physical trauma: serious or life-threatening physical injury, wound, or shock.

platelets: cell fragments that help form blood clots at wound sites; also called *thrombocytes.*

poison: a chemical that can harm the body if ingested, absorbed, or breathed in sufficiently high concentrations.

polymerase chain reaction (PCR): technique used to make billions of copies of a specific segment of DNA; allows very minute samples of DNA to be copied billions of times.

postmortem interval (PMI): the interval of time between when death occurs and the body is discovered.

pressure: the amount of force per unit area.

MONKEY BUSINESS IMAGES/SHUTTERSTOCK.COM

presumptive test: a test to screen evidence and narrow down the possible type of a substance.

primer: small piece of DNA used to begin and end replication during PCR.

pulp: softer connective tissue that composes the innermost portion of the tooth; contains nerves and blood vessels.

pupa (pupae, *pl*)**:** nonfeeding and relatively inactive developmental stage of some insects.

pyrolysis: decomposition of organic matter by heat in the absence of oxygen.

R

reagent: a substance used to produce a chemical reaction to detect, measure, or produce other substances.

reconstruction: in criminal investigations, the process of recreating the actions and circumstances based on examination and interpretation of evidence.

reference sample: a sample from a known source used for comparison, also referred to as *exemplar*.

R_f value: retention factor; in paper and thin layer chromatography, ratio of the distance a substance traveled to the distance the solvent traveled.

rigor mortis: the stiffening of the skeletal muscles after death.

root: the portion of the tooth that extends into the tooth socket and is covered with cementum.

S

scientific method: a series of logical steps to ensure careful and systematic collection, identification, organization, and analysis of information.

short tandem repeat (STR): short segment of DNA in which the same sequence of 2–6 base pairs is repeated many times; varying numbers of repeats found among individuals.

signature: something unusual or specific left at the crime scene by the perpetrator.

skid marks: marks left on the roadway by the tires when the driver of the vehicle applies the brakes suddenly.

staging: the intentional altering of a crime scene in order to disguise what really happened.

substrate control: a similar, but uncontaminated, sample; used for making comparisons.

suffocation: condition in which the amount of oxygen available to the lungs is quickly diminished.

suspect: an individual under investigation for his or her alleged involvement in a crime.

T

terrorism: the intentional use of force or violence to coerce or intimidate governments or other large organized groups.

therapeutic wound: a wound caused by incision in a medical setting.

MEL YATES/GETTY IMAGES

tolerance: in response to prolonged, heavy drinking,the body's need for progressively larger amounts of a chemical to cause the same levels of intoxication.

toxin: a type of poison produced naturally by living things.

Trojan horse: software designed with the intention to harm a computer or the information on/within the computer.

V

victim: person who has experienced harm, injuries, loss, or death.

victimology: the study of victims affected by crime, accidents, or natural disasters.

W

wavelength: the distance between crests, or peaks, of two consecutive waves.

worm: self-replicating malware program that spreads through a computer system by sending copies of itself to networked computers.

Y

yaw marks: skid marks in a curved path as a result of an out-of-control skid.

Appendix A

THE NATURE OF SCIENCE

WHAT IS SCIENCE?

Science is a method of explaining the natural world. Scientific knowledge is based on evidence gathered through careful study and testing. Science gives us more than just a better understanding of how things work. The knowledge gained through science affects society and individuals. Governments, communities, and businesses make decisions based on studies conducted by scientists. Many technologies that entertain, save lives, improve comfort, preserve the environment, and make work more efficient and profitable are applications of scientific understanding.

SCIENTIFIC PROCESSES

Scientists draw conclusions about the natural world through a scientific process of observation, experimentation, explanation, and peer review. Because scientists all over the world use a similar process to gather information, they are able to confirm and build on one another's work.

In general, scientists begin the scientific process by observing the world around them and asking questions. They might ask questions about how things work, what things are made of, what causes certain phenomena, or why things behave the way they do. These questions may stem from long-held mysteries, they may come from unsolved problems generated by previous scientific studies, or they may arise when new inventions or discoveries shed new light on unexplored areas. Whatever the inspiration, the scientific process begins with a question.

After choosing the question or questions a scientist wants to answer, he or she will begin the experimentation stage of the process. Experimentation is guided by the scientific method. The scientific method is a series of steps a scientist takes to find the answers to questions. The specific steps taken may vary in different fields, from one scientist to the next, or even in different experiments carried out by the same scientist. However, they always involve drawing logical conclusions based on observed, and often measurable, evidence. Figure 1 maps out one way that a scientist might carry out the scientific method.

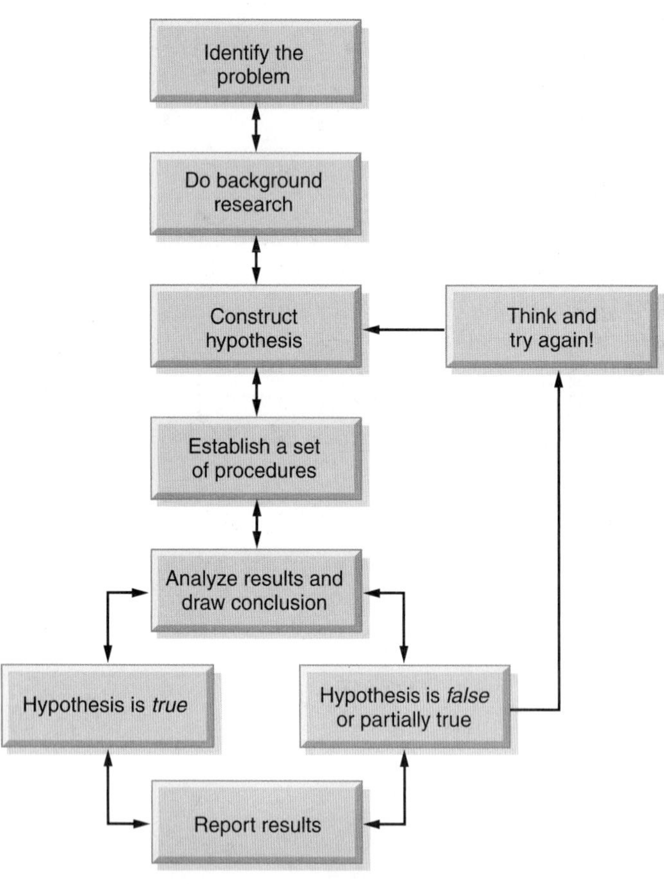

Figure 1. *The scientific method.*

SCIENTIFIC OBSERVATIONS AND INFERENCES

© ACE STOCK LIMITED/ALAMY

***Figure 2.* Observation:** *The lock on the front door of the residence is broken.* **Inference:** *The person who stole goods from the residence last night broke the lock.*

An *observation* is information about the world gathered through the senses. Observations can be qualitative—described in words or drawings. Scientists may use their senses of sight, touch, hearing, or smell to observe the immeasurable qualities of a specimen or study subject, such as its color, shape, scent, or behavior. They may even extend those senses using equipment such as microscopes or hand lenses to make qualitative observations.

Observations can also be quantitative—described by numbers. Counting the number of individuals in a group or the number of instances an event occurs in a specific time are two examples of quantitative observations. Quantitative observations are often measurements that scientists make using tools such as rulers, graduated cylinders, balances, and thermometers.

Observations differ from inferences. An *inference* is a conclusion drawn about the world based on previous knowledge or past experience. Scientists use inferences when constructing explanations of their observations. However, it is important to distinguish observation from inference when interpreting scientific conclusions. Figure 2 describes an observation and an inference a detective might make about a crime scene at a home burglary.

HYPOTHESES

The design of an experiment is guided by the construction of a hypothesis. A *hypothesis* is a statement that answers a testable scientific question. A scientist usually constructs the hypothesis from an educated guess based on background research, previous experimentation, and observation. A hypothesis does not have to be correct to be valid. That is, the quality of a hypothesis is not measured by how accurate it is, but by how testable it is.

Consider the following question: How does temperature affect the decay of a dead animal? A valid hypothesis would be: The higher the temperature, the more quickly a dead animal will decay. A scientist testing this hypothesis would set up a controlled experiment—a test in which all but one of the variables was the same for each test subject. If the data show that each dead animal kept at a greater temperature took less time to decay than each dead animal kept at a lower temperature, the data would support the hypothesis. The scientist would conclude that the hypothesis was true. If he found the opposite to be true, the results of the experiment would still be valid, because they were collected through methodical research and a sound experimental design. Additionally, even with a hypothesis that is found not to be supported by the data, the scientist has learned something about the natural world.

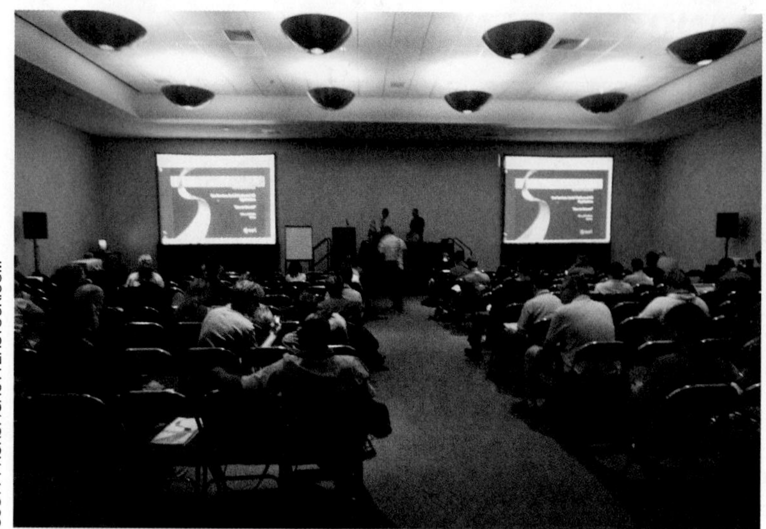

SCOTT PROKOP/SHUTTERSTOCK.COM

Figure 3. *Scientists present their findings at scientific conferences that often bring together researchers in a particular field from all over the world.*

PEER REVIEW

After analyzing their data and drawing conclusions, scientists present their work to other scientists. They do so by publishing papers in research journals and by presenting their data and conclusions at scientific conferences (see Figure 3). This allows scientists to review and compare each other's work. Peer review is an important stage of the scientific process. During this stage, other scientists examine the methods, analyses, and conclusions of an experiment. They ask for clarification, point out errors, or dispute inferences. They even perform their own experiments to confirm that the results are repeatable.

The process of communicating the results of an experiment provides an additional important purpose beyond confirmation: to add to the world's body of scientific knowledge. By reading about the experiments of others, scientists often develop new questions that may inspire new research. They also draw upon others' results and conclusions to help make inferences in their own experiments.

THEORIES

Once many scientists have collected data and drawn conclusions in a particular area of study, scientists might formulate a theory that explains the collective findings. In everyday conversation, people often use the word *theory* to mean a guess. However, in science, the term *theory* has a much more substantial meaning. A theory is an explanation that describes phenomena in the natural world based on extensive, scientifically gathered evidence.

Theories are different from hypotheses. A hypothesis is an answer to a scientific question. A theory, on the other hand, is much broader in scope. A theory is an explanation of natural phenomena. Also, generally, a hypothesis drives the construction of a single experiment, while a theory results from data gathered from many experiments—experiments that have confirmed or overturned many hypotheses.

As important as theories are to explaining the natural world, they are not permanent. They can be revised and even overturned when new, compelling evidence arises that calls into question their accuracy. Even theories that have withstood centuries of testing can be revised or discredited when new discoveries bring about changes in the way scientists understand the world. In this way, revision is a key part of the scientific process.

SCIENTIFIC MODELS

A scientific model is a representation of an object or phenomenon. Models are important tools that scientists use not only to explain their findings to others, but also to collect data and analyze it. For example, maps, like the one in Figure 4, are models that show the topography or geography of a region. They can show information that would be difficult to portray in photographs, graphs, or verbal descriptions.

Figure 4. *A map is a model of a place. This map shows the layout of the streets, buildings, and objects within and near a crime scene.*

There are many reasons a scientist might use a model instead of the actual object it represents. It may be less expensive or less environmentally harmful to perform tests on a model than on the original. Scientists use models of physical objects, such as the anatomical model in Figure 5, to test hypotheses without destroying or damaging the original. Furthermore, scientists might use a model when it is more dangerous to perform an experiment using real test subjects. Computer programs called *simulations* model systems and phenomena, such as severe storms or the conditions under which a passenger airplane flies. Finally, scientists might use models to examine objects that are too small or too large to see with the naked eye. For example, a model of the solar system and a model of an atom are scale models that are proportionately small or large versions of the original object.

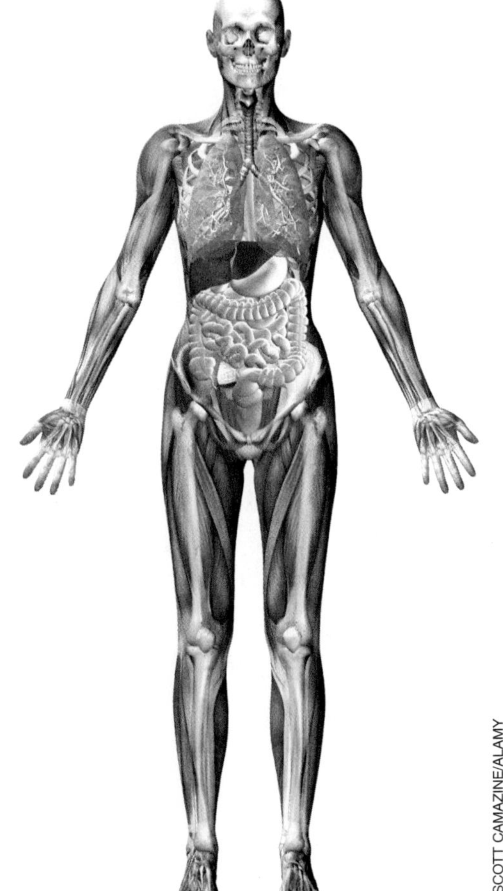

Figure 5. *An anatomical model is a representation of part or all of the body of a living thing.*

Review Questions

1. Which of the following is a scientific model?
   A. a sample of soil taken from a riverbed
   B. a family of cheetahs living at a public zoo
   C. a set of spheres whose arrangement represents the solar system
   D. a group of researchers which reviews the work of other scientists
2. Distinguish between an inference and an observation.
3. Why do scientists perform peer review?

Think Critically **4. How is a theory different from a hypothesis?**

CONVERSION TABLES

INTERNATIONAL SYSTEM OF UNITS (SI)

Because much of the scientific process is collaborative and science takes place in many different countries and in many different languages, scientists have agreed on a "common language" for describing measurements, or quantities. Scientists around the world report their findings using a standard system of units called the *International System of Units*, or *SI* (abbreviated from its name in French, *Le Système International d'Unités*). Many SI units are the same units found in the metric system—a system commonly used by governments and universities all over the world, though not commonly used in the United States.

SI units consist of a base unit and a prefix—the same prefixes used in the metric system. The SI base unit for each quantity is given in Figure 6, and a select list of prefixes is given in the table in Figure 7.

The prefixes allow scientists to describe quantities that are much larger or much smaller than the base unit with more manageable numerical values. For example, 6,100 meters is the same as 6.1 kilometers, and 0.0032 grams is the same as 3.2 milligrams. Also, the prefixes are based on multiples of 10, which makes conversions very easy. To convert among units of the same quantity, you simply move the decimal to the left or to the right by the appropriate number of decimal places.

Quantity	Base Unit	Symbol
Length	meter	m
Mass	kilogram	kg
Time	second	s
electric current	ampere	a
temperature	kelvin	K
Amount	mole	mol
luminous intensity	candela	cd

Figure 6. *SI base units.*

Prefix	Symbol	Meaning: "Multiply by..."
kilo-	k	1,000
hecto-	h	100
deka-	da	10
deci-	d	0.1
centi-	c	0.01
milli-	m	0.001

Figure 7. *Metric prefixes.*

CONVERTING BETWEEN UNITS

Most of the units you find in the grocery store or on road signs in the United States are not SI units. Instead, they are part of the U.S. customary system (also called "English units").

- To convert U.S. customary units into metric or SI units, multiply by the conversion factors given in Figure 8.
- To perform the opposite procedure (converting from U.S. customary units to metric units), divide by the conversion factor.

Quantity	English/U.S. Customary Unit "To convert from..."	Metric Unit "...to..."	Conversion Factor "...multiply by..."
Length	inch (in)	centimeter (cm)	2.54 cm/in
Length	feet (ft)	meter (m)	0.31 m/ft
Length	mile (mi)	kilometer (km)	1.61 km/mi
mass/weight	ounce (oz)	gram (g)	28.35 g/oz
mass/weight	pound (lb)	kilogram (kg)	0.45 kg/lb
Volume	fluid ounce (fl oz)	milliliter (mL)	29.57 mL/fl oz
Volume	gallon (gal)	liter (L)	3.79 L/gal

Figure 8. *Converting from English/U.S. customary units to metric units.*

TEMPERATURE CONVERSIONS

There are three common scales for describing temperature. The SI unit for temperature is the kelvin (K), based on the Kelvin scale. In the Kelvin scale, water freezes at 273.15 kelvin. Scientists also often use the Celsius scale to describe temperature, as do most people outside of the United States. In the Celsius scale, water freezes at 0°C. In the United States, weather reports and medical offices usually report temperature using the Fahrenheit scale. In the Fahrenheit scale, water freezes at 32°F.

- The following equation lets you convert from degrees Celsius to degrees Fahrenheit:

$$\left(\tfrac{9}{5} \times {}^{\circ}\mathrm{C}\right) + 32 = {}^{\circ}\mathrm{F}$$

- The following equation lets you to convert from degrees Fahrenheit to degrees Celsius:

$$({}^{\circ}\mathrm{F} - 32) \times \tfrac{5}{9} = {}^{\circ}\mathrm{C}$$

- The following equation lets you to convert from degrees Celsius to kelvin:

$${}^{\circ}\mathrm{C} + 273.15 = \mathrm{K}$$

Review Questions

1. How many meters are there in 323.4 kilometers?
   A. 0.3234 m
   B. 3.234 m
   C. 3,234 m
   D. 323,400 m
2. What is 14.2 ounces expressed in grams?
   A. 0.0142 g
   B. 0.501 g
   C. 403 g
   D. 420 g
3. Convert 0.455 km to centimeters.
4. Convert 22°C into °F.
5. Convert 8 miles to kilometers.
6. Explain why it is important for scientists to use a standard measurement system.

Think Critically

7. **What is interesting about the temperature 233.15 K? (Hint: Convert this temperature to degrees Celsius and degrees Fahrenheit.)**
8. **Convert 15 miles into meters.**

THE PERIODIC TABLE

WHAT IS THE PERIODIC TABLE?

There are more than 100 different elements that make up matter. Each element has a unique name and a set of characteristic properties. To keep track of the elements, scientists have developed the periodic table—a reference tool that shows the name, chemical symbol, atomic number, and atomic mass of each element.

Each element has its own box in the periodic table. Figure 9 is a key that describes the information given in each element's box. The top number is the element's unique atomic number—the number of protons each atom has in its nucleus. Below the atomic number is the chemical symbol, the one- or two-letter abbreviation that represents the element. Below the chemical symbol is the atomic mass—the average mass of one atom of the element—given in atomic mass units (amu). Finally, at the bottom of the box, the element name is listed.

GROUPS AND PERIODS

Besides being a handy reference for element information, the periodic table can be used to understand which elements have similar characteristics, or properties. The elements are arranged from left to right and top to bottom in the periodic table by increasing atomic number, and not by atomic mass. They are arranged this way for an important reason. In the 1800s and 1900s, scientists discovered that when arranged by increasing atomic number, the elements showed repeating patterns of properties. Elements in each column, or group, have similar chemical and physical properties. And, as you go from left to right across each row, or period, there are trends in how these properties change from one group to the next. The periodic table is shown on the next two pages (see Figure 10).

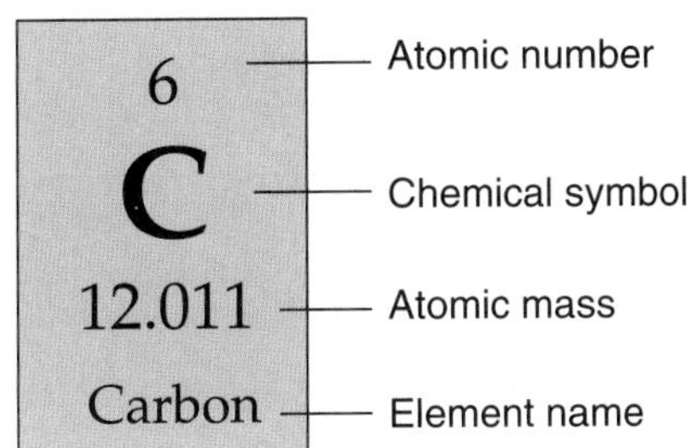

Figure 9. *The box on the periodic table for carbon shows that this element has 6 protons in its nucleus, is represented by the symbol C, and has an atomic mass of 12.011 amu.*

THE PERIODIC TABLE

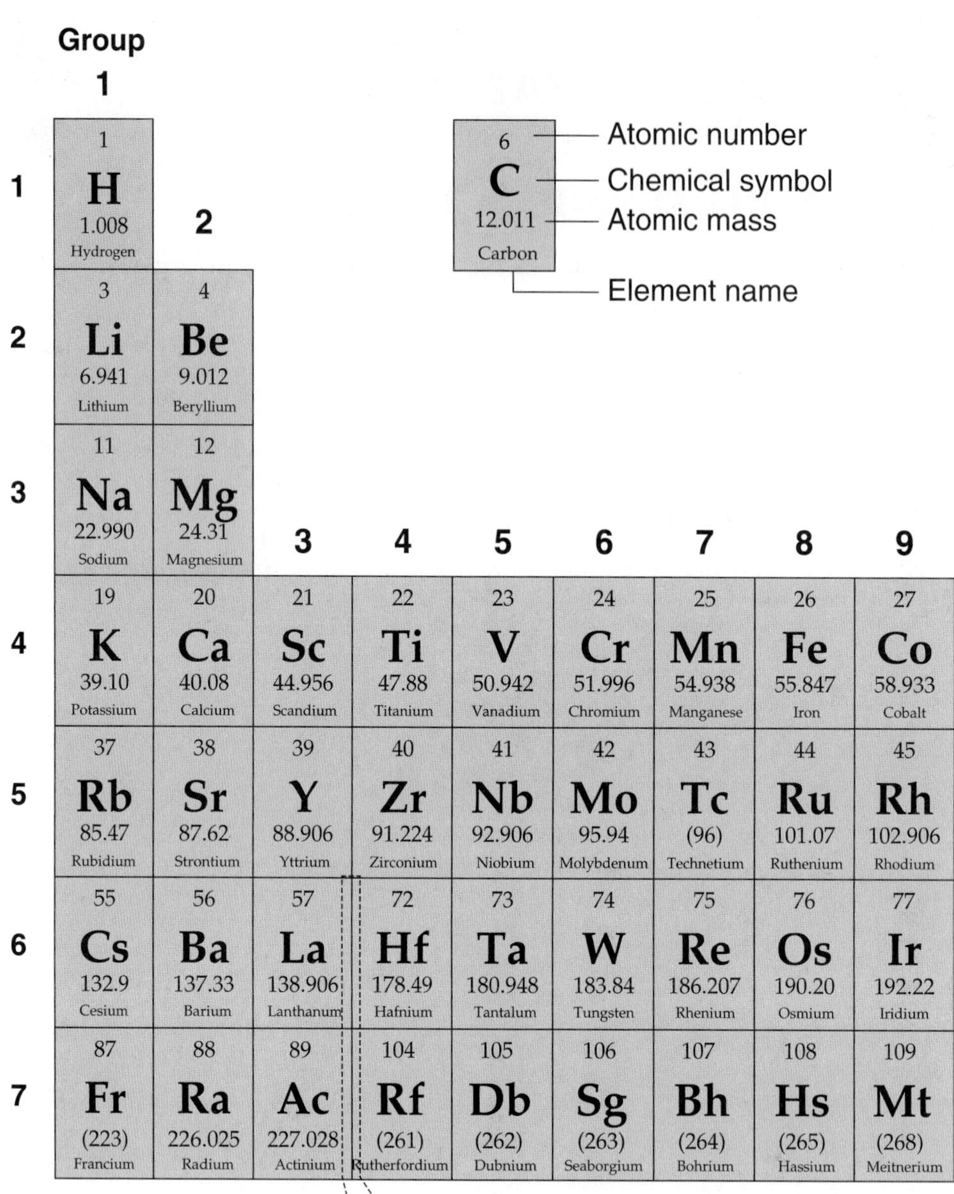

Figure 10. *The Periodic Table.*

10	11	12	13	14	15	16	17	18
								2 **He** 4.0026 Helium
			5 **B** 10.81 Boron	6 **C** 12.011 Carbon	7 **N** 14.007 Nitrogen	8 **O** 15.999 Oxygen	9 **F** 18.996 Fluorine	10 **Ne** 20.179 Neon
			13 **Al** 26.982 Aluminum	14 **Si** 28.086 Silicon	15 **P** 30.974 Phosphorus	16 **S** 32.066 Sulfur	17 **Cl** 35.453 Chlorine	18 **Ar** 39.948 Argon
28 **Ni** 58.69 Nickel	29 **Cu** 63.546 Copper	30 **Zn** 65.37 Zinc	31 **Ga** 69.72 Gallium	32 **Ge** 72.61 Germanium	33 **As** 74.922 Arsenic	34 **Se** 78.96 Selenium	35 **Br** 79.904 Bromine	36 **Kr** 83.80 Krypton
46 **Pd** 106.42 Palladium	47 **Ag** 107.868 Silver	48 **Cd** 112.4 Cadmium	49 **In** 114.82 Indium	50 **Sn** 118.71 Tin	51 **Sb** 121.763 Antimony	52 **Te** 127.60 Tellurium	53 **I** 126.904 Iodine	54 **Xe** 131.29 Xenon
78 **Pt** 195.06 Platinum	79 **Au** 106.967 Gold	80 **Hg** 200.6 Gold	81 **Tl** 204.383 Thallium	82 **Pb** 207.2 Lead	83 **Bi** 208.980 Bismuth	84 **Po** (209) Polonium	85 **At** (210) Astatine	86 **Rn** (222) Radon
110 **Ds** (269) Darmstadtium	111 **Rg** (272) Roentgenium	112 **Cn** (277) Copernicium	113 ***Uut** (284) Ununtrium	114 ***Uuq** (289) Ununquadium	115 ***Uup** (288) Ununpentium	116 ***Uuh** (293) Ununhexium		118 ***Uuo** (299) Ununoctium

*Official name not yet assigned
Parentheses indicate mass number of the most stable or most common isotope.

63 **Eu** 151.97 Europium	64 **Gd** 157.25 Gadolinium	65 **Tb** 158.925 Terbium	66 **Dy** 162.50 Dysprosium	67 **Ho** 164.930 Holmium	68 **Er** 167.26 Erbium	69 **Tm** 168.934 Thulium	70 **Yb** 173.04 Ytterbium	71 **Lu** 174.967 Lutetium
95 **Am** (243) Americium	96 **Cm** (247) Curium	97 **Bk** (247) Berkelium	98 **Cf** (251) Californium	99 **Es** (252) Einsteinium	100 **Fm** (257) Fermium	101 **Md** (258) Mendelevium	102 **No** (259) Nobelium	103 **Lr** (262) Lawrencium

PERIODIC TRENDS

One trend you can see on the periodic table is how metallic an element is. The elements in the groups on the left-hand side of the periodic table are metals. Metals tend to be shiny, good conductors, malleable (easily flattened with a hammer), and ductile (easily drawn into thin wires). On the right-hand side of the table, the elements are mostly nonmetals. Many nonmetals are brittle solids that are dull in appearance and poor conductors. Some are also gases at room temperature. Dividing the metals from the nonmetals is a zigzag line (see Figure 11). The elements on either side of this zigzag line, called *metalloids*, have many properties of metals and nonmetals.

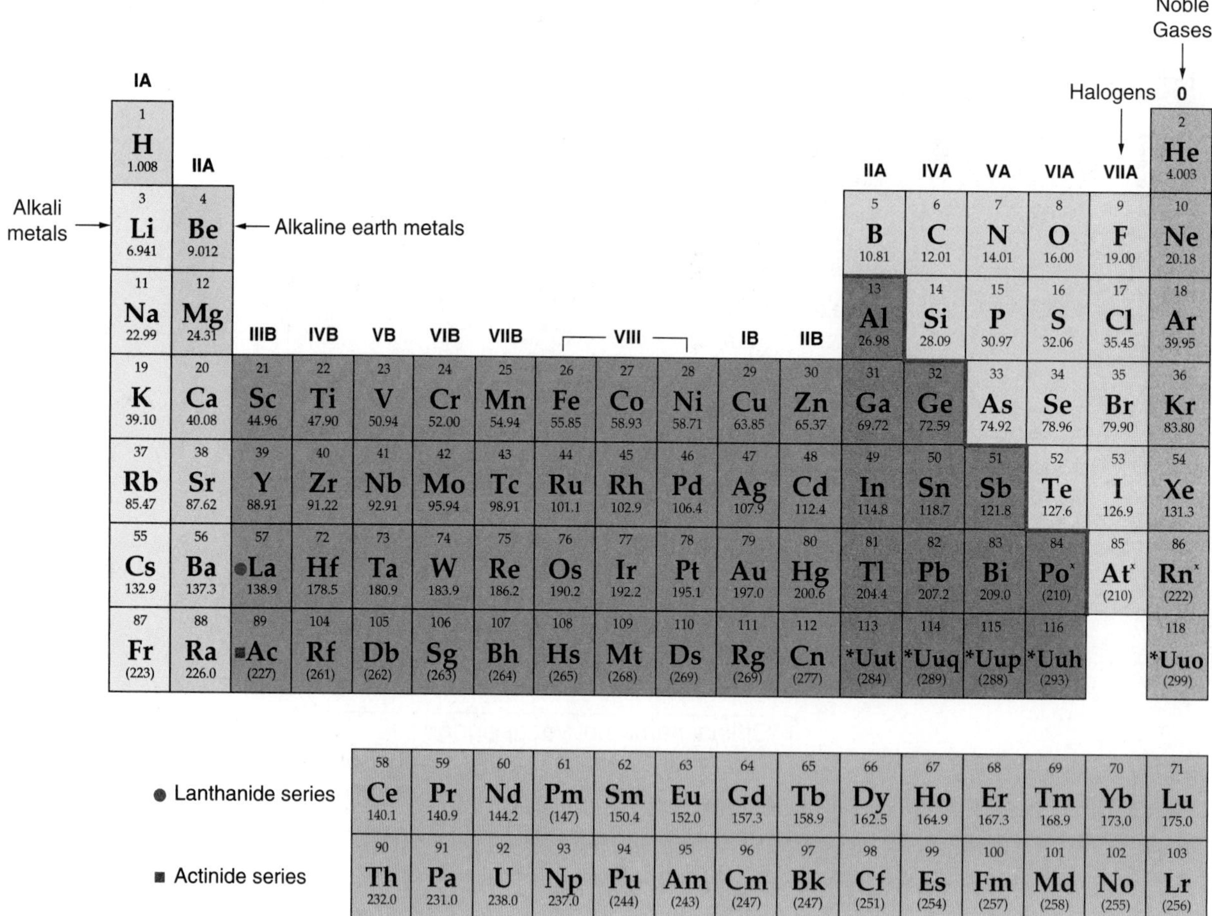

Figure 11. *The red line divides the metals from the nonmetals on the periodic table.*

NAMED GROUPS

Many of the groups of the periodic table are given special names, as shown in Figure 11. For example, the noble gases, which are in the far right column, are all inert gases—they rarely react with other elements, so they are usually found uncombined in nature. The alkali metals located in the column on the far left, on the other hand, are very reactive metals. They are rarely found uncombined in nature. When bonded to carbon atoms, the halogens (the elements in the column second from the far right) give carbon compounds special properties.

Review Questions

1. Which information about an element cannot be found in the periodic table?
   A. number of protons
   B. mass of an atom
   C. metallic nature
   D. solubility
2. What is true of the elements in a period as you go from left to right across the periodic table?
   A. They become more metallic.
   B. Their atomic masses decrease.
   C. Their atomic numbers increase.
   D. They become better conductors.
3. Neon lights are made of glass tubes filled with neon. What can you tell about neon from looking at the periodic table?
4. In what group on the periodic table do lithium (Li), Sodium (Na), and potassium (K) fall?

Think Critically **5. Consider the elements chlorine (Cl), argon (Ar), and bromine (Br). Which pair of elements is most likely to have similar properties? Explain how you know this.**

6. Consider the elements carbon (C), nitrogen (N), and neon (Ne). Which pair of elements is most likely to have similar properties? Explain how you know this.

Appendix D

CHEMICAL REACTIONS

WHAT IS A CHEMICAL REACTION?

Substances can react with one another to form new substances—a process called a *chemical reaction.* An example of a chemical reaction is when steel rusts. The iron in steel reacts with oxygen in air to form rust, or iron oxide. During a chemical reaction, bonds between atoms in the starting substances (called *reactants*) break and reform between different atoms to produce new substances (called *products*). Chemical equations represent reactants as chemical formulas on the left side of the equation and products as chemical formulas on the right side of the equation. An arrow representing chemical change separates reactants from the products in the equation. Scientists classify chemical reactions into several general types: synthesis, decomposition, replacement, acid-base, and redox.

PHOTOLIBRARY.COM

Figure 12. *The electrolysis of water is a decomposition reaction used to produce the oxygen gas that people who work in a submarine need to breathe. Electrolysis requires the input of energy in the form of electricity.*

SYNTHESIS

During a synthesis reaction, two or more reactants react to produce a single product. The formation of iron (II) sulfide from iron and sulfur is an example of synthesis. The following equation shows this reaction:

$$8Fe + S_8 \longrightarrow 8FeS$$

$$\text{iron} + \text{sulfur} \longrightarrow \text{iron (II) sulfide}$$

DECOMPOSITION

During a decomposition reaction, a single reactant breaks down to form two or more products. The electrolysis of water to form hydrogen gas and oxygen gas is an example of decomposition. Figure 12 shows the laboratory setup for this reaction. The following equation shows the balanced chemical reaction:

$$2H_2O \longrightarrow 2H_2 + O_2$$

$$\text{water} \longrightarrow \text{hydrogen gas} + \text{oxygen gas}$$

In the opposite reaction, water is synthesized from hydrogen and oxygen gas, releasing great amounts of energy. The following equation shows the balanced synthesis reaction:

$$2H_2 + O_2 \longrightarrow 2H_2O$$

$$\text{hydrogen gas} + \text{oxygen gas} \longrightarrow \text{water}$$

REPLACEMENT

In replacement reactions, one part of a chemical replaces another in a compound. There are two main types of replacement reactions: single-replacement and double-replacement.

During a single-replacement reaction, an uncombined element replaces an element or group of elements in a compound to produce a new compound and a new uncombined element. The reaction of magnesium with hydrochloric acid, shown in Figure 13, is an example of a single-replacement reaction. The following equation shows this reaction:

$$2Mg + 2HCl \longrightarrow H_2 + 2MgCl$$

magnesium + hydrochloric acid ⟶ hydrogen gas + magnesium chloride

During a double-replacement reaction, elements or ions in two different compounds switch places to form two new compounds. The reaction of silver nitrate with sodium chloride is an example of a double-replacement reaction. Silver chloride and sodium nitrate are the products. The following equation shows this reaction:

$$AgNO_3 + NaCl \longrightarrow AgCl + NaNO_3$$

silver nitrate + sodium chloride ⟶ silver chloride + sodium nitrate

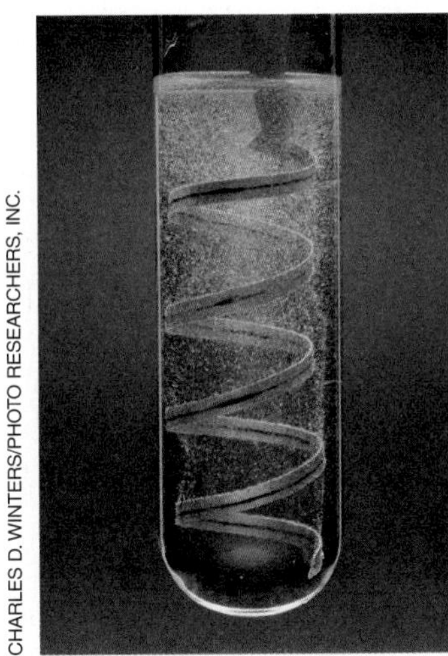

CHARLES D. WINTERS/PHOTO RESEARCHERS, INC.

Figure 13. *Magnesium reacts with hydrochloric acid in a single-replacement reaction that produces hydrogen gas and magnesium chloride.*

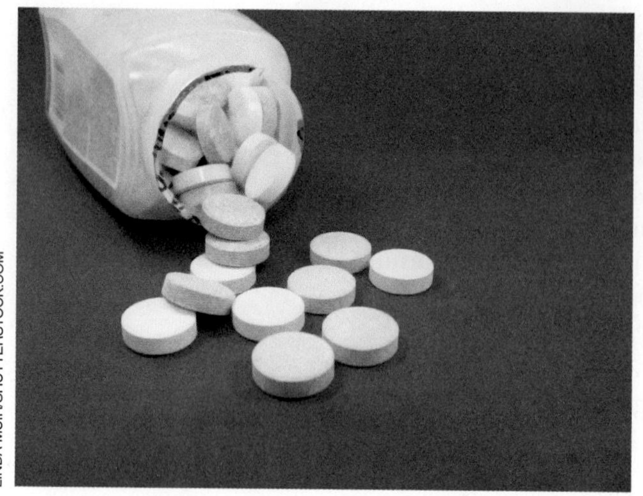
LINDA MUIR/SHUTTERSTOCK.COM

Figure 14. *Some antacids taken for indigestion contain a base, such as magnesium hydroxide. When the base reacts with stomach acid (hydrochloric acid), the acid-base reaction produces neutral compounds that are less irritating to the stomach. The following equation shows one such acid-base reaction with an antacid:* $Mg(OH)_2 + 2\ HCl \rightarrow MgCl_2 + 2H_2O$.

ACID-BASE REACTIONS

Acid-base reactions are a special kind of double-replacement reaction in which an acid reacts with a base to produce water and a salt. These are also sometimes called *neutralization reactions,* because the products are usually neutral—neither acidic nor basic. The reaction of hydrochloric acid and sodium hydroxide (a base) results in the neutral salt, sodium chloride, and water.

$$HCl + NaOH \longrightarrow NaCl + H_2O$$

hydrochloric acid + sodium hydroxide ⟶ sodium chloride + water

Figure 14 shows another example of an acid-base reaction.

REDOX REACTIONS

Redox reactions are also called *oxidation-reduction reactions.* During a redox reaction, one reactant is oxidized (it loses electrons), while the other reactant is reduced (it gains electrons). You can think of a redox reaction as the summary of two "half reactions." A half reaction shows how neutral atoms gain or lose electrons to form negative or positive ions. Look at the following two half reactions and the redox reaction that they combine to form.

$$4Al \longrightarrow 4Al^{3+} + 12e^-$$

4 aluminum atoms ⟶ 4 aluminum ions + 12 electrons

$$3O_2 + 12e^- \longrightarrow 6O^{2-}$$

3 oxygen molecules + 12 electrons → 6 oxide ions

$$4Al + 3O_2 \longrightarrow 2Al_2O_3$$

aluminum + oxygen gas ⟶ alumina

In this redox reaction, aluminum is oxidized, while oxygen is reduced, and alumina is formed.

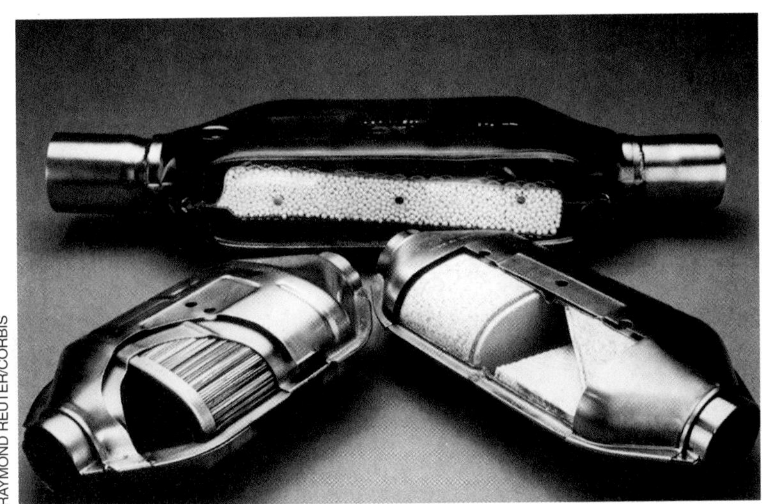

Figure 15. *Catalytic converters carry out redox reactions that convert harmful exhaust chemicals into less harmful chemicals to reduce the pollution of car exhaust.*

A car's catalytic converter provides an example of a useful redox reaction. Catalytic converters help carry out chemical reactions that reduce the amount of harmful gases in car exhaust (see Figure 15). One of the many reactions that take place in a catalytic converter is the decomposition of nitric oxide to oxygen and nitrogen gases. The following equations show the half reactions and the redox reaction that they combine to form.

$$2N^{2+} + 4e^- \longrightarrow N_2$$

2 nitrogen ions + 4 electrons ⟶ nitrogen molecule

$$2O^{2-} \longrightarrow O_2 + 4e^-$$

2 oxide ions ⟶ oxygen molecule + 4 electrons

$$2NO \longrightarrow O_2 + N_2$$

nitric oxide ⟶ oxygen gas + nitrogen gas

Review Questions

1. Zinc reacts with hydrochloric acid to produce zinc chloride and hydrogen gas according to the following equation:

$$Zn + 2HCl \longrightarrow ZnCl_2 + H_2$$

What kind of reaction is this?

A. synthesis
B. acid-base
C. single-replacement
D. double-replacement

2. Which of the following reactions is a double-replacement reaction?

A. $Pb(NO_3)_2 + 2KI \longrightarrow PbI_2 + 2KNO_3$
B. $Pb + O_2 \longrightarrow PbO_2$
C. $2Fe + 6NaBr \longrightarrow 2FeBr_3 + 6Na$
D. $Na_2CO_3 \longrightarrow Na_2O + CO_2$

3. Define the following types of chemical reactions: synthesis, decomposition, single-replacement, double-replacement, acid-base, and redox.

Think Critically **4. When living things die, their tissues start to rot. Rotting is a chemical reaction in which molecules break down to produce other products, including gases such as methane. What type of chemical reaction is likely common during the rotting process?**

SAFETY

To protect yourself and others, follow these safety guidelines when working in the laboratory or performing fieldwork (see Figure 16).

CONDUCT AND PERSONAL SAFETY

- Do not engage in horseplay or practical jokes in the laboratory.
- Do not ever run, especially while handling chemicals, equipment, or sharp objects.
- Do not perform experiments that have not been previously authorized by your teacher.
- Do not eat or drink in the laboratory, and do not bring open food or drink containers into the laboratory.
- Remove or tie back loose hair, jewelry, and clothing.
- Wash your hands after working in the laboratory and before eating or touching your eyes.
- Wear safety equipment, including goggles, apron, and gloves, as directed by your teacher.

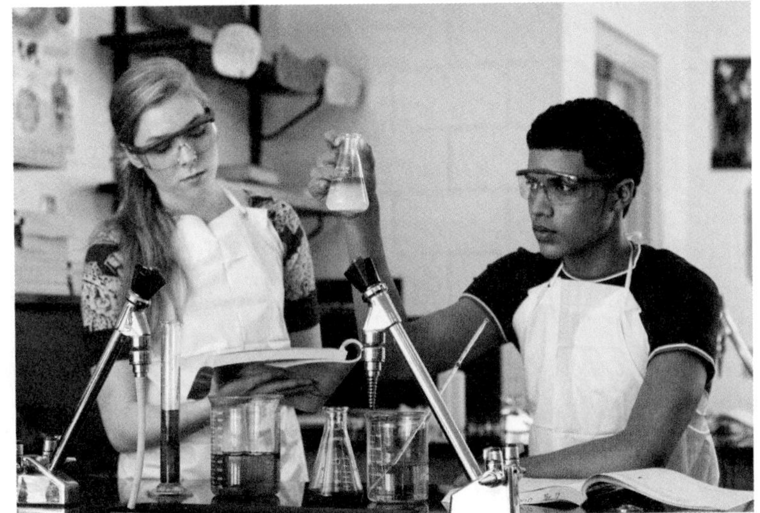

Figure 16. *When working with chemicals or anything that could splash into the eyes or onto clothing, it is important to wear goggles and an apron. Gloves protect your skin from chemicals. You should wear gloves whenever instructed to do so by the text or your teacher.*

LABORATORY PROCEDURE

- Follow all of the teacher's instructions.
- Read the directions for each laboratory or field activity before proceeding with your work.
- Keep your work area neat and uncluttered.
- Be sure you understand and follow all safety precautions described in the laboratory.

HANDLING EQUIPMENT, ORGANISMS, AND CHEMICALS

- Handle any living thing with respect and care. Wash your hands after touching plants or animals in the field or in the laboratory.
- Do not touch, taste, or directly smell any chemical. Use the wafting technique to detect the smell of a chemical, when told to do so. To waft, hold the container (usually a beaker) 5 or 6 inches in front of you in one hand. Move the other hand in a sweeping motion over the container to bring the scent toward you. Breathe normally.
- Notify your teacher if any chemical is spilled.
- Do not use chemicals that are unlabeled or mislabeled.
- Never heat flammable liquids with an open flame. Use a hot water bath instead.
- When diluting an acid, never add the water to the concentrated acid. Instead, add the acid to the water slowly.
- Conserve laboratory chemicals. Take only the amount of chemical you will need for a laboratory activity.
- Do not put any unused chemical back into its original container. Instead, dispose of it according to your teacher's instructions.
- Dispose of all cracked and broken glassware in the appropriate container. Also, dispose of all used razors, needles, and other sharp instruments in the appropriate container.
- When in doubt, ask your teacher how to dispose of chemicals and materials used in the laboratory.

Figure 17. *Use the emergency eye-wash station if chemicals have splashed into your eyes. Use the emergency shower if harmful chemicals have splashed onto or soaked into your clothing.*

EMERGENCIES

- In case of emergency, notify your teacher immediately.
- Know the location of the fire extinguisher, emergency shower, and emergency eye-wash station, and know how to use these pieces of safety equipment (see Figure 17).
- Know the location of all exits from the laboratory and from the building.

Review Questions

1. Which piece of safety equipment protects your eyes?
   A. apron
   B. emergency shower
   C. fire extinguisher
   D. goggles
2. Describe the proper way to smell a chemical.
3. Describe the proper procedure for diluting an acid.
4. Explain how reviewing all of the instructions related to the lab activity before performing the activity relates to safety.

Think Critically

5. **Your teacher has asked you to gently heat a flammable liquid. Name at least two safety procedures you should follow during this activity.**
6. **During a lab activity, your lab partner drops a flask and spills chemical on the lab table. The flask breaks. Outline the steps you should take in response.**

OCEAN/CORBIS

GRAPHING DATA

Trends are not always obvious in a list of numbers, which is why scientists usually collect and organize their data in graphs. Graphs allow scientists to analyze the data visually. They also help scientists share the results of their work in a display that is often easier to interpret than a table. Different graphs have different purposes and different types of data for which they are best suited.

BAR GRAPHS

A bar graph uses horizontal or vertical bars of different lengths or heights to represent values (see Figure 18). Bar graphs are best suited for data that is not continuous, such as comparing the final values of different experimental groups. Bars of different colors described in a key can show multiple values of the different groups (such as initial and final amounts or male and female subjects).

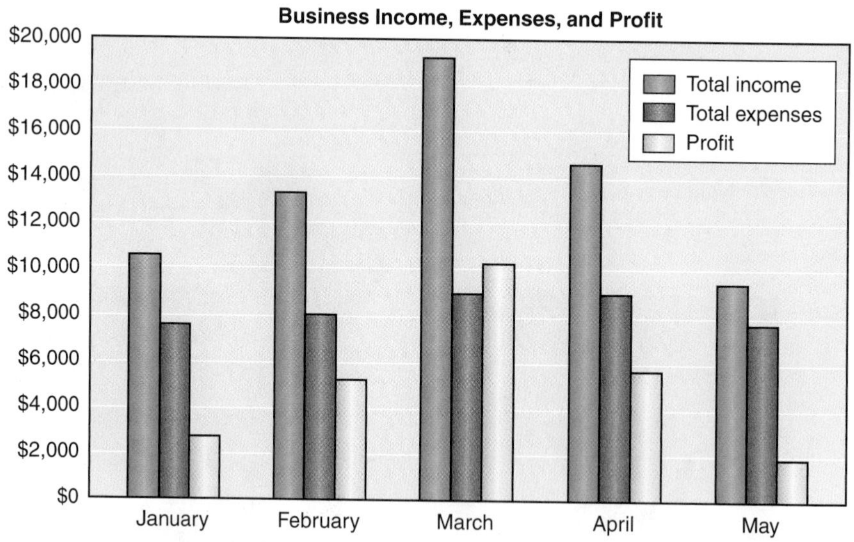

Figure 18. *This bar graph compares three monthly values for a business over a period of five months.*

LINE GRAPHS

A line graph uses a line to show how a dependent variable responds to an independent variable (see Figure 19). The independent variable is usually described by the *x*-axis (horizontal), while the dependent variable is usually described by the *y*-axis (vertical). Line graphs are best used for showing data that is continuous, such as data collected over time, data collected as temperature is increased, or data collected as a substance is added in increasing amounts.

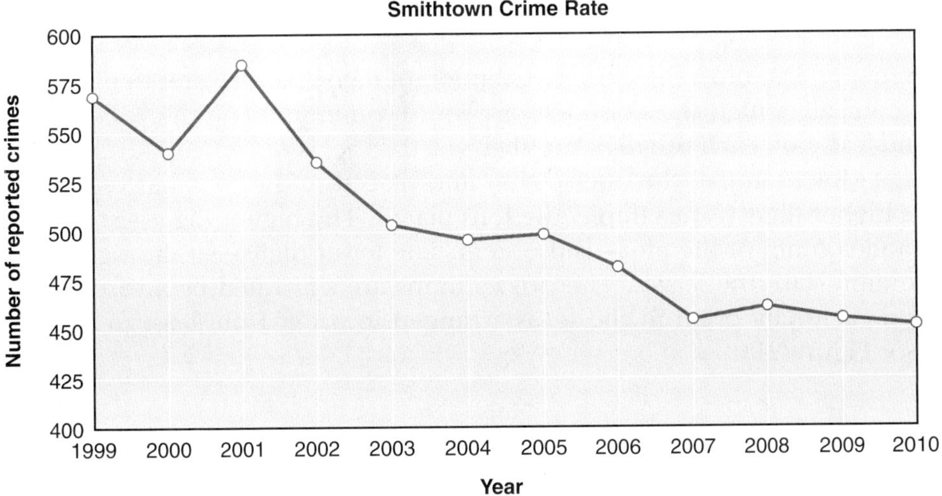

Figure 19. *This line graph shows continuous change over time in the number of crimes reported each year.*

CIRCLE GRAPHS

Data that are parts of a whole are often best displayed in a circle graph (see Figure 20). A circle graph, which is also called a *pie chart,* represents data points as wedges—or slices of the pie. The size of each wedge represents the percentage or fraction of a whole. Thus, a wedge that represents 30 percent is twice as big as a wedge that represents 15 percent. In a circle graph, the percentages must add up to 100 percent, and fractions must add up to 1.

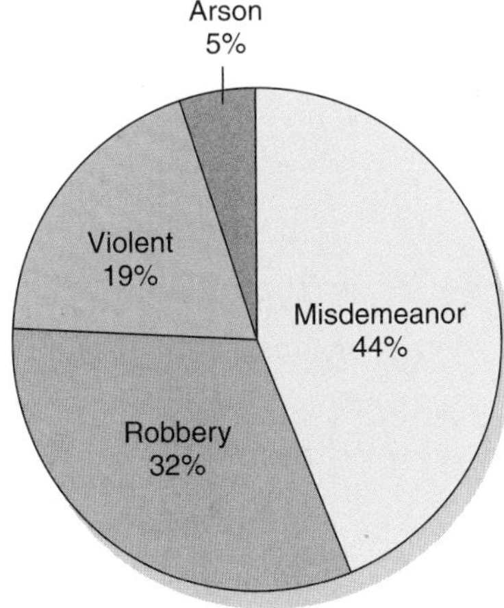

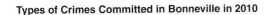

Figure 20. *This circle graph shows that the greatest percentage of crimes committed in 2010 were misdemeanors, while the smallest percentage were arson.*

Ages of Victims

1	8 9 9
2	0 1 1 1 2 2 4 6 7 9
3	0 3 3
4	1 7
5	
6	2

Figure 21. *This stem-and-leaf plot shows that the ages of victims are not randomly distributed—most of the victims are in their twenties.*

STEM-AND-LEAF PLOTS

Stem-and-leaf plots show the distribution of data in a set. That is, they show graphically which types of values are more common than others. In a stem-and-leaf plot, each two-digit number in a data set is represented by its lower digit (for example, the ones place) and placed in the row that represents its higher digit (for example, the tens place). The higher digit—in the left-hand column of the plot—is the stem. The lower digits—in the right-hand column—are the leaves. The two columns are separated by a vertical bar, and each data point in the set is arranged in order from least to greatest (see Figure 21).

HISTOGRAMS

Like stem-and-leaf plots, histograms show visually the distribution of data in a set. A histogram uses vertical bars of different heights to represent frequencies, or the number of times something has a certain characteristic. The frequency is represented on the *x*-axis, and the varying characteristic is represented on the *y*-axis. Each bar in the histogram stands for an interval, a group of data points that fall within a certain range of values for that characteristic (see Figure 22).

Shoe Sizes of Students in 12th Grade

Frequency: 0, 20, 40, 60, 80, 100, 120

Shoe size: Smaller than 6; 6 to 7½; 8 to 9½; 10 to 11½; 12 to 13½; 14 and larger

Figure 22. *According to this histogram, most of the students in twelfth grade have shoes that are sizes 8 to 9½ or sizes 10 to 11½.*

BOX-AND-WHISKER PLOTS

Sometimes it is useful to know how spread apart the data are, or whether most of the data fall to one side or another. Box-and-whisker plots can be useful in such cases. Scientists often summarize data sets with five statistics numbers:

- the lowest number in the set (the minimum),
- the middle number of the lowest half of the numbers in the set (the lower quartile),
- the middle number of the set (the median),
- the middle number of the highest half of the numbers in the set (the upper quartile), and
- the highest number (the maximum).

Box-and-whisker plots show these numbers visually. The two boxes in the plot together represent exactly half of the values in the data set—the half that are in the middle. Each whisker, or line with a hash mark at one end, represents the lower or upper quarter of the values in the set. The left-hand box spans from the lower quartile to the median, while the right-hand box spans from the median to the upper quartile. The left-hand whisker ends at the minimum, and the right-hand whisker ends at the maximum.

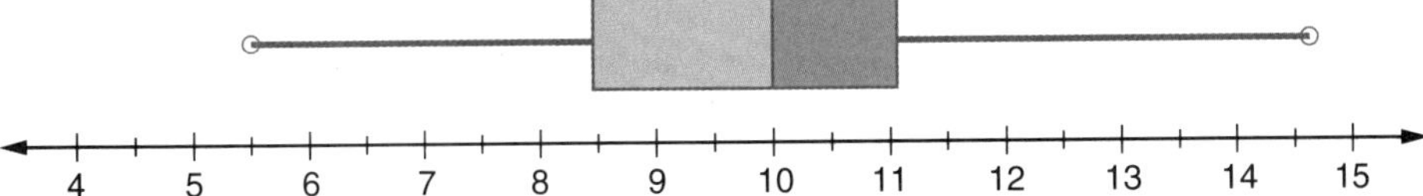

Figure 23. *This box-and-whisker plot shows the same data set as the histogram in Figure 22. However, with this kind of graph, you can tell that the smallest shoe size worn by a twelfth-grade student is 5½, and the largest shoe size is 14½.*

In Figure 23, the data used in Figure 22 was used to develop a box-and-whisker plot.

SCATTER PLOTS

Scatter plots use Cartesian coordinates to display data points that are made up of two values, such as those that can be written in the format (x, y). As with line graphs, the independent variable is usually described by the x-axis (horizontal), while the dependent variable is usually described by the y-axis (vertical).

Scatter plots allow scientists to determine whether there is a correlation between two variables, and if there is a correlation, the strength of that correlation. A random distribution of data points on a scatter plot shows no correlation. If the data points form a band, then there is a correlation. A thick band shows a weak correlation, while a narrow band shows a strong correlation. Figure 24 provides an example of a scatter plot.

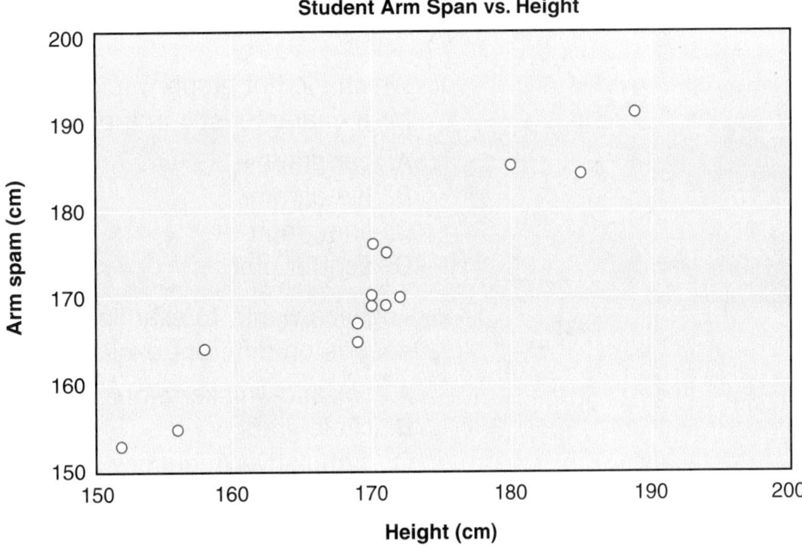

Figure 24. *The data points in this scatter plot form a narrow band, which shows that arm span and height are correlated.*

CUMULATIVE FREQUENCY (OGIVE) GRAPHS

A cumulative frequency graph, or ogive graph, is a line graph that shows a growing (or shrinking) total. The slope of the line between two data points represents the rate of accumulation for that interval (see Figure 25). The steeper the slope, the greater is the change in amount. This kind of graph is best suited for looking at trends in data where the total value is as important as the rate of change in that value, such as with a savings account.

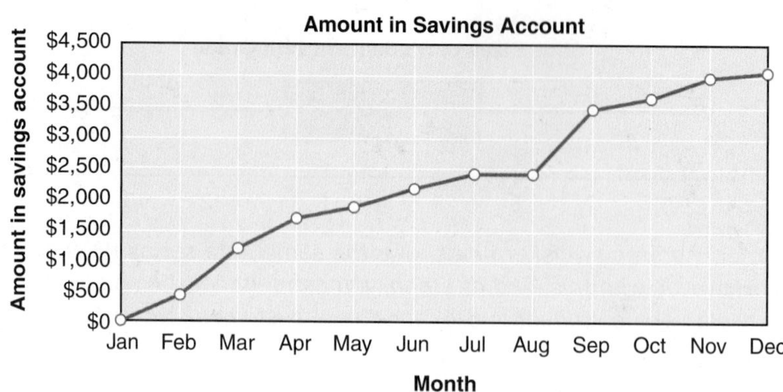

Figure 25. *This cumulative frequency graph shows the total amount of money kept in a savings account. An increase in amount appears as a sloped line between two data points.*

Review Questions

1. Which kind of graph would best show the distribution of frequencies for a characteristic in a population?
   A. bar graph
   B. histogram
   C. line graph
   D. scatter plot

2. Angelica wants to examine the portions of her total income that she spends on different expenses. Which kind of graph should she use?
   A. box-and-whisker plot
   B. circle graph
   C. cumulative frequency graph
   D. stem-and-leaf plot

3. Look at Figure 19 on page 493. In which year did the crime rate reach its highest for the time period shown?

Think Critically

4. Imagine that the cumulative frequency graph in Figure 25 describes the savings account of a man named Steve. At one point Steve lost his job. Then, a month later, he got a new job that paid him a greater salary. Which data in the graph match these events in Steve's life?

5. Juan and Erin are working on forensic entomology lab. As part of the investigation, they have been measuring the air temperature and humidity every day at 3:00 P.M. for the last two weeks. What kind of graph or graphs should they use to display their data? Justify your answer.

Index

C

D

G

H

I

U

V

W

Y

Z